QUANTUM FIELD THEORY AND CONDENSED MATTER

Providing a broad review of many techniques and their application to condensed matter systems, this book begins with a review of thermodynamics and statistical mechanics, before moving on to real- and imaginary-time path integrals and the link between Euclidean quantum mechanics and statistical mechanics. A detailed study of the Ising, gauge-Ising and *XY* models is included. The renormalization group is developed and applied to critical phenomena, Fermi liquid theory, and the renormalization of field theories. Next, the book explores bosonization and its applications to one-dimensional fermionic systems and the correlation functions of homogeneous and random-bond Ising models. It concludes with the Bohm–Pines and Chern–Simons theories applied to the quantum Hall effect. Introducing the reader to a variety of techniques, it opens up vast areas of condensed matter theory for both graduate students and researchers in theoretical, statistical, and condensed matter physics.

R. SHANKAR is the John Randolph Huffman Professor of Physics at Yale University, with a research focus on theoretical condensed matter physics. He has held positions at the Aspen Center for Physics, the American Physical Society, and the American Academy of Arts and Sciences. He has also been a visiting professor at several universities including MIT, Princeton, UC Berkeley, and IIT Madras. Recipient of both the Harwood Byrnes and Richard Sewell teaching prizes at Yale (2005) and the Julius Edgar Lilienfeld prize of the American Physical Society (2009), he has also authored several books: *Principles of Quantum Mechanics, Basic Training in Mathematics*, and *Fundamentals of Physics* I and II.

QUANTUM FIELD THEORY AND CONDENSED MATTER

An Introduction

R. SHANKAR
Yale University, Connecticut

CAMBRIDGE
UNIVERSITY PRESS

CAMBRIDGE
UNIVERSITY PRESS

University Printing House, Cambridge CB2 8BS, United Kingdom

One Liberty Plaza, 20th Floor, New York, NY 10006, USA

477 Williamstown Road, Port Melbourne, VIC 3207, Australia

314-321, 3rd Floor, Plot 3, Splendor Forum, Jasola District Centre, New Delhi – 110025, India

79 Anson Road, #06-04/06, Singapore 079906

Cambridge University Press is part of the University of Cambridge.

It furthers the University's mission by disseminating knowledge in the pursuit of education, learning, and research at the highest international levels of excellence.

www.cambridge.org
Information on this title: www.cambridge.org/9780521592109
DOI: 10.1017/9781139044349

First published 2017
3rd printing 2018

Printed and bound in Great Britain by Clays Ltd, Elcograf S.p.A.

A catalogue record for this publication is available from the British Library.

ISBN 978-0-521-59210-9 Hardback

Dedicated to
Michael Fisher, Leo Kadanoff, Ben Widom, and Ken Wilson
Architects of the modern RG

Contents

Preface

Condensed matter theory is a massive field to which no book or books can do full justice. Every chapter in this book is possible material for a book or books. So it is clearly neither my intention nor within my capabilities to give an overview of the entire subject. Instead I will focus on certain techniques that have served me well over the years and whose strengths and limitations I am familiar with.

My presentation is at a level of rigor I am accustomed to and at ease with. In any topic, say the renormalization group (RG) or bosonization, there are treatments that are more rigorous. How I deal with this depends on the topic. For example, in the RG I usually stop at one loop, which suffices to make the point, with exceptions like wave function renormalization where you need a minimum of two loops. For non-relativistic fermions I am not aware of anything new one gets by going to higher loops. I do not see much point in a scheme that is exact to all orders (just like the original problem) if in practice no real gain is made after one loop. In the case of bosonization I work in infinite volume from the beginning and pay scant attention to the behavior at infinity. I show many examples where this is adequate, but point to cases where it is not and suggest references. In any event I think the student should get acquainted with these more rigorous treatments after getting the hang of it from the treatment in this book. I make one exception in the case of the two-dimensional Ising model where I pay considerable attention to boundary conditions, without which one cannot properly understand how symmetry breaking occurs only in the thermodynamic limit.

This book has been a few years in the writing and as a result some of the topics may seem old-fashioned; on the other hand, they have stood the test of time.

Ideally the chapters should be read in sequence, but if that is not possible, the reader may have to go back to earlier chapters when encountering an unfamiliar notion.

I am grateful to the Aspen Center for Physics (funded by NSF Grant 1066293) and the Indian Institute of Technology, Madras for providing the facilities to write parts of this book.

Over the years I have drawn freely on the wisdom of my collaborators and friends in acquiring the techniques described here. I am particularly grateful to my long-standing collaborator Ganpathy Murthy for countless discussions over the years.

Of all the topics covered here, my favorite is the renormalization group. I have had the privilege of interacting with its founders: Michael Fisher, Leo Kadanoff, Ben Widom, and Ken Wilson. I gratefully acknowledge the pleasure their work has given me while learning it, using it, and teaching it.

In addition, Michael Fisher has been a long-time friend and role model – from his exemplary citizenship and his quest for accuracy in thought and speech, right down to the imperative to punctuate all equations. I will always remain grateful for his role in ensuring my safe passage from particle theory to condensed-matter theory.

This is also a good occasion to acknowledge the 40+ happy years I have spent at Yale, where one can still sense the legacy of its twin giants: Josiah Willard Gibbs and Lars Onsager. Their lives and work inspire all who become aware of them.

To the team at Cambridge University Press, my hearty thanks: Editor Simon Capelin for his endless patience with me and faith in this book, Roisin Munnelly and Helen Flitton for document handling, and my peerless manuscript editor Richard Hutchinson for a most careful reading of the manuscript and correction of errors in style, syntax, punctuation, referencing, and, occasionally, even equations! Three cheers for university presses, which exemplify what textbook publication is all about.

Finally I thank the three generations of my family for their love and support.

1

Thermodynamics and Statistical Mechanics Review

This is a book about techniques, and the first few chapters are about the techniques you have to learn *before* you can learn the *real* techniques.

Your mastery of these crucial chapters will be presumed in subsequent ones. So go though them carefully, paying special attention to exercises whose answers are numbered equations.

I begin with a review of basic ideas from thermodynamics and statistical mechanics. Some books are suggested at the end of the chapter [1–4].

In the beginning there was thermodynamics. It was developed before it was known that matter was made of atoms. It is notorious for its multiple but equivalent formalisms and its orgy of partial derivatives. I will try to provide one way to navigate this mess that will suffice for this book.

1.1 Energy and Entropy in Thermodynamics

I will illustrate the basic ideas by taking as an example a cylinder of gas, with a piston on top. The piston exerts some pressure P and encloses a volume V of gas. We say the system (gas) is in equilibrium when nothing changes at the macroscopic level. In equilibrium the gas may be represented by a point in the (P,V) plane.

The gas has an internal energy U. This was known even before knowing the gas was made of atoms. The main point about U is that it is a *state variable*: it has a unique value associated with every state, i.e., every point (P,V). It returns to that value if the gas is taken on a closed loop in the (P,V) plane.

There are two ways to change U. One is to move the piston and do some mechanical work, in which case

$$dU = -PdV \tag{1.1}$$

by the law of conservation of energy. The other is to put the gas on a hot or cold plate. In this case some heat δQ can be added and we write

$$dU = \delta Q, \tag{1.2}$$

which acknowledges the fact that heat is a form of energy as well. The first law of thermodynamics simply expresses energy conservation:

$$dU = \delta Q - PdV. \tag{1.3}$$

I use δQ and not dQ since Q is not a state variable. Unlike U, there is no unique Q associated with a point (P,V): we can go on a closed loop in the (P,V) plane, come back to the same state, but Q would have changed by the negative of the work done, which is the area inside the loop.

The second law of thermodynamics introduces another state variable, S, the *entropy*, which changes by

$$dS = \frac{\delta Q}{T} \tag{1.4}$$

when heat δQ is added reversibly, i.e., arbitrarily close to equilibrium. It is a state variable because it can be shown that

$$\oint dS = 0 \tag{1.5}$$

for a quasi-static cyclic process.

Since dU is independent of how we go from one point to another, we may as well assume that the heat was added reversibly, and write

$$dU = TdS - PdV, \tag{1.6}$$

which tells us that

$$U = U(S,V), \tag{1.7}$$

$$T = \left.\frac{\partial U}{\partial S}\right|_V, \tag{1.8}$$

$$-P = \left.\frac{\partial U}{\partial V}\right|_S. \tag{1.9}$$

For future use note that we may rewrite these equations as follows:

$$dS = \frac{1}{T}dU + \frac{P}{T}dV, \tag{1.10}$$

$$S = S(U,V), \tag{1.11}$$

$$\frac{1}{T} = \left.\frac{\partial S}{\partial U}\right|_V, \tag{1.12}$$

$$\frac{P}{T} = \left.\frac{\partial S}{\partial V}\right|_U. \tag{1.13}$$

These equations will be recalled shortly when we consider statistical mechanics, in which macroscopic thermodynamic quantities like energy or entropy emerge from a microscopic description in terms of the underlying atoms and molecules.

The function $U(S,V)$, called the fundamental relation, constitutes complete thermodynamics knowledge of the system.

This is like saying that the Hamiltonian function $H(x,p)$ constitutes complete knowledge of a mechanical system. However, we still need to find what H is for a particular situation, say the harmonic oscillator, by empirical means.

As an example, let us consider n moles of an ideal gas for which it is known from experiments that

$$U(S,V) = C\left[\frac{e^{S/nR}}{V}\right]^{2/3}, \tag{1.14}$$

where $R = 8.31\,\mathrm{J\,mol^{-1}\,K^{-1}}$ is the universal gas constant and C is independent of S and V. From the definition of P and T in Eqs. (1.8) and (1.9), we get the *equations of state*:

$$P = -\left.\frac{\partial U}{\partial V}\right|_S = \frac{2}{3}\frac{U}{V}, \tag{1.15}$$

$$T = \left.\frac{\partial U}{\partial S}\right|_V = \frac{2}{3nR}U. \tag{1.16}$$

The two may be combined to give the more familiar $PV = nRT$.

1.2 Equilibrium as Maximum of Entropy

The second law of thermodynamics states that when equilibrium is disturbed, the entropy of the universe will either increase or remain the same. Equivalently,

S is a maximum at equilibrium.

But I have emphasized that S, like U, is a state variable defined only in equilibrium! How can you maximize a function defined only at its maximum?

What this statement means is this. Imagine a box of gas in equilibrium. It has a volume V and energy U. Suppose the box has a conducting piston that is held in place by a pin. The piston divides the volume into two parts of size $V_1 = \alpha V$ and $V_2 = (1-\alpha)V$. They are at some common temperature $T_1 = T_2 = T$, but not necessarily at a common pressure. The system is forced into equilibrium despite this due to a constraint, the pin holding the piston in place. The entropy of the combined system is just the sum:

$$S = S_1(V_1) + S_2(V_2) = S_1(\alpha V) + S_2((1-\alpha)V). \tag{1.17}$$

Suppose we now let the piston move. It may no longer stay in place and α could change. Where will it settle down? We are told it will settle down at the value that maximizes S:

$$0 = dS = dS_1 + dS_2 \tag{1.18}$$

$$= \frac{\partial S_1}{\partial V_1}dV_1 + \frac{\partial S_2}{\partial V_2}dV_2 \tag{1.19}$$

$$= \left(\frac{P_1}{T_1} - \frac{P_2}{T_2} \right) d\alpha\ V \tag{1.20}$$

$$0 = \left(\frac{P_1 - P_2}{T} \right) d\alpha, \tag{1.21}$$

which is the correct physical answer: in equilibrium, when S is maximized, the pressures will be equal.

So the principle of maximum entropy means that when a system held in equilibrium by a constraint becomes free to explore new equilibrium states due to the removal of the constraint, it will pick the one which maximizes S. In this example, where its options are parametrized by α, it will pick

$$\frac{\partial S(\alpha)}{\partial \alpha} = 0 \stackrel{\text{def}}{=} \text{equilibrium}. \tag{1.22}$$

A subtle point: Suppose initially we had $\alpha = 0.1$, and finally $\alpha = 0.5$. In this experiment, only $S(\alpha = 0.1)$ and $S(\alpha = 0.5)$ are equilibrium entropies. However, we could define an $S(\alpha)$ by making any α into an equilibrium state by restraining the piston at that α and letting the system settle down. It is this $S(\alpha)$ that is maximized at the new equilibrium.

1.3 Free Energy in Thermodynamics

Temporarily, let V be fixed, so that $U = U(S)$ and

$$T = \frac{dU}{dS}. \tag{1.23}$$

Since $U = U(S)$, this gives us T as a function of S:

$$T = T(S). \tag{1.24}$$

Assuming that this relation can be inverted to yield

$$S = S(T), \tag{1.25}$$

let us construct a function $F(T)$, called the *free energy*, as follows:

$$F(T) = U(S(T)) - TS(T). \tag{1.26}$$

Look at the T-derivative of $F(T)$:

$$\frac{dF}{dT} = \frac{dU(S(T))}{dT} - S(T) - T\frac{dS(T)}{dT} \tag{1.27}$$

$$= \frac{dU}{dS} \cdot \frac{dS}{dT} - S(T) - T\frac{dS(T)}{dT} \tag{1.28}$$

$$= -S(T) \quad \text{using Eq. (1.8)}. \tag{1.29}$$

Thus we see that while U was a function of S, with T as the derivative, F is a function of T, with $-S$ as its derivative. Equation (1.26), which brings about this exchange of roles, is an example of a *Legendre transformation*.

Let us manipulate Eq. (1.26) to derive a result that will be invoked soon:

$$F = U - ST \tag{1.30}$$

$$U = F + ST \tag{1.31}$$

$$= F - T\frac{dF}{dT} \quad \text{using Eq. (1.29).} \tag{1.32}$$

If we bring back V, which was held fixed so far, and repeat the analysis, we will find that

$$F = F(T,V), \tag{1.33}$$

$$-S = \left.\frac{\partial F}{\partial T}\right|_V, \tag{1.34}$$

$$-P = \left.\frac{\partial F}{\partial V}\right|_T, \tag{1.35}$$

$$dF = -SdT - PdV. \tag{1.36}$$

The following recommended exercise invites you to find $F(T,V)$ for an ideal gas by carrying out the Legendre transform.

Exercise 1.3.1 *For an ideal gas, start with Eq. (1.14) and show that*

$$T(S,V) = \frac{\partial U}{\partial S} = \frac{2}{3nR}U \tag{1.37}$$

to obtain U as a function of T. Next, construct $F(T,V) = U(T) - S(T,V)T$; to get $S(T,V)$, go back to Eq. (1.14), write S in terms of U, and then U in terms of T, and show that

$$F(T,V) = \frac{3nRT}{2}\left[(1+\ln C) - \ln\frac{3nRT}{2} - \frac{2}{3}\ln V\right]. \tag{1.38}$$

Verify that the partial derivatives with respect to T and V give the expected results for the entropy and pressure of an ideal gas.

1.4 Equilibrium as Minimum of Free Energy

Knowledge of $F(T,V)$ is as complete as the knowledge of $U(S,V)$. Often $F(T)$ is preferred, since it is easier to control its independent variable T than the entropy S which enters $U(S)$. What does it mean to say that $F(V,T)$ has the same information as $U(V,S)$?

Start with the principle defining the equilibrium of an isolated system as the *maximum of S at fixed U* (when a constraint is removed). It can equally well be stated as the *minimum of U at fixed S*. After the Legendre transformation from U to F, is there an equivalent principle that determines equilibrium? If so, what is it?

It is that when a constraint is removed in a system in equilibrium with a reservoir at fixed T, *it will find a new equilibrium state that minimizes* F.

Let us try this out for a simple case. Imagine the same box of gas as before with an immovable piston that divides the volume into $V_1 = \alpha V$ and $V_2 = (1-\alpha)V$, but in contact with a reservoir at T. The two parts of the box are at the same temperature by virtue of being in contact with the reservoir, but at possibly different pressures. If we now remove the constraint, the pin holding the piston in place, where will the piston come to rest? The claim above is that the piston will come to rest at a position where the free energy $F(\alpha)$ is maximized. Let us repeat the arguments leading to Eq. (1.21) with S replaced by F:

$$0 = dF = dF_1 + dF_2 \tag{1.39}$$

$$= \frac{\partial F_1}{\partial V_1} dV_1 + \frac{\partial F_2}{\partial V_2} dV_2 \tag{1.40}$$

$$0 = \left(\frac{-P_1 + P_2}{T} \right) d\alpha \tag{1.41}$$

$$P_1 = P_2, \tag{1.42}$$

which is the correct answer.

This completes our review of thermodynamics. We now turn to statistical mechanics.

1.5 The Microcanonical Distribution

Statistical mechanics provides the rational, microscopic foundations of thermodynamics in terms of the underlying atoms (and molecules). There are many equivalent formulations, depending on what is held fixed: the energy, the temperature, the number of particles, and so forth.

Consider an isolated system in equilibrium. It can be in any one of its possible *microstates* – states in which it is described in maximum possible detail. In the classical case this would be done by specifying the coordinate x and momentum p of every particle in it, while in the quantum case it would be the energy eigenstate of the entire system (energy being the only conserved quantity).

In statistical mechanics one abandons a microscopic description in favor of a statistical one, being content to give the probabilities for measuring various values of macroscopic quantities like pressure. One usually computes the average, and sometimes the fluctuations around the average.

Consider a system that can be in one of many microstates. Let the state labeled by an index i occur with probability p_i, and in this state an observable O has a value $O(i)$. The average of O is

$$\langle O \rangle = \sum_i p_i O(i). \tag{1.43}$$

One measure of fluctuations is the mean-squared deviation:

$$(\Delta O)^2 = \sum_i (O(i) - \langle O \rangle)^2 p_i = \langle O^2 \rangle - \langle O \rangle^2. \tag{1.44}$$

To proceed, we need p_i, the probability that the system will be found in a particular microstate i. This is given by the fundamental postulate of statistical mechanics: *A macroscopic isolated system in thermal equilibrium is equally likely to be found in any of its accessible microstates.* This "equal weight" probability distribution is called the *microcanonical distribution.*

A central result due to Boltzmann identifies the entropy of the isolated system to be

$$S = k \ln \Omega, \tag{1.45}$$

where Ω is the number of different microscopic states or *microstates* of the system compatible with its known macroscopic properties, such as its energy, volume, and so on. Boltzmann's constant $k = 1.38 \times 10^{-23}\,\mathrm{J\,K^{-1}}$ is related to the macroscopically defined gas constant R, which enters

$$PV = nRT, \tag{1.46}$$

and Avogadro's number $N_A \simeq 6 \times 10^{23}$ as follows:

$$R = N_A k. \tag{1.47}$$

I will illustrate Eq. (1.45) by applying it to an ideal gas of non-interacting atoms, treated classically. I will compute

$$S(U,V,N) = k \ln \Omega(U,V,N), \tag{1.48}$$

where $\Omega(U,V,N)$ is the number of states in which every atomic coordinate of the N atoms lies inside the box of volume V and the momenta are such that the sum of the individual kinetic energies adds up to U.

For pedagogical reasons the complete dependence of S on N will not be computed here. We will find, however, that as long as N is fixed, the partial derivatives of S with respect to U and V can be evaluated with no error, and these will confirm beyond any doubt that Boltzmann's S indeed corresponds to the one in thermodynamics, by reproducing $PV = NkT$ and $U = \frac{3}{2}NkT$.

First, consider the spatial coordinates. Because each atom is point-like in our description, its position is a point. If we equate the number of possible positions to the number of points inside the volume V, the answer will be infinite, no matter what V is! So what one does is divide the box mentally into tiny cells of volume a^3, where a is some tiny number determined by our desired accuracy in specifying atomic positions in practice. Let us say we choose $a = 10^{-6}$ m. In a volume V, there will be V/a^3 cells indexed by $i = 1, 2, \ldots, V/a^3$. We label the atoms A, B, ..., and say in which cell each one lies. If A is in cell $i = 20$ and B in cell $i = 98\,000$, etc., that's one microscopic arrangement or microstate.

We can assign them to other cells, and if we permute them, say with $A \to B \to C \to D \to A$, that is counted as another arrangement (except when two exchanged atoms are in the same cell). Thus, when the gas is restricted to volume V, and each of the N atoms has V/a^3 possible cell locations, the number of positional configurations is

$$\Omega_V = \left[\frac{V}{a^3}\right]^N, \tag{1.49}$$

and the entropy associated with all possible positions is

$$S_V = k \ln \left[\frac{V}{a^3}\right]^N = Nk \ln \frac{V}{a^3}. \tag{1.50}$$

Notice that S_V depends on the cell size a. If we change a, we will change S_V by a constant, because of the $\ln a^3$ term. This is unavoidable until quantum mechanics comes in to specify a unique cell size. However, *changes* in S_V, which alone are defined in classical statistical mechanics, will be unaffected by the varying a.

But Eq. (1.50) is incomplete. The state of the atom is not given by just its location, but also its momentum $\boldsymbol{p}$. Thus, Ω_V above should be multiplied by a factor $\Omega_p(U)$ that counts the number of momentum states open to the gas at a given value of U. Again, one divides the possible atomic momenta into cells of some size b^3. Whereas the atoms could occupy any spatial cell in the box independently of the others, now they can only assume momentum configurations in which the total kinetic energy of the gas adds up to a given fixed U. Thus, the formula to use is

$$\Omega = \left[\frac{V}{a^3}\right]^N \times \Omega_p(U), \tag{1.51}$$

$$S(U,V) = Nk \ln \frac{V}{a^3} + k \ln \Omega_p(U). \tag{1.52}$$

Now for the computation of $\Omega_p(U)$. The energy of an ideal gas is entirely kinetic and independent of the particle positions. The internal energy is (for the allowed configuration with every atom inside the box)

$$U = \sum_{i=1}^{N} \frac{1}{2} m |v_i|^2 = \sum_{i=1}^{N} \frac{|\boldsymbol{p}_i|^2}{2m} = \sum_{i=1}^{N} \frac{p_{ix}^2 + p_{iy}^2 + p_{iz}^2}{2m}, \tag{1.53}$$

where $\boldsymbol{p} = m\boldsymbol{v}$ is the momentum.

Let us now form a vector $\boldsymbol{P}$ with $3N$ components,

$$\boldsymbol{P} = (p_{1x}, p_{1y}, p_{1z}, p_{2x}, \ldots, p_{Nz}), \tag{1.54}$$

which is simply the collection of the 3 components of the N momentum vectors $\boldsymbol{p}_i$. If we renumber the components of $\boldsymbol{P}$ with an index $j = 1, \ldots, 3N$,

$$\boldsymbol{P} = (P_1, P_2, \ldots, P_{3N}), \tag{1.55}$$

that is to say,

$$P_1 = p_{1x}, P_2 = p_{1y}, P_3 = p_{1z}, P_4 = p_{2x}, \ldots, P_{3N} = p_{Nz}, \tag{1.56}$$

we may write

$$U = \sum_{j=1}^{3N} \frac{P_j^2}{2m}. \tag{1.57}$$

Regardless of their position, the atoms can have any momentum as long as the components satisfy Eq. (1.57). So we must see how many possible momenta exist obeying this condition. The condition may be rewritten as

$$\sum_{j=1}^{3N} P_j^2 = 2mU. \tag{1.58}$$

This is the equation for a hypersphere of radius $\mathcal{R} = \sqrt{2mU}$ in $3N$ dimensions. By dimensional analysis, a sphere of radius $\mathcal{R}$ in d dimensions has an area that goes as $\mathcal{R}^{d-1}$. In our problem, $\mathcal{R} = \left[\sqrt{2mU}\right]$ and $d = 3N - 1 \simeq 3N$. If we divide the individual momenta into cells of size b^3, which, like a^3, is small but arbitrary, the total number of states allowed to the gas behaves as

$$\Omega(V,U) = V^N U^{3N/2} f(m,N,a,b), \tag{1.59}$$

where we have focused on the dependence on U and V and lumped the rest of the dependence on m, a, b, and N in the unknown function $f(m,N,a,b)$. We do not need f because we just want to take

$$S = k \ln \Omega = k\left[N \ln V + \frac{3N}{2} \ln U\right] + k \ln f(m,N,a,b) \tag{1.60}$$

and find its V and U partial derivatives, to which f makes no contribution. These derivatives are

$$\left.\frac{\partial S}{\partial V}\right|_U = \frac{kN}{V}, \tag{1.61}$$

$$\left.\frac{\partial S}{\partial U}\right|_V = \frac{3kN}{2U}. \tag{1.62}$$

If $S = k \ln \Omega$ were indeed the S of thermodynamics, it should obey

$$\left.\frac{\partial S}{\partial V}\right|_U = \frac{P}{T}, \tag{1.63}$$

$$\left.\frac{\partial S}{\partial U}\right|_V = \frac{1}{T}. \tag{1.64}$$

We will see that assuming this indeed gives us the correct ideal gas equation:

$$\frac{kN}{V} = \frac{P}{T} \quad \text{which is just } PV = NkT; \tag{1.65}$$

$$\frac{3kN}{2U} = \frac{1}{T} \quad \text{which is just } U = \tfrac{3}{2}NkT. \tag{1.66}$$

Thus we are able derive these *equations of state* of the ideal gas from the Boltzmann definition of entropy. Going forward, remember that the (inverse) temperature is the derivative of Boltzmann's entropy with respect to energy, exactly as in thermodynamics:

$$\left.\frac{\partial S}{\partial U}\right|_V = \frac{1}{T}. \tag{1.67}$$

Dividing both sides by k, we obtain another important variable, β:

$$\beta = \left.\frac{\partial \ln \Omega(U)}{\partial U}\right|_V = \frac{1}{kT}. \tag{1.68}$$

With more work we could get the full N-dependence of Ω as well. It has interesting consequences, but I will not go there, leaving it to you to pursue the topic on your own.

While

$$S = k \ln \Omega \tag{1.69}$$

is valid for any thermodynamic system, computing Ω is generally impossible except for some idealized models, like the ideal gas.

Finally, consider two systems that are independent. Then

$$\Omega = \Omega_1 \times \Omega_2, \tag{1.70}$$

i.e., the number of options open to the two systems is the product of the numbers open to each. This ensures that the total entropy, S, is additive:

$$S = S_1 + S_2. \tag{1.71}$$

1.6 Gibbs's Approach: The Canonical Distribution

In contrast to Boltzmann, who gave a statistical description of isolated systems with a *definite energy* U, Gibbs wanted to describe systems at a *definite temperature* by virtue of being in thermal equilibrium with a heat reservoir at a fixed T. For example, the system could be a gas, confined to a heat-conducting box, placed inside a gigantic oven at that T. By definition, the temperature of the reservoir remains fixed no matter what the system does.

What is the probability p_i that the system will be in microstate i? We do not need a new postulate for this. The system and reservoir may exchange heat but the two of them are isolated from everything else and have a total conserved energy U_0. *So, together, they obey the microcanonical distribution: every microstate of the joint system is equally likely.* Let $\Omega_R(U)$ be the number of microstates the reservoir can be in if its energy is U. Consider two cases: one where the system is in a *particular* microstate i of energy $E(i)$ and the reservoir is in any one of $\Omega_R(U_0 - E(i))$ microstates, and the other where the system is in a particular microstate j with energy $E(j)$ and the reservoir is in any one of $\Omega_R(U_0 - E(j))$ microstates. What can we say about the probabilities $p(i)$ and $p(j)$ for these two outcomes? Since the system is in a definite microstate in either case (i or j), the number of states the entire system can have is just the corresponding $\Omega_R(U)$. From the microcanonical postulate, the ratio of probabilities for the two cases is:

$$\frac{p(i)}{p(j)} = \frac{\Omega_R(U_0 - E(i)) \times 1}{\Omega_R(U_0 - E(j)) \times 1}, \tag{1.72}$$

where the 1's denote the one microstate (i or j) open to the system under the conditions specified.

It is natural to work with the logarithm of Ω. Since $E(i) \ll U_0$, we Taylor expand $\ln \Omega_R$:

$$\ln \Omega_R(U_0 - E(i)) = \ln \Omega_R(U_0) - \beta E(i) + \cdots, \text{ where} \tag{1.73}$$

$$\beta = \left.\frac{\partial \ln \Omega_R(U)}{\partial U}\right|_{U_0} = \frac{1}{kT}. \tag{1.74}$$

We drop higher derivatives because they correspond to the rate of change of β or $1/kT$ as the system moves up and down in energy, and this is zero because the T of the reservoir, by definition, is unaffected by what the system does.

Exponentiating both sides of Eq. (1.73) and invoking Eq. (1.72),

$$\frac{p(i)}{p(j)} = \frac{\Omega_R(U_0)e^{-\beta E(i)}}{\Omega_R(U_0)e^{-\beta E(j)}} = \frac{e^{-\beta E(i)}}{e^{-\beta E(j)}}; \tag{1.75}$$

that is to say, the relative probability of the system being in state i of energy $E(i)$ is $e^{-\beta E(i)}$. This is the *canonical distribution* and $e^{-\beta E(i)}$ is called the *Boltzmann weight* of configuration i.

Now $p(i)$, the *absolute probability* obeying

$$\sum_i p_i = 1 \tag{1.76}$$

(which I denote by the same symbol), is given by rescaling each p_i by the sum over i:

$$p(i) = \frac{e^{-\beta E(i)}}{\sum_i e^{-\beta E(i)}} = \frac{e^{-\beta E(i)}}{Z}, \tag{1.77}$$

where we have defined the *partition function*

$$Z = \sum_i e^{-\beta E(i)}. \tag{1.78}$$

While Eq. (1.78) for Z holds for both classical and quantum problems, there is a big difference in the work involved in evaluating it, as illustrated by the following example.

Consider a harmonic oscillator of frequency ω_0. In the classical case, given the energy in terms of the coordinate x and momentum p,

$$E(x,p) = \frac{p^2}{2m} + \frac{1}{2}m\omega_0^2 x^2, \tag{1.79}$$

we need to evaluate the integral

$$Z(\beta) = \int_{-\infty}^{\infty}\int_{-\infty}^{\infty} dp\, dx\, e^{-\beta\left(\frac{p^2}{2m}+\frac{1}{2}m\omega_0^2x^2\right)}. \tag{1.80}$$

(Constants like the cell sizes a or b which will not ultimately affect physical quantities are suppressed in this Z.)

Exercise 1.6.1 *Evaluate $Z(\beta)$ for the classical oscillator.*

In the quantum case, given the Hamiltonian operator in terms of the corresponding operators X and P,

$$H(X,P) = \frac{P^2}{2m} + \frac{1}{2}m\omega_0^2 X^2, \tag{1.81}$$

we need to *first* solve its eigenvalue problem, take its eigenvalues

$$E_n = \left(n + \frac{1}{2}\right)\hbar\omega_0, \tag{1.82}$$

and *then* do the sum

$$Z = \sum_{n=0}^{\infty} e^{-\beta(n+\frac{1}{2})\hbar\omega_0}. \tag{1.83}$$

Exercise 1.6.2 *Evaluate $Z(\beta)$ for the quantum oscillator.*

In the quantum case the sum over energy eigenvalues makes Z the trace of $e^{-\beta H}$ in the energy basis. Since the trace is basis independent, we can switch to another basis, say the eigenkets $|x\rangle$ of X:

$$Z = \sum_i e^{-\beta E(i)} = \mathrm{Tr}\, e^{-\beta H} = \int_{-\infty}^{\infty} dx \langle x|e^{-\beta H(X,P)}|x\rangle. \tag{1.84}$$

This version in terms of X-eigenstates leads to the path integral, to be described later in this chapter.

Let us look at the sum (in the classical or quantum cases)

$$Z(\beta) = \sum_i e^{-\beta E(i)}. \tag{1.85}$$

If the sum is evaluated in closed form, and we have the partition function $Z(\beta)$, we can extract all kinds of statistical properties of the energy of the system. For example, $\langle E \rangle$, the weighted average of energy, is

$$\langle E \rangle = \frac{\sum_i E(i) e^{-\beta E(i)}}{Z} = \frac{1}{Z} \frac{\partial Z}{\partial(-\beta)} = -\frac{\partial \ln Z}{\partial \beta} = U. \tag{1.86}$$

This weighted average $\langle E \rangle$ is also denoted by U since this is what corresponds to the *internal energy* U of thermodynamics. I will switch between U and $\langle E \rangle$ depending on the context.

Exercise 1.6.3 *Evaluate $\langle E \rangle$ for the classical and quantum oscillators. Show that they become equal in the appropriate limit.*

Since $\ln Z$ occurs frequently, let us give it a name:

$$\ln Z = -\beta F, \qquad \text{or} \qquad Z = e^{-\beta F}. \tag{1.87}$$

This notation is intentional because F may be identified with the free energy of thermodynamics. Recall that in statistical mechanics,

$$\langle E \rangle \stackrel{\text{def}}{=} U = -\frac{d \ln Z}{d\beta} = \frac{\partial [\beta F]}{\partial \beta}. \tag{1.88}$$

On the other hand, we have seen in Eq. (1.32) that in thermodynamics,

$$U = F + ST = F - T\frac{dF}{dT} = F + \beta \frac{dF}{d\beta} = \frac{d[\beta F]}{d\beta}, \tag{1.89}$$

in agreement with Eq. (1.88).

1.7 More on the Free Energy in Statistical Mechanics

Having seen that the F that appears in statistical mechanics corresponds to the F in thermodynamics, let us see what additional insights statistical mechanics provides.

Consider C_V, the *specific heat at constant volume*:

$$C_V = \left.\frac{dQ}{dT}\right|_V \tag{1.90}$$

$$= \left.\frac{dU}{dT}\right|_V \quad (\text{since } PdV = 0) \tag{1.91}$$

$$= -\frac{1}{kT^2} \frac{\partial^2 [\beta F]}{\partial \beta^2} \tag{1.92}$$

$$= -T \frac{\partial^2 F}{\partial T^2}. \tag{1.93}$$

Consider next the second derivative of βF appearing in Eq. (1.92). If we take one more $(-\beta)$-derivative of Eq. (1.86), we obtain

$$-\frac{\partial^2[\beta F]}{\partial\beta^2} = \frac{\sum_i (E(i))^2 e^{-\beta E(i)}}{Z} - \frac{\left(\sum_i E(i)e^{-\beta E(i)}\right)^2}{Z^2}$$
$$= \langle E^2\rangle - \langle E\rangle^2 = \langle (E-\langle E\rangle)^2\rangle, \tag{1.94}$$

which measures the *fluctuations in energy around the average*. It follows that the specific heat is a direct measure of the fluctuations in internal energy in the canonical distribution.

Statistical mechanics has managed to give a microscopic basis for the entropy of a system of definite energy U via Boltzmann's formula,

$$S = k\ln\Omega(U), \tag{1.95}$$

where

$$\Omega(U) = \sum_i 1, \tag{1.96}$$

where the sum on i is over every allowed state of the prescribed energy U. (I have suppressed the dependence of S on V and N.)

The partition function

$$Z = \sum_i e^{-\beta E_i} \tag{1.97}$$

is a similar sum, but over states with all energies, and weighted by $e^{-\beta E_i}$. Does its logarithm correspond to anything in thermodynamics? The answer, as claimed earlier, is that

$$\ln Z = -\beta F, \quad \text{or } F = -kT\ln Z, \tag{1.98}$$

where F is the free energy. This will now be confirmed in greater detail.

Since, in thermodynamics,

$$dF = -SdT - PdV, \tag{1.99}$$

it must be true of the putative free energy defined by $\ln Z = -\beta F$ that

$$\frac{\partial F}{\partial V} = -P, \tag{1.100}$$
$$\frac{\partial F}{\partial T} = -S. \tag{1.101}$$

Let us verify this. First consider

$$-\frac{\partial F}{\partial V} = \frac{\partial[kT\ln Z]}{\partial V} \tag{1.102}$$
$$= kT\frac{1}{Z}\sum_i \frac{\partial e^{-E_i/kT}}{\partial V} \tag{1.103}$$

$$= \sum_i \frac{e^{-E_i/kT}}{Z} \left[-\frac{dE_i}{dV} \right] \tag{1.104}$$

$$= \sum_i p_i P_i \tag{1.105}$$

$$= P, \tag{1.106}$$

where p_i is the probability of being in state i and $\left[-\frac{dE_i}{dV}\right] = P_i$ is the pressure if the system is in state i. It is the weighted average of P_i that emerges as the P of thermodynamics. As in the case of U, thermodynamics deals with the averaged quantities of statistical mechanics and ignores the fluctuations.

Next, consider

$$-\frac{\partial F}{\partial T} = \frac{\partial [kT \ln Z]}{\partial T} \tag{1.107}$$

$$= k \ln Z + kT \frac{1}{Z} \sum_i e^{-E_i/kT} \frac{E_i}{kT^2} \tag{1.108}$$

$$= k \sum_i \frac{e^{-E_i/kT}}{Z} \left[\ln Z + \frac{E_i}{kT} \right] \tag{1.109}$$

$$= -k \sum_i p_i \ln p_i, \tag{1.110}$$

where $p_i = e^{-E_i/kT}/Z$ is the absolute probability of being in state i.

This does not at all look like the S of thermodynamics, or even Boltzmann's S. I will now argue that, nonetheless,

$$S = -k \sum_i p_i \ln p_i \tag{1.111}$$

indeed represents entropy.

First consider a case when all the allowed energies are equal to some E^*. If there are $\Omega(E^*)$ of them, each has the same probability of being realized:

$$p_i = 1/\Omega(E^*) \ \forall\, i, \tag{1.112}$$

and

$$S = -k \sum_{i=1}^{\Omega(E^*)} \frac{1}{\Omega(E^*)} \ln \frac{1}{\Omega(E^*)} \tag{1.113}$$

$$= k \ln \Omega(E^*). \tag{1.114}$$

Thus, Eq. (1.111) appears to be the natural extension of $\Omega = k \ln \Omega(E)$ for a system with fixed energy to a system that can sample all energies with the Boltzmann probability p_i.

But here is the decisive proof that Eq. (1.111) indeed describes the familiar S. It consists of showing that this S obeys

$$TdS = \delta Q. \tag{1.115}$$

To follow the argument, we need to know how heat and work are described in the language of the partition function. Consider

$$U = \sum_i p_i E_i \tag{1.116}$$

$$dU = \sum_i (dp_i E_i + p_i dE_i) \tag{1.117}$$

$$= \delta Q - PdV. \tag{1.118}$$

The identification in the last line of the two terms in the penultimate line comes from asking how we can change the energy of a gas. If we fix the volume and heat (or cool) it, the energy levels remain fixed but the probabilities change. This is the heat added:

$$\delta Q = \sum_i (dp_i E_i). \tag{1.119}$$

If we very, very slowly let the gas contract or expand, so that no transitions take place, the particles will rise and fall with the levels as the levels move up or down. This corresponds to the work:

$$\sum_i p_i dE_i = \sum_i p_i \frac{dE_i}{dV} \tag{1.120}$$

$$= -\sum_i p_i P_i dV \tag{1.121}$$

$$= -PdV. \tag{1.122}$$

Armed with this, let us return to S. We find, starting with

$$S = -k \sum_i p_i \ln p_i, \text{ that} \tag{1.123}$$

$$TdS = -kT \sum_i \left[dp_i \ln p_i + \frac{p_i}{p_i} dp_i \right] \tag{1.124}$$

$$= kT \sum_i dp_i \left[\frac{E_i}{kT} + \ln Z \right] \text{ using } \textstyle\sum_i dp_i = 0 \tag{1.125}$$

$$= \sum_i dp_i E_i \text{ using } \textstyle\sum_i dp_i = 0 \text{ again} \tag{1.126}$$

$$= \delta Q \text{ based on Eq. (1.119).} \tag{1.127}$$

This concludes the proof that the F in $F = -kT \ln Z$ indeed corresponds to the free energy of thermodynamics, but with a much deeper microscopic underpinning.

The system in contact with the heat bath and described by Z need not be large. It could even be a single atom in a gas. The fluctuations of the average energy relative to the average need not be small in general. However, for large systems a certain simplification arises.

Consider a very large system. Suppose we group all states of energy E (and hence the same Boltzmann weight) and write Z as a sum over energies rather than states; we obtain

$$Z = \sum_E e^{-\beta E} \Omega(E), \tag{1.128}$$

where Ω is the number of states of energy E. For a system with N degrees of freedom, $\Omega(E)$ grows very fast in N (as E^N) while $e^{-\beta E}$ falls exponentially. The product then has a very sharp maximum at some E^*. The maximum of

$$e^{-\beta E}\Omega(E) = e^{-(\beta E - \ln \Omega(E))} \tag{1.129}$$

is the minimum of $\beta E - \ln \Omega(E)$, which occurs when its E-derivative vanishes:

$$\beta = \left. \frac{d \ln \Omega(E)}{dE} \right|_{E^*}, \tag{1.130}$$

which is the familiar result that in the most probable state the temperatures of the two parts in thermal contact, the reservoir and the system, are equal. (This assumes the system is large enough to have a well-defined temperature. For example, it cannot contain just ten molecules. But then the preceding arguments would not apply anyway.)

We may write

$$Z = A e^{-\beta(E^* - kT \ln \Omega(E^*))}, \tag{1.131}$$

where A is some prefactor which measures the width of the peak and is typically some power of N. If we take the logarithm of both sides we obtain (upon dropping $\ln A$ compared to the ln of the exponential),

$$\ln Z = -\beta F = \beta(E^* - kT \ln \Omega(E^*)) \equiv \beta(E^* - S(E^*)T), \tag{1.132}$$

where the *entropy* of the system is

$$S = k \ln \Omega(E^*). \tag{1.133}$$

In the limit $N \to \infty$, the maximum at E^* is so sharp that the system has essentially that one energy which we may identify also with the mean $\langle E \rangle = U$ and write

$$F = U - ST. \tag{1.134}$$

The free energy thus captures the struggle between energy and entropy. At low temperatures a few states at the lowest E dominate Z because the Boltzmann factor favors them. At higher temperatures, their large numbers (entropy) allow states at high energy to overcome the factor $e^{-\beta E}$ which suppresses them individually.

1.8 The Grand Canonical Distribution

We conclude with one more distribution, the *grand canonical distribution*. Consider a system that is in contact with an ideal reservoir with which it can exchange heat as well as particles. Let the total energy be E_0 and total number of particles be N_0. In analogy with the canonical distribution, we will now be dealing with the following Taylor series when the system has N particles and is in a state i of energy $E(i)$:

$$\ln[\Omega(E_0 - E(i), N_0 - N)] = \ln\Omega(E_0, N_0) - \beta E(i) + \beta\mu N + \cdots \tag{1.135}$$

where the *chemical potential* μ is defined as

$$\mu = -\frac{1}{\beta}\left.\frac{\partial \ln\Omega(E_R, N_R)}{\partial N_R}\right|_{E_0, N_0}. \tag{1.136}$$

The absolute probability of the system having N particles and being in an N-particle state i of energy $E(i)$ is

$$p(E(i), N) = \frac{e^{-\beta(E(i)-\mu N)}}{Z}, \tag{1.137}$$

where the *grand canonical partition function* is

$$Z = \sum_N \sum_{i(N)} e^{-\beta(E(i)-\mu N)}, \tag{1.138}$$

where $i(N)$ is any N-particle state. The average number of particles in the system is given by

$$\langle N \rangle = \frac{1}{\beta}\left.\frac{\partial \ln Z}{\partial \mu}\right|_{\beta}. \tag{1.139}$$

The average energy U is determined by the relation

$$-\frac{d\ln Z}{d\beta} = \langle U \rangle - \mu\langle N \rangle. \tag{1.140}$$

The system can be found having any energy and any number of particles, neither being restricted in the sum in Z. However, if the system is very large, it can be shown that the distribution will have a very sharp peak at a single energy E^* and a single number of particles N^*, depending on β and μ. In other words, for large systems all ensembles become equivalent to the microcanonical ensemble. However, the ones with unrestricted sums over E and N are more convenient to work with.

Exercise 1.8.1 (A must-do problem) *Apply the grand canonical ideas to a system which is a* quantum state *of energy ε that may be occupied by fermions (bosons). Show that in the two cases*

$$\langle N \rangle = n_{F/B} = \frac{1}{e^{\beta(\varepsilon-\mu)} \pm 1} \tag{1.141}$$

by evaluating Z and employing Eq. (1.139). These averages are commonly referred to with lower case symbols as n_F and n_B respectively.

References and Further Reading

[1] H. B. Callen, *Thermodynamics and an Introduction to Thermostatistics*, Wiley, 2nd edition (1965).
[2] F. Reif, *Fundamentals of Thermal and Statistical Physics*, McGraw-Hill (1965).
[3] C. Kittel and H. Kromer, *Thermal Physics*, W. H. Freeman, 2nd edition (1980).
[4] R. K. Pathria and P. D. Beale, *Statistical Mechanics*, Academic Press, 3rd edition (2011).

2

The Ising Model in $d = 0$ and $d = 1$

We are now going to see how the canonical ensemble of Gibbs is to be applied to certain model systems. The problems of interest to us will involve many (possibly infinite) degrees of freedom, classical or quantum, possibly interacting with each other and with external fields, and residing in any number of dimensions d.

We begin with the simplest example and slowly work our way up.

2.1 The Ising Model in $d = 0$

The system in our first example has just two degrees of freedom, called s_1 and s_2. In addition, the degrees of freedom are the simplest imaginable. They are *Ising spins*, which means they can take only two values: $s_i = \pm 1$. The configuration of the system is given by the pair of values (s_1, s_2).

To compute the partition function Z, we need the energy of the system for every configuration (s_1, s_2). It is assumed to be

$$E(s) = -Js_1s_2 - B(s_1 + s_2). \tag{2.1}$$

This energy models a system made of magnetic moments. Although in general the moments can point in any direction, in the Ising model they point up or down one axis, say the z-axis, along which we are applying an external magnetic field B. The constant J has units of energy, and a constant that multiplies B (and has units of magnetic moment) to give the second terms the units of energy has been suppressed (set equal to unity.)

In our illustrative example we assume $J > 0$, the *ferromagnetic* case, where the energy favors aligned spins. The *antiferromagnetic* case, $J < 0$, favors antiparallel spins. In both cases, we choose $B > 0$, so that the spins like to align with the external field B.

Let us evaluate the partition function for this case:

$$Z = \sum_{s_1, s_2} e^{Ks_1s_2 + h(s_1+s_2)}, \tag{2.2}$$

$$K = \beta J, \quad h = \beta B. \tag{2.3}$$

It is readily seen that

$$Z(K,h) = 2\cosh(2h) \cdot e^{K} + 2e^{-K}. \tag{2.4}$$

At very low T or very high β, the state of lowest energy should dominate the sum and we expect the spins to be aligned with B and also with each other. At very high T and vanishing β, all four states should get equal weight and the spins should fluctuate independently of each other and the applied field. Let us see if all this follows from Eq. (2.4).

First, consider

$$M = \frac{s_1 + s_2}{2}, \tag{2.5}$$

the *average magnetization* of the system in any given configuration. (This average is over the spins in the system, and not the thermal average. It can be defined at every instant. Had there been N spins, we would have divided by N to obtain the average magnetization.) It takes on different values in different states in the sum. The weighted or *thermal* average of M, denoted by $\langle M \rangle$, is given by

$$\langle M \rangle = \frac{\sum_{s_1,s_2} \frac{1}{2}(s_1 + s_2) e^{Ks_1 s_2 + h(s_1+s_2)}}{Z} \tag{2.6}$$

$$= \frac{1}{2Z} \frac{\partial Z(K,h)}{\partial h} \tag{2.7}$$

$$= \frac{1}{2} \frac{\partial \ln Z(K,h)}{\partial h}. \tag{2.8}$$

In terms of the free energy $F(K,h)$, defined as before by

$$Z = e^{-\beta F}, \tag{2.9}$$

we have

$$\langle M \rangle = \frac{1}{2} \frac{\partial[-\beta F(K,h)]}{\partial h}. \tag{2.10}$$

In our problem,

$$-\beta F = \ln\left(2\cosh(2h) \cdot e^{K} + 2e^{-K}\right), \tag{2.11}$$

so that

$$\langle M \rangle = \frac{\sinh 2h}{\cosh 2h + e^{-2K}}. \tag{2.12}$$

As expected, $\langle M \rangle \to 1$ as $h, K \to \infty$, and $\langle M \rangle \to h$ as $h, K \to 0$.

Suppose we want the thermal average of a *particular* spin, say s_1. We should then add a *source term* $h_1 s_1 + s_2 h_2$ in the Boltzmann weight which couples each spin s_i to its own independent field h_i, so that

$$Z = \sum_{s_1,s_2} e^{Ks_1 s_2 + h_1 s_1 + h_2 s_2} = e^{-\beta F(K,h_1,h_2)}. \tag{2.13}$$

Now you can check that

$$\frac{\partial[-\beta F(K,h_1,h_2)]}{\partial h_i} = \frac{\partial \ln Z}{\partial h_i} = \frac{1}{Z}\frac{\partial Z}{\partial h_i} = \langle s_i \rangle. \tag{2.14}$$

Taking the mixed derivative with respect to h_1 and h_2, we obtain

$$\frac{\partial^2[-\beta F(K,h_1,h_2)]}{\partial h_1 \partial h_2} = \frac{\partial}{\partial h_1}\left(\frac{1}{Z}\frac{\partial Z}{\partial h_2}\right) \tag{2.15}$$

$$= \frac{1}{Z}\frac{\partial^2 Z}{\partial h_1 \partial h_2} - \frac{1}{Z^2}\frac{\partial Z}{\partial h_1}\frac{\partial Z}{\partial h_2} \tag{2.16}$$

$$= \langle s_1 s_2 \rangle - \langle s_1 \rangle\langle s_2 \rangle \equiv \langle s_1 s_2 \rangle_c, \tag{2.17}$$

which is called the *connected correlation function*. It differs from $\langle s_1 s_2 \rangle$, called the *correlation function*, in that $\langle s_1 \rangle\langle s_2 \rangle$, the product of single spin averages, is subtracted.

The significance of the two correlation functions will be taken up in Section 3.3.

Exercise 2.1.1 *Prove Eq. (2.17).*

If there are four spins coupled to $h_1, \ldots, h_4$, we can show that

$$\left.\frac{\partial^4[-\beta F(K,h_1,h_2,h_3,h_4)]}{\partial h_1 \partial h_2 \partial h_3 \partial h_4}\right|_{h=0}$$

$$= \langle s_1 s_2 s_3 s_4 \rangle_c \tag{2.18}$$

$$= \langle s_1 s_2 s_3 s_4 \rangle - \langle s_1 s_2 \rangle\langle s_3 s_4 \rangle - \langle s_1 s_3 \rangle\langle s_2 s_4 \rangle - \langle s_1 s_4 \rangle\langle s_2 s_3 \rangle. \tag{2.19}$$

Since $h=0$ in the end, there will be no correlators with an odd number of spins.

Exercise 2.1.2 *Prove Eq. (2.19).*

Equations (2.17) and (2.19) are valid even if s is replaced by some other variable (say with continuous values) coupled to the corresponding "magnetic" field in the Boltzmann weight, because nowhere in their derivation did we use the fact that $s = \pm 1$.

How would we use the sources to compute thermal averages if in the actual problem h is uniform or zero? We would evaluate the h_i-derivatives and *then* set $h_i = h$ or 0 $\forall i$ in Eqs. (2.13) and (2.17). Thus, for example, if there were a uniform external field h we would write

$$\langle s_1 s_2 \rangle - \langle s_1 \rangle\langle s_2 \rangle \equiv \langle s_1 s_2 \rangle_c = \left.\frac{\partial^2[-\beta F(K,h_1,h_2)]}{\partial h_1 \partial h_2}\right|_{h_i = h\ \forall\ i}. \tag{2.20}$$

We will mostly be working with a uniform field h, which might be zero.

It turns out that we can also obtain $\langle s_1 s_2 \rangle$ by taking a K-derivative:

$$\langle s_1 s_2 \rangle = \frac{\partial[-\beta F(K,h)]}{\partial K} \tag{2.21}$$

$$= \frac{e^K \cosh 2h - e^{-K}}{e^K \cosh 2h + e^{-K}}. \tag{2.22}$$

But now we interpret it as measuring the *average interaction energy*, which *happens* to coincide with the correlation of neighboring spins. In a model with more spins, correlations of spins that are not neighbors, say s_5 and s_{92}, cannot be obtained by taking the K-derivative. The distinction between average interaction energy and generic spin correlations is unfortunately blurred in our toy model with just two spins.

Exercise 2.1.3 *Derive Eq. (2.22) and discuss its K dependence at $h = 0$.*

We are taking K and h as the independent parameters. Sometimes we may revert to B, J, and T if we want to know the averages (or their derivatives) as functions of these basic variables.

Higher derivatives with respect to K or h (for uniform h) give additional information about fluctuations about the mean just as in Eq. (1.94), where the second β-derivative gave the fluctuation in energy. Thus,

$$\chi = \frac{1}{2}\frac{\partial\langle M\rangle}{dh} = \frac{1}{4}\frac{\partial^2[-\beta F]}{dh^2} = \langle M^2\rangle - \langle M\rangle^2 \tag{2.23}$$

measures the fluctuation of the magnetization about its mean. It is called the *magnetic susceptibility* because it tells us the rate of change of the average magnetization with the applied field. (The canonical χ is the derivative of $\langle M\rangle$ with respect to B and not h. The χ defined above differs by a factor of β. There are situations where the difference matters, and you will be alerted.)

For later use, let us note that with N spins,

$$\chi = \frac{1}{N}\frac{\partial\langle M\rangle}{dh} = \frac{1}{N^2}\frac{\partial^2[-\beta F]}{dh^2} = \langle M^2\rangle - \langle M\rangle^2. \tag{2.24}$$

2.2 The Ising Model in $d = 1$

By a $d = 1$ system we mean one that extends over the entire infinite line. A system with a finite number of spins, no matter in how many spatial dimensions, is referred to as zero dimensional.

Given that any real system is finite, one may ask why we should focus on infinite systems. The answer is that systems with, say, 10^{23} degrees of freedom look more like infinite systems than finite ones in the following sense. Consider magnetism. We know phenomenologically that below some Curie temperature T_C iron will magnetize. But it can be shown rigorously that a finite system of Ising spins can never magnetize: if it is polarized

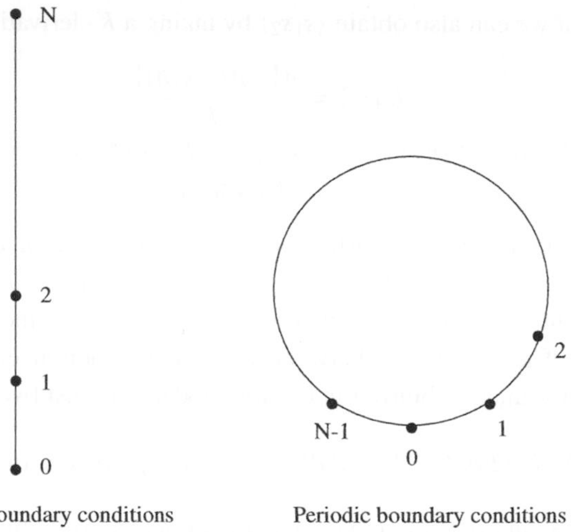

Figure 2.1 Ising models in $d = 1$ with open and periodic boundary conditions. In the latter, site N is not shown since it is the same as site 0.

one way (say up the z-axis) for some time, it can jump to the opposite polarization after some time, so that on average it is not magnetized. We reconcile this rigorous result with reality by noting that the time to flip can be as large as the age of the universe, so that in human time scales magnetization is possible. The nice thing about infinite systems is that remote possibilities are rendered impossible, making them a better model of real life.

We first begin with the Ising model defined on a lattice of $N+1$ equally spaced points numbered by an index $i = 0, 1, \ldots, N$, as shown in the left part of Figure 2.1. We will eventually take the *thermodynamic limit* $N \to \infty$. The reason for lining up the lattice points on a vertical axis is that later it will become the time axis for a related problem and it is common to view time as evolving from the bottom of the page to the top.

By "Ising model" I will always mean the nearest-neighbor Ising model, in which each spin interacts with its nearest neighbor, unless otherwise stated. In $d = 1$, the energy of a given configuration is

$$E = -J \sum_{0}^{N-1} s_i s_{i+1}. \tag{2.25}$$

We initially set the external field $B = 0 = h$.

The partition function is

$$Z = \sum_{s_i = \pm 1} \exp\left[\sum_{i=0}^{N-1} K(s_i s_{i+1} - 1)\right], \tag{2.26}$$

where $K = \beta J > 0$. The additional, spin-independent constant of $-K$ is added to every site for convenience. This will merely shift βF by NK.

Let us first keep s_0 fixed at one value and define a relative-spin variable:

$$t_i = s_i s_{i+1}. \tag{2.27}$$

Given s_0 and all the t_i, we can reconstruct the state of the system. So we can write

$$Z = \sum_{t_i=\pm 1} \exp\left[\sum_{i=0}^{N-1} K(t_i - 1)\right] = \sum_{t_i=\pm} \prod_{i=0}^{N-1} e^{K(t_i-1)}. \tag{2.28}$$

Since the exponential factorizes into a product over i, we can do the sums over each t_i and obtain (after appending a factor of 2 for the two possible choices of s_0)

$$Z = 2(1+e^{-2K})^N. \tag{2.29}$$

One is generally interested in the free energy per site in the *thermodynamic limit* $N \to \infty$:

$$f(K) = -\lim_{N\to\infty} \frac{1}{N} \ln Z. \tag{2.30}$$

This definition of f differs by a factor β from the traditional one.

We see that

$$-f(K) = \ln(1+e^{-2K}) \tag{2.31}$$

upon dropping $(\ln 2)/N$ in the thermodynamic limit. Had we chosen to fix s_0 at one of the two values, the factor 2 would have been missing in Eq. (2.29) but there would have been no difference in Eq. (2.31) for the free energy per site. Boundary conditions are unimportant in the thermodynamic limit in this sense.

Consider next the correlation function between two spins located at sites i and $j > i$:

$$\langle s_j s_i \rangle = \frac{\sum_{s_k} s_j s_i \exp\left[\sum_k K(s_k s_{k+1} - 1)\right]}{Z}. \tag{2.32}$$

As the name suggests, this function measures how likely s_i and s_j are, on average, to point in the same direction. In any ferromagnetic system, the average will be positive for a pair of neighboring spins since the Boltzmann weight is biased toward parallel values. Surely if a spin can influence its neighbors to be parallel to it, they in turn will act similarly on *their* neighbors. So we expect that a given spin will tend to be correlated not just with its immediate neighbors, but with those even further away. The correlation function $\langle s_j s_i \rangle$ is a quantitative measure of this. (In an antiferromagnet, there will be correlations as well, but now the correlation function will alternate in sign.)

If in addition there is an external magnetic field, it will enhance this correlation further. We will return to this point later.

Using the fact that $s_i^2 \equiv 1$, we can write, for $j > i$,

$$s_j s_i = s_i s_j = s_i s_{i+1} s_{i+1} s_{i+2} \cdots s_{j-1} s_j = t_i t_{i+1} \cdots t_{j-1}. \tag{2.33}$$

Thus,

$$\langle s_j s_i \rangle = \langle t_i \rangle \langle t_{i+1} \rangle \cdots \langle t_{j-1} \rangle. \tag{2.34}$$

The answer factorizes over i since the Boltzmann weight in Eq. (2.28) factorizes over i when written in terms of t_i. The average for any one t is easy:

$$\langle t \rangle = \frac{1e^{0 \cdot K} - 1e^{-2K}}{e^{0 \cdot K} + e^{-2K}} = \tanh K, \tag{2.35}$$

so that, finally,

$$\langle s_j s_i \rangle = (\tanh K)^{j-i} = \exp\left[(j-i)\ln\tanh K\right]. \tag{2.36}$$

By choosing $i > j$ and repeating the analysis, we can see that in general

$$\langle s_j s_i \rangle = (\tanh K)^{|j-i|} = \exp\left[|j-i|\ln\tanh K\right]. \tag{2.37}$$

At any finite K, since $\tanh K < 1$,

$$\langle s_j s_i \rangle \to 0 \text{ as } |j-i| \to \infty. \tag{2.38}$$

Only at $T = 0$ or $K = \infty$ (when $\tanh K = 1$) does the correlation not decay exponentially but remain flat.

The average $\langle s_j s_i \rangle$ depends on just the *difference* in coordinates, a feature called *translation invariance*. This is not a generic result for a finite chain, but a peculiarity of this model. In general, in a problem of $N + 1$ points (for any finite N), correlations between two spins will generally depend on where the two points are in relation to the ends. On the other hand, in all models, when $N \to \infty$ we expect translational invariance to hold for correlations of spins far from the ends.

To have translational invariance in a finite system, we must use *periodic boundary conditions*, so the system has the shape of a ring with the point labeled N identified with that labeled 0, as in the right half of Figure 2.1. Now every point is equivalent to every other. Correlation functions will now depend only on the difference between the two coordinates but they will not decay monotonically with separation because as one point starts moving away from the other, it eventually starts approaching the first point from the other side! Thus the correlation function will be a sum of two terms, one of which grows as $j - i$ increases to values of order N. However, if we promise never to consider separations comparable to N, this complication can be ignored; see Exercise 3.3.6. (Our calculation of correlations in terms of t_i must be amended in the face of periodic boundary conditions to ensure that the sum over t_i is restricted to configurations for which the product of all the t_i over the ring equals unity.)

2.3 The Monte Carlo Method

We have seen that in $d = 1$ the computation of Z and thermal averages like $\langle s_i s_j \rangle$ can be done *exactly*. We cannot do this in general, especially for $d > 1$. Then we turn to other, approximate, methods. Here I describe one which happens to be a *numerical method* called the *Monte Carlo method*. Consider

$$\langle s_i s_j \rangle = \frac{\sum_C s_i s_j e^{-E(C)/kT}}{\sum_C e^{-E(C)/kT}} \equiv \sum_C s_i s_j(C) p(C), \tag{2.39}$$

where C is any configuration of the entire system, (i.e., a complete list of what every spin is doing), $E(C)$ and $p(C)$ are the corresponding energies and absolute probabilities, and $s_i s_j(C)$ is the value of $s_i s_j$ when the system is in configuration C. The system need not be in $d = 1$.

The Monte Carlo method is a trick by which huge multidimensional integrals (or sums) can be done on a computer. In our problem, to do a (Boltzmann) weighted sum over configurations, there is a trick (see Exercise 2.3.1 for details) by which we can make the computer generate configurations with the given Boltzmann probability $p(C)$. In other words, each configuration of spins will occur at a rate proportional to its Boltzmann weight. As the computer churns out these configurations C, we can ask it to remember the value of $s_i s_j$ in that C. Since the configurations are already weighted, the *arithmetic average* of these numbers is the *weighted* thermal average we seek. From this correlation function we can deduce the correlation length ξ. By the same method we can measure the average magnetization, or any other quantity of interest. Clearly this can be done for any problem with a real, positive, Boltzmann weight, not just Ising models, and in any dimension.

Exercise 2.3.1 explains one specific algorithm called the Metropolis method for generating these weighted configurations. You are urged to do it if you want to get a feeling for the basic idea.

The numerical method has its limitations. For example, if we want the thermodynamic limit, we cannot directly work with an infinite number of spins on any real computer. Instead, we must consider larger and larger systems and see if the measured quantities converge to a clear limit.

Exercise 2.3.1 *The* Metropolis method *is a way to generate configurations that obey the Boltzmann distribution. It goes as follows:*

1. *Start with some configuration i of energy* E_i.
2. *Consider another configuration j obtained by changing some degrees of freedom (spins in our example).*
3. *If* $E_j < E_i$ *jump to j with unit probability, and if* $E_j > E_i$, *jump with a probability* $e^{-\beta(E_j - E_i)}$.

If you do this for a long time, you should find that the configurations appear with probabilities obeying

$$\frac{p(i)}{p(j)} = e^{-\beta(E_i - E_j)}. \tag{2.40}$$

To show this, consider the rate equation, whose meaning should be clear:

$$\frac{dp(i)}{dt} = -p(i)\sum_j R(i \to j) + \sum_j p(j)R(j \to i), \tag{2.41}$$

where $R(i \to j)$ is the rate of jumping from i to j and vice versa. In equilibrium or steady state, we will have $\frac{dp(i)}{dt} = 0$. One way to kill the right-hand side is by detailed balance, *to ensure that the contribution from every value of j is separately zero, rather than just the sum. It is not necessary, but clearly sufficient. In this case,*

$$\frac{p(i)}{p(j)} = \frac{R(j \to i)}{R(i \to j)}. \tag{2.42}$$

Show that if you use the rates specified by the Metropolis algorithm, you get Eq. (2.40). (Hint: First assume $E_j > E_i$ and then the reverse.) Given a computer that can generate a random number between 0 and 1, how will you accept a jump with probability $e^{-\beta(E_j - E_i)}$?

3

Statistical to Quantum Mechanics

We have thus far looked at the classical statistical mechanics of a collection of Ising spins in terms of their partition function. We have seen how to extract the free energy per site and correlation functions of the $d = 1$ Ising model.

We will now consider a second way to do this. While the results are, of course, going to be the same, the method exposes a deep connection between classical statistical mechanics in $d = 1$ and the quantum mechanics of a *single* spin-$\frac{1}{2}$ particle. This is the simplest way to learn about a connection that has a much wider range of applicability: the classical statistical mechanics of a problem in d dimensions may be mapped on to a quantum problem in $d - 1$ dimensions. The nature of the quantum variable will depend on the nature of the classical variable. In general, the allowed values of the classical variable ($s = \pm 1$ in our example) will correspond to the maximal set of simultaneous eigenvalues of the operators in the quantum problem (σ_z in our example). Subsequently we will study a problem where the classical variable called x lies in the range $-\infty < x < \infty$ and will correspond to the eigenvalues of the familiar position operator X that appears in the quantum problem. The correlation functions will become the expectation values in the ground state of a certain *transfer matrix*.

The different sites in the $d = 1$ lattice will correspond to different, discrete, times in the life of the quantum degree of freedom.

By the same token, a quantum problem may be mapped into a statistical mechanics problem in one-higher dimensions, essentially by running the derivation backwards. In this case the resulting partition function is also called the *path integral* for the quantum problem. There is, however, no guarantee that the classical partition function generated by a legitimate quantum problem (with a Hermitian Hamiltonian) will always make physical sense: the corresponding Boltzmann weights may be negative, or even complex!

In the quantum mechanics we first learn, time t is a real parameter. For our purposes we also need to get familiar with quantum mechanics in which time takes on purely imaginary values $t = -i\tau$, where τ is real. It turns out that it is possible to define such a *Euclidean quantum mechanics*. The adjective "Euclidean" is used because if t is imaginary, the invariant $x^2 - c^2t^2$ of Minkowski space becomes the Pythagoras sum $x^2 + c^2\tau^2$ of Euclidean space. It is Euclidean quantum mechanics that naturally emerges from classical statistical mechanics. We begin with a review of the key ideas from quantum mechanics.

An excellent source for this chapter and a few down the road is a paper by J. B. Kogut [1] (see also [2]).

3.1 Real-Time Quantum Mechanics

I now review the similarities and differences between real-time and Euclidean quantum mechanics.

In real-time quantum mechanics, the state vectors obey Schrödinger's equation:

$$i\hbar\frac{d|\psi(t)\rangle}{dt} = H|\psi(t)\rangle. \tag{3.1}$$

Given any initial state $|\psi(0)\rangle$ we can find its future evolution by solving this equation. For time-independent Hamiltonians the future of any initial state may be found once and for all in terms of the *propagator* $U(t)$, which obeys

$$i\hbar\frac{dU(t)}{dt} = HU(t), \tag{3.2}$$

as follows:

$$|\psi(t)\rangle = U(t)|\psi(0)\rangle. \tag{3.3}$$

A formal solution to $U(t)$ is

$$U(t) = e^{-i\frac{H}{\hbar}t}. \tag{3.4}$$

Since $H^\dagger = H$,

$$U^\dagger U = I, \tag{3.5}$$

which means U preserves the norm of the state.

If you knew all the eigenvalues and eigenvectors of H you could write

$$U(t) = \sum_n |n\rangle\langle n|e^{-i\frac{E_n}{\hbar}t}, \quad \text{where} \tag{3.6}$$

$$H|n\rangle = E_n|n\rangle. \tag{3.7}$$

If H is simple, say a 2×2 spin Hamiltonian, we may be able to exponentiate H as in Eq. (3.4) or do the sum in Eq. (3.6), but if H is a differential operator in an infinite-dimensional space, the formal solution typically remains formal. I will shortly mention a few exceptions where we can write down U explicitly and in closed form in the X-basis spanned by $|x\rangle$.

In the $|x\rangle$ basis the matrix elements of $U(t)$ written in terms of the energy eigenfunctions $\psi_n(x) = \langle x|n\rangle$ follow from Eq. (3.6):

$$\langle x'|U(t)|x\rangle \equiv U(x',x;t) = \sum_n \psi_n(x')\psi_n^*(x)e^{-i\frac{E_n}{\hbar}t}, \tag{3.8}$$

and denote the amplitude that a particle known to be at position x at time $t = 0$ will be detected at x' at time t. This means that as $t \to 0$, we must expect $U(x,x' : t) \to \delta(x - x')$. Indeed, we see this in Eq. (3.8) by using the completeness condition of the eigenfunctions.

For a free particle of mass m, the states $|n\rangle$ are just plane wave momentum states and the sum over n (which becomes an integral over momenta) can be carried out to yield the explicit expression

$$U(x',x;t) = \sqrt{\frac{m}{2\pi i\hbar t}} \exp\left[\frac{im(x'-x)^2}{2\hbar t}\right]. \tag{3.9}$$

Remarkably, even for the oscillator of frequency ω (where the propagator involves an infinite sum over products of Hermite polynomials) we can write in closed form

$$U(x',x;t) = \sqrt{\frac{m\omega}{2\pi i\hbar \sin\omega t}} \exp\left[\frac{im\omega}{2\hbar \sin\omega t}((x^2 + x'^2)\cos\omega t - 2xx')\right]. \tag{3.10}$$

These two results can be obtained from path integrals with little effort.

So much for the Schrödinger picture. In the Heisenberg picture, operators have time dependence. The Heisenberg operator $\Omega(t)$ is related to the Schrödinger operator Ω as per

$$\Omega(t) = U^\dagger(t)\Omega U(t). \tag{3.11}$$

We define the *time-ordered Green's function* as follows:

$$iG(t) = \langle 0|\mathcal{T}(\Omega(t)\Omega(0))|0\rangle, \tag{3.12}$$

where $|0\rangle$ is the ground state of H and $\mathcal{T}$ is the *time-ordering symbol* that puts the operators that follow it in order of increasing time, reading from right to left:

$$\mathcal{T}(\Omega(t_1)\Omega(t_2)) = \theta(t_2 - t_1)\Omega(t_2)\Omega(t_1) + \theta(t_1 - t_2)\Omega(t_1)\Omega(t_2). \tag{3.13}$$

These Green's functions are unfamiliar in elementary quantum mechanics, but central in quantum field theory and many-body theory. We will see the time-ordered product emerge naturally and describe correlations in statistical mechanics, in, say, the Ising model. For now, I ask you to learn about them in faith.

In condensed matter physics (and sometimes in field theory as well) one often wants a generalization of the above ground state expectation value to one at finite temperature β when the system can be in any state of energy E_n with the probability given by the Boltzmann weight:

$$iG(t,\beta) = \frac{\sum_n e^{-\beta E_n}\langle n|\mathcal{T}(\Omega(t)\Omega(0))|n\rangle}{\sum_n e^{-\beta E_n}} \tag{3.14}$$

$$= \frac{\mathrm{Tr}\left[e^{-\beta H}\mathcal{T}(\Omega(t)\Omega(0))\right]}{\mathrm{Tr}\, e^{-\beta H}}. \tag{3.15}$$

In the computation of the response to external probes (how much current flows in response to an applied electric field?) one works with a related quantity,

$$iG_R(t,\beta) = \frac{\mathrm{Tr}\left[e^{-\beta H}\theta(t)\left[\Omega(t),\ \Omega(0)\right]\right]}{\mathrm{Tr}\, e^{-\beta H}}, \tag{3.16}$$

called the *retarded Green's function.*

In this review, we began with the Schrödinger equation which specified the continuous time evolution governed by H and integrated it to get the propagator which gives the time evolution over finite times. Suppose instead we had been brought up on a formalism in which the starting point was $U(t)$, as is the case in Feynman's version of quantum mechanics. How would the notion of H and the Schrödinger equation have emerged?

We would have asked how the states changed over short periods of time and argued that since $U(0) = I$, for short times ε it must assume the form

$$U(\varepsilon) = I - \frac{i\varepsilon}{\hbar}H + \mathcal{O}(\varepsilon^2), \tag{3.17}$$

where the $\hbar$ is inserted for dimensional reasons and the i to ensure that unitarity of U (to order ε) implies that H is Hermitian. From this expansion would follow the Schrödinger equation:

$$\frac{d|\psi\rangle}{dt} = \lim_{\varepsilon\to 0}\frac{(U(\varepsilon) - U(0))|\psi(t)\rangle}{\varepsilon} = -\frac{i}{\hbar}H|\psi(t)\rangle. \tag{3.18}$$

3.2 Imaginary-Time Quantum Mechanics

Now it turns out to be very fruitful to consider what happens if we let time assume purely imaginary values,

$$t = -i\tau, \tag{3.19}$$

where τ is real.

Formally, this means solving the *imaginary-time Schrödinger equation*

$$-\hbar\frac{d|\psi(\tau)\rangle}{d\tau} = H|\psi(\tau)\rangle. \tag{3.20}$$

The Hamiltonian is the same as before; only the parameter has changed from t to τ. The propagator

$$U(\tau) = e^{-\frac{H}{\hbar}\tau} \tag{3.21}$$

is Hermitian and not unitary.

We can write an expression for $U(\tau)$ as easily as for $U(t)$:

$$U(\tau) = \sum |n\rangle\langle n| e^{-\frac{1}{\hbar}E_n\tau}, \tag{3.22}$$

in terms of the *same eigenstates*

$$H|n\rangle = E_n|n\rangle. \tag{3.23}$$

The main point is that *even though the time is now imaginary, the eigenvalues and eigenfunctions that enter into the formula for $U(\tau)$ are the usual ones*. Conversely, if we knew $U(\tau)$, we could extract the former (Exercise 3.2.1).

We can get $U(\tau)$ from $U(t)$ by setting $t = -i\tau$. Thus, for the oscillator,

$$U(x,x';\tau) = \sqrt{\frac{m\omega}{2\pi\hbar\sinh\omega\tau}}\exp\left[-\frac{m\omega}{2\hbar\sinh\omega\tau}((x^2+x'^2)\cosh\omega\tau - 2xx')\right]. \tag{3.24}$$

Unlike $U(t)$, which is unitary, $U(\tau)$ is Hermitian, and the dependence on energies in Eq. (3.22) is a decaying and not oscillating exponential. As a result, upon acting on any initial state for a long time, $U(\tau)$ kills all but its ground state projection:

$$\lim_{\tau\to\infty} U(\tau)|\psi(0)\rangle \to |0\rangle\langle 0|\psi(0)\rangle e^{-\frac{1}{\hbar}E_0\tau}. \tag{3.25}$$

This means that in order to find $|0\rangle$, we can take any state and hit it with $U(\tau\to\infty)$. In the case that $|\psi(0)\rangle$ has no overlap with $|0\rangle, |1\rangle, \ldots, |n_0\rangle$, U will asymptotically project along $|n_0+1\rangle$.

Exercise 3.2.1

(i) Consider $U(x,x';\tau)$ for the oscillator as $\tau\to\infty$ and read off the ground-state wavefunction and energy. Compare to what you learned as a child. Given this, try to pull out the next state from the subdominant terms.

(ii) Set $x = x' = 0$ and extract the energies from $U(\tau)$ in Eq. (3.24). Why are some energies missing?

The Heisenberg operators $\Omega(\tau)$, which are defined in terms of Schrödinger operators Ω,

$$\Omega(\tau) = e^{\frac{H}{\hbar}\tau}\Omega e^{-\frac{H}{\hbar}\tau}, \tag{3.26}$$

have a feature that is new: two operators Ω and $\Omega^\dagger$ which are adjoints at $\tau = 0$ do not evolve into adjoints at later times:

$$\Omega^\dagger(\tau) = e^{\frac{H}{\hbar}\tau}\Omega^\dagger e^{-\frac{H}{\hbar}\tau} \neq [\Omega(\tau)]^\dagger = e^{-\frac{H}{\hbar}\tau}\Omega^\dagger e^{\frac{H}{\hbar}\tau}. \tag{3.27}$$

We will be interested in (imaginary) time-ordered products,

$$\mathcal{T}(\Omega(\tau_2)\Omega(\tau_1)) = \theta(\tau_2-\tau_1)\Omega(\tau_2)\Omega(\tau_1) + \theta(\tau_1-\tau_2)\Omega(\tau_1)\Omega(\tau_2), \tag{3.28}$$

and their expectation values,

$$G(\tau_2-\tau_1) = -\langle 0|\mathcal{T}(\Omega(\tau_2)\Omega(\tau_1))|0\rangle, \tag{3.29}$$

in the ground state or all states weighted by the Boltzmann factor:

$$G(\tau_2 - \tau_1, \beta) = -\frac{\mathrm{Tr}\, e^{-\beta H} \mathcal{T}(\Omega(\tau_2)\Omega(\tau_1))}{\mathrm{Tr}\, e^{-\beta H}}. \tag{3.30}$$

3.3 The Transfer Matrix

Now for the promised derivation of the Euclidean quantum problem underlying the classical $d = 1$ Ising model.

The word "quantum" conjures up Hilbert spaces, operators, eigenvalues, and so on. In addition, there is quantum dynamics driven by the Hamiltonian H or the propagator, or $U(\tau)$ in imaginary time. (Real-time quantum mechanics will not come from statistical mechanics because there is no $i = \sqrt{-1}$ in sight.) Let us see how these entities appear, starting with the Ising partition function.

It all begins with the idea of a *transfer matrix* invented by H. A. Kramers and G. H. Wannier [3].

Consider the $d = 1$ Ising model for which

$$Z = \sum_{s_i} \prod_i e^{K(s_i s_{i+1} - 1)}. \tag{3.31}$$

Each exponential factor $e^{K(s_i s_{i+1}-1)}$ is labeled by two discrete indices (s_i and s_{i+1}) which can take two values each. Thus there are four numbers. Let us introduce a 2×2 transfer matrix whose rows and columns are labeled by these spins and whose matrix elements equal the Boltzmann weight associated with that pair of neighboring spins (which we call s and s' to simplify notation):

$$T_{s's} = e^{K(s's-1)}. \tag{3.32}$$

In other words,

$$T_{++} = T_{--} = 1, \; T_{+-} = T_{-+} = \exp(-2K), \tag{3.33}$$

so

$$T = \begin{pmatrix} 1 & e^{-2K} \\ e^{-2K} & 1 \end{pmatrix} = I + e^{-2K}\sigma_1, \tag{3.34}$$

where σ_1 is the first Pauli matrix. Notice that T is real and Hermitian.

It follows from Eq. (3.31) that

$$Z = \sum_{s_i, i=1\ldots N-1} T_{s_N s_{N-1}} \cdots T_{s_2 s_1} T_{s_1 s_0} \tag{3.35}$$

involves the repeated matrix product of T with itself, and Z reduces to

$$Z = \langle s_N | T^N | s_0 \rangle \tag{3.36}$$

for the case of fixed boundary conditions (which we will focus on) where the first spin is fixed at s_0 and the last at s_N. If we sum over the end spins (free boundary conditions),

$$Z = \sum_{s_0 \; s_N} \langle s_N | T^N | s_0 \rangle. \tag{3.37}$$

If we consider periodic boundary conditions where $s_0 = s_N$ and sum over these equal spins, we obtain a trace:

$$Z = \mathrm{Tr}\, T^N. \tag{3.38}$$

As a warm-up, we will now use this formalism to show the insensitivity of the free energy per site to boundary conditions in the thermodynamic limit. Consider, for example, fixed boundary conditions with the first and last spins being s_0 and s_N. If we write

$$T = \lambda_0 |0\rangle\langle 0| + \lambda_1 |1\rangle\langle 1|, \tag{3.39}$$

where $|i\rangle$, λ_i $[i = 0, 1]$ are the orthonormal eigenvectors and eigenvalues of T, then

$$T^N = \lambda_0^N |0\rangle\langle 0| + \lambda_1^N |1\rangle\langle 1|. \tag{3.40}$$

Now we invoke the *Perron–Frobenius theorem*, which states that

A square matrix with positive non-zero entries will have a non-degenerate largest eigenvalue and a corresponding eigenvector with strictly positive components.

The matrix T is such a matrix for any finite K.

Assuming λ_0 is the bigger of the two eigenvalues,

$$T^N \lim_{N\to\infty} \simeq \lambda_0^N \left[|0\rangle\langle 0| + \mathcal{O}\left(\frac{\lambda_1}{\lambda_0}\right)^N \right] \tag{3.41}$$

and

$$Z = \langle s_N | T^N | s_0 \rangle \simeq \langle s_N | 0 \rangle \langle 0 | s_0 \rangle \lambda_0^N \left(1 + \mathcal{O}\left(\frac{\lambda_1}{\lambda_0}\right)^N \right), \tag{3.42}$$

and the free energy per site in the infinite volume limit,

$$-f = \lim_{N\to\infty} \left[\ln \lambda_0 + \frac{1}{N} \ln(\langle s_N | 0 \rangle \langle 0 | s_0 \rangle) + \cdots \right] = \ln \lambda_0 \tag{3.43}$$

is clearly independent of the boundary spins as long as $\langle 0 | s_0 \rangle$ and $\langle s_N | 0 \rangle$ do not vanish. This is assured by the Perron–Frobenius theorem which states that all the components of the dominant eigenvector will be positive.

Exercise 3.3.1 *Check that the free energy per site is the same as above for periodic boundary conditions starting with Eq. (3.38).*

Let us return to Eq. (3.34):

$$T = I + e^{-2K}\sigma_1.$$

Consider the identity

$$e^{K^*\sigma_1} = \cosh K^* + \sinh K^*\sigma_1 \tag{3.44}$$

$$= \cosh K^*(I + \tanh K^*\sigma_1), \tag{3.45}$$

where K^* is presently unrelated to K; in particular, it is not the conjugate. If we choose K^* as a function of K such that

$$\tanh K^*(K) = e^{-2K}, \tag{3.46}$$

then

$$T = \frac{e^{K^*(K)\sigma_1}}{\cosh K^*(K)}. \tag{3.47}$$

I will often drop the $\cosh K^*$ *in the denominator since it drops out of all averages.* I will also suppress the fact that K^* is a function of K and simply call it K^*.

One calls K^* the *dual* of K. Remarkably, if you invert Eq. (3.46), you will find (Exercise 3.3.2) that you get the same functional form:

$$\tanh K = e^{-2K^*}. \tag{3.48}$$

Thus, K^* is the dual of K as well. Observe that when one is small the other is large, and vice versa.

Exercise 3.3.2 *Show that if you invert Eq. (3.46) you find* $\tanh K = e^{-2K^*}$.

For later reference, let us note that in the present case, the eigenvalues of T are

$$\lambda_0 = e^{K^*}, \qquad \lambda_1 = e^{-K^*}, \qquad \frac{\lambda_1}{\lambda_0} = e^{-2K^*}, \tag{3.49}$$

and the corresponding eigenvectors are

$$|0\rangle, |1\rangle = \frac{1}{\sqrt{2}}\begin{pmatrix} 1 \\ \pm 1 \end{pmatrix}. \tag{3.50}$$

Consider next the correlation function $\langle s_j s_i \rangle$ for $j > i$. I claim that if the boundary spins are fixed at s_0 and s_N,

$$\langle s_j s_i \rangle = \frac{\langle s_N | T^{N-j}\sigma_3 T^{j-i}\sigma_3 T^i | s_0 \rangle}{\langle s_N | T^N | s_0 \rangle}. \tag{3.51}$$

To see the correctness of this, look at the numerator. Retrace our derivation by introducing a complete set of σ_3 eigenstates between every factor of T. Reading from right to left, we get just the sum over Boltzmann weights till we get to site i. There, the σ_3 acting on its eigenstate gives s_i, the value of the spin there. Then we proceed as usual to j, pull out s_j and go to the Nth site. The denominator is just Z.

Let us rewrite Eq. (3.51) another way. Define *Heisenberg operators* by

$$\sigma_3(n) = T^{-n}\sigma_3 T^n. \tag{3.52}$$

We use the name "Heisenberg operators" because *the site index n plays the role of discrete integer-valued time, and T is the time-evolution operator or propagator for one unit of Euclidean time.*

In terms of these Heisenberg operators,

$$\langle s_j s_i\rangle = \frac{\langle s_N|T^N\sigma_3(j)\sigma_3(i)|s_0\rangle}{\langle s_N|T^N|s_0\rangle}. \tag{3.53}$$

(As an aside, observe that dropping the $\cosh K^*$ in Eq. (3.47) does not affect $\langle s_j s_i\rangle$, because it cancels out.)

Consider now the limit as $N \to \infty$, with i and j *fixed at values far from the end points* (labeled 0 and N), so that $N-j$ and i are large. We may approximate:

$$T^\alpha \simeq |0\rangle\langle 0|\lambda_0^\alpha \qquad \alpha = N, N-j, i. \tag{3.54}$$

In this limit, we have, from Eq. (3.51),

$$\langle s_j s_i\rangle = \frac{\langle s_N|0\rangle\langle 0|\lambda_0^{N-j}\sigma_3 T^{j-i}\sigma_3\lambda_0^i|0\rangle\langle 0|s_0\rangle}{\langle s_N|0\rangle\lambda_0^N\langle 0|s_0\rangle} = \langle 0|\sigma_3(j)\sigma_3(i)|0\rangle, \tag{3.55}$$

and the dependence on the boundary has dropped out. For the case $i > j$, we will get the operators in the other order. In general, then,

$$\langle s_j s_i\rangle = \langle 0|\mathcal{T}(\sigma_3(j)\sigma_3(i))|0\rangle, \tag{3.56}$$

where the *time-ordering symbol* $\mathcal{T}$ will order the operators with their arguments increasing from right to left:

$$\mathcal{T}(\sigma_3(j)\sigma_3(i)) = \theta(j-i)(\sigma_3(j)\sigma_3(i)) + \theta(i-j)(\sigma_3(i)\sigma_3(j)). \tag{3.57}$$

The reason for associating i and j with time will be explained later in this chapter. For now, notice how naturally the time-ordered product arises.

We will pursue the evaluation of this correlation function using the eigenvectors of T. But first, let us replace $\sigma_3(j)$ by the unit operator in the above derivation to obtain the mean magnetization as

$$\langle s_i\rangle = \langle 0|\sigma_3|0\rangle = \langle s,\rangle \tag{3.58}$$

independent of i as long as it is far from the ends, so that Eq. (3.54) applies. In our example, $|0\rangle$ is the eigenket of σ_1 so that there is no mean magnetization. The only exception is at zero temperature or zero K^*: now $T = I$, the eigenvalues are equal, and we can form linear combinations corresponding to either of the fully ordered (up or down) σ_3 eigenstates. The Perron–Frobenius theorem does not apply since some entries of T are not > 0.

Let us return to Eq. (3.56). Even though it appears that everything depends on just the ground state, knowledge of all states is required even in the infinite volume limit to evaluate the correlation. Going to Eq. (3.56) for the case $j > i$, let us insert the complete set of (two) eigenvectors of T between the Pauli matrices. When we insert $|0\rangle\langle 0|$ we get $\langle s\rangle^2$, the square of the magnetization. Even though it vanishes in this case ($\langle 0|\sigma_3|0\rangle = 0$ since $|0\rangle$ is an eigenstate of σ_1), let us keep the term and move it to the left-hand side, to get the *connected correlation function*

$$\begin{aligned}\langle s_j s_i\rangle_c \equiv \langle s_j s_i\rangle - \langle s\rangle^2 &= \langle 0|T^{-j}\sigma_3(0)T^{j-i}|1\rangle\langle 1|\sigma_3(0)T^i|0\rangle \\ &= \left(\frac{\lambda_1}{\lambda_0}\right)^{j-i} |\langle 0|\sigma_3|1\rangle|^2 \\ &= e^{-2K^*(j-i)}|\langle 0|\sigma_3|1\rangle|^2 \text{ [recall Eq. (3.49)]} \\ &= e^{-2K^*|j-i|}|\langle 0|\sigma_3|1\rangle|^2 \qquad (3.59)\\ &= [\tanh K]^{|j-i|}\,|\langle 0|\sigma_3|1\rangle|^2, \qquad (3.60)\end{aligned}$$

where we have replaced $j-i$ by $|j-i|$ because this is what we will find if we repeat the analysis for the case $i > j$ and recalled $e^{2K^*} = \tanh K$ in the last step.

Observe that, asymptotically, $\langle s_j s_i\rangle \to \langle s\rangle^2$ in the limit $|j-i| \to \infty$.

We are going to define the *correlation length* ξ as follows. For general models (not necessarily of Ising spins or in $d = 1$) we will find (except at isolated points, or in very special models):

$$\langle s_i s_j\rangle_c \lim_{|j-i|\to\infty} \to \frac{e^{-|j-i|/\xi}}{|j-i|^{d-2+\eta}}, \qquad (3.61)$$

where η is some number and $|j-i|$ stands for the distance between the two spins. Since the exponential decay dominates the power law asymptotically, the correlation length may be extracted from $\langle s_i s_j\rangle_c$ as follows:

$$\xi^{-1} = \lim_{|j-i|\to\infty}\left[-\frac{\ln\langle s_i s_j\rangle_c}{|j-i|}\right]. \qquad (3.62)$$

Exercise 3.3.3 *Verify that when the correlation function of Eq. (3.61) is inserted into Eq. (3.62), the power law prefactor drops out in the determination of ξ.*

In our problem $\langle s\rangle = 0$, and so

$$\langle s_i s_j\rangle = \langle s_i s_j\rangle_c = \exp\left[|j-i|\ln\tanh K\right],$$

which allows us to read off ξ by inspection:

$$\xi^{-1} = -\ln\tanh K = 2K^*, \text{ using Eq. (3.48).} \qquad (3.63)$$

(Exponential decay for all separations and not just when $|j-i| \to \infty$ is again peculiar to our model and stems from the fact that it is in one spatial dimension and the Ising spin can take only two values.)

To understand why ξ is defined in terms of $\langle s_i s_j \rangle_c$, we need to consider a problem where $\langle s \rangle \neq 0$, so let us turn on a magnetic field $h > 0$. The transfer matrix T, whose entries are Boltzmann weights in this field, would still have all its entries positive, so that, by the Perron–Frobenius theorem, will still have $\lambda_0 > \lambda_1$; an equation similar to Eq. (3.59) would follow and tell us again that it is $\langle s_i s_j \rangle_c$ that determines ξ. If $\langle s_i s_j \rangle_c$ gives us ξ, what does

$$\langle s_i s_j \rangle = \langle s_i s_j \rangle_c + \langle s \rangle^2 \tag{3.64}$$

tell us, and what do its two parts mean?

By definition, $\langle s_i s_j \rangle$ is the average value of the product of two spins. If they fluctuate independently, this average will be zero, since the product is as likely to be $+1$ as -1. But there are two reasons why their behavior is correlated and the average is non-zero.

The first is the background field $h > 0$ which permeates all of space and encourages spins to point in its own direction. This produces a kind of correlation that exists even in a model with $K = 0$, which we will first consider. In this model, the probability distributions for s_i and s_j factorize into two independent distributions with $\langle s_i s_j \rangle = \langle s \rangle^2$, *independent of the distance* $|i-j|$. This part of $\langle s_i s_j \rangle$ coming from a non-zero product of averages does not signify any *direct* correlation between spins: the distance between them does not enter it, and each spin has no idea the other is even present (because $K = 0$); it is simply correlated with the external field.

Suppose we now turn on $K > 0$. There are two effects. First, the aligned spins (little magnets that they are) will generate, via the K term, an additional uniform internal field parallel to h leading to an enhancement of $\langle s \rangle$. This added correlation among averages is just more of what we saw in the $K = 0$ case.

What we are looking for are additional correlations in the *fluctuations on top of the average* $\langle s \rangle$. This suggests we introduce a new correlation function:

$$\langle s_j s_i \rangle_{\text{new}} = \langle (s_j - \langle s \rangle)(s_i - \langle s \rangle) \rangle. \tag{3.65}$$

Happily, this is just what the connected part is:

$$\langle (s_j - \langle s \rangle)(s_i - \langle s \rangle) \rangle = \langle s_j s_i \rangle - \langle s_j \rangle \langle s \rangle - \langle s_i \rangle \langle s \rangle + \langle s \rangle^2 = \langle s_j s_i \rangle_c, \tag{3.66}$$

where we have used translation invariance: $\langle s_i \rangle = \langle s_j \rangle = \langle s \rangle$. In contrast to the product of the averages, the connected correlation function depends on the separation $|i-j|$ and defines the correlation length.

Once more with feeling: The full correlation $\langle s_j s_i \rangle$ is a sum of $\langle s \rangle^2$, which can be measured by observing just one spin at a time, and $\langle s_j s_i \rangle_c$, which necessarily requires simultaneous observation of two spins.

The Ising example illustrates a general physical principle called *clustering*, which says that the connected correlation between any two variables A and B must die asymptotically, i.e., that the average of their product must become the product of their averages:

$$\langle A_j B_i \rangle \lim_{|j-i|\to\infty} \to \langle A_j \rangle \langle B_i \rangle. \tag{3.67}$$

(In a translationally invariant system, the site labels i and j are not needed on the right-hand side.) Clustering is the statement that the joint probability distribution for the two variables must asymptotically factorize into a product of individual distributions.

We will refer back to this physically motivated principle when we study phase transitions. There we will run into some mathematically allowed solutions that will be rejected because they do not obey clustering.

3.3.1 The Hamiltonian

We have seen that the transfer matrix T plays the role of a time-evolution operator, given that the Heisenberg operators that arose naturally were defined as

$$\sigma_3(j) = T^{-j}\sigma_3 T^j. \tag{3.68}$$

If we are to identify T with $U(\tau) = e^{-H\tau}$, what is τ? It is one step on our discrete time lattice. Using that step as the unit of time (i.e., setting the time-lattice spacing to unity), we now introduce a Hamiltonian H by the following definition:

$$T = e^{-H}. \tag{3.69}$$

The H above is dimensionless because the unit of time has been set to unity by our choice of time units.

Given that T is symmetric and real, so is H, so it shares its orthogonal eigenvectors $|0\rangle, |1\rangle$ with T. Here is a summary:

$$T = e^{K^*\sigma_1}, \tag{3.70}$$

$$H = -K^*\sigma_1, \tag{3.71}$$

$$|0\rangle \; : \; T|0\rangle = \lambda_0|0\rangle = e^{K^*}|0\rangle \text{ and } H|0\rangle = E_0|0\rangle = -K^*|0\rangle \,, \tag{3.72}$$

$$|1\rangle \; : \; T|1\rangle = \lambda_1|1\rangle = e^{-K^*}|1\rangle \text{ and } H|1\rangle = E_1|1\rangle = K^*|1\rangle, \tag{3.73}$$

$$f = -\ln\lambda_0 = -K^* = E_0. \tag{3.74}$$

In the present case it was obvious by inspection that $H = -K^*\sigma_1$ because T had the form of an exponential. In generic cases (we will see one soon), where it is a product of exponentials of non-commuting operators, the explicit form of H may be hard or impossible to work out, but it surely exists, and has the properties mentioned above. Later we will see under what circumstances a simple H can be associated with T. In any event, the asymptotic behavior as the Euclidean time $N \to \infty$ is dominated by the largest eigenstate of T or the ground state of H.

Now for the fact that H is dimensionless, like everything else so far. The lattice sites were indexed by a number, but how far apart are they? This was not specified and need not be since we can measure all distances in that unit. If these spins really lived on a lattice with spacing, say, 1 Å, sites numbered $i = 12$ and $j = 23$ would really be 11 Å apart. The correlation length $\xi = (-\ln \tanh K)^{-1}$ would be that many Å. In other words, the correlation length has been rendered dimensionless by dividing by the lattice spacing. Likewise, the Ising coupling J has been rendered dimensionless by forming the combination $K = J/kT$. The dimensionless free energy per site we called f is really the free energy per site divided by kT. Likewise, H is some Hamiltonian rendered dimensionless by measuring it in some (unspecified) energy units.

Occasionally we will introduce corresponding entities with the right dimensions, but keep using the same symbols.

Exercise 3.3.4 *Show that f in Eq. (3.74) agrees with Eq. (2.31) upon subtracting* $\ln\cosh K^*$ *and using the definition of K^*.*

Let us note that:

- The connected correlation depends only on ratios of the eigenvalues of T and falls exponentially with distance with a coefficient $2K^*$. Now, $2K^*$ is just the gap to the first excited state of the Hamiltonian H defined by $T = e^{-H}$, which in our example is $-K^*\sigma_1$. The result

$$\xi^{-1} = E_1 - E_0 \equiv m \tag{3.75}$$

  is very general. The reason one uses the symbol m for the gap (called the *mass gap*) is that in a field theory the lowest energy state above the ground state (vacuum) has a single particle at rest with energy m (in units where $c = 1$). The correlations fall with Euclidean time as $e^{-m\tau}$ as $\tau \to \infty$.
- The connected correlation function in Eq. (3.59) is determined by the matrix element of the operator in question (σ_3) between the ground state and the next excited state. This is also a general feature. If this matrix element vanishes, we must go up in the levels till we find a state that is connected to the ground state by the action of the operator. (In this problem, we know $|\langle 0|\sigma_3|1\rangle|^2 = 1$ since σ_3 is the spin-flip operator for the eigenstates of σ_1.)

This simple example has revealed most of the general features of the problem. The only difference is that for a bigger transfer matrix, the sum over states will have more than two (possibly infinite) terms. Thus, the correlation function will be a sum of decaying exponentials, and a unique correlation length will emerge only asymptotically when the smallest mass gap dominates.

3.3.2 Turning on h

We must now bring in the magnetic field by adding a term $h\sum_i s_i$ to the exponent in Eq. (2.26). The transfer matrix

$$T = e^{K^*\sigma_1} e^{h\sigma_3} \equiv T_K T_h \tag{3.76}$$

reproduces the Boltzmann weight, but is not symmetric (Hermitian). By rewriting

$$h\sum_i s_i = \frac{1}{2}h\sum_i (s_i + s_{i+1}), \tag{3.77}$$

we obtain

$$T = T_h^{1/2} T_K T_h^{1/2}, \tag{3.78}$$

which is symmetric.

Again, we would like to write $T = e^{-H}$, but evaluating this H requires some work because to combine exponentials of non-commuting operators we have to use the Baker–Campbell–Hausdorff (BCH) formula

$$e^A e^B = e^{A+B+\frac{1}{2}[A,B]\cdots}, \tag{3.79}$$

where higher-order repeated commutators are shown by the unending string of dots. Now, in this 2×2 case we can finesse this, but this will not be the case when we consider problems involving larger or even infinite-dimensional operators like X and P.

In the present case, suppose all exponentials in $T_h^{1/2} T_K T_h^{1/2}$ have been combined into a single exponential, which we will call $-H$. It will be a 2×2 matrix function of $K = J/(kT)$, $h = B/(kT)$. Thus the temperature of the classical model enters H as a parameter. As the classical system of Ising spins gets hot and cold, the parameters of the Hamiltonian vary.

Exercise 3.3.5 (Very important) *Find the eigenvalues of T in Eq. (3.78) and show that there is degeneracy only for $h = K^* = 0$. Why does this degeneracy not violate the Perron–Frobenius theorem? Show that the magnetization is*

$$\langle s\rangle = \sinh h / \left(\sqrt{\sinh^2 h + e^{-4K}}\right) \tag{3.80}$$

in the thermodynamic limit.

Exercise 3.3.6 *Consider the correlation function for the $h = 0$ problem with periodic boundary conditions and write it as a ratio of two traces. Saturate the denominator with the largest eigenket, but keep both eigenvectors in the numerator and show that the answer is invariant under $j - i \leftrightarrow N - (j - i)$. Using the fact that σ_3 exchanges $|0\rangle$ and $|1\rangle$ should speed things up. Argue that as long as $|j - i|$ is much smaller than N, only one term is needed.*

Exercise 3.3.7 *Recall the remarkable fact that the correlation function $\langle s_j s_i \rangle$ in the $h = 0$ Ising model was translation invariant in the finite open chain with one end fixed at s_0. Derive this result using the transfer matrix formalism as follows: Explicitly evaluate $\sigma_3(j)$ by evaluating $T^{-j}\sigma_3 T^j$ in terms of σ_3 and σ_1. Show that $\sigma_3(j)\sigma_3(i)$ is a function only of $j - i$ by using some identities for hyperbolic functions. Keep going until you explicitly have the correlation function. It might help to use $\sum_{s_N} |s_N\rangle = (I + \sigma_1)|s_0\rangle$. (This means the following: We want to sum s_N over its two values because this is an open chain. The two values must equal s_0 and $-s_0$ whatever we choose for the fixed s_0. These two values are generated from s_0 by the action of I and σ_1 respectively.)*

3.4 Classical to Quantum Mapping: The Dictionary

We have seen that the partition function and correlation functions of the classical Ising problem can be restated in terms of a two-dimensional Hilbert space, i.e., the kinematical homeland of a quantum spin-$\frac{1}{2}$ degree of freedom. Here, I collect the various correspondences that emerged so that they will be more easily remembered as we advance to more complicated versions of the classical–quantum correspondence.

The Schrödinger operators of this theory are σ_1 and σ_3. The operator σ_2 does not enter since there are no complex numbers in the classical problem. This reminds us that the quantum problem that emerges need not be the most general one of its kind. The Ising spin s is associated with the eigenvalues of σ_3.

The transfer matrix T plays the role of the Euclidean time-evolution operator for one unit of some discrete time $\Delta\tau$:

$$T \Longleftrightarrow U(\Delta\tau). \tag{3.81}$$

We have been using units in which

$$\Delta\tau = 1. \tag{3.82}$$

The $N+1$ points in the spatial lattice, numbered by $\left[j = 0, \ldots, N\right]$, correspond to $N+1$ instants of Euclidean time $\tau = j\Delta\tau$. The finite Ising chain corresponds to time evolution over a finite time.

Next, the partition function with $s_0 = s_i$ and $s_N = s_f$ (where i and f stand for "initial" and "final"),

$$\langle s_N = s_f | T^N | s_0 = s_i \rangle \Longleftrightarrow \langle s_f | U(N\Delta\tau) | s_i \rangle, \tag{3.83}$$

corresponds to the matrix element of the propagator U for imaginary time $N\Delta\tau$ between the states $\langle s_f|$ and $|s_i\rangle$.

The Heisenberg operators are related to the Schrödinger operators as expected:

$$\sigma_3(j) = T^{-j}\sigma_3 T^j \Longleftrightarrow U^{-1}(j\Delta\tau)\sigma_3 U(j\Delta\tau) = \sigma_3(\tau = j\Delta\tau). \tag{3.84}$$

If we formally define a Hamiltonian H by

$$T = e^{-H\Delta\tau} = e^{-H} \quad \text{(if we set the time step } \tau = 1\text{)}, \tag{3.85}$$

then the dominant eigenvector of T is the ground state eigenvector $|0\rangle$ of H. In the $d = 1$ Ising case, H was a simple operator, $-K^*\sigma_1$. In general, the T coming from a sensible statistical mechanics problem will be a nice operator, but its logarithm H need not be. (By nice, I mean this: Suppose, instead of just one quantum spin, we had a line of them. Then T will usually involve products of typically two or four operators that are close by in this line, while H may involve products of arbitrary numbers of operators arbitrarily far apart.)

The correlation function of the Ising model in the thermodynamic limit $N \to \infty$ is the ground state expectation value of the time-ordered product of the corresponding Heisenberg operators:

$$\langle s_j s_i \rangle \Longleftrightarrow \langle 0|\mathcal{T}(\sigma_3(j)\sigma_3(i))|0\rangle. \tag{3.86}$$

References and Further Reading

[1] J. B. Kogut, Reviews of Modern Physics, **51**, 659 (1979).
[2] M. Kardar, *Statistical Physics of Fields*, Cambridge University Press (2003).
[3] H. A. Kramers and G. H. Wannier, Physical Review, **60**, 252 (1941).

4

Quantum to Statistical Mechanics

Now I will show you how we can go back from a problem in Euclidean quantum mechanics to one in classical statistical mechanics, by essentially retracing the path that led to the transfer matrix.

Why bother? Because even if it is the same road, the perspective is different if you travel in the other direction.

In the process, we would also have managed to map the problem of *quantum statistical mechanics* to a classical problem. The reason is that quantum statistical mechanics is a subset of Euclidean quantum mechanics. Let us see why.

The central object in quantum statistical mechanics is the partition function:

$$Z_Q = \mathrm{Tr}\, e^{-\beta H} = \mathrm{Tr}\, e^{-\frac{\beta\hbar}{\hbar}H} = \mathrm{Tr}\, U(\tau = \beta\hbar), \tag{4.1}$$

where H is the quantum Hamiltonian, with the right dimensions. I have inserted the $\hbar$ so we can make the following important assertion: *the quantum partition function Z_Q is just the trace of the imaginary-time-evolution operator for a time $\beta\hbar$.*

It follows that in the study of a more general object, namely

$$U(f,i;\tau) = \langle f|U(\tau)|i\rangle$$

for any τ and any initial and final state, the problem of Z_Q constitutes a subset that corresponds to choosing $\tau = \beta\hbar$ and tracing over the diagonal elements. Indeed, one of the reasons for developing Euclidean quantum mechanics was the study of Z_Q.

4.1 From U to Z

Let us then begin with the fundamental entity in the quantum problem: the matrix element of the propagator $U(\tau)$ between initial and final states $|i\rangle$ and $|f\rangle$,

$$U(f,i;\tau) = \langle f|U(\tau)|i\rangle. \tag{4.2}$$

To simplify the notation I will use units in which $\hbar = 1$ in this and the next section. You can put back the $\hbar$ at any time by dimensional analysis. Later, I may revert to showing $\hbar$ explicitly. Such behavior is common in the field and you should get used to it.

We are going to see how it becomes a classical statistical mechanics model and with what parameters.

4.2 A Detailed Example from Spin $\frac{1}{2}$

Let us focus on a concrete example of a quantum problem, the one governed by the Hamiltonian

$$H = -B_1\sigma_1 - B_3\sigma_3, \tag{4.3}$$

which describes a spin-$\frac{1}{2}$ in a magnetic field $\mathbf{B} = (\mathbf{i}B_1 + \mathbf{k}B_3)$ in the x–z plane. The numbers B_1 and B_3 include the field, the magnetic moment, and $\hbar$.

Let us now write

$$U(\tau) = \left[U\left(\frac{\tau}{N}\right)\right]^N, \quad \text{where} \tag{4.4}$$

$$U\left(\frac{\tau}{N}\right) \equiv T_\varepsilon = e^{-\varepsilon H} \text{ with} \tag{4.5}$$

$$\varepsilon = \frac{\tau}{N}. \tag{4.6}$$

We are simply factoring the evolution operator for time τ into a product of N evolution operators for time $\frac{\tau}{N} = \varepsilon$. *Bear in mind that at this stage ε is not necessarily small.*

Putting in $N-1$ intermediate sums over σ_3 eigenstates and renaming $|i\rangle$ and $|f\rangle$, appropriately for the spin problem, as $|s_0\rangle$ and $|s_N\rangle$ respectively, we obtain

$$U(s_N, s_0) = \sum_{s_i} \prod_{i=0}^{N-1} \langle s_{i+1}|e^{-\varepsilon H}|s_i\rangle, \tag{4.7}$$

which makes it a partition function of an Ising chain of $N+1$ spins (whose ends are fixed at s_0 and s_N) and whose transfer matrix has elements

$$T_{s_{i+1}s_i} = \langle s_{i+1}|e^{-\varepsilon H}|s_i\rangle. \tag{4.8}$$

To figure out the parameters of the classical Ising chain we equate the general expression for the Boltzmann weight of the Ising problem to the matrix elements of the quantum problem:

$$e^{K(s's-1)+\frac{h}{2}(s'+s)+c} = \langle s'|e^{-\varepsilon H}|s\rangle = \langle s'|e^{\varepsilon(B_1\sigma_1+B_3\sigma_3)}|s\rangle. \tag{4.9}$$

We have chosen simpler labels s and s' just for the extraction of K, h, and a spin-independent constant c. Exercise 4.2.1 explains how the form in the exponential on the left-hand side arises, while Exercise 4.2.2 goes into how one solves for K, h, and c in terms of ε, B_1, and B_3.

Exercise 4.2.1 *Write $T_{ss'} = e^{R(s,s')}$, and expand the function R in a series, allowing all possible powers of s and s'. Show that $R(s,s') = Ass' + B(s+s') + C$ follows, given that*

$s^2 = s'^2 = 1$ and both $T_{ss'}$ and R are symmetric under $s \leftrightarrow s'$. (We go from here to the left-hand side of Eq. (4.9) by writing $C = c - K$. This is not mandatory and simply a matter of convention.)

Exercise 4.2.2 (Important)

(i) Solve Eq. (4.9) for K, h, and c in terms of B_1, B_3, and ε by choosing $s = s' = \pm 1$ and $s = -s'$, and show that

$$\tanh h = \frac{B_3}{B} \tanh \varepsilon B, \tag{4.10}$$

$$e^{-2K+c} = \frac{B_1}{B} \sinh \varepsilon B, \tag{4.11}$$

$$e^{2c}(1 - e^{-4K}) = 1, \tag{4.12}$$

where $\mathbf{B} = (\mathbf{i}B_1 + \mathbf{k}B_3)$ is the field appearing in H. (First prove and then use $e^{\varepsilon \boldsymbol{\sigma} \cdot \mathbf{B}} = \cosh \varepsilon B + \frac{\mathbf{B}}{B} \cdot \boldsymbol{\sigma} \sinh \varepsilon B$.) As expected, the three parameters c, K, and h originating from two independent parameters εB_1 and εB_3 will necessarily be constrained. This is evident from Eq. (4.12), which is a restatement of det $\left[e^{\varepsilon \boldsymbol{\sigma} \cdot \mathbf{B}}\right] = 1$, which in turn reflects the tracelessness of the Pauli matrices.

(ii) Show, by going to the eigenbasis of H, that its tracelessness implies its exponential has unit determinant.

(iii) Explain what goes wrong with Eq. (4.11) if B_1 is negative.

(iv) Why do we not include a σ_2 term in H in mapping to the Ising model even though it is Hermitian?

There are two things to bear in mind for later use. Even if you did not solve Exercise 4.2.2, notice one thing from Eq. (4.9): the classical parameters depend on ε, the time slice τ/N. Thus, as you vary the number of time slices or Ising spins, *you change the parameters, keeping the partition function Z constant* since it always signifies the amplitude to go from s_0 to s_N in time τ no matter how we slice the interval.

When we pick a value of N, say 10, we are dividing the continuous interval $[0, \tau]$ into N discrete parts and viewing the propagation amplitude from s_0 to s_N as a partition function for $N + 1$ spins with the ones at the end fixed at s_0 and s_N.

Those who start out with the quantum propagator call the equivalent partition function a *path integral*. The name arises as follows. Imagine the Ising spin at some point $0 \le n \le N$ (which labels Euclidean time) as an arrow pointing to the left or right, perpendicular to the time axis. View the tip of the arrow as the location of a particle which has only two locations: ± 1, as in Figure 4.1. Every term in the sum over spin configurations in Z corresponds to some "trajectory" the particle took in the time interval $[0, \tau]$. This is the "path" in "path integral," and the sum over spins in Z (which in some cases could be an integral) is the "integral" over all possible paths with these end points. If we were doing the quantum partition function, we would have chosen $\tau = \beta$ (remember $\hbar = 1$), and the

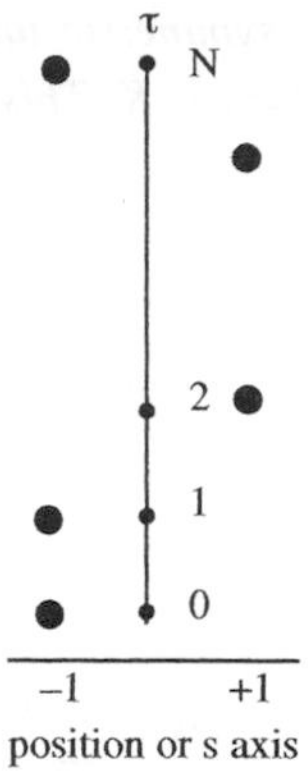

Figure 4.1 One term in the Ising partition function as a trajectory or path of a particle. Its position at time $\tau = n\varepsilon$ equals the value of spin s_n (measured perpendicular to the time axis). Since we are doing a trace, the particle starts and ends at the same location, which happens to be (-1) in the figure.

particle locations (spins) at the start and finish would have been equal and summed over as part of the trace.

Some people refer to the classical partition arising from inserting intermediate states as a *sum over histories*.

The path described above is not very much of a path for two reasons: the particle has only two locations, and the points on its path are defined only at times $\tau_n = n\varepsilon = \frac{n}{N}\tau, \ n = 0, 1, \ldots, N$.

There is not much we can do about the first objection as long as we work with a spin-$\frac{1}{2}$ particle. Soon we will turn to the quantum mechanics of a particle described by a quantum Hamiltonian $H(X,P)$ and repeatedly use the complete set of position eigenstates, which will run over a continuous infinity of locations $-\infty < x < \infty$.

We will now fix the second problem of discrete times by going to the limit $N \to \infty$, in which the interval $\varepsilon = \tau/N$ between adjacent lattice points goes to zero. This is a natural thing to do since τ is a continuous variable in the quantum problem and one is interested in what is happening at all times in the interval $[0, \tau]$, not just the N points that appear in the slicing process. *In Euclidean quantum mechanics the lattice in time is an artifact that must be banished at the end by taking the limit $N \to \infty$ or $\varepsilon = \tau/N \to 0$.* To be concrete, consider a typical situation where we are interested in the correlation function

$$G(\tau_2 - \tau_1) = \frac{\mathrm{Tr}\, e^{-H\tau}\left[\mathcal{T}\sigma_3(\tau_2)\sigma_3(\tau_1)\right]}{\mathrm{Tr}\, e^{-H\tau}}. \tag{4.13}$$

If we map into an Ising model with parameters determined by Eq. (4.9) and periodic boundary conditions (because of the trace), the correlation function $G(\tau_2 - \tau_1)$ will be just $\langle s_{i_2} s_{i_1} \rangle$ in the Ising model, where site labels i_1 and i_2 are given by $i_1\varepsilon = \tau_1$ and $i_2\varepsilon = \tau_2$.

As a check, we may now want to do a Monte Carlo calculation on a computer to evaluate $\langle s_{i_2} s_{i_1} \rangle$ as explained in Section 2.3. If τ_1 and τ_2 are not one of the N points $\tau = n\varepsilon = n\tau/N$, we are out of luck. Since we cannot be sure which τ's will be of interest, we need to make the N points dense in the interval $[0, \tau]$.

This is the limit one uses in the path integral and it will be understood.

In this limit we find a simplification. Recall that for finite N, it was quite a chore to find the parameters of the Ising model, i.e., the elements of the transfer matrix T_ε. This was because T_ε was the exponential of H and *it took as much effort to map the quantum problem to the classical model as to solve the former!* But in the $\varepsilon \to 0$ limit,

$$U(\tau/N) = T_\varepsilon \to I - \varepsilon H + \mathcal{O}(\varepsilon^2), \tag{4.14}$$

so that the matrix elements of the transfer matrix T_ε are essentially those of H.

We are looking at a T that is infinitesimally close to the identity I.

Let us work out the details for the case $H = -(B_1\sigma_1 + B_3\sigma_3)$. Once again, we need to match

$$e^{K(s's-1)+\frac{h}{2}(s'+s)+c} = \langle s'|e^{-\varepsilon H}|s\rangle = \langle s'|I + \varepsilon B_1\sigma_1 + \varepsilon B_3\sigma_3)|s\rangle \tag{4.15}$$

to find the parameters of the Ising model when $\varepsilon \to 0$. We find:

$$T_{++} = e^{h+c} = 1 + \varepsilon B_3, \tag{4.16}$$

$$T_{--} = e^{-h+c} = 1 - \varepsilon B_3, \tag{4.17}$$

$$T_{+-} = e^{-2K+c} = \varepsilon B_1. \tag{4.18}$$

From the product of the first two equations, we deduce that

$$e^{2c} = 1 + \mathcal{O}(\varepsilon^2), \tag{4.19}$$

which means $c = 0$ to this order. The other parameters follow:

$$h = \varepsilon B_3, \tag{4.20}$$

$$e^{-2K} = \varepsilon B_1. \tag{4.21}$$

These agree with the exact results in Eqs. (4.10)–(4.12) when $\varepsilon \to 0$. Observe that h is very small and K is very large.

Exercise 4.2.3 *Verify that the exact results in Eqs. (4.10)–(4.12) from Exercise 4.2.2 reduce to Eqs. (4.20) and (4.21) and $c = 0$ to $\mathcal{O}(\varepsilon)$.*

4.3 The τ-Continuum Limit of Fradkin and Susskind

We just saw how in the limit when the Euclidean time step $\varepsilon = \frac{\tau}{N} \to 0$, the quantum problem mapped into a classical Ising problem with a special set of parameters in which e^{-2K} and h were infinitesimal. These parameters corresponded to a transfer matrix that was close to the identity, which in turn was because now T_ε was generating time evolution over a truly infinitesimal time ε. In other words, in the large parameter space $[h, K]$ of the Ising model there is a tiny corner with

$$e^{-2K} = \varepsilon B_1 \qquad \varepsilon \to 0, \tag{4.22}$$

$$h = \varepsilon B_3 \qquad \varepsilon \to 0, \tag{4.23}$$

in which $T = I - \varepsilon H + \mathcal{O}(\varepsilon^2)$ with

$$H = -B_1\sigma_1 - B_3\sigma_3. \tag{4.24}$$

It follows that if we are interested in Ising models in this corner we can go backwards and map it to the eigenvalue problem of the quantum H above, which is a lot easier than the eigenvalue problem of the transfer matrix.

Coming from the Ising side we must now take the view that K is a number we are free to dial and that it determines the product of ε and B_1. There is some freedom in assigning values to them separately. We will choose B_1 to be some fixed number with units of energy – it can be a GeV or a μeV, it does not matter. Once the choice is made, K determines ε in these units and we must let $K \to \infty$ to ensure $\varepsilon \to 0$. We see that h determines B_3 in these units. In short, K and h determine ε, the time between slices, and the field B_3 in units set by B_1.

The choice of the classical model's parameters that leads to a T close to the identity (and makes it generate infinitesimal change from one time slice to the next in the quantum interpretation) was invented and dubbed the *τ-continuum limit* by Fradkin and Susskind [1], who went on to exploit it to great effect.

A general feature of the τ-continuum limit is that couplings involving spins at one time (like h) will be infinitesimal, while couplings between spins at neighboring times (like K) will be large and ferromagnetic, so that e^{-2K} or K^* is infinitesimal. We understand that a large K strongly discourages changes in spin, so that T has tiny off-diagonal elements. But why does h have to be infinitesimal? Because all deviations from the identity have to be infinitesimal so we may extract an ε-independent $-H$ as $\lim_{\varepsilon\to 0}(T - I)/\varepsilon$.

The utility of the τ-continuum limit will become abundantly clear in later chapters, but here is the partial answer to why one would ever focus on a limited region wherein the τ-continuum limit maps the classical problem to the eigenvalue problem of a simple H. One focus of this book will be *phase transition*, in which a system undergoes qualitative change, say from being magnetized to unmagnetized, as a function of parameters in the Boltzmann weight. The transition will occur on a hypersurface in parameter space. Some aspects of the transition will be *universal*, i.e, independent of where we are on the surface, be it at generic values of the parameters or in the τ-continuum limit. In this case it will prove much easier to extract the universal features in the τ-continuum limit.

4.4 Two $N \to \infty$ Limits and Two Temperatures

The meaning of $N \to \infty$ is different when we take the limit of continuous time as compared to taking the thermodynamic limit as in Section 2.2. The difference comes from what is held constant.

- If we work with a *fixed* transfer matrix T that describes a classical model with some fixed parameters K and h and take $N \to \infty$, we are taking the *thermodynamic limit* in which the system is infinitely large. In this limit, T is dominated by the eigenvector with the largest eigenvalue, which we called λ_0. If we write $T = e^{-H}$, it is dominated by the ground state of H.
- If we choose $T_\varepsilon = U(\tau/N) = U(\varepsilon)$, the parameters of T_ε vary with N in such a way that the partition function $Z = \mathrm{Tr}\, T^N$ or $Z = \langle s_N | T_\varepsilon^N | s_0 \rangle$ is independent of N. The system has finite extent in the τ direction, but the points in that interval are becoming dense as $N \to \infty$. In this limit the physics is not dominated by a single eigenvector of $U(\tau)$; that happens only for $\tau \to \infty$.
- If we want a quantum system at zero temperature, i.e., $\beta \to \infty$, we must let $\tau = \beta\hbar \to \infty$. This is when the ground state of the quantum system dominates. In this limit the number of points in any fixed τ interval and the length of the overall interval in τ are both diverging. This poses a double problem for numerical (Monte Carlo) work since the number of spins the computer has to keep track of diverges in this double sense. One then has to extrapolate the numerical results obtained from a large but necessarily finite value of N to the continuum/zero-temperature limits.

When we map a quantum partition function to a classical Ising model, there are *two different temperatures*. On the one hand, we have β, the quantum temperature, which controls the length ($\beta\hbar$) of the system in the τ direction, which is also the spatial extent of the Ising model. On the other hand we have the temperature of the Ising model (hidden inside, say, $K = J/kT$), which varies with the parameters of the quantum problem (such as B_1) as well as the value of $\varepsilon = \hbar\beta/N$.

References and Further Reading

[1] E. Fradkin and L. Susskind, Physical Review D, **17**, 2637 (1978).

5
The Feynman Path Integral

Now we turn to the path integral that was first studied by Feynman [1,2] in his treatment of a particle in one dimension whose points are labeled by $-\infty < x < \infty$ and for which the resolution of the identity is

$$I = \int_{-\infty}^{\infty} |x\rangle\langle x|\, dx. \tag{5.1}$$

We will first review the real-time case and then quickly go to the case $t = -i\tau$. We will also consider a related path integral based on another resolution of the identity.

We will reinstate $\hbar$ since the Feynman approach has so much to say about the classical limit $\hbar \to 0$.

Except for the fact that the Hilbert space is infinite dimensional, the game plan is the same: chop up $U(t)$ into a product of N factors of $U(t/N)$, insert the resolution of the identity $N-1$ times, and take the limit $N \to \infty$.

5.1 The Feynman Path Integral in Real Time

Let us assume that the Hamiltonian is time independent and has the form

$$H = \frac{P^2}{2m} + V(X). \tag{5.2}$$

First, it is evident that we may write

$$\exp\left[-\frac{i\varepsilon}{\hbar}(P^2/2m + V(X))\right] \simeq \exp\left[-\frac{i\varepsilon}{2m\hbar}P^2\right] \cdot \exp\left[-\frac{i\varepsilon}{\hbar}V(X)\right], \tag{5.3}$$

because the commutators in

$$e^A e^B = e^{A+B+\frac{1}{2}[A,B]+\cdots}$$

are proportional to higher powers of ε, which is going to 0. While all this is fine if A and B are finite-dimensional matrices with finite matrix elements, it is clearly more delicate for operators in Hilbert space which could have large or even singular matrix elements. We will simply assume that in the limit $\varepsilon \to 0$ the $\simeq$ sign in Eq. (5.3) will become the equality sign for the purpose of computing any reasonable physical quantity. We could also have split the $V(X)$ term into two equal parts and put one on each side of the P terms in symmetric form, just as we wrote the magnetic field term in the Ising model as $T_h^{1/2} T_K T_h^{1/2}$, but we will not, for this will add to the clutter with no corresponding gains if we work to this order in ε.

So we have to compute

$$\langle x' | \underbrace{e^{-\frac{i\varepsilon}{2m\hbar}P^2} \cdot e^{-\frac{i\varepsilon}{\hbar}V(X)} \cdot e^{-\frac{i\varepsilon}{2m\hbar}P^2} \cdot e^{-\frac{i\varepsilon}{\hbar}V(X)} \cdots}_{N \text{ times}} |x\rangle. \tag{5.4}$$

The next step is to introduce the resolution of the identity Eq. (5.1) between every two adjacent factors of $U(t/N)$. Let me illustrate the outcome by considering $N = 3$. We find (upon renaming x, x' as x_3, x_0 for the usual reasons),

$$\begin{aligned} U(x_3, x_0; t) = \int \prod_{n=1}^{2} dx_n \, &\langle x_3 | e^{-\frac{i\varepsilon}{2m\hbar}P^2} e^{-\frac{i\varepsilon}{\hbar}V(X)} |x_2\rangle \\ &\times \langle x_2 | e^{-\frac{i\varepsilon}{2m\hbar}P^2} e^{-\frac{i\varepsilon}{\hbar}V(X)} |x_1\rangle \langle x_1 | e^{-\frac{i\varepsilon}{2m\hbar}P^2} e^{-\frac{i\varepsilon}{\hbar}V(X)} |x_0\rangle. \end{aligned} \tag{5.5}$$

Consider now the evaluation of the matrix element

$$\langle x_n | e^{-\frac{i\varepsilon}{2m\hbar}P^2} \cdot e^{-\frac{i\varepsilon}{\hbar}V(X)} |x_{n-1}\rangle. \tag{5.6}$$

When the right-most exponential operates on the ket to its right, the operator X gets replaced by the eigenvalue x_{n-1}. Thus,

$$\langle x_n | e^{-\frac{i\varepsilon}{2m\hbar}P^2} \cdot e^{-\frac{i\varepsilon}{\hbar}V(X)} |x_{n-1}\rangle = \langle x_n | e^{-\frac{i\varepsilon}{2m\hbar}P^2} |x_{n-1}\rangle \, e^{-\frac{i\varepsilon}{\hbar}V(x_{n-1})}. \tag{5.7}$$

Consider now the remaining matrix element. It is simply the free particle propagator from x_{n-1} to x_n in time ε. We know from before, or from doing Exercise 5.1.1, that

$$\langle x_n | e^{\frac{-i\varepsilon}{2m\hbar}P^2} |x_{n-1}\rangle = \left[\frac{m}{2\pi i\hbar\varepsilon}\right]^{1/2} \exp\left[\frac{im(x_n - x_{n-1})^2}{2\hbar\varepsilon}\right]. \tag{5.8}$$

You may ask why we do not expand the exponential $e^{-\frac{i\varepsilon}{2m\hbar}P^2}$ in Eq. (5.7) out to order ε as we did with $e^{\varepsilon B_1 \sigma_1}$ evaluated between eigenstates of σ_3. The reason is that the momentum operator has singular matrix elements between $\langle x_n|$ and $|x_{n-1}\rangle$, and the bra and ket have singular dot products. You can see from Eq. (5.8) that the exact answer does not have a series expansion in ε.

Exercise 5.1.1 *Derive Eq. (5.8) by introducing a resolution of the identity in terms of momentum states between the exponential operator and the position eigenket on the left-hand side of Eq. (5.8). That is, use*

$$I = \int_{-\infty}^{\infty} \frac{dp}{2\pi\hbar} |p\rangle\langle p|, \tag{5.9}$$

where the plane wave states have a wavefunction given by

$$\langle x|p\rangle = e^{ipx/\hbar}, \tag{5.10}$$

which explains the measure for the p integration.

Resuming our derivation, we now have

$$\begin{aligned}
&\langle x_n|e^{-\frac{i\varepsilon}{2m\hbar}P^2}\cdot e^{-\frac{i\varepsilon}{\hbar}V(X)}|x_{n-1}\rangle \\
&= \left[\frac{m}{2\pi i\hbar\varepsilon}\right]^{1/2}\exp\left[\frac{im(x_n-x_{n-1})^2}{2\hbar\varepsilon}\right]e^{-\frac{i\varepsilon}{\hbar}V(x_{n-1})}.
\end{aligned} \tag{5.11}$$

(Had we used the $V(X)$ term symmetrically we would have had $V(x_{n-1}) \to \frac{1}{2}(V(x_{n-1}) + V(x_n))$ in the above.) Collecting all such factors (there are just two more in this case with $N = 3$), we can readily see that, for general N,

$$\begin{aligned}
U(x_N, x_0; t) &= \left(\frac{m}{2\pi i\hbar\varepsilon}\right)^{1/2}\left[\int\prod_{n=1}^{N-1}\left(\frac{m}{2\pi i\hbar\varepsilon}\right)^{1/2}dx_n\right] \\
&\times\exp\left[\sum_{n=1}^{N}\frac{im(x_n-x_{n-1})^2}{2\hbar\varepsilon}-\frac{i\varepsilon}{\hbar}V(x_{n-1})\right].
\end{aligned} \tag{5.12}$$

If we rewrite

$$\begin{aligned}
&\exp\left[\sum_{n=1}^{N}\left[\frac{im(x_n-x_{n-1})^2}{2\hbar\varepsilon}-\frac{i\varepsilon}{\hbar}V(x_{n-1})\right]\right] \\
&= \exp\frac{i}{\hbar}\varepsilon\sum_{n=1}^{N}\left[\frac{m(x_n-x_{n-1})^2}{2\varepsilon^2}-V(x_{n-1})\right],
\end{aligned} \tag{5.13}$$

we obtain the continuum version as $\varepsilon \to 0$:

$$U(x, x'; t) = \int[\mathcal{D}x]\exp\left[\frac{i}{\hbar}\int_0^t\mathcal{L}(x,\dot{x})dt\right], \tag{5.14}$$

where

$$\int[\mathcal{D}x] = \lim_{N\to\infty}\left(\frac{m}{2\pi i\hbar\varepsilon}\right)^{1/2}\int\left[\prod_{n=1}^{N-1}\left(\frac{m}{2\pi i\hbar\varepsilon}\right)^{1/2}dx_n\right] \tag{5.15}$$

and the *Lagrangian* $\mathcal{L}(x,\dot{x})$ is the familiar function from classical mechanics:

$$\mathcal{L}(x,\dot{x}) = \frac{1}{2}m\dot{x}^2 - V(x) = K - V, \tag{5.16}$$

where K is the kinetic energy and V is the potential energy.

The continuum notation is really a schematic for the discretized version of Eq. (5.12) that preceded it, and which actually defines what one means by the path integral. It is easy to make many mistakes if one forgets this. In particular, there is no reason to believe that replacing differences by derivatives is always legitimate. For example, in this problem, in a time ε the x typically changes by $\mathcal{O}(\varepsilon^{1/2})$ and not $\mathcal{O}(\varepsilon)$, making the velocity singular. The continuum version is, however, very useful to bear in mind since it exposes some aspects of the theory that would not be so transparent otherwise. It is also very useful for getting the picture at the semiclassical level and for finding whatever connection there is between the macroscopic world of smooth paths and the quantum world.

The quantity

$$S[x(t)] = \int_{t_1}^{t_2} \mathcal{L}dt, \tag{5.17}$$

which is the time integral of the Lagrangian over a path $x(t)$, is called the *action* for the path $x(t)$. It is a function of the path $x(t)$, which itself is a function. It is called a *functional*. The path integral is often called a *functional integral*. We may write the path integral schematically as

$$U(x',x;t) = \int [\mathcal{D}x]e^{\frac{i}{\hbar}S}. \tag{5.18}$$

This is how Feynman manages to express the quantum mechanical amplitude to go from a point x to x' in a time t as a sum over paths with equal unimodular weight $e^{\frac{i}{\hbar}S}$ for every classical path connecting the end points in spacetime. Why, then, does the world look classical? The answer is that as we sum over all crazy paths, the rapidly oscillating exponentials cancel out, until we approach a path for which the action is an extremum. Now S remains nearly constant and the unimodular numbers $e^{\frac{i}{\hbar}S}$ add up in phase to something sizable.

The stationary points of S are the points obeying the Euler–Lagrange equation

$$\frac{d}{dt}\left(\frac{\partial \mathcal{L}}{\partial \dot{x}}\right) = \frac{\partial \mathcal{L}}{\partial x}, \tag{5.19}$$

which is just $F = ma$ derived now from the *principle of least action*.

Paths that contribute coherently must have an action within roughly $\hbar$ of the action of the classical path. Outside this region, the paths cancel out. In the limit $\hbar \to 0$, we may then forget all but the classical path and its coherent neighbors and write

$$U(x',x;t) \simeq A\, e^{\frac{i}{\hbar}S_c}, \tag{5.20}$$

where S_c is the action of the classical path $x_c(t)$ and A is a prefactor that reflects the sum over the nearby paths. In the limit $\hbar \to 0$ one focuses on the exponential factors and pays little attention to A. However, with some work one can calculate A by systematically including the fluctuations around $x_c(t)$. This is called the *semiclassical approximation*. It works not because $\hbar$ actually tends to 0 (for $\hbar$ is what it is), but because we are interested in a particle that is getting so macroscopic that any tiny change in its path causes huge changes in the action measured *in units of* $\hbar$.

There are many problems where the answer is a sum over configurations and there is a number (the counterpart of $1/\hbar$) in front of the corresponding "action" that is very large, allowing one to replace the sum over configurations by the contribution from the stationary point of S. These are also called semiclassical or *saddle point approximations*.

The preceding treatment relies on the fact that the complex exponential $e^{i\frac{S}{\hbar}}$ behaves like a real one in that its main contribution comes from where the action is stationary: it is a mathematical fact that a rapidly oscillating exponential is as bad as a rapidly falling one. When we go to imaginary-time path integrals, things will be a lot simpler: we will be working with a real exponential, $e^{-S_{\mathrm{E}}/\hbar}$, which will be dominated by the minima (not just stationary points) of the Euclidean action S_{E} for obvious reasons.

So here is a toy model of a saddle point calculation for a real exponential that allows us to get a feeling for what happens with both (real- and imaginary-time) path integrals. Consider a one-dimensional integral

$$Z(\alpha) = \int_{-\infty}^{\infty} dx e^{-\alpha F(x)} \tag{5.21}$$

where $\alpha \to \infty$. In this case, the dominant contribution will come where F is smallest. Assume it has only one minimum at $x = x_0$, and that

$$F(x) = F(x_0) + F'(x_0)(x - x_0) + F''(x_0)\frac{(x - x_0)^2}{2} + \cdots, \tag{5.22}$$

where the primes on F denote its derivatives. Since x_0 is a minimum, $F'(x_0) = 0$. We will drop the higher-order terms in the Taylor series, because they make subdominant contributions as $\alpha \to \infty$, and simply do the Gaussian integral, called the *fluctuation integral*. Thus,

$$Z(\alpha \to \infty) \to e^{-\alpha F(x_0)} \int_{-\infty}^{\infty} dx e^{-\frac{1}{2}\alpha F''(x_0)(x - x_0)^2} = \sqrt{\frac{2\pi}{\alpha F''(x_0)}} e^{-\alpha F(x_0)}. \tag{5.23}$$

The leading dependence on α is in the exponential and is called the saddle point contribution, and it dominates the prefactor, coming from the fluctuation integral.

Exercise 5.1.2 *To show that the higher derivatives can be ignored in the limit $\alpha \to \infty$, first shift the origin to $x = x_0$ and estimate the average value of the new x^2 in terms of α. Argue that a factor $e^{\alpha x^3}$ will be essentially equal to 1 in the region where the Gaussian integral has any real support.*

The above toy integral captures the essential ideas of the saddle point approximation in path integrals except for one – the integral over paths is really an integral over an infinite number of coordinates. Sitting at a stationary point in *path space*, namely at the classical path, we can move away in an infinite number of directions which correspond to the infinite number of ways of slightly deforming the classical path. All these fluctuations have to be included by doing that many Gaussian integrals.

There are some cases when we do not have to apologize for dropping higher derivatives of F: that is when F has no cubic or higher terms in its Taylor expansion. In that case Eq. (5.23) becomes exact and $F''(x_0) = F''$, the x-independent number that multiplies $\frac{1}{2}x^2$.

These features are true even in an actual path integral, not just the toy model above. In all cases where the Lagrangian has no terms of cubic or higher order in x or $\dot{x}$, we may write

$$Z = Ae^{-i\frac{S_c}{\hbar}}, \tag{5.24}$$

where S_c is the action of the classical path and A is a factor obtained from the Gaussian integration over fluctuations about the classical path.

Consider a free particle whose action is clearly quadratic. The classical path connecting $(x,0)$ to (x',t) is a straight line in spacetime with constant velocity $v = (x'-x)/t$, and so

$$S_c = \int_0^t \frac{m}{2} v^2 dt' = \frac{m}{2} \frac{(x'-x)^2}{t}, \tag{5.25}$$

and

$$U(x',x;t) = A(t) \exp\left[\frac{im(x'-x)^2}{2\hbar t}\right], \tag{5.26}$$

where to find $A(t)$, we need to do a Gaussian *functional integral* around the straight path. However, if we combine two facts:

- in the limit $t \to 0$, $U(x',x;t)$ must approach $\delta(x-x')$;
- $\delta(x-x') = \lim_{\Delta\to 0}\left(\frac{1}{\pi\Delta^2}\right)^{1/2} \exp\left[-\frac{(x-x')^2}{\Delta^2}\right]$ (even for complex Δ^2);

we may deduce that

$$A(t) = \left[\frac{m}{2\pi\hbar i t}\right]^{1/2}. \tag{5.27}$$

Exercise 5.1.3 (Important) *Consider the oscillator. First solve for a trajectory that connects the spacetime points $(x,0)$ to (x',t) by choosing the two free parameters in its solution. (Normally you choose them given the initial position and velocity; now you want the solution to go through the two end points in spacetime.) Then find its action and show that*

$$U(x',x;t) = A(t) \exp\left[\frac{im\omega}{2\hbar \sin\omega t}\left[(x^2 + x'^2)\cos\omega t - 2x'x\right]\right]. \tag{5.28}$$

If you want $A(t)$ you need to modify the trick used for the free particle, since the exponential is not a Gaussian in $(x-x')$. Note, however, that if you choose $x=0$, it becomes a Gaussian in x' which allows you to show that $A(t) = (m\omega/2\pi i\hbar \sin\omega t)^{1/2}$. That the answer for the fluctuation integral A does not depend on the classical path, especially the end points x or

x', is a property of the quadratic action. It amounts in our toy model to the fact that if $F(x)$ is quadratic, that is $F = ax^2 + bx + c$, then $F'' = 2a$ for all x, including the minimum.

5.2 The Feynman Phase Space Path Integral

The path integral derived above is called the *configuration space path integral*, or simply *the* path integral. We now consider another one. Let us go back to

$$\langle x_N|\underbrace{e^{-\frac{i\varepsilon}{2m\hbar}P^2}\cdot e^{-\frac{i\varepsilon}{\hbar}V(X)}\cdot e^{-\frac{i\varepsilon}{2m\hbar}P^2}\cdot e^{-\frac{i\varepsilon}{\hbar}V(X)}\cdots}_{N\text{ times}}|x_0\rangle. \tag{5.29}$$

Let us now introduce resolutions of the identity between *every exponential* and the next. We need two versions,

$$I = \int_{-\infty}^{\infty} dx|x\rangle\langle x|, \tag{5.30}$$

$$I = \int_{-\infty}^{\infty} \frac{dp}{2\pi\hbar}|p\rangle\langle p|, \tag{5.31}$$

where the plane wave states have a wavefunction given by

$$\langle x|p\rangle = e^{ipx/\hbar}. \tag{5.32}$$

Let us first set $N = 3$ and insert three resolutions of the identity in terms of p-states and two in terms of x-states with x and p resolutions alternating. This gives us

$$U(x_3,x_0,t) = \int [\mathcal{D}p\mathcal{D}x]\ \langle x_3|e^{-\frac{i\varepsilon}{2m\hbar}P^2}|p_3\rangle\langle p_3|e^{-\frac{i\varepsilon}{\hbar}V(X)}|x_2\rangle\langle x_2|e^{-\frac{i\varepsilon}{2m\hbar}P^2}|p_2\rangle$$
$$\times\langle p_2|e^{-\frac{i\varepsilon}{\hbar}V(X)}|x_1\rangle\langle x_1|e^{-\frac{i\varepsilon}{2m\hbar}P^2}|p_1\rangle\langle p_1|e^{-\frac{i\varepsilon}{\hbar}V(X)}|x_0\rangle, \tag{5.33}$$

where

$$\int [\mathcal{D}p\mathcal{D}x] = \underbrace{\int_{-\infty}^{\infty}\int_{-\infty}^{\infty}\int_{-\infty}^{\infty}\int_{-\infty}^{\infty}\cdots\int_{-\infty}^{\infty}}_{2N-1\text{ times}}\prod_{n=1}^{N}\frac{dp_n}{2\pi\hbar}\prod_{n=1}^{N-1}dx_n \tag{5.34}$$

with $N = 3$. Evaluating all the matrix elements of the exponential operators is trivial since each operator can act on the eigenstate to its right and get replaced by the eigenvalue. Collecting all the factors (a strongly recommended exercise for you), we obtain, for general N, the result

$$U(x,x',t) = \int [\mathcal{D}p\mathcal{D}x]\exp\left[\sum_{i=1}^{N}\left(\frac{-i\varepsilon}{2m\hbar}p_n^2 + \frac{i}{\hbar}p_n(x_n - x_{n-1}) - \frac{i\varepsilon}{\hbar}V(x_{n-1})\right)\right], \tag{5.35}$$

with $[\mathcal{D}p\mathcal{D}x]$ as in Eq. (5.34).

In the limit $N \to \infty$ i.e., $\varepsilon \to 0$, we write schematically in continuous time (upon multiplying and dividing the middle term by ε) the following continuum version:

$$U(x,x',t) = \int [\mathcal{D}p\mathcal{D}x] \exp\left[\frac{i}{\hbar}\int_0^t [p\dot{x} - \mathcal{H}(x,p)]dt\right], \tag{5.36}$$

where $\mathcal{H} = p^2/2m + V(x)$ and $(x(t), p(t))$ are now written as functions of a continuous variable t. This is the *phase space path integral* for the propagator. The continuum version is very pretty (with the Lagrangian in the exponent, but expressed in terms of (x,p)), but is only a schematic for the discretized version preceding it.

In our problem, since p enters the Hamiltonian quadratically, it is possible to integrate out all the N variables p_n. Going back to the discretized form, we isolate the part that depends on just p's and do the integrals:

$$\prod_1^N \int_{-\infty}^{\infty} \frac{dp_n}{2\pi\hbar} \exp\left[-i\frac{\varepsilon}{2m\hbar}p_n^2 + \frac{i}{\hbar}p_n(x_n - x_{n-1})\right]$$
$$= \prod_1^N \left(\frac{m}{2\pi i\hbar\varepsilon}\right)^{1/2} \exp\left[\frac{im(x_n - x_{n-1})^2}{2\hbar\varepsilon}\right]. \tag{5.37}$$

If we now bring in the x-integrals we find that this gives us exactly the configuration space path integral, as it should.

If p does not enter the Hamiltonian in a separable quadratic way, it will not be possible to integrate it out and get a path integral over just x, since we do not know how to do non-Gaussian integrals. In that case we can only write down the phase space path integral.

5.3 The Feynman Path Integral for Imaginary Time

Consider the matrix element

$$U(x,x';\tau) = \langle x|U(\tau)|x'\rangle. \tag{5.38}$$

We can write down a path integral for it following exactly the same steps as for real time. The final answer in continuum notation is

$$\langle x|U(\tau)|x'\rangle = \int [\mathcal{D}x] \exp\left[-\frac{1}{\hbar}\int_0^\tau \mathcal{L}_E(x,\dot{x})d\tau\right], \tag{5.39}$$

$$\int [\mathcal{D}x] = \lim_{N\to\infty}\left(\frac{m}{2\pi\hbar\varepsilon}\right)^{1/2} \prod_0^{N-1}\left(\frac{m}{2\pi\hbar\varepsilon}\right)^{1/2} dx_i, \tag{5.40}$$

$$\mathcal{L}_E = \frac{m}{2}\left(\frac{dx}{d\tau}\right)^2 + V(x), \tag{5.41}$$

where $\varepsilon = \tau/N$ and $\mathcal{L}_E$ is called the *Euclidean Lagrangian*. Notice that $\mathcal{L}_E$ is the *sum* of the Euclidean kinetic energy and real-time potential energy. *Thus a particle obeying the Euclidean equations of motion will see the potential turned upside down. This will be exploited later.*

The *Euclidean action* is defined in the obvious way in terms of the Euclidean Lagrangian:

$$S_{\mathrm{E}} = \int \mathcal{L}_{\mathrm{E}} d\tau. \tag{5.42}$$

We have emphasized that the continuum form of the path integral is a shorthand for the discrete version. It is true here also, but of all the path integrals, the Euclidean is the best behaved. Rapidly varying paths with large kinetic energy are suppressed by the falling (rather than rapidly oscillating) exponential factor. Likewise, the path of least action dominates the sum over paths not because of stationary phase, but because of simple numerical dominance: $e^{-\frac{S_{\mathrm{E}}}{\hbar}}$ is biggest when S_{E} is smallest.

5.4 Classical ↔ Quantum Connection Redux

We saw that in order to have the right limit in continuous time, the classical Ising model must have its parameters in a special τ-continuum regime, such as $e^{-2K} \to \varepsilon$, $h \to \varepsilon$. We want to derive similar conditions on the parameters of a classical model where the fluctuating variable is continuous and can take on any real value.

Consider a classical system with $N+1$ sites and a degree of freedom x_n at each site. A concrete example is a one-dimensional crystal where the atom labeled n is located at $x = na + x_n$, with na being the equilibrium position and x_n the *deviation* from it. The variables at the end of the chain, called x_0 and x_N, are fixed. Then

$$Z = \int_{-\infty}^{\infty} \prod_{1}^{N-1} dx_i e^{-\frac{1}{kT}E(x_0,\ldots,x_N)}, \tag{5.43}$$

where E is the energy function and we have written β in terms of the more familiar temperature variable as $\beta = \frac{1}{kT}$. The reason for neglecting the kinetic energy of the atoms may be found in Exercise 5.4.1.

Let E/kT have the form

$$\frac{E}{kT} = \sum_{1}^{N-1} \left[K_1(x_n - x_{n-1})^2 + K_2 x_n^2 \right], \tag{5.44}$$

where the first term favors a fixed separation ($x_n = x_{n-1}$ corresponds to spacing a) and the second one provides a quadratic potential that discourages each x_n from wandering off its neutral position $x_n = 0$ or $x = na$.

Exercise 5.4.1 *Include the kinetic energy $p^2/2m$ of the atoms in the Boltzmann weight and do the p integrals as part of Z. Obtain an overall factor $(\sqrt{2\pi mkT})^{N-1}$ and show that the statistical properties of the x's are still given by the Z in Eq. (5.44).*

If we compare this to the discretized imaginary-time Feynman path integral for the oscillator,

$$U(x_0,x_N,\tau)=\int_{-\infty}^{\infty}\prod_{1}^{N-1}dx_i\exp\left[-\frac{1}{\hbar}\sum_{1}^{N-1}\varepsilon\left(\frac{m}{2}\frac{(x_n-x_{n-1})^2}{\varepsilon^2}+\frac{m\omega^2}{2}x_n^2\right)\right], \tag{5.45}$$

we see the following generalizations to infinite-dimensional spaces of the Ising spin-$\frac{1}{2}$ correspondence:

- The Feynman path integral from x_0 to x_N is identical in form to a classical partition function of a system of $N+1$ coordinates x_n with the boundary condition that the first and last be fixed at x_0 and x_N. The variables x_n are interpreted as intermediate state labels of the quantum problem (in the repeated resolution of the identity) and as the classical variables summed over in the partition function.
- The role of the Euclidean action S_{E} in the Feynman integral is played by the energy in the partition function.
- The role of $\hbar$ is played by T. In particular, as either variable goes to zero, the sum over configurations is dominated by the minimum of action or energy and fluctuations are suppressed.
- The parameters in the classical and quantum problems can be mapped onto each other. For example, $K_1=m/2\hbar\varepsilon$ and $K_2=m\omega^2\varepsilon/2\hbar$. This is the τ-continuum limit, with very strong coupling (K_1) between variables in neighboring times and very weak coefficients (K_2) for terms at a given time.
- Since $\varepsilon\to 0$ in the quantum problem, the parameters of the classical problem must take some limiting values ($K_1\to\infty$ and $K_2\to 0$ in a special way, namely the τ-continuum limit) to really be in correspondence with the quantum problem with continuous time evolution generated by $H=P^2/2m+m\omega^2X^2/2$.
- The single quantum degree of freedom is traded for a one-dimensional array of classical degrees of freedom. This is a general feature: the dimensionality goes up by one as we go from the quantum to the classical problem. For example, a one-dimensional array of *quantum* oscillators would map onto the partition function of a two-dimensional array of classical variables. The latter array would be labeled by the time slice n as well as the quantum oscillator whose intermediate state label it stands for.

5.5 Tunneling by Euclidean Path Integrals

We now consider an application of the Euclidean formalism. We have seen how the path integral can be approximated by the contribution from the classical path in the limit $\hbar\to 0$. This procedure does not work for tunneling amplitudes across barriers since we cannot find a classical path that goes over the barrier. On the other hand, in the Euclidean dynamics the potential is turned upside down, and what is forbidden in Minkowski space is suddenly allowed in the Euclidean region!

Here is a problem that illustrates this point and many more. Consider a particle in a double-well potential,

$$V(x) = A^2(x^2 - a^2)^2. \tag{5.46}$$

The classical minima are at

$$x = \mp a. \tag{5.47}$$

Figure 5.1 shows the potential in Minkowski and Euclidean space for the case $A = a = 1$. Notice that in the Euclidean problem the double well has been inverted into a double hill.

What is the ground state of the system? The classical ground state is doubly degenerate: the particle can be sitting at either of the two minima. In the semiclassical approximation, we can broaden these out to Gaussians that are ground states $|\pm a\rangle$ in the harmonic-oscillator-like potential around each minimum at $x = \pm a$. This will shift each degenerate ground state by $\hbar\omega$, where ω measures the curvature of the potential near the minimum. We can go to higher-order approximations that recognize that the bottom of the well is not exactly quadratic, and shift the ground state energies by higher powers of $\hbar$. However, none of this will split the degeneracy of the ground states since whatever we find at the left minimum we will find at the right by symmetry under reflection. Lifting of the degeneracy will happen only if we take into account tunneling between the two wells. So we study this problem in the following stripped-down version. First, we drop all but the degenerate ground states $|\pm a\rangle$. (The Gaussians centered around the two minima are not quite orthogonal. Assume they have been orthogonalized by a Gram–Schmidt procedure.) The approximate Hamiltonian looks like this in this subspace:

$$H = \begin{pmatrix} E_0 & 0 \\ 0 & E_0 \end{pmatrix}. \tag{5.48}$$

Let us shift our reference energy so that $E_0 = 0$.

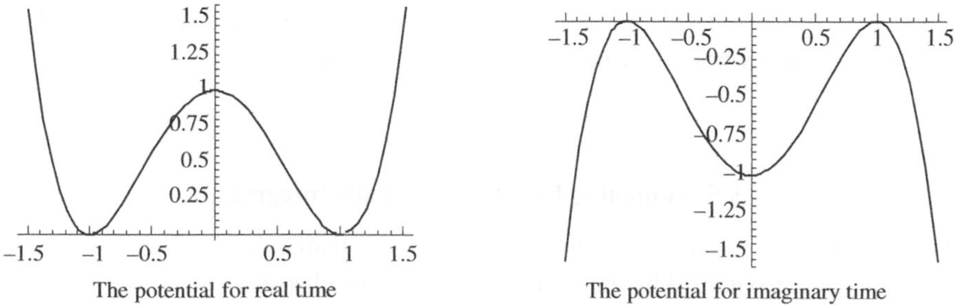

Figure 5.1 The potential with minima at $x = \pm 1$ gets inverted to one with maxima at $x = \pm 1$ as we go from real to imaginary time. Thus, the two minima separated by a barrier in real time become maxima connected by a classical path in imaginary time, allowing a saddle point calculation.

There are no off-diagonal matrix elements. If this were an exact result, it should mean that if a particle starts out in one well it will never be found at the other. But we know from the wavefunction approach that if it starts at one side, it can tunnel to the other. This means that there is effectively a non-zero off-diagonal matrix element $H_{+-} = H_{-+} = \langle a|H|-a\rangle$ in this basis. *The challenge is to find that element in the semiclassical approximation.* Once we find it, it is evident that the energy levels will be split into

$$E = \pm H_{+-} \tag{5.49}$$

and the eigenstates will be $|S/A\rangle$, the symmetric and antisymmetric combinations of $|\pm a\rangle$.

The problem is that there is no classical path connecting the two sides of the well due to the barrier. We will finesse that by considering the imaginary-time dynamics in which the potential gets inverted and $x = \pm a$ are connected by a classical path.

Consider

$$\langle a|U(\tau)|-a\rangle = \langle a|e^{-\frac{1}{\hbar}H\tau}|-a\rangle. \tag{5.50}$$

In this discussion of tunneling, $U(\tau)$ is the propagator from $-\tau/2$ to $\tau/2$ and not from 0 to τ. The term linear in τ gives us the off-diagonal matrix element that we are after:

$$\langle a|e^{-\frac{1}{\hbar}H\tau}|-a\rangle = 0 - \frac{1}{\hbar}\tau\langle a|H|-a\rangle + \mathcal{O}(\tau^2). \tag{5.51}$$

The point to bear in mind is that even though we are considering a Euclidean transition amplitude, *the Hamiltonian H is the same as in real time.*

We shall calculate $\langle a|e^{-\frac{1}{\hbar}H\tau}|-a\rangle$ by the semiclassical approximation to the Euclidean path integral and extract the approximate matrix element H_{+-} as the coefficient of the term linear in τ. I am not saying that τ has to be small, only that the coefficient of the linear term is the matrix element we want.

We shall find that in the semiclassical approximation the term linear in τ assumes the form

$$\langle a|e^{-\frac{1}{\hbar}H\tau}|-a\rangle \simeq \tau e^{-\frac{1}{\hbar}S_{\rm cl}}, \tag{5.52}$$

where $S_{\rm cl}$ is the Euclidean action for the classical path connecting the left hill to the right and numerical prefactors other than τ have been suppressed.

Comparing this to

$$\langle a|e^{-\frac{1}{\hbar}H\tau}|-a\rangle = 0 - \frac{1}{\hbar}\tau\langle a|H|-a\rangle + \mathcal{O}(\tau^2), \tag{5.53}$$

we infer that

$$\langle a|H|-a\rangle \simeq e^{-\frac{1}{\hbar}S_{\rm cl}} \tag{5.54}$$

up to factors that do not affect the exponential, which dominates.

The key point is that in the double-hill potential of Euclidean mechanics the classical ground states are not separated by a barrier but a valley, so that there will be no problem finding a classical path going from one hilltop to the other.

The Euclidean equations of motion are the same as the real-time ones, *except for the reversal of the potential.* Thus there will be a conserved energy E_e given by

$$E_e = \frac{m}{2}\left(\frac{dx}{d\tau}\right)^2 - V(x). \tag{5.55}$$

Using this, we can solve for the trajectory by quadrature:

$$\int_{x_1}^{x_2} \frac{\sqrt{m}dx}{\sqrt{2(E_e + V(x))}} = \int_{t_1}^{t_2} d\tau. \tag{5.56}$$

Now we want the tunneling from the state $|-a\rangle$ to the state $|a\rangle$. These are not eigenstates of position, but Gaussians centered at $x = \mp a$. We shall, however, calculate the amplitude to tunnel from the *position eigenstate* $x = -a$ to the position eigenstate $x = a$. Except for the overlaps $\langle x = a|a\rangle$ and $\langle -a|x = -a\rangle$, this is the same as $\langle a|U| - a\rangle$. These overlaps know nothing about the tunneling barrier. They will constitute undetermined prefactors in front of the exponential dependence on the barrier that we are after.

Let us consider the trajectory that has $E_e = 0$. It is given by doing the above integral with $E_e = 0$:

$$x(\tau) = a \tanh\left[\sqrt{\frac{2}{m}} Aa(\tau - \tau_0)\right], \tag{5.57}$$

where τ_0 is a constant not determined by the equation of motion or the boundary condition $x \to \pm a$ as $\tau \to \pm\infty$. Notice that in this trajectory, the particle starts out at the left *maximum* (Figure 5.1) at $\tau \to -\infty$ and rolls down the hill and up to the top of the right maximum, reaching it only as $\tau \to \infty$. Physically, it takes for ever since the particle must start from rest at the left end to have zero Euclidean energy. Only at infinite times does the tanh take its limiting value of $\pm a$, and only in this limit is τ_0 a free parameter. Normally a second-order equation should have no free parameters left once we have specified two points in spacetime that it passes through. The loophole now is that the two times involved are infinite, and the finite τ_0 gets buried under them.

The action for the above solution (combining the general result $\mathcal{L}_E = T + V$ with the result $T = V$ for the zero-energy solution) is

$$S_{cl} = \int (T + V)d\tau = \int 2T d\tau = \int m\dot{x}\dot{x} d\tau = \int_{-a}^{a} p(x)dx = \int_{-a}^{a} \sqrt{2mV(x)}dx, \tag{5.58}$$

and the tunneling amplitude

$$e^{-\frac{1}{\hbar}\int_{-a}^{a}\sqrt{2mV(x)}dx}$$

is in agreement with the tunneling result in the Schrödinger approach.

Notice that the action does not depend on τ_0. This means that for every τ_0 we have a path with exactly this action. We must sum over all of them to get

$$\langle a|U(\tau)|-a\rangle \simeq \tau e^{-\frac{1}{\hbar}S_{\rm cl}}. \tag{5.59}$$

Comparing it to

$$\langle a|e^{-\frac{1}{\hbar}H\tau}|-a\rangle \simeq 0 - \frac{1}{\hbar}\tau\langle a|H|-a\rangle + \mathcal{O}(\tau^2), \tag{5.60}$$

we read off

$$H_{+-} \simeq -e^{-\frac{1}{\hbar}S_{\rm cl}}, \tag{5.61}$$

where once again we have dropped all prefactors except for the sign, which is important. (All Euclidean transition amplitudes are positive since the functional is positive. The minus sign comes from the definition $U(\tau) = e^{-\frac{1}{\hbar}H\tau}$.)

It is now clear that with H_{-+} negative, the new eigenstates and energies are as follows:

$$|S\rangle = \sqrt{\frac{1}{2}}[|+a\rangle + |-a\rangle], \qquad E_S = -e^{-\frac{1}{\hbar}S_{\rm cl}}, \tag{5.62}$$

$$|A\rangle = \sqrt{\frac{1}{2}}[|+a\rangle - |-a\rangle], \qquad E_A = e^{-\frac{1}{\hbar}S_{cl}}. \tag{5.63}$$

Some of you may be uneasy with using the infinite-τ solution since it means the prefactor τ in Eq. (5.60) is infinite. So let us consider a slightly different solution: the particle leaves at $-a'$, a little bit to the right of the left maximum $-a$, and reaches a symmetric point $+a'$ after a large but finite time τ. Now τ is finite, but the solution has a fixed parameter $\tau_0 = 0$, in agreement with the expectation that a second-order equation obeying two boundary conditions at two finite times has no residual parameters. The choice $\tau_0 = 0$ follows from symmetry: we want the tanh to go to $\pm a'$ at $\pm\tau/2$. So how do we get the τ as a prefactor?

Notice that tunneling takes place in a short time near $\tau = 0$, when the tanh switches sign, while the total tunneling time τ can be made as large as we like by starting as close to the top as we want. Pictorially, the particle takes a long time to roll off the top, but once it gets going, it rolls down very quickly to a point close to the other end point. For this reason, this solution is called an *instanton*, a term coined by 't Hooft: except for the brief "instant" when tunneling takes place, the system is essentially in one of its classical ground states. If we draw a new trajectory in which the tunneling takes place in the same time interval, but is centered around a different time (but far from the end points), this too will be close to being a minimum of the action. (If it has exactly the same action in the limit $\tau \to \infty$, it must have nearly the same action for arbitrarily large τ.) In other words, even the finite-time solution has many companions, all of nearly the same action, but different tunneling instants τ_0. Since they all have nearly the same action, the effect is to multiply the answer by τ. (If you limit τ_0 to $-0.01\frac{\tau}{2} < \tau_0 < 0.01\frac{\tau}{2}$ to stay off the edges, nothing changes: as long as you grant me a prefactor *proportional to* τ, I can extract the matrix element of H, and the 0.01 will get absorbed in the other prefactors which are anyway dwarfed by the exponential $e^{-S_{cl}/\hbar}$.)

So we have argued for a prefactor of τ which came from considering a fluctuation about the classical solution with $\tau_0 = 0$, a fluctuation which moved the tunneling instant. We were forced to consider this particular fluctuation since the action did not change due to it (for large τ). The flatness of the action in this "direction" in path space means that the curvature $F'' \simeq 0$ for doing the Gaussian fluctuation integral vanishes. The integral of a Gaussian with no F'' to damp it yields an answer that diverges as the range of integration, namely τ. We can, however, ignore the finite Gaussian integrals over the rest of the fluctuations since they cost finite action (have $F'' > 0$).

More generally, if a saddle point has other companion solutions of the same action because of a symmetry, the prefactor equal to the volume of the symmetry transformations must be appended by hand. Trying to obtain these as Gaussian fluctuations will lead to divergences because the corresponding F'' will be zero. The preceding example corresponded to time translation symmetry, and the symmetry volume was just τ. All these and more aspects of tunneling are thoroughly explained in the books by Rajaraman [3] and by Coleman [4] listed at the end of this chapter.

5.6 Spontaneous Symmetry Breaking

Why are we interested in a term that vanishes exponentially fast as $\hbar \to 0$ when we ignored all the perturbative corrections to the states $|\pm a\rangle$ which vanished as finite powers of $\hbar$? The reason is that the exponentially small term is the leading term in the *splitting* of the two classically degenerate ground states.

But there is another very significant implication of the tunneling calculation. This has to do with the phenomenon of *spontaneous symmetry breaking*, which will now be described.

Consider a Hamiltonian which has a symmetry, say under parity. *If the lowest-energy state of the problem is itself not invariant under the symmetry, we say symmetry is spontaneously broken.*

Spontaneous symmetry breaking occurs quite readily in classical mechanics. Consider the single-well oscillator. The Hamiltonian is invariant under parity. The ground state is a particle sitting at the bottom of the well. This state respects the symmetry: the effect of parity on this state gives back the state. Now consider the double well with minima at $x = \pm a$. There are two lowest-energy configurations available to the particle: sitting still at the bottom of either well. No matter which choice it makes, it breaks the symmetry. The breakdown is spontaneous in that there was nothing in the Hamiltonian that tilted the scales. The particle has made a choice based on accidents of initial conditions. Let us note the twin signatures of symmetry breaking: there is more than one ground state, and these states are not invariant under the symmetry (some observable, not invariant under the symmetry, has a non-zero value), but instead get mapped into each other by the symmetry operation.

Now consider the quantum case of the double well, *but with an infinite barrier between the wells* (I mean a barrier across which tunneling is impossible either in the path integrals or wavefunction approach, so a delta function spike is not such a barrier). Once again the particle has two choices, these being Gaussian-like functions centered at the two troughs: $|\pm a\rangle$. They show the twin features of symmetry breaking: they are degenerate and non-invariant under parity ($\langle X\rangle \neq 0$). But here is a twist. In quantum theory a particle can be in two places at the same time. In particular, we can form the combinations of these degenerate eigenvectors,

$$|S/A\rangle = \frac{[|+a\rangle \pm |-a\rangle]}{\sqrt{2}}, \tag{5.64}$$

$$\Pi|S/A\rangle = \pm|S/A\rangle, \tag{5.65}$$

which are eigenstates of parity. Indeed, in quantum theory the relation

$$[\Pi, H] = 0 \tag{5.66}$$

guarantees that such parity eigenstates *can* be formed. But *should* they be formed? The answer is negative in this problem due to the infinite barrier. The reason is this: Suppose the particle in question is sighted on one side during a measurement. Then *there is no way for its wavefunction to develop any support in the other side*. (One says the motion is not ergodic.) Even in quantum theory, barrier penetration is forbidden if the barrier is infinite. This means in particular that the symmetric and antisymmetric functions will never be realized by any particle that has ever been seen on either side. The correct thing to do then is to build a Hilbert space of functions with support on just one side. That every state so built has a degenerate partner in the inaccessible well across the barrier is academic. The particle will not even know a parallel universe just like its own exists. Real life will not be symmetric in such a problem and the symmetric and antisymmetric wavefunctions (with zero $\langle X\rangle$) represent unrealizable situations. Symmetry *is* spontaneously broken.

Now for the more typical problem with a finite barrier. In this case, a particle once seen on the left side *can* later be seen on the right side, and vice versa. Symmetric and antisymmetric wavefunctions are physically sensible and we can choose energy eigenstates that are also parity eigenstates. These states will no longer be degenerate. In normal problems, the symmetric state, or more generally the state with eigenvalue unity for the symmetry operation, the one invariant under the symmetry operation, will be the unique ground state. Recall that in the oscillator problem the ground state not only had definite parity, it was invariant under parity. Likewise, in the hydrogen atom the ground state not only has definite angular momentum, the angular momentum is zero and is invariant under rotations. However, in both these problems there was no multiplicity of classical ground states and no real chance of symmetry breakdown. (The oscillator had just one classical ground state at the bottom of the well, and the hydrogen atom had one infinitely deep within the Coulomb well.) What the instanton calculation tells us is that the double well, despite having two classical ground states that break symmetry, has, in the quantum theory, a unique, symmetric, ground state.

Thus, even though the tunneling calculation was very crude and approximate, it led to a very profound conclusion: the symmetry of the Hamiltonian is the symmetry of the ground state; symmetry breaking does not take place in the double-well problem.

This concept of symmetry restoration by tunneling (which in turn is tied to the existence of classical Euclidean solutions with finite action going from one putative degenerate ground state to another) is very deep, and plays a big role in many problems. In quantum chromodynamics, the theory of quarks and gluons, one did not initially realize that the minimum one had assumed was unique for years was one of an infinite family of degenerate minima until an instanton (of finite action) connecting the two classical minima was found and interpreted by 't Hooft and others [5–8]. We discuss a simpler example to illustrate the generality of the notion: a particle in a periodic potential $V(x) = 1 - \cos 2\pi x$. The minima are at $x = n$, where n is any integer. The symmetry of the problem is the discrete translation $x \to x+1$. The approximate states, $|n\rangle$, which are Gaussians centered around the classical minima, break the symmetry and are converted to each other by T, the operator that translates $x \to x+1$:

$$T|n\rangle = |n+1\rangle. \tag{5.67}$$

However, adjacent classical minima are connected by a non-zero tunneling amplitude of the type we just calculated and H has off-diagonal amplitudes between $|n\rangle$ and $|n \pm 1\rangle$. (There are also solutions describing tunneling to next-nearest-neighbor minima, but these have roughly double the action of the nearest-neighbor tunneling process and lead to an off-diagonal matrix element that is roughly the square of the one due to nearest-neighbor tunneling.) Suppose the one-dimensional world were finite and formed a closed ring of size N, so that there were N degenerate classical minima. These would evolve into N non-degenerate levels (the analogs of $|S/A\rangle$) due to the mixing caused by tunneling. The ground state would be a symmetric combination:

$$|S\rangle = \frac{1}{\sqrt{N}} \sum_{1}^{N} |n\rangle. \tag{5.68}$$

The details are left to the following exercise.

Exercise 5.6.1 (Very important) *Assume that*

$$H = \sum_{1}^{N} E_0 |n\rangle\langle n| - t(|n\rangle\langle n+1| + |n+1\rangle\langle n|) \tag{5.69}$$

describes the low-energy Hamiltonian of a particle in a periodic potential with minima at integers n. The integers n go from 1 to N since it is assumed that the world is a ring of length N, so that the $(N+1)$th point is the first. Thus, the problem has symmetry under translation by one site despite the finite length of the world. The first term in H represents *the energy of the Gaussian state centered at $x = n$. The second represents the tunneling to adjacent minima with tunneling amplitude t. Consider the state*

$$|\theta\rangle = \frac{1}{\sqrt{N}} \sum_{1}^{N} e^{in\theta} |n\rangle. \tag{5.70}$$

Show that it is an eigenstate of T. Find the eigenvalue. Use the condition $T^N = I$ to restrict the allowed values of θ and make sure that we still have just N states. Show that $|\theta\rangle$ is an eigenstate of H with eigenvalue $E(\theta) = E_0 - 2t\cos\theta$. Consider $N = 2$ and regain the double-well result. (You might have some trouble with a factor of two in front of the $\cos\theta$ term. Remember that in a ring with just two sites, each site is both ahead and behind the other, and H couples them twice.)

Will the ground state always be invariant under the symmetry operation that commutes with H? The answer is yes, as long as the barrier height is finite, or more precisely, as long as there is a finite action solution to the Euclidean equations of motion linking classical minima. This is usually the case for quantum mechanics of finite number of degrees of freedom with finite parameters in the Hamiltonian. On the other hand, if $V_0 \to \infty$ in the periodic potential, there really will be N degenerate minima with particles living in any one minimum, trapped there for ever. In quantum field theory, where there are infinitely many degrees of freedom, even if the parameters are finite, the barrier is often infinitely high if all degrees of freedom try to jump over a barrier. Spontaneous symmetry breaking is common there.

5.6.1 Instantons in the $d = 1$ Ising Model

Look at the Hamiltonian for this model in the σ_3 basis:

$$H = -\begin{pmatrix} 0 & K^* \\ K^* & 0 \end{pmatrix}. \tag{5.71}$$

Suppose the off-diagonal term is absent, that is, $K^* = \tanh K^* = e^{-2K} = T = 0$. The two diagonal entries describe states of fixed $\sigma_3 = \pm 1$, whose common energy has been set to $E_0 = 0$. These states will not evolve in real or Euclidean time. There are two equivalent ways to understand this. Given that H is diagonal, the Euclidean Schrödinger equation will preserve an eigenstate of H except for the factor $e^{-E_0\tau} = 1$. Equivalently, in the Euclidean sum over paths a configuration that starts out with the first spin (s_0) up (or down) will have all subsequent spins also up (or down) because any spin flip costs energy and the corresponding Boltzmann weight when $T = 0$ or $K = \infty$ is 0.

If we now pass from $K^* = 0$ to $K^* \to 0$, we may employ a "semiclassical" approximation in which the sum over paths is dominated by the one of least action, because $K^*, T \to 0$ is like $\hbar \to 0$.

The "instanton" that connects, say, the up spin ground state to the down spin ground state (the analogs of $x = \pm a$ in the double well) is a configuration in which the spin is up until some time when it flips to down and thereafter stays down. The energy or action cost is $2K$ (there is one broken bond at the "domain wall' between up and down spins), and the weight is $e^{-2K} = \tanh K^* \simeq K^*$. Just as in the instanton calculation of the double well, the spin flip can occur at any time in the interval of width τ, and in the same notation, the identification $-H_{+-}\tau = e^{-2K}\tau$ follows.

This tunneling restores symmetry at any non-zero temperature T.

5.7 The Classical Limit of Quantum Statistical Mechanics

Consider a single particle of mass m in a potential $V(x)$. The quantum partition function, see Eqs. (5.39)–(5.41), is:

$$Z(\beta) = \int dx \int_x^x [\mathcal{D}x] \exp\left[-\frac{1}{\hbar}\int_0^{\beta\hbar}\left[\frac{m}{2}\left(\frac{dx}{d\tau}\right)^2 + V[x(\tau)]\right]d\tau\right], \tag{5.72}$$

where the limits on the functional integral remind us to consider paths starting and ending at the same point x, which is then integrated over because we are performing a trace. Consider the limit $\beta\hbar \to 0$, either due to high temperatures or vanishing $\hbar$ (the classical limit). Look at any one value of x. We need to sum over paths that start at x, go somewhere, and come back to x in a very short time $\beta\hbar$. If the particle wanders off a distance Δx, the kinetic energy is of order (dropping factors of order unity) $m(\Delta x/\beta\hbar)^2$ and the Boltzmann factor is

$$\simeq e^{-\frac{1}{\hbar}m(\Delta x/\beta\hbar)^2\beta\hbar}, \tag{5.73}$$

from which it follows that

$$\Delta x \simeq \sqrt{\frac{\beta}{m}}\hbar. \tag{5.74}$$

If the potential does not vary over such a length scale (called the *thermal wavelength* – see Exercise 5.7.1), we can approximate it by a constant equal to its value at the starting point x and write

$$\begin{aligned} Z(\beta) &\simeq \int dx e^{-\beta V(x)} \int_x^x [\mathcal{D}x] \exp\left[-\frac{1}{\hbar}\int_0^{\beta\hbar}\left[\frac{m}{2}\left(\frac{dx}{d\tau}\right)^2\right]d\tau\right] \\ &= \int dx e^{-\beta V(x)} \sqrt{\frac{m}{2\pi\hbar\beta\hbar}}, \end{aligned} \tag{5.75}$$

where in the last step we have used the fact that with $V(x)$ pulled out, the functional integral is just the amplitude for a free particle to go from x to x in time $\beta\hbar$. How does this compare with classical statistical mechanics? There, the sum over states is replaced by an integral over phase space:

$$Z = A \int dx \int dp \, \exp\left[-\beta\left(\frac{p^2}{2m} + V(x)\right)\right], \tag{5.76}$$

where the arbitrary prefactor A reflects one's freedom to multiply Z by a constant without changing anything physical since Z is a sum over relative probabilities and any prefactor will drop out in any averaging process. Equivalently, it corresponds to the fact that the number of classical states in a region $dxdp$ of phase space is not uniquely defined. If we do the p integral and compare to the classical limit of the path integral we see that quantum theory fixes

$$A = \frac{1}{2\pi\hbar} \tag{5.77}$$

in accordance with the uncertainty principle, which associates an area of order $\Delta X \Delta P \simeq \hbar$ in phase space with each quantum state.

Exercise 5.7.1 *Consider a particle at temperature T, with mean energy of order kT. Assuming all the energy is kinetic, estimate its momentum and convert to the de Broglie wavelength. Show that this gives us a number of the order of the thermal wavelength. This is the minimum size over which the particle can be localized.*

References and Further Reading

[1] R. P. Feynman and A. R. Hibbs, *Quantum Mechanics and Path Integrals*, McGraw Hill (1965).
[2] R. Shankar, *Principles of Quantum Mechanics*, Springer, 2nd edition (1994). This has a concise treatment of path integrals.
[3] R. Rajaraman, *Instantons and Solitons: An Introduction to Instantons and Solitons in Quantum Field Theory*, North Holland (1982).
[4] S. R. Coleman, *Aspects of Symmetry*, Cambridge University Press (1985).
[5] G. 't Hooft, Phys. Rev. Lett., **37**, 8 (1976).
[6] R. Jackiw and C. Rebbi, Phys. Rev. Lett., **37**, 172 (1976).
[7] R. Jackiw and C. Rebbi, Phys. Rev. D **14**, 517 (1976).
[8] C. Callan, R. Dashen, and D. Gross, Phys. Lett. B **63**, 334 (1976).

6

Coherent State Path Integrals for Spins, Bosons, and Fermions

There are two threads that run through this chapter. One is the path integrals for problems with no classical limit: spins and fermions. These have a finite-dimensional Hilbert space for each particle: for spin S there are the $2S+1$ eigenstates of S_z, and for the fermions' two states, occupied and empty. The second theme is one of using an *over-complete basis*, a basis with far more states than minimally needed, in the resolution of the identity. For example, in the case of spin we will use not the $2S+1$ eigenstates of S_z, but *coherent states* labeled by points (θ, ϕ) on the unit sphere. For the bosons, instead of using the eigenstates $|n\rangle$ of the harmonic oscillator, we will use a continuous infinity of states labeled by a complex number z. These over-complete states cannot, of course, be mutually orthogonal, and this poses some challenges. However, the sum over paths can now be more readily visualized given the continuous labels for intermediate states.

For fermions we will use coherent states that are labeled neither by real numbers nor complex numbers, but strange entities called *Grassmann numbers* (to be defined in due course). We need to learn what they are, how to manipulate them, how to integrate over them, and finally how to obtain the resolution of the identity in terms of states labeled by them.

Although the procedure we follow is the familiar one – chop the time-evolution operator U into N pieces (with $N \to \infty$) and introduce resolutions of the identity N times, using the label of the intermediate states to define a "path" – new complications arise that make this chapter longer than you would expect.

6.1 Spin Coherent State Path Integral

Consider a spin S degree of freedom. The Hilbert space is $(2S+1)$-dimensional. Choosing S_z eigenstates as our basis, we can write the propagator $U(t)$ as a sum over configurations by using the following resolution of the identity:

$$I = \sum_{-S}^{S} |S_z\rangle\langle S_z|. \tag{6.1}$$

The intermediate states will have discrete labels (as in the Ising model).

We consider here an alternate scheme in which an over-complete basis is used. Consider the *spin coherent state*

$$|\Omega\rangle \equiv |\theta,\phi\rangle = U[R(\Omega)]\,|SS\rangle, \tag{6.2}$$

where $|\Omega\rangle$ denotes the state obtained by rotating the normalized, fully polarized state $|SS\rangle$ by an angle θ around the x-axis and then by ϕ around the z-axis using the unitary rotation operator $U[R(\Omega)]$.

Given that

$$\langle SS|\mathbf{S}|SS\rangle = \mathbf{k}S, \tag{6.3}$$

it is clear (say by considering $U^\dagger \mathbf{S} U$) that

$$\langle\Omega|\mathbf{S}|\Omega\rangle = S(\mathbf{i}\sin\theta\cos\phi + \mathbf{j}\sin\theta\sin\phi + \mathbf{k}\cos\theta). \tag{6.4}$$

Our spin operators are not defined with an $\hbar$. Thus, for spin 1, the eigenvalues of S_z are $0, \pm 1$.

The coherent state is one in which the spin operator has a nice expectation value: equal to a classical spin of length S pointing along the direction of Ω. *It is not an eigenvector of the spin operator.* This is not expected anyway, since the three components of spin do not commute. Higher powers of the spin operators do not have expectation values equal to the corresponding powers of the classical spin. For example, $\langle\Omega|S_x^2|\Omega\rangle \neq S^2\sin^2\theta\cos^2\phi$. However, the difference between this wrong answer and the right one is of order S. Generally, the nth power of the spin operator will have an expectation value equal to the nth power of the expectation value of that operator plus corrections that are of order S^{n-1}. If S is large, the corrections may be ignored. This is typically when one usually uses these states.

Let us now examine the equation

$$\langle\Omega_2|\Omega_1\rangle = \left(\cos\frac{\theta_2}{2}\cos\frac{\theta_1}{2} + e^{i(\phi_1-\phi_2)}\sin\frac{\theta_2}{2}\sin\frac{\theta_1}{2}\right)^{2S}. \tag{6.5}$$

The result is obviously true for $S = \frac{1}{2}$, given that the up spinor along the direction θ ϕ is

$$|\Omega\rangle \equiv |\theta\phi\rangle = \cos\frac{\theta}{2}\left|\frac{1}{2},\frac{1}{2}\right\rangle + e^{i\phi}\sin\frac{\theta}{2}\left|\frac{1}{2},-\frac{1}{2}\right\rangle. \tag{6.6}$$

As for higher spin, imagine $2S$ spin-$\frac{1}{2}$ particles joining to form a spin-S state. There is only one direct product state with $S_z = S$: where all the spin-$\frac{1}{2}$'s are pointing up. Thus the normalized fully polarized state is

$$|SS\rangle = \left|\frac{1}{2},\frac{1}{2}\right\rangle \otimes \left|\frac{1}{2},\frac{1}{2}\right\rangle \otimes \cdots \otimes \left|\frac{1}{2},\frac{1}{2}\right\rangle. \tag{6.7}$$

If we now rotate this state, it becomes a tensor product of rotated states and, when we form the inner product on the left-hand side of Eq. (6.5), we obtain the right-hand side.

The resolution of the identity in terms of these states is

$$I = \frac{2S+1}{4\pi}\int d\Omega|\Omega\rangle\langle\Omega|, \tag{6.8}$$

where $d\Omega = d\cos\theta d\phi$. The proof can be found in the references. You are urged to do the following exercise that deals with $S = \frac{1}{2}$.

Exercise 6.1.1 *Prove Eq. (6.8) for $S = \frac{1}{2}$ by carrying out the integral over Ω using Eq. (6.6).*

6.2 Real-Time Path Integral for Spin

When we work out the path integral we will get a product of factors like the following:

$$\cdots\langle\Omega(t+\varepsilon)|I - \frac{i\varepsilon}{\hbar}H(\mathbf{S})|\Omega(t)\rangle\cdots \tag{6.9}$$

We work to order ε. Since H already has a factor of ε in front of it, we set

$$\langle\Omega(t+\varepsilon)|\left[-\frac{i\varepsilon}{\hbar}H(\mathbf{S})\right]|\Omega(t)\rangle \simeq -\frac{i\varepsilon}{\hbar}\langle\Omega(t)|H(\mathbf{S})|\Omega(t)\rangle \equiv -i\varepsilon\mathcal{H}(\Omega). \tag{6.10}$$

If the Hamiltonian is linear in S, we simply replace the quantum spin operator by the classical vector pointing along θ,ϕ and, if not, we can replace the operator by the suitable expectation value in the state $|\Omega(t)\rangle$. This is what we called $\hbar\mathcal{H}(\Omega)$ in the preceding equation.

Next, we turn to the product

$$\langle\Omega(t+\varepsilon)|\Omega(t)\rangle \simeq 1 - i\varepsilon S(1-\cos\theta)\dot{\phi} \simeq e^{iS(\cos\theta-1)\dot{\phi}\varepsilon}, \tag{6.11}$$

where we have expanded Eq. (6.5) to first order in $\Delta\theta$ and $\Delta\phi$.

Exercise 6.2.1 *Derive Eq. (6.11).*

This gives us the following representation of the propagator in the continuum limit:

$$\langle\Omega_f|U(t)|\Omega_i\rangle = \int [\mathcal{D}\Omega]\,\exp\left[i\int_{t_1}^{t_2}[S\cos\theta\dot{\phi} - \mathcal{H}(\Omega)]dt\right], \tag{6.12}$$

where a total derivative in ϕ has been dropped and $\left[\int\mathcal{D}\Omega\right]$ is the measure with all factors of π in it.

Even by the lax standards of continuum functional integrals we have hit a new low. First of all, we have replaced differences by derivatives as if the paths are smooth. In the configuration path integral, the factor that provided any kind of damping on the variation in the coordinate from one time slice to the next was the kinetic energy term $\exp\left[im(x'-x)^2/2\hbar\varepsilon\right]$. *In the present problem there is no such term.* There is no reason why the difference in Ω from one time to another should be treated as a small quantity.

Thus, although the discretized functional integral is never wrong (since all we use is the resolution of the identity), any further assumptions about the smallness of the change in Ω from one time to the next are suspect. There is one exception. Suppose $S \to \infty$. Then we see from Eq. (6.5) that the overlap is unity if the two states are equal and falls rapidly if they are different. (It is easier to consider the case $\phi_2 = \phi_1$.) This is usually the limit ($S \to \infty$) in which one uses this formalism.

We now consider two simple applications. First, let

$$H = \hbar S_z. \tag{6.13}$$

We know that the allowed eigenvalues are $\hbar(-S, -S+1, \ldots, S)$. Let us derive this from the continuum path integral.

Given that $\langle\Omega|H|\Omega\rangle = \hbar S\cos\theta$, it follows that $\mathcal{H} = S\cos\theta$, and that the functional integral is

$$\left[\int \mathcal{D}\cos\theta \mathcal{D}\phi\right] \exp\left[iS\int(\cos\theta\dot{\phi} - \cos\theta)dt\right]. \tag{6.14}$$

We note that:

- This is a phase space path integral with $\cos\theta$ as the *momentum conjugate* to ϕ!
- Phase space is compact here (the unit sphere), as compared to the problem of a particle moving on a sphere for which the configuration space is compact but momenta can be arbitrarily large and the phase space is infinite in extent.
- The spin S plays the role of $1/\hbar$.
- The Hamiltonian for the dynamics is $\cos\theta$ since we pulled the S out to the front. This means in particular that $\cos\theta$ is a constant of motion, i.e., the orbits will be along a fixed latitude.

Recall the Wentzel–Kramers–Brillouin (WKB) quantization rule

$$\oint p\,dq = 2\pi n\hbar \tag{6.15}$$

for a problem with no turning points. In our problem, $p = \cos\theta$ is just the conserved energy E. Of all the classical orbits along constant latitude lines, the ones chosen by WKB obey

$$\oint E\,d\phi = 2\pi n S^{-1} \tag{6.16}$$

since S^{-1} plays the role of $\hbar$. The allowed energies are

$$E_n = \frac{n}{S} \qquad [-S \leq n \leq S]. \tag{6.17}$$

There is exactly enough room in this compact phase space for $2S+1$ orbits, and the allowed values of E translate into the allowed values of H when we reinstate the factor of $\hbar S$ that was pulled out along the way.

So, we got lucky with this problem. In general, if H is more complicated we cannot hope for much luck unless S is large. Now, you may ask why we bother with this formalism given that the spins of real systems are very small. Here is at least one reason, based on a problem I am familiar with: In nuclear physics one introduces a *pseudospin* formalism in which the proton is called spin up and the neutron is called spin down. A big nucleus can have a large pseudospin, say 25. The Hamiltonian for the problem can be written in terms of the pseudospin operators, and they can be 50×50 matrices. Finding the energy levels analytically is hopeless, but we can turn the large S in our favor by doing a WKB quantization using the appropriate H.

Coherent states are also very useful in the study of interacting quantum spins. For example, in the one-dimensional Heisenberg model, the Hamiltonian is a sum of dot products of nearest-neighbor spin operators on a line of points. Since each spin operator appears linearly, the Hamiltonian in the action is just the quantum H with $\mathbf{S}$ replaced by a classical vector of length S. Even though the spin is never very large in these problems, one studies the large-S limit to get a feeling for the subject and to make controlled approximations in $1/S$.

6.3 Bosonic Coherent States

Let us start by recalling the familiar harmonic oscillator with

$$H = \left(a^\dagger a + \frac{1}{2}\right)\hbar\omega, \tag{6.18}$$

$$\left[a, a^\dagger\right] = 1, \tag{6.19}$$

$$H|n\rangle = E_n|n\rangle \quad (n = 0, 1, \ldots), \tag{6.20}$$

$$E_n = \left(n + \frac{1}{2}\right)\hbar\omega. \tag{6.21}$$

The fact that the energy levels are uniformly spaced allows one to introduce the notion of quanta. Rather than saying the oscillator was in the nth state, we could say (dropping the zero-point energy) that there was one quantum level or state of energy $\hbar\omega$ and there were n quanta in it. This is how phonons, photons, and so forth are viewed, and it is a seminal idea.

That the level could be occupied by any number of quanta means they are bosons. Indeed, our perception of a classical electric or magnetic field or sound waves in solids is thanks to this feature.

The bosonic coherent state $|z\rangle$ is defined as follows:

$$|z\rangle = e^{a^\dagger z}|0\rangle = \sum_{n=0}^{\infty} \frac{z^n}{\sqrt{n!}}|n\rangle. \tag{6.22}$$

The state $|z\rangle$ is an eigenstate of a:

$$a|z\rangle = z|z\rangle. \tag{6.23}$$

The adjoint gives us

$$\langle z|a^\dagger = \langle z|z^*. \tag{6.24}$$

Exercise 6.3.1 *Verify Eq. (6.23).*

Using the result

$$e^A e^B = e^B e^A e^{[A,B]}, \tag{6.25}$$

where $[A,B]$ is a c-number, it can be shown that

$$\langle z_2|z_1\rangle = e^{z_2^* z_1}. \tag{6.26}$$

We finally need the following resolution of the identity:

$$I = \int \frac{dz dz^*}{2\pi i} e^{-z^* z} |z\rangle\langle z| \tag{6.27}$$

$$= \int_{-\infty}^{\infty}\int_{-\infty}^{\infty} \frac{dxdy}{\pi} e^{-z^* z} |z\rangle\langle z| \tag{6.28}$$

$$= \int_0^{\infty}\int_0^{2\pi} \frac{rdrd\theta}{\pi} e^{-z^* z} |z\rangle\langle z|. \tag{6.29}$$

The verification is left to the following exercise.

Exercise 6.3.2 *Verify the resolution of the identity in Eq. (6.29). Start with $|z\rangle$ and $\langle z|$ written in terms of $|n\rangle$ and $\langle m|$ (Eq. (6.22) and its adjoint), and do the angular and then radial integrals to show that I reduces to $\sum_n |n\rangle\langle n|$. Recall the definition of the Γ function.*

We are now ready to write down the path integral. Start with

$$\langle z_f|e^{-\frac{i}{\hbar}:H(a^\dagger,a):t}|z_i\rangle, \tag{6.30}$$

where $:H(a^\dagger,a):$ is the *normal-ordered* Hamiltonian with all creation operators to the left and destruction operators to the right. (If the given Hamiltonian is not normal ordered, it must first be normal ordered using the commutation relations of a and $a^\dagger$.)

By chopping the time t into N equal parts of infinitesimal size $\varepsilon = \frac{t}{N}$, we will run into the following string upon repeated insertion of the identity:

$$\cdots|z_{n+1}\rangle\langle z_{n+1}|\left(I - \frac{i\varepsilon}{\hbar} : H(a^\dagger,a) : |z_n\rangle\langle z_n|\frac{dz_n dz_n^* e^{-z_n^* z_n}}{2\pi i}\right)\left(I - \frac{i\varepsilon}{\hbar}\cdots\right)\cdots$$

$$= \cdots|z_{n+1}\rangle\frac{dz_n dz_n^*}{2\pi i}\exp\left[(z_{n+1}^* - z_n^*)z_n - \frac{i\varepsilon}{\hbar} : H(z_n^*, z_n) :\right]\langle z_n|\cdots$$

I have used the fact that a and $a^\dagger$ in $:H:$ are acting on their eigenstates to the right and left, and that given the ε in front of $:H:$, we may set $z_{n+1} = z_n$ inside it.

By putting in factors of ε, $\hbar$ as needed, we arrive at the following continuum version with the usual caveats and in the standard notation:

$$\langle z_f|e^{-\frac{i}{\hbar}:H(a^\dagger,a):t}|z_i\rangle = \int_{z_i}^{z_f^*} \left[\mathcal{D}z^*\mathcal{D}z\right]\exp\left[\frac{i}{\hbar}\int_0^t \left(i\hbar z^*\frac{dz}{dt} - :H(z^*,z):\right)dt\right]. \tag{6.31}$$

I have been sloppy about integration by parts and the end points. If you ever need to use this path integral you should learn the details. My main goal was to make you aware of this path integral and pave the way for the fermion path integral, which bears a strong formal resemblance to it.

6.4 The Fermion Problem

Before jumping into path integrals for fermions, let us get acquainted with the operator formalism, which may be novel to some of you. This will be introduced in the simplest possible way: through a *fermionic oscillator*. It has only one level, which can contain one or no quanta due to the Pauli principle. There can be no macroscopic field associated with this state, which is why the fermion problem is unfamiliar to us at first.

We now develop the theory of a fermionic oscillator. We will study its spectrum and thermodynamics first. We will then proceed to reproduce all these results in the path integral approach. The route is familiar in some ways: chop up the time-evolution operator U into N factors, one for each time slice, and keep inserting the resolution of the identity.

It is in the resolution of the identity that new issues arise, and we cannot get to the path integral until they are resolved. Our plan is to use an over-complete basis of *fermionic coherent states*, which are eigenstates of Ψ. However, the corresponding eigenvalues are neither real nor complex, but altogether new beasts. We will meet them shortly.

Since fermions, in particular electrons or quarks, are ubiquitous, the Fermi oscillator plays a central role in physics. We will now discuss non-interacting fermions in the operator approach. Interactions can be handled by perturbation theory. Following this, we will see how to pass from the operator approach to one based on path integrals using the resolution of identity in terms of Fermionic coherent states.

6.5 Fermionic Oscillator: Spectrum and Thermodynamics

We start by writing down the Hamiltonian:

$$H_0 = \Omega_0\, \Psi^\dagger\Psi. \tag{6.32}$$

What distinguishes this problem from the bosonic one are the *anticommutation relations*:

$$\{\Psi^\dagger,\Psi\} = \Psi^\dagger\Psi + \Psi\Psi^\dagger = 1, \tag{6.33}$$

$$\{\Psi,\Psi\} = \{\Psi^\dagger,\Psi^\dagger\} = 0. \tag{6.34}$$

The last equation tells us that

$$\Psi^{\dagger 2} = \Psi^2 = 0. \tag{6.35}$$

This equation will be used all the time without explicit warning. We shall see that it represents the Pauli principle forbidding double occupancy. The *number operator*,

$$N = \Psi^\dagger \Psi, \tag{6.36}$$

obeys

$$N^2 = \Psi^\dagger \Psi \Psi^\dagger \Psi = \Psi^\dagger (1 - \Psi^\dagger \Psi)\Psi = \Psi^\dagger \Psi = N. \tag{6.37}$$

Thus, the eigenvalues of N can only be 0 or 1. The corresponding normalized eigenstates obey

$$N|0\rangle = 0|0\rangle, \tag{6.38}$$

$$N|1\rangle = 1|1\rangle. \tag{6.39}$$

We will now prove that

$$\Psi^\dagger|0\rangle = |1\rangle, \tag{6.40}$$

$$\Psi|1\rangle = |0\rangle. \tag{6.41}$$

As for the first,

$$N\Psi^\dagger|0\rangle = \Psi^\dagger\Psi\Psi^\dagger|0\rangle = \Psi^\dagger(1 - \Psi^\dagger\Psi)|0\rangle = \Psi^\dagger|0\rangle, \tag{6.42}$$

which shows that $\Psi^\dagger|0\rangle$ has $N = 1$. Its norm is unity:

$$||\Psi^\dagger|0\rangle||^2 = \langle 0|\Psi\Psi^\dagger|0\rangle = \langle 0|(1 - \Psi^\dagger\Psi)|0\rangle = \langle 0|0\rangle = 1. \tag{6.43}$$

It can similarly be shown that $\Psi|1\rangle = |0\rangle$.

There are no other vectors in the Hilbert space; any attempts to produce more states are thwarted by $\Psi^2 = \Psi^{\dagger 2} = 0$. In other words, the Pauli principle rules out more vectors – the state is either empty or singly occupied.

Thus the Fermi oscillator Hamiltonian

$$H_0 = \Omega_0 \Psi^\dagger \Psi \tag{6.44}$$

has eigenvalues 0 and Ω_0.

We will work not with H_0, but with

$$H = H_0 - \mu N, \tag{6.45}$$

where μ is the *chemical potential*. For the oscillator, since

$$H = (\Omega_0 - \mu)\Psi^\dagger \Psi, \tag{6.46}$$

this merely amounts to measuring all energies relative to the chemical potential. The role of the chemical potential will be apparent soon.

Let us now turn to thermodynamics. The central object here is the *grand partition function* defined to be

$$Z = \mathrm{Tr}\, e^{-\beta(H_0 - \mu N)}, \tag{6.47}$$

where the trace is over any complete set of eigenstates. If we use the N basis, this sum is trivial:

$$Z = 1 + e^{-\beta(\Omega_0 - \mu)}, \tag{6.48}$$

where the two terms correspond to $N = 0$ and $N = 1$. All thermodynamic quantities can be deduced from this function. For example, it is clear from Eq. (6.47) that the mean occupation number is

$$\langle N \rangle = \frac{1}{\beta Z}\frac{\partial Z}{\partial \mu} = \frac{1}{\beta}\frac{\partial \ln Z}{\partial \mu} = \frac{1}{e^{\beta(\Omega_0 - \mu)} + 1} \equiv n_F, \tag{6.49}$$

in agreement with the results from Exercise 1.8.1.

At zero temperature, we find, from Eq. (6.49),

$$\langle N \rangle = \theta(\mu - \Omega_0); \tag{6.50}$$

i.e., the fermion is present if its energy is below the chemical potential and absent if it is not. At finite temperatures the mean number varies more smoothly with μ.

6.6 Coherent States for Fermions

As mentioned at the outset, for the repeated resolution of the identity we will use fermion coherent states $|\psi\rangle$, which are eigenstates of the destruction operator

$$\Psi|\psi\rangle = \psi|\psi\rangle. \tag{6.51}$$

The eigenvalue ψ is a peculiar object: if we act once more with Ψ we find

$$\psi^2 = 0, \tag{6.52}$$

since $\Psi^2 = 0$. Any ordinary variable whose square is zero is itself zero. But this ψ is no ordinary variable, it is a *Grassmann variable*. These variables *anticommute* with each other and with all fermionic creation and destruction operators. (They will therefore commute with a string containing an even number of such operators.) That is how they are defined. The variable ψ is rather abstract and defined by its anticommuting nature. There are no big or small Grassmann variables. You will get used to them, and even learn to love them just as you did the complex numbers, despite some initial resistance.

There are treacherous minus signs lurking around. For example, ψ does not commute with all the state vectors. If we postulate that $\psi|0\rangle = |0\rangle\psi$, then it follows that $\psi|1\rangle = -|1\rangle\psi$ because $|1\rangle = \Psi^\dagger|0\rangle$ and ψ anticommutes with Ψ and $\Psi^\dagger$.

We now write down the coherent state. It is

$$|\psi\rangle = |0\rangle - \psi|1\rangle, \tag{6.53}$$

where ψ is a Grassmann number. This state obeys:

$$\Psi|\psi\rangle = \Psi|0\rangle - \Psi\psi|1\rangle \tag{6.54}$$
$$= 0 + \psi\Psi|1\rangle \tag{6.55}$$
$$= \psi|0\rangle \tag{6.56}$$
$$= \psi(|0\rangle - \psi|1\rangle) \tag{6.57}$$
$$= \psi|\psi\rangle, \tag{6.58}$$

where we have appealed to the fact that ψ anticommutes with Ψ and that $\psi^2 = 0$. If we act on both sides of Eq. (6.58) with Ψ, the left vanishes due to $\Psi^2 = 0$ and the right due to $\psi^2 = 0$.

It may similarly be verified that

$$\langle\overline{\psi}|\Psi^\dagger = \langle\overline{\psi}|\overline{\psi}, \tag{6.59}$$

where

$$\langle\overline{\psi}| = \langle 0| - \langle 1|\overline{\psi} = \langle 0| + \overline{\psi}\langle 1|. \tag{6.60}$$

Please note two points. First, the coherent state vectors are not the usual vectors from a complex vector space since they are linear combinations with Grassmann coefficients. Second, $\overline{\psi}$ is not in any sense the complex conjugate of ψ, and $\langle\overline{\psi}|$ is not the adjoint of $|\psi\rangle$. You should therefore be prepared to see a change of Grassmann variables in which ψ and $\overline{\psi}$ undergo totally unrelated transformations. We might even choose to call $\bar{\psi}$ η. The only reason behind calling it $\bar{\psi}$ is to remind us that in a theory where every operator Ψ is accompanied by its adjoint $\Psi^\dagger$, for every label ψ there is another independent label $\bar{\psi}$.

The inner product of two coherent states is

$$\langle\overline{\psi}|\psi\rangle = (\langle 0| - \langle 1|\overline{\psi})(|0\rangle - \psi|1\rangle) \tag{6.61}$$
$$= \langle 0|0\rangle + \langle 1|\overline{\psi}\psi|1\rangle \tag{6.62}$$
$$= 1 + \overline{\psi}\psi \tag{6.63}$$
$$= e^{\overline{\psi}\psi}. \tag{6.64}$$

Compare this to

$$\langle z_1|z_2\rangle = e^{z_1^* z_2} \tag{6.65}$$

for bosons.

Any function of a Grassmann variable can be expanded as follows:

$$F(\psi) = F_0 + F_1 \psi, \tag{6.66}$$

there being no higher powers possible.

6.7 Integration over Grassmann Numbers

Before we can do path integrals, we have to learn to do integrals. We will now *define* integrals over Grassmann numbers. These have no geometric significance (as areas or volumes) and are formally defined. We just have to know how to integrate 1 and ψ, since that takes care of all possible functions. Here is the complete table of Grassmann integrals:

$$\int \psi d\psi = 1, \tag{6.67}$$

$$\int 1 d\psi = 0. \tag{6.68}$$

The integral is postulated to be *translationally invariant* under a shift by another Grassmann number η:

$$\int F(\psi + \eta) d\psi = \int F(\psi) d\psi. \tag{6.69}$$

This agrees with the expansion Eq. (6.66) if we set

$$\int \eta \, d\psi = 0. \tag{6.70}$$

In general, for a collection of Grassmann numbers $(\psi_1, \ldots, \psi_N)$ we postulate that

$$\int \psi_i d\psi_j = \delta_{ij}. \tag{6.71}$$

There are no limits on these integrals. Integration is assumed to be a linear operation. The differential $d\psi$ is also a Grassmann number. Thus, $\int d\psi \psi = -1$. The integrals for $\overline{\psi}$ or any other Grassmann variable are identical. These integrals are simply assigned these values. They are very important since we see for the first time ordinary numbers on the right-hand side. Anything numerical we calculate in this theory goes back to these integrals.

Jacobians for Grassmann change of variables are the inverses of what you expect. Start with

$$\chi = a\phi, \tag{6.72}$$

where a is an ordinary number. Now demand that since χ is as good a Grassmann number as any, it too must obey

$$1 = \int \chi d\chi = \int a\phi \frac{d\chi}{d\phi} d\phi = a\frac{d\chi}{d\phi} \int \phi d\phi = a\frac{d\chi}{d\phi}, \tag{6.73}$$

where the Jacobian is a constant that can be pulled out of the integral. All this implies that

$$\frac{d\chi}{d\phi} = \frac{1}{a}, \quad \text{and not} \tag{6.74}$$

$$\frac{d\chi}{d\phi} = a \quad \text{as you might expect.} \tag{6.75}$$

The funny business occurs because we want both the old and new variables to have a unit integral. This can happen only if the differentials transform oppositely to the variable.

More generally, under a linear transformation

$$\phi_i = \sum_j M_{\mathbf{ij}}\chi_j \tag{6.76}$$

the differentials transform as

$$d\phi_i = \sum_j d\chi_j M^{-1}_{\mathbf{ji}}. \tag{6.77}$$

This ensures that the new variables have the same integrals as the old:

$$\int \phi_i d\phi_j = \delta_{ij} \quad \text{if} \quad \int \chi_i d\chi_j = \delta_{ij}. \tag{6.78}$$

A result we will use often is this:

$$\int \overline{\psi}\psi d\psi d\overline{\psi} = 1. \tag{6.79}$$

If the differentials or variables come in any other order there can be a change of sign. For example, we will also invoke the result

$$\int \overline{\psi}\psi d\overline{\psi} d\psi = -1. \tag{6.80}$$

Let us now consider some Gaussian integrals.

Exercise 6.7.1 *You are urged to show the following:*

$$\int e^{-a\overline{\psi}\psi} d\overline{\psi} d\psi = a, \tag{6.81}$$

$$\int e^{-\overline{\psi}M\psi}[d\overline{\psi}d\psi] = \det M, \tag{6.82}$$

where in the second formula M is a 2-by-2 matrix, ψ is a column vector with entries ψ_1 and ψ_2, $\overline{\psi}$ a row vector with entries $\overline{\psi}_1$ and $\overline{\psi}_2$, and $[d\overline{\psi}d\psi] = d\overline{\psi}_1 d\psi_1 d\overline{\psi}_2 d\psi_2$. (This result is true for matrices of any size, as will be established shortly for Hermitian M.) To prove these, simply expand the exponential and do the integrals.

Consider next the "averages" over the Gaussian measure:

$$\langle \overline{\psi}\psi \rangle = \frac{\int \overline{\psi}\psi e^{-a\overline{\psi}\psi} d\overline{\psi}d\psi}{\int e^{-a\overline{\psi}\psi} d\overline{\psi}d\psi} = -\frac{1}{a} = -\langle \psi\overline{\psi} \rangle. \tag{6.83}$$

The proof is straightforward and left as an exercise.

Exercise 6.7.2 *Prove Eqs. (6.81), (6.82) (for the 2×2 case), and (6.83).*

Exercise 6.7.3 *Verify Eq. (6.78).*

After working out specific examples, we now consider the general problem with two sets of Grassmann variables

$$\psi = [\psi_1 \cdots \psi_N] \quad \text{and} \quad \bar{\psi} = \left[\bar{\psi}_1 \cdots \bar{\psi}_N\right] \tag{6.84}$$

and a Gaussian action:

$$S = -\bar{\psi} M \psi = -\sum_{i,j} \bar{\psi}_i M_{ij} \psi_j. \tag{6.85}$$

Let us assume that M is Hermitian and show the *extremely important result*:

$$\int e^{-\bar{\psi} M \psi} [\mathcal{D}\overline{\psi}\mathcal{D}\psi] = \det M, \tag{6.86}$$

where

$$[\mathcal{D}\overline{\psi}\mathcal{D}\psi] = \prod_{1}^{N} d\overline{\psi}_i d\psi_i. \tag{6.87}$$

To this end, begin with the usual orthonormal eigenvectors of M with complex components:

$$MV_n = \lambda_n V_n, \quad V_n^\dagger M = V_n^\dagger \lambda_n, \quad V_n^\dagger V_m = \delta_{mn}, \tag{6.88}$$

and expand as follows:

$$\psi = \sum_n \chi_n V_n, \quad \bar{\psi} = \sum_n \bar{\chi}_n V_n^\dagger, \tag{6.89}$$

where χ_n and $\bar{\chi}_n$ are Grassmann numbers. We then find that

$$Z = \exp\left[-\sum_n \lambda_n \bar{\chi}_n \chi_n\right] [d\overline{\chi} d\chi] = \prod_n \lambda_n = \det M, \tag{6.90}$$

using Eq. (6.81) for the Grassmann integral over a single pair $(\bar{\psi}, \psi)$ and the fact (not proven here) that under a unitary transformation of Grassmann numbers the Jacobian is unity.

6.8 Resolution of the Identity and Trace

We need two more results before we can write down the path integral. The first is the resolution of the identity:

$$I = \int |\psi\rangle\langle\overline{\psi}|e^{-\overline{\psi}\psi} d\overline{\psi} d\psi. \tag{6.91}$$

Compare this to the bosonic version,

$$I = \int |z\rangle\langle z|e^{-z^*z} \frac{dz^* dz}{2\pi i}. \tag{6.92}$$

In the following proof of this result we will use all the previously described properties and drop terms that are not going to survive integration. (Recall that only $\overline{\psi}\psi = -\psi\overline{\psi}$ has a non-zero integral.)

$$\begin{aligned}
\int |\psi\rangle\langle\overline{\psi}|e^{-\overline{\psi}\psi} d\overline{\psi} d\psi &= \int |\psi\rangle\langle\overline{\psi}|(1-\overline{\psi}\psi) d\overline{\psi} d\psi \\
&= \int (|0\rangle - \psi|1\rangle)(\langle 0| - \langle 1|\overline{\psi})(1-\overline{\psi}\psi) d\overline{\psi} d\psi \\
&= \int (|0\rangle\langle 0| + \psi|1\rangle\langle 1|\overline{\psi})(1-\overline{\psi}\psi) d\overline{\psi} d\psi \\
&= |0\rangle\langle 0| \int (-\overline{\psi}\psi) d\overline{\psi} d\psi + |1\rangle\langle 1| \int \psi\overline{\psi} d\overline{\psi} d\psi \\
&= I.
\end{aligned} \tag{6.93}$$

The final result we need is that for any bosonic operator Ω (an operator made of an even number of Fermi operators) the trace is given by

$$\mathrm{Tr}\,\Omega = \int \langle -\overline{\psi}|\Omega|\psi\rangle e^{-\overline{\psi}\psi} d\overline{\psi} d\psi. \tag{6.94}$$

The proof is very much like the one just given and is left as an exercise.

Exercise 6.8.1 *Prove Eq. (6.94).*

Exercise 6.8.2 (The Grassmann delta function) *Let η, ψ, and χ be Grassmann variables obeying $\int \psi d\psi = 1$ and $\int 1 d\psi = 0$, etc. Show that*

$$\int e^{(\eta-\chi)\psi} F(\eta) d\psi d\eta = F(\chi) \tag{6.95}$$

so that

$$\int e^{(\eta-\chi)\psi} d\psi = \delta(\eta - \chi). \tag{6.96}$$

Hint: Write $F(\eta) = F_0 + F_1\eta$ and expand everything in sight, keeping only terms that survive integration over η and ψ.

6.9 Thermodynamics of a Fermi Oscillator

Consider the partition function for a single oscillator, which we have studied in the operator approach. We will reproduce the results using path integrals. We begin by writing the trace in terms of coherent states:

$$Z = \mathrm{Tr}\, e^{-\beta(\Omega_0-\mu)\Psi^\dagger\Psi} \tag{6.97}$$

$$= \int \langle -\bar{\psi}| e^{-\beta(\Omega_0-\mu)\Psi^\dagger\Psi} |\psi\rangle e^{-\bar{\psi}\psi} d\bar{\psi} d\psi. \tag{6.98}$$

You cannot simply replace $\Psi^\dagger$ and Ψ by $-\bar{\psi}$ and ψ respectively in the exponential. This is because when we expand out the exponential, not all the Ψ's will be acting to the right on their eigenstates and neither will all $\Psi^\dagger$'s be acting to the left on their eigenstates. (Remember that we are now dealing with operators, not Grassmann numbers. The exponential will have an infinite number of terms in its expansion.) We need to convert the exponential to its *normal-ordered form* in which all the creation operators stand to the left and all the destruction operators to the right. Luckily we can write down the answer by inspection:

$$e^{-\beta(\Omega_0-\mu)\Psi^\dagger\Psi} = 1 + (e^{-\beta(\Omega_0-\mu)} - 1)\Psi^\dagger\Psi, \tag{6.99}$$

whose correctness we can verify by considering the two possible values of $\Psi^\dagger\Psi$. (Alternatively, you can expand the exponential and use the fact that $N^k = N$ for any non-zero k.) Now we may write

$$Z = \int \langle -\bar{\psi}| 1 + (e^{-\beta(\Omega_0-\mu)} - 1)\Psi^\dagger\Psi |\psi\rangle e^{-\bar{\psi}\psi} d\bar{\psi} d\psi \tag{6.100}$$

$$= \int \langle -\bar{\psi}|\psi\rangle (1 + (e^{-\beta(\Omega_0-\mu)} - 1)(-\bar{\psi}\psi)) e^{-\bar{\psi}\psi} d\bar{\psi} d\psi \tag{6.101}$$

$$= \int (1 - (e^{-\beta(\Omega_0-\mu)} - 1)\bar{\psi}\psi) e^{-2\bar{\psi}\psi} d\bar{\psi} d\psi \tag{6.102}$$

$$= 1 + e^{-\beta(\Omega_0-\mu)}, \tag{6.103}$$

as expected. While this is the right answer, and confirms the correctness of all the Grassmann integration and minus signs, it is not the path integral approach. We finally turn to that.

6.10 Fermionic Path Integral

After this lengthy preparation we are ready to map the quantum problem of fermions to a path integral. Let us begin with

$$Z = \mathrm{Tr}\, e^{-\beta H}, \tag{6.104}$$

where H is a normal-ordered operator $H(\Psi^\dagger, \Psi)$. We write the exponential as follows:

$$e^{-\beta H} = \lim_{N\to\infty} (e^{-\frac{\beta}{N}H})^N \tag{6.105}$$

$$= \underbrace{(1-\varepsilon H)\cdots(1-\varepsilon H)}_{N \text{ times}}, \qquad \varepsilon = \beta/N; \tag{6.106}$$

take the trace as per Eq. (6.94) by integrating over $\overline{\psi}_0\psi_0$; and introduce the resolution of the identity $N-1$ times:

$$Z = \int \langle -\overline{\psi}_0|(1-\varepsilon H)|\psi_{N-1}\rangle e^{-\overline{\psi}_{N-1}\psi_{N-1}} \langle \overline{\psi}_{N-1}|(1-\varepsilon H)|\psi_{N-2}\rangle e^{-\overline{\psi}_{N-2}\psi_{N-2}}$$
$$\times \langle \psi_{N-2}|\cdots|\psi_1\rangle e^{-\overline{\psi}_1\psi_1} \langle \overline{\psi}_1|(1-\varepsilon H)|\psi_0\rangle e^{-\overline{\psi}_0\psi_0} \prod_{i=0}^{N-1} d\overline{\psi}_i d\psi_i. \tag{6.107}$$

Note that $\varepsilon = \beta/N$ really has units of time/$\hbar$. I will set $\hbar = 1$ for the rest of this section.

Next, we legitimately make the replacement

$$\langle \overline{\psi}_{i+1}|1-\varepsilon H(\Psi^\dagger,\Psi)|\psi_i\rangle = \langle \overline{\psi}_{i+1}|1-\varepsilon H(\overline{\psi}_{i+1},\psi_i)|\psi_i\rangle$$
$$= e^{\overline{\psi}_{i+1}\psi_i} e^{-\varepsilon H(\overline{\psi}_{i+1},\psi_i)}, \tag{6.108}$$

where in the last step we are anticipating the limit of infinitesimal ε. Let us now *define* an additional pair of variables (not to be integrated over)

$$\overline{\psi}_N = -\overline{\psi}_0, \tag{6.109}$$

$$\psi_N = -\psi_0. \tag{6.110}$$

The first of these equations allows us to replace the left-most bra in Eq. (6.107), $\langle -\overline{\psi}_0|$, by $\langle \overline{\psi}_N|$. The reason for introducing ψ_N will be clear when we get to Eq. (6.147).

Putting together all the factors (including the overlap of coherent states), we end up with

$$Z = \int \prod_{i=0}^{N-1} e^{\overline{\psi}_{i+1}\psi_i} e^{-\varepsilon H(\overline{\psi}_{i+1},\psi_i)} e^{-\overline{\psi}_i\psi_i} d\overline{\psi}_i d\psi_i$$
$$= \int \prod_{i=0}^{N-1} \exp\left[\left[\frac{(\overline{\psi}_{i+1}-\overline{\psi}_i)}{\varepsilon}\psi_i - H(\overline{\psi}_{i+1},\psi_i)\right]\varepsilon\right] d\overline{\psi}_i d\psi_i$$
$$\simeq \int e^{S(\bar{\psi},\psi)} \left[\mathcal{D}\overline{\psi}\mathcal{D}\psi\right], \quad \text{where} \tag{6.111}$$

$$S = \int_0^\beta \left(\overline{\psi}(\tau)\left(-\frac{\partial}{\partial\tau}\right)\psi(\tau) - H(\overline{\psi}(\tau),\psi(\tau))\right) d\tau. \tag{6.112}$$

This formula is worth memorizing: the fermionic action is the sum of the $\bar{\psi}(-\partial_\tau)\psi$ and $-H(\bar{\psi}(\tau), \psi(\tau))$, where H is the normal-ordered quantum Hamiltonian with $\Psi^\dagger \to \bar{\psi}$, $\Psi \to \psi$.

For the fermionic oscillator we have been studying,

$$S = \int_0^\beta \overline{\psi}(\tau)\left(-\frac{\partial}{\partial \tau} - \omega_0 + \mu\right)\psi(\tau)d\tau. \tag{6.113}$$

The steps leading to the continuum form of the action Eq. (6.112) need some explanation. With all the factors of ε in place we do seem to get the continuum expression in the last formula. However, the notion of replacing differences by derivatives is purely symbolic for Grassmann variables. There is no sense in which $\overline{\psi}_{i+1} - \overline{\psi}_i$ is small; in fact, the objects have no numerical values. What this really means here is the following: In a while we will trade $\psi(\tau)$ for $\psi(\omega)$ related by Fourier transformation. At that stage we will replace $-\frac{\partial}{\partial \tau}$ by $i\omega$ while the exact answer is $e^{i\omega} - 1$. If we do not make this replacement, the Grassmann integral, when evaluated in terms of ordinary numbers, will give exact results for anything one wants to calculate, say the free energy. With this approximation, only quantities insensitive to high frequencies will be given correctly. The free energy will come out wrong, but the correlation functions will be correctly reproduced. (This is because the latter are given by derivatives of the free energy and these derivatives make the integrals sufficiently insensitive to high frequencies.) Notice also that we are replacing $H(\overline{\psi}_{i+1}, \psi_i) = H(\overline{\psi}(\tau+\varepsilon), \psi(\tau))$ by $H(\overline{\psi}(\tau), \psi(\tau))$ in the same spirit.

Consider now the average $\langle \psi(\tau_1)\bar{\psi}(\tau_2)\rangle$ with respect to the partition function Eq. (6.113),

$$\langle \psi(\tau_1)\bar{\psi}(\tau_2)\rangle$$
$$= \frac{\int \psi(\tau_1)\bar{\psi}(\tau_2)\prod_{i=0}^{N-1} \exp\left[\left[\frac{(\overline{\psi}_{i+1}-\overline{\psi}_i)}{\varepsilon}\psi_i - H(\overline{\psi}_{i+1}, \psi_i)\right]\varepsilon\right] d\overline{\psi}_i d\psi_i}{\int \prod_{i=0}^{N-1} \exp\left[\left[\frac{(\overline{\psi}_{i+1}-\overline{\psi}_i)}{\varepsilon}\psi_i - H(\overline{\psi}_{i+1}, \psi_i)\right]\varepsilon\right] d\overline{\psi}_i d\psi_i}.$$

We are now going to express this average in operator language by running our derivation backwards to drive home the correspondence. First, consider the case $\tau_1 > \tau_2$. Let us move these two variables $\psi(\tau_1)\bar{\psi}(\tau_2)$ to the appropriate time slices, encountering no minus signs as we will run into an even number of Grassmann variables on the way. Then we undo our derivation, going from Grassmann variables back to fermionic intermediate states. We may then replace $\psi(\tau_1)$ and $\bar{\psi}(\tau_2)$ by the corresponding operators Ψ and $\Psi^\dagger$, since these operators, acting to the right and left respectively on the coherent states, yield precisely these Grassmann numbers as eigenvalues. Thus we will find, for $\tau_1 > \tau_2$,

$$\langle \psi(\tau_1)\bar{\psi}(\tau_2)\rangle = \frac{\mathrm{Tr}\left[e^{-H(\beta-\tau_1)}\Psi e^{-H(\tau_1-\tau_2)}\Psi^\dagger e^{-H\tau_2}\right]}{\mathrm{Tr}\, e^{-\beta H}}. \tag{6.114}$$

You can verify the correctness of the expression in the numerator by inserting intermediate states and getting a path integral with the two Grassmann variables incorporated.

Thus, for $\tau_1 > \tau_2$ we find

$$\langle\psi(\tau_1)\bar{\psi}(\tau_2)\rangle = \frac{\mathrm{Tr}\left[e^{-\beta H}\Psi(\tau_1)\Psi^\dagger(\tau_2)\right]}{\mathrm{Tr}\, e^{-\beta H}} \tag{6.115}$$

upon invoking

$$\Psi(\tau_i) = e^{H\tau_i}\Psi e^{-H\tau_i}, \tag{6.116}$$

and likewise for $\Psi^\dagger$.

If, on the other hand, $\tau_1 < \tau_2$, we reorder the Grassmann numbers first using

$$\psi(\tau_1)\bar{\psi}(\tau_2) = -\bar{\psi}(\tau_2)\psi(\tau_1), \tag{6.117}$$

and repeat the process of replacing them by the corresponding operators to obtain [the fermionic analog of Eq. (3.55)]:

$$\langle\psi(\tau_1)\bar{\psi}(\tau_2)\rangle = \frac{\mathrm{Tr}\left[e^{-\beta H}|\mathcal{T}(\Psi(\tau_1)\Psi^\dagger(\tau_2))\right]}{\mathrm{Tr}\, e^{-\beta H}}, \tag{6.118}$$

where the time-ordered product now stands for

$$\mathcal{T}(\Psi(\tau_1)\Psi^\dagger(\tau_2)) = \theta(\tau_1 - \tau_2)\Psi(\tau_1)\Psi^\dagger(\tau_2) - \theta(\tau_2 - \tau_1)\Psi^\dagger(\tau_2)\Psi(\tau_1). \tag{6.119}$$

Observe the minus sign when the operator order is reversed. As a result, you may verify that operators within the time-ordered product may be permuted provided we change the sign:

$$\mathcal{T}(\Psi(\tau_1)\Psi^\dagger(\tau_2)) = -\mathcal{T}(\Psi^\dagger(\tau_2)\Psi(\tau_1)). \tag{6.120}$$

In other words, inside the $\mathcal{T}$ product, all fermionic operators simply anticommute, with no canonical anticommutator terms. This rule generalizes in an obvious way for products of many fermionic operators.

6.10.1 Finite-Temperature Green's Function

Let us explore the key properties of the finite-temperature Green's function

$$G(\tau = \tau_1 - \tau_2) = -\frac{\mathrm{Tr}\left[e^{-\beta H}\mathcal{T}(\Psi(\tau_1)\Psi^\dagger(\tau_2))\right]}{\mathrm{Tr}\ e^{-\beta H}} \tag{6.121}$$

$$= -\langle\psi(\tau_1)\bar{\psi}(\tau_2)\rangle = +\langle\bar{\psi}(\tau_2)\psi(\tau_1)\rangle, \tag{6.122}$$

where the minus sign in the first equation is a convention, the secret handshake of members of the club.

Since $G(\tau)$ plays a crucial role in many-body perturbation theory, I will now demonstrate its essential properties.

Translation invariance, the fact that G depends only on $\tau_1 - \tau_2$, may be readily verified by using the cyclicity of the trace in Eq. (6.114).

Although in the finite temperature problem the imaginary times lie in the interval $0 \leq \tau_1 \leq \beta$ and $0 \leq \tau_2 \leq \beta$ (I am using $\hbar = 1$ here), the *difference* τ lies in the interval $-\beta \leq \tau \leq \beta$.

We will now establish a key antiperiodicity property of G:

$$G(\tau - \beta) = -G(\tau), \quad 0 < \tau \leq \beta. \tag{6.123}$$

The proof uses the cyclicity of the trace and the sign change that accompanies permutation of operators within the $\mathcal{T}$ product [Eq. (6.120)], and goes as follows. First, let us choose $\tau_1 = \beta$ and $\tau_2 > 0$, and remember that $\beta > \tau_2$ in the $\mathcal{T}$ product. Then

$$ZG(\beta, \tau_2) = -\mathrm{Tr}\left[e^{-\beta H}(e^{\beta H}\Psi e^{-\beta H})(e^{H\tau_2}\Psi^\dagger e^{-H\tau_2})\right] \tag{6.124}$$

$$= -\mathrm{Tr}\left[e^{-\beta H}\Psi^\dagger(\tau_2)\Psi(0)\right] \tag{6.125}$$

$$= -\mathrm{Tr}\left[e^{-\beta H}\mathcal{T}(\Psi^\dagger(\tau_2)\Psi(0))\right] \tag{6.126}$$

$$= +\mathrm{Tr}\left[e^{-\beta H}\mathcal{T}\Psi(0)(\Psi^\dagger(\tau_2))\right] \tag{6.127}$$

$$= -ZG(0, \tau_2), \tag{6.128}$$

so that G changes sign when $\tau = \tau_1 - \tau_2 > 0$ is reduced by β.

Exercise 6.10.1 *If you are unhappy that one of the points was chosen to be at β, you may consider the more generic case $\frac{1}{2}\beta < \tau_1 < \beta$ and $0 < \tau_2 < \frac{1}{2}\beta$. Now reduce τ_1 by $\beta/2$ and raise τ_2 by $\beta/2$ so that the new values still stay in the allowed range $[0, \beta]$ and verify that G changes sign.*

We will now consider the Fourier transformation of $G(\tau)$. Since a change by β flips its sign, it is periodic in an interval $[-\beta, \beta]$. So we may employ the following Fourier series:

$$G(\tau) = \sum_{m=-\infty}^{\infty} e^{-i\omega_m \tau} G(\omega_m), \quad \text{where} \tag{6.129}$$

$$\omega_m = \frac{2\pi m}{2\beta} = \frac{m\pi}{\beta}, \tag{6.130}$$

$$G(\omega_m) = \frac{1}{2\beta}\int_{-\beta}^{\beta} G(\tau) e^{i\omega_m \tau} d\tau. \tag{6.131}$$

But there is more to $G(\tau)$ than periodicity in 2β, there is antiperiodicity under a shift β, which we will now use to write the integral between $[-\beta,\ 0]$ in terms of an integral between $[0,\ \beta]$. In terms of $\bar{\tau} = \tau + \beta$, we may write

$$\frac{1}{2\beta}\int_{-\beta}^{0} G(\tau)e^{i\omega_m \tau} d\tau = \frac{1}{2\beta}\int_{0}^{\beta} G(\bar{\tau} - \beta) e^{-i\omega_m \beta} e^{i\omega_m \bar{\tau}} d\bar{\tau}$$

$$= -1 \cdot \frac{1}{2\beta} \cdot e^{-i\omega_m \beta} \int_{0}^{\beta} G(\bar{\tau}) e^{i\omega_m \bar{\tau}} d\bar{\tau}.$$

Since $e^{i\omega_m\beta} = (-1)^m$, we find that if m is even, the integral cancels the one in the range $[0,\beta]$, while if $m = (2n+1)$, they are equal. Thus we arrive at the following relations that are used extensively:

$$G(\omega_n) = \frac{1}{\beta}\int_0^\beta G(\tau)e^{i\omega_n\tau}d\tau, \tag{6.132}$$

$$\omega_n = \frac{(2n+1)\pi}{\beta}, \tag{6.133}$$

$$G(\tau) = \sum_n e^{-i\omega_n\tau}G(\omega_n), \tag{6.134}$$

$$\int_0^\beta e^{i\omega_n\tau}e^{-i\omega_m\tau}d\tau = \frac{e^{i(\omega_n-\omega_m)\beta}-1}{i(\omega_n-\omega_m)} = \beta\delta_{mn}. \tag{6.135}$$

These ω_n are called the *Matsubara frequencies*.

6.10.2 $G(\tau)$ *for a Free Fermion*

Consider a free fermion for which

$$H = (\Omega_0 - \mu)\Psi^\dagger\Psi, \tag{6.136}$$

where we have included the chemical potential in the Hamiltonian. The Euclidean equation of motion for the Heisenberg operator $\Psi(\tau)$ is

$$\frac{d\Psi(\tau)}{d\tau} = [H, \Psi(\tau)] = -(\Omega_0 - \mu)\Psi(\tau), \tag{6.137}$$

with a solution

$$\Psi(\tau) = e^{-(\Omega_0-\mu)\tau}\Psi. \tag{6.138}$$

Similarly, one finds

$$\Psi^\dagger(\tau) = e^{(\Omega_0-\mu)\tau}\Psi^\dagger. \tag{6.139}$$

Choosing $\tau_1 = \tau$ and $\tau_2 = 0$ without any loss of generality (due to translational invariance),

$$\begin{aligned} G(\tau) &= -\theta(\tau)\frac{\mathrm{Tr}\left[e^{-\beta H}(\Psi(\tau)\Psi^\dagger(0))\right]}{Z} + \theta(-\tau)\frac{\mathrm{Tr}\left[e^{-\beta H}(\Psi^\dagger(0)\Psi(\tau))\right]}{Z} \\ &= -\theta(\tau)e^{-(\Omega_0-\mu)\tau}(1-n_F(\Omega_0-\mu)) + \theta(-\tau)e^{-(\Omega_0-\mu)\tau}n_F(\Omega_0-\mu), \end{aligned} \tag{6.140}$$

where

$$n_F(\Omega_0 - \mu) = \frac{\mathrm{Tr}\left[e^{-\beta H}\Psi^\dagger\Psi\right]}{Z} = \frac{1}{e^{\beta(\Omega_0-\mu)}+1} \tag{6.141}$$

is the thermally averaged fermion occupation number.

From this, we readily find that

$$G(\omega_n) = -\frac{1}{\beta}\int_0^\beta e^{i\omega_n\tau} e^{-(\Omega_0-\mu)\tau}(1-n_F(\Omega_0-\mu))$$
$$= \frac{1}{\beta}\frac{1}{i\omega_n-(\Omega_0-\mu)}, \tag{6.142}$$

the details of which are left as an exercise.

Exercise 6.10.2 *Derive Eq. (6.140).*

Exercise 6.10.3 *Perform the integral leading to Eq. (6.142).*

In the zero-temperature limit, $G(\tau)$ reduces to

$$G(\tau) = -\theta(\tau)e^{-(\Omega_0-\mu)\tau} \quad (\mu < \Omega_0) \tag{6.143}$$
$$= +\theta(-\tau)e^{-(\Omega_0-\mu)\tau} \quad (\mu > \Omega_0). \tag{6.144}$$

6.10.3 Fermion Path Integral in Frequency Space

Let us return to the path integral Eq. (6.113). We want to change variables from $\psi(\tau)$ and $\bar{\psi}(\tau)$ to their Fourier transforms. From Eqs. (6.109) and (6.110), we see that these functions are antiperiodic in the interval $[0,\beta]$. Thus they too will be expanded in terms of the Matsubara frequencies

$$\omega_n = \frac{(2n+1)\pi}{\beta}. \tag{6.145}$$

So we write as we did with G:

$$\overline{\psi}(\tau) = \sum_n e^{i\omega_n\tau}\overline{\psi}(\omega_n), \tag{6.146}$$
$$\psi(\tau) = \sum_n e^{-i\omega_n\tau}\psi(\omega_n). \tag{6.147}$$

We have chosen the Fourier expansions as if ψ and $\overline{\psi}$ were complex conjugates, which they are not. This choice, however, makes the calculations easier.

The inverse transformations are

$$\psi(\omega_n) = \frac{1}{\beta}\int_0^\beta \psi(\tau)e^{i\omega_n\tau}d\tau, \tag{6.148}$$
$$\overline{\psi}(\omega_n) = \frac{1}{\beta}\int_0^\beta \overline{\psi}(\tau)e^{-i\omega_n\tau}d\tau, \tag{6.149}$$

where we again use the orthogonality property

$$\frac{1}{\beta}\int_0^\beta e^{i(\omega_m-\omega_n)\tau}d\tau = \delta_{mn}. \tag{6.150}$$

In anticipation of what follows, let us note that if $\beta \to \infty$, it follows from Eq. (6.145) that when n increases by unity, ω_n changes by $d\omega = 2\pi/\beta$ so that

$$\frac{1}{\beta}\sum_n \to \int \frac{d\omega}{2\pi}. \tag{6.151}$$

The action for the Fermi oscillator in Eq. (6.113) transforms as follows under Fourier transformation:

$$S = \int_0^\beta \overline{\psi}(\tau)\left(-\frac{\partial}{\partial\tau} - (\Omega_0 - \mu)\right)\psi(\tau)d\tau \tag{6.152}$$

$$= \beta\sum_m \overline{\psi}(\omega_n)\left(i\omega_n - (\Omega_0 - \mu)\right)\psi(\omega_n). \tag{6.153}$$

In the limit $\beta \to \infty$, when ω_n becomes a continuous variable ω, if we introduce rescaled Grassmann variables

$$\bar{\psi}(\omega) = \beta\bar{\psi}(\omega_n), \qquad \psi(\omega) = \beta\psi(\omega_n) \tag{6.154}$$

and use Eq. (6.151), we find that the action is

$$S = \int_{-\infty}^{\infty}\frac{d\omega}{2\pi}\overline{\psi}(\omega)(i\omega - \Omega_0 + \mu)\psi(\omega). \tag{6.155}$$

Using the fact that the Jacobian for the change from $(\bar{\psi}(\tau), \psi(\tau))$ to $(\bar{\psi}(\omega), \psi(\omega))$ is unity (which I state without proof here), we end up with the path integral that will be invoked often:

$$Z = \int \exp\left[\int_{-\infty}^{\infty}\frac{d\omega}{2\pi}\overline{\psi}(\omega)(i\omega - \Omega_0 + \mu)\psi(\omega)\right]\left[\mathcal{D}\overline{\psi}(\omega)\mathcal{D}\psi(\omega)\right]. \tag{6.156}$$

Although β has disappeared from the picture, it will appear as $2\pi\delta(0)$, which we know stands for the total time. (Recall Fermi's golden rule calculations.) An example will follow shortly.

Let us first note that the frequency space correlation function is related to the integral over just a single pair of variables [Eq. (6.83)] and is given by:

$$\langle\overline{\psi}(\omega_1)\psi(\omega_2)\rangle$$

$$= \frac{\int \overline{\psi}(\omega_1)\psi(\omega_2)\exp\left[\int_{-\infty}^{\infty}\frac{d\omega}{2\pi}\overline{\psi}(\omega)(i\omega - \Omega_0 + \mu)\psi(\omega)\right]\left[\mathcal{D}\overline{\psi}(\omega)\mathcal{D}\psi(\omega)\right]}{\int \exp\left[\int_{-\infty}^{\infty}\frac{d\omega}{2\pi}\overline{\psi}(\omega)(i\omega - \Omega_0 + \mu)\psi(\omega)\right]\mathcal{D}\overline{\psi}(\omega)\mathcal{D}\psi(\omega)}$$

$$= \frac{2\pi\delta(\omega_1 - \omega_2)}{i\omega_1 - \Omega_0 + \mu}. \tag{6.157}$$

In particular,

$$G(\omega) = \langle \overline{\psi}(\omega)\psi(\omega)\rangle = \frac{2\pi\delta(0)}{i\omega - \Omega_0 + \mu} = \frac{\beta}{i\omega - \Omega_0 + \mu}. \tag{6.158}$$

Note that $G(\omega) = \beta^2 G(\omega_m)$ because of the rescaling [Eq. (6.154)].

Exercise 6.10.4 *Try to demonstrate Eqs. (6.157) and (6.158). Note first of all that unless $\omega_1 = \omega_2$, we get zero since only a $\overline{\psi}\psi$ pair has a chance of having a non-zero integral. This explains the δ-function. As for the 2π, go back to the stage where we had a sum over frequencies and not an integral, i.e., go against the arrow in Eq. (6.151), and use it in the exponent of Eq. (6.157).*

Let us now calculate the mean occupation number $\langle N\rangle$:

$$\langle N\rangle = \frac{1}{\beta Z}\frac{\partial Z}{\partial \mu} \tag{6.159}$$

$$= \frac{1}{\beta}\int_{-\infty}^{\infty}\frac{d\omega}{2\pi}\langle \overline{\psi}(\omega)\psi(\omega)\rangle \tag{6.160}$$

$$= \int_{-\infty}^{\infty}\frac{d\omega}{2\pi}\frac{e^{i\omega o^+}}{i\omega - \Omega_0 + \mu} \tag{6.161}$$

$$= \theta(\mu - \Omega_0), \tag{6.162}$$

as in the operator approach.

We had to introduce the factor $e^{i\omega o^+}$ into the ω integral. We understand this as follows: If we had done the calculation using time τ instead of frequency ω, we would have calculated the average of $\Psi^\dagger\Psi$. This would automatically have turned into $\overline{\psi}(\tau+\varepsilon)\psi(\tau)$ when introduced into the path integral since the coherent state bra to the left of the operator would have come from the next time slice compared to the ket at the right. (Remember how $H(\Psi^\dagger,\Psi)$ turned into $H(\overline{\psi}(i+1)\psi(i))$.) The integral over ω was not convergent, varying as $d\omega/\omega$. It was therefore sensitive to the high frequencies and we had to intervene with the factor $e^{i\omega o^+}$. This factor allows us to close the contour in the upper half-plane. If $\mu > \Omega_0$, the pole of the integrand lies in that half-plane and makes a contribution. If not, we get zero. In correlation functions that involve integrals that have two or more powers of ω in the denominator and are hence convergent, we will not introduce this factor.

6.11 Generating Functions $Z(J)$ and $W(J)$

So far we have encountered a variety of correlation functions, namely averages of products of several variables weighted by some Boltzmann weight or action. The products can involve any number of variables, and in each case the correlation functions will be functions of that many variables. I will now introduce you to a compact way to subsume *all* the correlation functions into one single function called the *generating function*.

We will discuss generating functions for the following variables:

- Ising spins
- real bosonic variables
- complex bosonic variables
- Grassmann variables.

6.11.1 Ising Correlators

In this case we have been using a generating function without being aware of the terminology. Recall that by coupling the spins to an external field $h = (h_1 \cdots h_N)$ we obtained a partition function

$$Z(h_1 \cdots h_N) \equiv Z(h) = \sum_{s_i} \exp\left[-E(K, s_i) + \sum_i h_i s_i\right], \tag{6.163}$$

where $E(K, s_i)$ is the energy as a function of the spins and some parameters, denoted by K. The derivatives of $Z(h)$ yielded the averages, such as

$$\langle s_i \rangle = \frac{1}{Z}\frac{\partial Z}{\partial h_i}, \tag{6.164}$$

$$\langle s_i s_j \rangle = \frac{1}{Z}\frac{\partial^2 Z}{\partial h_i \partial h_j}, \tag{6.165}$$

and so on. If we want the external field to be zero we may take these derivatives and then set $h = 0$.

The derivatives of the function $W(h)$ defined by

$$Z(h) = e^{-W(h)} = e^{-\beta F} \tag{6.166}$$

give the *connected* correlation functions. For example, recalling Eqs. (2.17) and (2.19),

$$-\frac{\partial^2 W}{\partial h_i \partial h_j} = \langle s_i s_j \rangle_c = \langle s_i s_j \rangle - \langle s_i \rangle \langle s_j \rangle, \tag{6.167}$$

$$\begin{aligned} -\frac{\partial^4 W}{\partial h_i \partial h_j \partial h_k \partial h_l} &= \langle s_i s_j s_k s_l \rangle_c \\ &= \langle s_i s_j s_k s_l \rangle - \left[\langle s_i s_j \rangle \langle s_k s_l \rangle + \langle s_i s_k \rangle \langle s_j s_l \rangle + \langle s_i s_l \rangle \langle s_j s_k \rangle\right]. \end{aligned} \tag{6.168}$$

Once again, the derivatives are usually evaluated at $h = 0$ in a problem with no real external field. In Eq. (6.168) I am assuming for simplicity that there is no spontaneous symmetry breaking: when $h = 0$, averages of odd powers of s such as $\langle s_i \rangle$ or $\langle s_i s_j s_k \rangle$ are zero.

6.11.2 Real Scalar Variables

Imagine we are interested in a collection of N real variables assembled into a vector $|X\rangle$ with components x_i:

$$|X\rangle \leftrightarrow [x_1, x_2, \ldots, x_N], \qquad -\infty < x_i < +\infty, \tag{6.169}$$

with a partition function

$$Z(J) = \prod_{i=1}^{N} \int_{-\infty}^{\infty} dx_i \, e^{-S_0(X) + \langle J|X\rangle} = e^{-W(J)}, \tag{6.170}$$

where S_0 does not involve the source $|J\rangle$, which is also a vector like $|X\rangle$:

$$|J\rangle \leftrightarrow [J_1, J_2, \ldots, J_N], \tag{6.171}$$

$$\langle J|X\rangle = \sum_{i=1}^{N} J_i x_i. \tag{6.172}$$

(I will use h as the source if I want to emphasize that X is a magnetic variable. The symbol J for the generic source is common in field theory.)

Everything I said for Ising correlators holds here with the substitution

$$s_i \to x_i, \qquad h_i \to J_i. \tag{6.173}$$

In particular,

$$-\left.\frac{\partial^2 W}{\partial J_i \partial J_j}\right|_{J=0} = \langle x_i x_j \rangle_c = \langle x_i x_j \rangle \quad (\text{I assume } \langle x_i \rangle = 0) \tag{6.174}$$

$$\begin{aligned} -\left.\frac{\partial^4 W}{\partial J_i \partial J_j \partial J_k \partial J_l}\right|_{J=0} &= \langle x_i x_j x_k x_l \rangle_c \\ &= \langle x_i x_j x_k x_l \rangle - \langle x_i x_j \rangle \langle x_k x_l \rangle - \langle x_i x_k \rangle \langle x_j x_l \rangle - \langle x_i x_l \rangle \langle x_j x_k \rangle. \end{aligned} \tag{6.175}$$

Something very simple happens if $S_0(X)$ is quadratic in X and the integral is a Gaussian: *All connected correlators vanish except the two-point function* $\langle x_i x_j \rangle$. I will now show this, and then extract a fundamental consequence called *Wick's theorem.*

I assume you know that for a single variable,

$$\int_{-\infty}^{\infty} e^{-\frac{1}{2}mx^2 + Jx} = \sqrt{\frac{2\pi}{m}} \exp\left[\frac{J^2}{2m}\right]. \tag{6.176}$$

Consider a quadratic action

$$S(J) = -\frac{1}{2}\langle X|M|X\rangle + \langle J|X\rangle, \tag{6.177}$$

where M is a real symmetric matrix with orthonormal eigenvectors:

$$M|\alpha\rangle = m_\alpha|\alpha\rangle, \tag{6.178}$$

$$\langle\alpha|\beta\rangle = \delta_{\alpha\beta}. \tag{6.179}$$

Let us expand X and J as follows:

$$|X\rangle = \sum_\alpha x_\alpha|\alpha\rangle, \tag{6.180}$$

$$|J\rangle = \sum_\alpha J_\alpha|\alpha\rangle, \quad \text{so that} \tag{6.181}$$

$$\langle J|X\rangle = \sum_\alpha J_\alpha x_\alpha. \tag{6.182}$$

The Jacobian for the orthogonal transformation from x_i to x_α is unity.

The action separates in these new variables:

$$-\frac{1}{2}\langle X|M|X\rangle + \langle J|X\rangle = \sum_\alpha\left(-\frac{1}{2}x_\alpha^2 m_\alpha + J_\alpha x_\alpha\right), \tag{6.183}$$

and $Z(J)$ factorizes into a product of Gaussian integrals:

$$Z(J) = \prod_\alpha \int_{-\infty}^{\infty} dx_\alpha \exp\left[-\frac{1}{2}m_\alpha x_\alpha^2 + J_\alpha x_\alpha\right] \tag{6.184}$$

$$= \prod_\alpha \sqrt{\frac{2\pi}{m_\alpha}}\, e^{\frac{1}{2}J_\alpha \frac{1}{m_\alpha} J_\alpha} \tag{6.185}$$

$$= e^{\frac{1}{2}\langle J|M^{-1}|J\rangle} \cdot \frac{(2\pi)^{N/2}}{\sqrt{\det M}}, \tag{6.186}$$

where $\det M$ is the determinant of M. This is a central result. A related one is

$$W(J) = -\frac{1}{2}\langle J|M^{-1}|J\rangle + \frac{1}{2}\ln\det M + \text{constants} \tag{6.187}$$

$$= -\frac{1}{2}\sum_{i,j} J_i(M^{-1})_{ij}J_j + J\text{-independent terms}. \tag{6.188}$$

Thus, the two-point function is

$$\langle x_i x_j\rangle = -\left.\frac{\partial^2 W}{\partial J_i \partial J_j}\right|_{J=0} = (M^{-1})_{ij} \equiv G_{ij}, \tag{6.189}$$

where I have followed tradition to denote by G the two-point function. The connected four-point function vanishes since $W(J)$ has only two J's to differentiate:

$$0 = -\left.\frac{\partial^4 W}{\partial J_i \partial J_j \partial J_k \partial J_l}\right|_{J=0} \tag{6.190}$$

$$= \langle x_i x_j x_k x_l \rangle_c \tag{6.191}$$

$$= \langle x_i x_j x_k x_l \rangle - \langle x_i x_j \rangle \langle x_k x_l \rangle - \langle x_i x_k \rangle \langle x_j x_l \rangle - \langle x_i x_l \rangle \langle x_j x_k \rangle, \tag{6.192}$$

which means

$$\langle x_i x_j x_k x_l \rangle = \langle x_i x_j \rangle \langle x_k x_l \rangle + \langle x_i x_k \rangle \langle x_j x_l \rangle + \langle x_i x_l \rangle \langle x_j x_k \rangle, \tag{6.193}$$

which is Wick's theorem. It expresses the four-point function in a Gaussian theory as a sum over products of two-point functions whose arguments are obtained by pairing the indices (i,j,k,l) in three independent ways. (You might think there are 24 such terms, but they reduce to 3 using $M^{-1}_{ij} = M^{-1}_{ji}$.) In terms of G_{ij} in Eq. (6.189),

$$\langle x_i x_j x_k x_l \rangle = G_{ij} G_{kl} + G_{ik} G_{jl} + G_{il} G_{jk}. \tag{6.194}$$

The generalization to higher-point correlators is obvious.

If the action is non-Gaussian and has, say, a quartic term, we may expand the exponential in a power series and evaluate the averages of the monomials term by term, using Wick's theorem. This is how one develops a perturbation theory for interacting systems.

Of interest to us will be problems where the statistical variable is not a vector $|X\rangle$ with a discrete index for its components x_i but a *field* $\phi(x)$, where x, now a label and not a variable, locates points in d dimensions. (Thus, in three dimensions we would have $x = \boldsymbol{r}$.) Nothing much changes except for replacing sums by integrals:

$$Z(J) = \prod_x \int d\phi(x) \exp\left[-\frac{1}{2}\int\int dx dy \phi(x) M(x,y) \phi(y) + \int dx J(x) \phi(x)\right] \tag{6.195}$$

$$= \exp\left[\frac{1}{2}\int\int dx dy J(x) M^{-1}(x,y) J(y)\right] \cdot \frac{1}{\sqrt{\det M}} \cdot \text{const.}, \tag{6.196}$$

and of course

$$\langle \phi(x_1)\phi(x_2)\rangle = M^{-1}(x_1,x_2) \equiv G(x_1,x_2), \tag{6.197}$$

$$\langle \phi(x_1)\phi(x_2)\phi(x_3)\phi(x_4)\rangle = G(x_1,x_2)G(x_3,x_4) + G(x_1,x_3)G(x_2,x_4) + G(x_1,x_4)G(x_2,x_3). \tag{6.198}$$

Here is a concrete example that will be invoked later on:

$$M(x,y) = \frac{1}{2}\left(-\nabla_x^2 + r_0\right)\delta(x-y), \tag{6.199}$$

$$S_0 = \int dx\phi(x)\left(-\frac{1}{2}\nabla^2 + \frac{1}{2}r_0\right)\phi(x), \tag{6.200}$$

$$Z_0 = \int e^{-S_0(\phi)} \prod_x d\phi(x). \tag{6.201}$$

The correlator is

$$G(x,y) = \langle x|\frac{1}{-\nabla^2 + r_0}|y\rangle. \tag{6.202}$$

By introducing a complete set of momentum states to the right of $\langle x|$ and to the left of $|y\rangle$, we end up with

$$G(x,y) = G(x-y) = \int dk \frac{e^{ik(x-y)}}{k^2 + r_0}, \tag{6.203}$$

where it is understood that $k(x-y)$ is a d-dimensional dot product and dk the d-dimensional integration measure (possibly including powers of 2π).

6.11.3 Complex Scalar Variables

Next, consider the case where the variables are z and its conjugate z^*. As you may be new to this kind of problem, let us first get used to the kind of correlation functions that arise and then ask how the generating function is to be defined.

Consider a toy partition function:

$$Z = \int_{-\infty}^{\infty} \frac{dzdz^*}{2\pi i} e^{-mz^*z}. \tag{6.204}$$

In practice, we switch to x and y, the real and imaginary parts of the integration variables, and use

$$\frac{dzdz^*}{2\pi i} \to \frac{dxdy}{\pi} = \frac{rdrd\theta}{\pi}. \tag{6.205}$$

The result is

$$Z = \int_{-\infty}^{\infty} \frac{dxdy}{\pi} e^{-m(x^2+y^2)} = \frac{1}{m}. \tag{6.206}$$

The only average we will need involves the complex conjugate pairs z and z^*:

$$\langle z^*z\rangle = \frac{\int_{-\infty}^{\infty} \frac{dzdz^*}{2\pi i} z^*ze^{-az^*z}}{\int_{-\infty}^{\infty} \frac{dzdz^*}{2\pi i} e^{-az^*z}} = \frac{1}{a}, \tag{6.207}$$

which can be most easily verified in polar coordinates.

The other two bilinears have zero average:

$$\langle zz\rangle = \langle z^* z^*\rangle = 0, \tag{6.208}$$

because the action and measure are invariant under

$$z \to z e^{i\theta}, \; z^* \to z^* e^{-i\theta}, \tag{6.209}$$

while the bilinears are not. This result may be verified using polar coordinates.

If there are more variables and M is a Hermitian matrix there is the obvious generalization:

$$Z = \int \left[\mathcal{D}z\mathcal{D}^\dagger z\right] e^{-z^\dagger M z}, \quad \text{where} \tag{6.210}$$

$$\left[\mathcal{D}z\mathcal{D}z^\dagger\right] = \prod_{i=1}^{N} \int \frac{dz_i dz_i^*}{2\pi i}, \tag{6.211}$$

$$z = [z_1 \cdots z_N], \quad z^\dagger = \left[z_1^* \cdots z_N^*\right], \tag{6.212}$$

$$Z = \prod_i \frac{1}{m_i} = \frac{1}{\det M}. \tag{6.213}$$

I have skipped the steps that invoke the complete set of eigenvectors and eigenvalues m_i of M.

As for correlations, consider the simple case where there are two sets of variables:

$$\langle z_i^* z_j\rangle = \frac{\int_{-\infty}^{\infty} \frac{dz_1 dz_1^*}{2\pi i}\frac{dz_2 dz_2^*}{2\pi i} z_i^* z_j e^{-m_1 z_1^* z_1 - m_2 z_2^* z_2}}{\int_{-\infty}^{\infty} \frac{dz_1 dz_1^*}{2\pi i}\frac{dz_2 dz_2^*}{2\pi i} e^{-m_1 z_1^* z_1 - m_2 z_2^* z_2}} = \frac{\delta_{ij}}{m_i} \equiv \langle \bar{i}j\rangle. \tag{6.214}$$

You may verify Wick's theorem in the same notation:

$$\langle z_i^* z_j z_k^* z_l\rangle = \langle \bar{i}j\rangle\langle \bar{k}l\rangle + \langle \bar{i}l\rangle\langle \bar{k}j\rangle. \tag{6.215}$$

This result makes sense: it demands that for the answer to be non-zero, the fields must come in complex conjugate pairs. Since this can happen in two ways, the result is a sum of two terms. The generalization to more variables and longer strings is obvious.

The generating function requires two sources: a complex column vector $\boldsymbol{J}$ and its adjoint $\boldsymbol{J}^\dagger$:

$$Z(\boldsymbol{J},\boldsymbol{J}^\dagger) = e^{-W(\boldsymbol{J}\boldsymbol{J}^\dagger)} = \int \left[\mathcal{D}z\mathcal{D}^\dagger z\right] e^{-z^\dagger M z + \boldsymbol{J}^\dagger z + z^\dagger \boldsymbol{J}}. \tag{6.216}$$

The correlations of a string of z_i's and z_j^*'s are found by taking derivatives of W with respect to the corresponding J_i^* and J_j respectively.

Consider now the case of a complex field ϕ in four dimensions with action

$$S_0(\phi,\phi^*) = \int \phi^*(\mathbf{k})k^2\phi(\mathbf{k})\frac{d^4k}{(2\pi)^4} \quad \text{and} \tag{6.217}$$

$$Z = \int [\mathcal{D}\phi\mathcal{D}\phi^*]e^{-S_0}, \quad \text{where} \tag{6.218}$$

$$[\mathcal{D}\phi\mathcal{D}\phi^*] = \prod_k \frac{d\mathrm{Re}\,\phi(\mathbf{k})d\mathrm{Im}\,\phi(\mathbf{k})}{\pi} = \prod_k \frac{d\phi^*(\mathbf{k})d\phi(\mathbf{k})}{2\pi i}. \tag{6.219}$$

This is called the *Gaussian model*. The corresponding functional integral is a product of ordinary Gaussian integrals, one for each $\mathbf{k}$. This makes it possible to express all the correlation functions in terms of averages involving a single Gaussian integral. *The only averages that do not vanish are products of an even number of variables, wherein each $\phi(\mathbf{k})$ is accompanied by its complex conjugate*. This is because the action and measure are invariant under

$$\phi(\mathbf{k}) \to \phi(\mathbf{k})e^{i\theta}, \quad \phi^*(\mathbf{k}) \to \phi^*(\mathbf{k})e^{-i\theta}, \tag{6.220}$$

where θ can be different for different $\mathbf{k}$'s.

In view of the above, the *two-point function* in our Gaussian model is

$$\langle\phi^*(\mathbf{k}_1)\phi(\mathbf{k}_2)\rangle = \frac{(2\pi)^4\delta^4(\mathbf{k}_1-\mathbf{k}_2)}{k^2} \tag{6.221}$$

$$\equiv (2\pi)^4\delta^4(\mathbf{k}_1-\mathbf{k}_2)G(k_1) \tag{6.222}$$

$$\equiv \langle\bar{2}1\rangle, \tag{6.223}$$

and likewise

$$\langle\phi^*(\mathbf{k}_4)\phi^*(\mathbf{k}_3)\phi(\mathbf{k}_2)\phi(\mathbf{k}_1)\rangle = \langle\bar{4}2\rangle\langle\bar{3}1\rangle + \langle\bar{4}1\rangle\langle\bar{3}2\rangle. \tag{6.224}$$

This is a case of Wick's theorem for complex bosons. For the case of $2n$ fields, the answer is a sum over all possible pairings, each term in the sum being a product of n two-point functions. The result follows from the preceding discussion, upon making the change from Kronecker deltas to Dirac delta functions in Eq. (6.215) to take into account the fact that the action in the Gaussian model is an integral (over $\mathbf{k}$) rather than a sum over variable labels.

For the generating function one must introduce the "source term"

$$S_0 \to S_0 + \int \left[J^*(k)\phi(k) + J(k)\phi^*(k)\right]\frac{d^4k}{(2\pi)^4}.$$

6.11.4 Grassmann Variables

The entire machinery of sources can be applied to Grassmann integrals, with the expected complication of a Grassmann source and even more minus signs. First, we define

$$Z(J,\bar{J}) = \int \exp\left[S(\bar{\psi},\psi) + \bar{J}\psi + \bar{\psi}J\right]\left[\mathcal{D}\bar{\psi}\mathcal{D}\psi\right], \tag{6.225}$$

$$\bar{J}\psi = \sum_i \bar{J}_i\psi_i, \tag{6.226}$$

$$\bar{\psi}J = \sum_i \bar{\psi}_i J_i, \tag{6.227}$$

$$\left[\mathcal{D}\bar{\psi}\mathcal{D}\psi\right] = \prod_i d\bar{\psi}_i d\psi_i, \tag{6.228}$$

where S is the action without sources and $(\bar{J},J)$ are Grassmann numbers that anticommute with all Grassmann numbers.

Notice that I integrate $e^{+S(\bar{\psi},\psi)}$ over ψ and $\bar{\psi}$, whereas for bosons I always integrated $e^{-S(\phi)}$. For bosons, the minus sign indicates that paths with large action are exponentially suppressed. For fermions, the action itself is a Grassmann number, which is neither big nor small, and the sign in front of S can be anything. I chose e^{+S} to make this point. (Later, in discussing bosonization, which relates bosonic and fermionic theories, I will revert to using e^{-S} for both to optimize the notation.)

You are invited to verify that

$$\langle \bar{\psi}_\beta \psi_\alpha \rangle = -\frac{1}{Z}\frac{\partial^2 Z}{\partial J_\beta \partial \bar{J}_\alpha}. \tag{6.229}$$

To verify this, you must:

- Expand the source terms to linear order in the exponents.
- Take derivatives, remembering that $\frac{\partial}{\partial J}$ and $\frac{\partial}{\partial \bar{J}}$ commute with all even terms like $\bar{\psi}J$ and anticommute with odd terms like themselves or the $\bar{\psi}$ in $\bar{\psi}J$.
- Remember that $\langle \bar{\psi}_\beta \psi_\alpha \rangle = -\langle \psi_\alpha \bar{\psi}_\beta \rangle$.

Similarly, you can get the four-point function as the fourth derivative with respect to sources.

In terms of $W(J,\bar{J})$ defined by

$$Z(J,\bar{J}) = e^{-W(J,\bar{J})}, \tag{6.230}$$

the connected two-point function is

$$\left.\frac{\partial^2 W}{\partial J_\beta \partial \bar{J}_\alpha}\right|_{J,\bar{J}=0} = \langle \bar{\psi}_\beta \psi_\alpha \rangle_{\mathrm{c}} = \langle \bar{\psi}_\beta \psi_\alpha \rangle \equiv \langle \bar{\beta}\alpha \rangle, \tag{6.231}$$

where we have used the fact that

$$\langle \bar{\psi}_\beta \rangle = \langle \psi_\alpha \rangle = 0 \text{ when } (J, \bar{J}) = 0. \tag{6.232}$$

The connected four-point function is related to the four-point function as follows, again with some minus signs and in compact notation:

$$-\left.\frac{\partial^4 W}{\partial J_\alpha \partial J_\beta \partial \bar{J}_\gamma \partial \bar{J}_\delta}\right|_{J,\bar{J}=0} = \langle \bar{\alpha}\bar{\beta}\gamma\delta \rangle_{\mathrm{c}} \tag{6.233}$$

$$= \langle \bar{\alpha}\bar{\beta}\gamma\delta \rangle - \left[\langle \bar{\alpha}\delta \rangle \langle \bar{\beta}\gamma \rangle - \langle \bar{\alpha}\gamma \rangle \langle \bar{\beta}\delta \rangle \right]. \tag{6.234}$$

The rules for the minus signs in what is subtracted to form the connected part are as follows: $\langle \bar{\alpha}\delta \rangle \langle \bar{\beta}\gamma \rangle$ has a positive coefficient since the indices can be brought to this order from $\bar{\alpha}\bar{\beta}\gamma\delta$ by an even number of Grassmann exchanges, while $\langle \bar{\alpha}\gamma \rangle \langle \bar{\beta}\delta \rangle$ comes with a minus sign because it involves an odd number of exchanges.

Here is the central result that subsumes everything you need to know about the Gaussian Grassmann action:

$$\int e^{-\bar{\psi} M \psi + \bar{J}\psi + \bar{\psi} J} \left[d\bar{\psi} d\psi \right] = \det(M) e^{\bar{J} M^{-1} J}, \tag{6.235}$$

which may be proved by making the following translation of variables:

$$\psi \to \psi + M^{-1} J, \qquad \bar{\psi} \to \bar{\psi} + \bar{J} M^{-1}. \tag{6.236}$$

Thus, for a Gaussian action

$$W(J, \bar{J}) = -\ln \det(M) - \bar{J} M^{-1} J, \tag{6.237}$$

it follows that

$$\langle \bar{\psi}_\beta \psi_\alpha \rangle_{\mathrm{c}} = \frac{\partial^2 W}{\partial J_\beta \partial \bar{J}_\alpha} = -M^{-1}_{\alpha\beta}, \tag{6.238}$$

and that

$$\langle \bar{\alpha}\bar{\beta}\gamma\delta \rangle_{\mathrm{c}} = 0, \tag{6.239}$$

since we can differentiate W only twice. Thus, *only the connected two-point correlation function is non-zero in a fermionic Gaussian theory, as in the bosonic theories*. For the four-point function, Eqs. (6.234) and (6.233) imply that

$$\langle \bar{\alpha}\bar{\beta}\gamma\delta \rangle = \langle \bar{\alpha}\delta \rangle \langle \bar{\beta}\gamma \rangle - \langle \bar{\alpha}\gamma \rangle \langle \bar{\beta}\delta \rangle. \tag{6.240}$$

This is just Wick's theorem, which expresses higher-point correlation functions in terms of products of two-point functions, as applied to fermions.

References and Further Reading

For this chapter and further use I suggest:

[1] E. Fradkin, *Field Theories of Condensed Matter Systems*, Addison Wesley (1991).
[2] M. Kardar, *Statistical Physics of Fields*, Cambridge University Press (2007).
[3] S. Sachdev, *Quantum Phase Transitions*, Cambridge University Press (1999).
[4] J. Kogut, Reviews of Modern Physics, **51**, 659 (1979).
[5] R. Shankar, *Principles of Quantum Mechanics*, Springer (1994).
[6] T. Giamarchi, *Quantum Physics in One Dimension*, Oxford University Press (2004).
[7] A. Fetter and J. D. Walecka, *Quantum Theory of Many-Particle systems*, Dover (1971).

7
The Two-Dimensional Ising Model

7.1 Ode to the Model

The two-dimensional Ising model is a necessary rite of passage in our transition from basics to hardcore topics. Like the harmonic oscillator in quantum mechanics, it is the easiest example of a completely solvable problem. By studying it one can learn many valuable lessons, notably about perturbative methods, the concept of duality, and the exact modeling of a phase transition.

Phase transitions will occupy much of this book, and we will return to study them in detail. For now, let us focus on the magnetic transition in the Ising model. The most detailed book on this subject is the one by B. McCoy and T. T. Wu [1].

On a square lattice with N columns and M rows, we define the model by

$$Z = \sum_{s_i} \exp\left[K \sum_{\langle i,j \rangle} s_i s_j \right], \tag{7.1}$$

where $K = J/kT$ and the symbol $\langle i,j \rangle$ means that sites i and j are nearest neighbors, as shown in Figure 7.1. There are many options at the edges: open boundary conditions in which the spins at the edges have no neighbors in one direction, periodic boundary conditions along one direction, which makes the system a cylinder, or along both directions, which makes it a torus. For now, let us just say that M and N and the number of sites $\mathcal{N} = MN$ are huge and we are nowhere near the ends. There are $2\mathcal{N}$ bonds on a square lattice with $\mathcal{N}$ sites because each site has four bonds emanating from it, but each bond is counted twice, once at each of its end points.

Consider the extreme limits. As $K = J/kT \to \infty$ or $T \to 0$, the spins will be all up or all down, the system will be magnetized, and $\langle M \rangle$, the average spin per site, will be at its maximum of ± 1. Let us pick $\langle M \rangle = +1$. As $K \to 0$ or $T \to \infty$, the Boltzmann weight will be 1 for all configurations, the spins will fluctuate independently, and $\langle M \rangle$ will vanish. The graph of $\langle M(T) \rangle$ will thus start out at $+1$ and decrease as we heat the system. It should be zero at $T = \infty$. One possibility is that it does not reach zero until we reach $T = \infty$. If, however, it vanishes at some finite $T = T_c$ and remains zero thereafter, we have a phase transition. There must be a singularity at T_c, since a non-trivial analytic function cannot

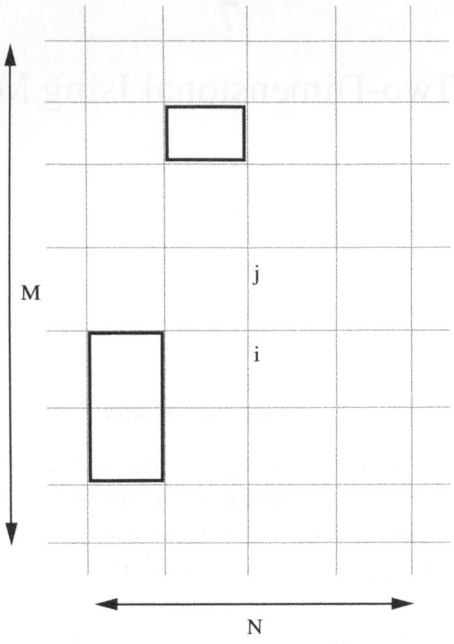

Figure 7.1 The square lattice with M rows and N columns. The site j is a nearest neighbor of i; there are three more. The solid rectangles correspond to terms at fourth and sixth orders in the $\tanh K$ (high-T) expansion.

identically vanish beyond some point. This singularity or non-analyticity is the signature of a phase transition. The free energy and its derivatives will also be singular at this point.

If we are to see this singular behavior analytically, we must go to the thermodynamic limit. Let us understand why. For any finite system, Z is a finite sum over positive terms, and its logarithm (the free energy F) will be analytic at any real T or K. But, Z could vanish arbitrarily close to the real axis, and in the limit of infinite system size these zeros could pinch the real axis, producing a singularity. However, we cannot simply compute F in this limit since it is extensive in system size like the energy, and will not approach a limit. However, under normal conditions f, the *free energy per site* will have a limit and this is what we are after. In the $d = 1$ Ising model we were able to obtain f, but it did not exhibit a finite-T phase transition; the system was unmagnetized at all $T > 0$. The $d = 2$ case is the celebrated example with a finite-T transition, displayed rigorously in Onsager's solution. Prior to Onsager it was not universally accepted that the innocuous sum over Boltzmann weights could reproduce a phenomenon as complex as a phase transition.

Before we plunge into the exact solution, let us learn some approximate methods that work in generic situations.

7.2 High-Temperature Expansion

This is a perturbative expansion around the point $K = 0$ when the Boltzmann weight is unity for all states, the spins do not talk to each other, and $Z = 2^{\mathcal{N}}$. For any one bond we may write

$$e^{Ks_is_j} = \cosh K + s_is_j \sinh K = \cosh K(1 + s_is_j \tanh K), \tag{7.2}$$

a result that follows from $(s_is_j)^2 = 1$. So

$$Z(K) = \sum_{s_i} \prod_{\text{bond } ij} \cosh K(1 + s_is_j \tanh K). \tag{7.3}$$

Each bond can contribute either a 1 or a $\tanh K$, and there are $2^{\mathcal{N}}$ terms in the product over bonds. The leading term in the $\tanh K$ expansion has a 1 from every bond and contributes $2^{\mathcal{N}}(\cosh K)^{2\mathcal{N}}$. The sum over 1 at each site gives us the $2^{\mathcal{N}}$, while the product of $\cosh K$ over the $2\mathcal{N}$ bonds gives the rest. The next term has a $\tanh K$ from one bond and a 1 from the others. There are $2\mathcal{N}$ such terms. These $2\mathcal{N}$ terms do not survive the sum over s_i since $\sum_{s=\pm 1} s = 0$ and we have two such free or dangling spins at each end. To order $\tanh^2 K$, we still get nothing: if the bonds share no sites, we have four spin sums that vanish, and if they do share a site we have a spin sum over the other two that vanish. The first non-zero contribution is at order $\tanh^4 K$, when we pick four bonds that form a square, as shown by the dark square in Figure 7.1. Now the spins at each corner appear twice (since there are two bonds from the square incident at each site), and we now get a contribution $2^{\mathcal{N}}(\cosh K)^{2\mathcal{N}} \cdot \mathcal{N} \tanh^4 K$ to $Z(K)$. The factor of $\mathcal{N}$ comes from the number of squares we can have, and this equals $\mathcal{N}$ because we can label each square by the site at its lower left-hand corner. The series so far looks as follows:

$$\frac{Z(K)}{2^{\mathcal{N}}(\cosh K)^{2\mathcal{N}}} = 1 + \mathcal{N} \tanh^4 K + 2\mathcal{N} \tanh^6 K + \cdots, \tag{7.4}$$

where the derivation of the $\tanh^6 K$ term is left to Exercise 7.2.1. We expect this series to work for small K.

Exercise 7.2.1 *Derive the sixth-order term in Eq. (7.4).*

The high-temperature series takes the form

$$\frac{Z(K)}{2^{\mathcal{N}}(\cosh K)^{2\mathcal{N}}} = \sum_{\text{closed loops}} C(L) \tanh^L K, \tag{7.5}$$

where $C(L)$ is the number of closed loops of length L we can draw on the lattice without covering any bond more than once.

The free energy per site is, to this order,

$$-\frac{f}{kT} = \frac{1}{\mathcal{N}} \ln Z = \ln\left[2\cosh^2 K\right] + \frac{1}{\mathcal{N}} \ln(1 + \mathcal{N} \tanh^4 K + \cdots)$$
$$= \ln\left[2\cosh^2 K\right] + \tanh^4 K + \cdots \tag{7.6}$$

using $\ln(1+x) = x + \cdots$. It is significant but not obvious that as we go to higher orders, we will keep getting a limit for f that is independent of $\mathcal{N}$. For example, if you consider the case of two disjoint elementary squares that contribute with factor $\tanh^8 K$, there will be $\mathcal{N}(\mathcal{N}-5)/2$ of them since the two squares cannot share an edge or be on top of each other. In addition, there are single loops with perimeter 8. Upon taking the logarithm, the $\mathcal{N}^2/2$ part of this cancels against the square of the $\mathcal{N}$ term due to the elementary square when $\ln(1+x) = x - \frac{1}{2}x^2 + \cdots$ is expanded. For more practice, do Exercise 7.2.2.

Exercise 7.2.2 *Show that to order* $\tanh^8 K$,

$$-\frac{f}{kT} = \ln\left[2\cosh^2 K\right] + \tanh^4 K + 2\tanh^6 K + \frac{9}{2}\tanh^8 K + \cdots \tag{7.7}$$

7.3 Low-Temperature Expansion

Consider now the regime near $T = 0$ or $K = \infty$. The spins will tend to be aligned in one direction, say up. The Boltzmann weight is e^K on each of the $2\mathcal{N}$ bonds. We say the bonds are all unbroken. If a spin is now flipped down, the four bonds linking it to its four nearest neighbors will be broken and the Boltzmann factor will be reduced by e^{-8K}. Since the spin flip can occur in any of $\mathcal{N}$ sites, we have

$$Z = e^{2\mathcal{N}K}(1 + \mathcal{N}e^{-8K} + \cdots). \tag{7.8}$$

Let us now consider flipping two spins. The energy cost depends on their relative locations. It is lowest if they are neighbors: the bond connecting them is unbroken, but the six other bonds linking them to all other neighbors will be broken. There are N ways to pick the first spin and two ways to pick the second: to its north or east. (Positions to the south and west are not needed since we will then be double counting.) Thus we have

$$\frac{Z}{e^{2\mathcal{N}K}} = (1 + \mathcal{N}e^{-8K} + 2\mathcal{N}e^{-12K} + \cdots). \tag{7.9}$$

We expect this series to work for small e^{-2K} or large K.

The obvious way to represent these configurations in the low-temperature expansion is to show flipped spins, as in Figure 7.2. A cleverer way due to Kramers and Wannier [2] is to surround the flipped spins by a contour that is made up of perpendicular bisectors of the broken bonds. The dotted lines in the figure correspond to the dual lattice.

If we create an island of spins pointing opposite to the majority, it costs an energy proportional to the perimeter of the island, as reflected in the Boltzmann factor e^{-2KL}.

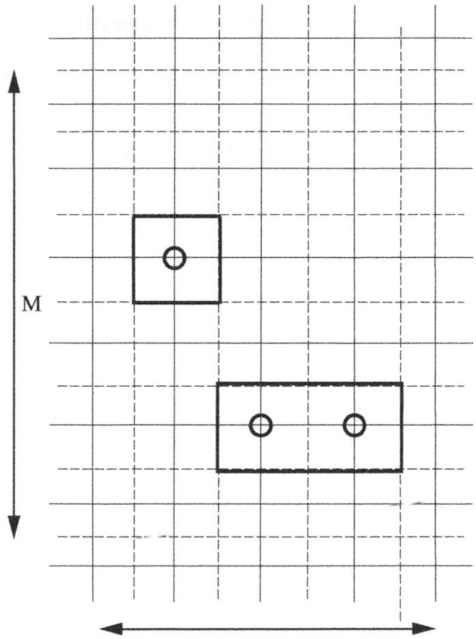

Figure 7.2 The low-temperature expansion in which only spins flipped relative to a fully aligned state are shown as tiny circles. These are then surrounded by bonds on a dual lattice of dotted lines. Note that the two such closed figures are topologically identical in shape and multiplicity to the leading ones in the high-temperature $\tanh K$ expansion.

Compare this to the $d = 1$ case where the cost of an island is just e^{-2K} regardless of its size. (The "perimeter" of this island is made of just the two end points.) This is why the $d = 1$ system loses its magnetization at any non-zero T. In $d = 2$, we can estimate the critical temperature by asking when large islands will go unsuppressed. Imagine laying out a loop of length L. At each stage we can move in three directions, since going back is not an option because each bond can be covered only once. Ignoring the condition that we end up where we began, and that we cannot run into other loops starting in other places and so on, we roughly get a factor $3^L e^{-2KL}$, so that loops of arbitrarily large size are no longer suppressed when we reach

$$e^{(-2K_c+\ln 3)L} = 1, \text{ or} \qquad K = K_c = 0.5493, \tag{7.10}$$

which you can compare to the exact result $K_c = 0.4407$ (to four places).

We can also do a similar analysis for the high-temperature series to estimate K_c:

$$(\tanh K_c)^L \cdot 3^L \simeq 1 \tag{7.11}$$

$$\tanh K_c = \frac{1}{3}, \text{ or} \qquad K = K_c = 0.3466. \tag{7.12}$$

The correct answer is seen to lie between these two estimates.

7.4 Kramer–Wannier Duality

Let us now note that the low-T expansion resembles the high-T expansion on the *dual lattice*: the lattice whose edges are the perpendicular bisectors of the original ones and whose sites are located at the center of each square element (plaquette) in the original lattice. You will agree that up to the order considered, the diagrams for the high- and low-temperature series have the same shapes (unit squares, 2×1 rectangles, etc.) and multiplicity ($\mathcal{N}$, $2\mathcal{N}$, etc.).

They do not, however, have the same weights. So we do the following: Since K is a dummy variable in Eq. (7.9), let us replace it by K^* to obtain

$$\frac{Z(K^*)}{e^{2\mathcal{N}K^*}} = (1 + \mathcal{N}e^{-8K^*} + 2\mathcal{N}e^{-12K^*} + \cdots). \tag{7.13}$$

So far, K^* is just a dummy variable. Let us now choose, for each K, a *dual temperature* $K^*(K)$ such that

$$e^{-2K^*(K)} = \tanh K. \tag{7.14}$$

Now the two series in Eqs. (7.4) and (7.13) agree numerically to the order shown. It can be shown that the agreement is good to all orders, implying the *self-duality* relation

$$\frac{Z(K)}{2^{\mathcal{N}}(\cosh K)^{2\mathcal{N}}} = \frac{Z(K^*)}{e^{2\mathcal{N}K^*}}. \tag{7.15}$$

Using

$$\sinh 2K \cdot \sinh 2K^* = 1, \tag{7.16}$$

one can rewrite Eq. (7.15) more symmetrically as

$$\frac{Z(K)}{(\sinh 2K)^{\frac{\mathcal{N}}{2}}} = \frac{Z(K^*)}{(\sinh 2K^*)^{\frac{\mathcal{N}}{2}}}. \tag{7.17}$$

Exercise 7.4.1 *Prove Eqs. (7.16) and (7.17).*

If K is small, then in order to satisfy Eq. (7.14) $K^*(K)$ has to be large, which is fine, as the low-temperature expansion works for large values of its argument. What is remarkable is that the thermodynamics of the model at low and high energies are related even though the physical properties are very different: one side has magnetization and one does not. *Self-duality*, relating the model at weak coupling (at a small value of a parameter, K in our example) to the *same* model at strong coupling (large K values) is quite rare. It is more common to encounter simply *duality*, in which one model at strong coupling is related to *another* model at weak coupling.

Recall from Chapter 1 that the inverse relation of Eq.(7.14) is

$$e^{-2K} = \tanh K^*. \tag{7.18}$$

In other words, K^* as a function of K coincides with K as a function of K^*. Consequently, the dual of the dual is the original K:

$$(K^*)^* = K. \tag{7.19}$$

(A trivial example of a function that is its own inverse is $y = \frac{1}{x}$, which implies $x = \frac{1}{y}$. The relation between K and K^* is, of course, much more interesting.)

Kramers and Wannier used duality to find the critical temperature of the Ising model as follows: Equation (7.17) implies that in the thermodynamic limit any singularity at some K, such as at a phase transition, implies one at $K^*(K)$. If, however, we assume that there is just one transition, it must occur at a critical value K_c that is its own dual:

$$K_c^* = K_c, \quad \text{or} \quad e^{-2K_c} = \tanh K_c, \quad \text{or} \quad e^{-2K_c} = \sqrt{2} - 1, \quad \text{or} \quad K_c = 0.4407\ldots \tag{7.20}$$

But bear in mind that there can be, and there *are*, problems where there are two phase transitions at critical points related by duality, with nothing interesting going on at the self-dual point itself.

We conclude this section with some remarks on the *anisotropic Ising model* with couplings K_x and K_τ in the two directions. We label the second direction by τ rather than y since it will play the role of imaginary time when we use the transfer matrix. By comparing high- and low-T expansions we find that the dual couplings K_x^{d} and K_τ^{d} are given by

$$e^{-2K_x^{\mathrm{d}}} = \tanh K_\tau, \qquad e^{-2K_\tau^{\mathrm{d}}} = \tanh K_x. \tag{7.21}$$

Now recall that we have defined the dual X^* of any real number by the symmetric relations

$$e^{-2X^*} = \tanh X, \qquad e^{-2X} = \tanh X^*. \tag{7.22}$$

Let us use this to trade the tanh's in Eq. (7.21) for exponentials of the duals to obtain exponentials on both sides:

$$e^{-2K_x^{\mathrm{d}}} = e^{-2K_\tau^*}, \qquad e^{-2K_\tau^{\mathrm{d}}} = e^{-2K_x^*}, \tag{7.23}$$

which allows us to read off the coordinates dual to any K_x and K_τ:

$$K_x^{\mathrm{d}} = K_\tau^*, \qquad K_\tau^{\mathrm{d}} = K_x^*. \tag{7.24}$$

To summarize, duality maps points in the (K_x, K_τ) plane as follows:

$$(K_x, K_\tau) \xrightarrow[\text{duality}]{} (K_\tau^*, K_x^*). \tag{7.25}$$

The self-dual points obey

$$(K_x, K_\tau) = (K_\tau^*, K_x^*). \tag{7.26}$$

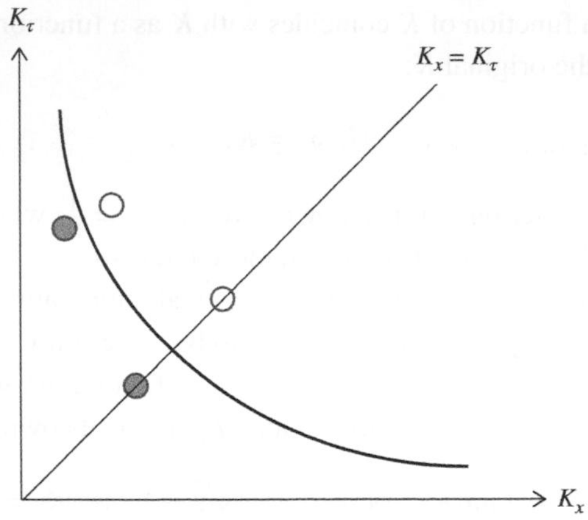

Figure 7.3 The situation in the K_x–K_τ plane. The isotropic points lie on the 45° line. The critical points lie on $\sinh 2K_x \sinh 2K_\tau = 1$. Also shown are a pair of points (solid circles) and their duals (open circles).

The condition obtained by matching the first coordinates,

$$K_x = K_\tau^*, \tag{7.27}$$

is the same as we get from equating the second coordinates, as can be seen by taking the dual of both sides. The self-dual points lie on a one-dimensional curve,

$$\sinh 2K_x \sinh 2K_\tau = 1. \tag{7.28}$$

The system is critical on this whole line and the isotropic point lies on the 45° line at $\sinh^2 2K_c = 1$ or $K_c = 0.4407\ldots$

For later use, note that no matter how small or large K_τ is, there is a critical value of K_x that goes with it, and vice versa. Figure 7.3 illustrates the situation in the K_x–K_τ plane.

Exercise 7.4.2 *Establish Eq. (7.21) by comparing high- and low-T expansions. Sketch the locus of self-dual points. Find the duals to a few points with (K_x, K_τ) both small, both large, and one large and one small. Show that the self-dual points obey* $\sinh 2K_x \sinh 2K_\tau = 1$.

7.5 Correlation Function in the tanh Expansion

Consider the thermal average $\langle s_i s_f \rangle$ between spins at some "initial" site i and "final" site f, named that way for a good reason. Let us employ the high-temperature tanh expansion. By definition,

$$\langle s_i s_f \rangle = \frac{\sum_s s_i s_f \prod_{\text{bonds}} (1 + s_m s_n \tanh K)}{\sum_s \prod_{\text{bonds}} (1 + s_m s_n \tanh K)}, \tag{7.29}$$

where m and n are neighbors connected by a bond, and common factors have been canceled between numerator and denominator. The first non-zero term in the numerator occurs when we have bonds starting at i and ending at f, yielding a product $s_i s_i s_{i+1} s_{i+1} s_{i+2} \cdots s_{f-1} s_f s_f$, where $i+1$ is the neighbor to i. Now there are no free Ising spins that can be summed over to give zero. In the simple case where the points lie on the same axis, this product will occur with a factor $(\tanh K)^{|i-f|}$, where $|i-f|$ is the distance between the end points. The first non-zero term in the denominator is just 1, so that

$$\langle s_i s_f \rangle = (\tanh K)^{|i-f|}(1+\cdots), \tag{7.30}$$

where the ellipses denote higher-order corrections from longer paths joining i and f, as well as contributions from closed paths in the numerator and denominator. If the points are not along the same axis, $|i-j|$ will be simply the Manhattan distance and there will be many paths of the shortest length; the 1 in the brackets above will be replaced by this multiplicity. This most important dependence is, of course, in the exponential prefactor $\exp(\ln \tanh K|i-f|)$.

Exercise 7.5.1 *Use the* $\tanh K$ *expansion for the N-site* $d=1$ *Ising model with periodic boundary conditions to compute* Z *as well as* $\langle s_i s_f \rangle$.

References and Further Reading

[1] B. McCoy and T. T. Wu, *The Two-Dimensional Ising Model*, Harvard University Press (1973).
[2] H. A. Kramers and G. H. Wannier, Phys. Rev. **60**, 252 (1941).

8

Exact Solution of the Two-Dimensional Ising Model

Here I will outline how the Ising model can be solved exactly. Some details will be left as exercises, and some uninteresting and easily reinstated additive constants in the free energy will be dropped. I will, however, pay close attention to boundary conditions on a finite lattice.

The original epoch-making solution of this model was due to Onsager [1]. This work was simplified by Shultz, Mattis, and Lieb [2], and it is this version we will study here.

The plan is to first write down the transfer matrix T and show how it can be diagonalized by introducing fermionic variables. The actual diagonalization will be carried out in the τ-continuum limit wherein $T = e^{-\tau H}$ and H is a simple, readily diagonalized, Hamiltonian.

8.1 The Transfer Matrix in Terms of Pauli Matrices

We want the transfer matrix corresponding to

$$Z = \sum_{s} \exp\left[\sum_{i} K_x s_i s_{i+x} + K_\tau (s_i s_{i+\tau} - 1)\right], \tag{8.1}$$

where $i+x$ and $i+\tau$ are neighbors of site i in the x and τ directions, and where I have subtracted the 1 from $s_i s_{i+\tau}$ so I can borrow some results from the $d=1$ Ising model. The transfer matrix for a lattice with N columns is

$$T = \frac{\exp\left[\sum_{n=1}^{N} K_\tau^* \sigma_1(n)\right]}{\left[\cosh K_\tau^*\right]^N} \cdot \exp\left[\sum_{n=1}^{N} K_x \sigma_3(n)\sigma_3(n+1)\right] \tag{8.2}$$

$$\equiv V_1 V_3, \tag{8.3}$$

where $\sigma_1(n)$ and $\sigma_3(n)$ are the usual Pauli matrices at site n.

To check this result, sandwich this T between $\langle s_1' \cdots s_n' \cdots s_N'|$ and $|s_1 \cdots s_n \cdots s_N\rangle$, where $(s_1' \cdots s_n' \cdots s_N')$ and $(s_1 \cdots s_n \cdots s_N)$ label the eigenvalues of $\sigma_3(n)$, $n = 1,\ldots,N$. Acting to the right on their own eigenstates, the $\sigma_3(n)$'s in V_3 will turn into s_n, yielding the Boltzmann weight associated with the horizontal bonds of the row containing the s_n's. As for V_1, its matrix elements factorize into a product over the sites n which make a contribution

$$\langle s_n'|\frac{e^{K_\tau^*\sigma_1(n)}}{\cosh K_\tau^*}|s_n\rangle = \langle s_n'|(1+\sigma_1(n)\tanh K_\tau^*)|s_n\rangle \tag{8.4}$$

$$= (\delta_{s_n s_n'} + \tanh K_\tau^* \delta_{s_n,-s_n'}) \tag{8.5}$$

$$= (\delta_{s_n s_n'} + e^{-2K_\tau}\delta_{s_n,-s_n'}), \tag{8.6}$$

which is the Boltzmann weight due to the vertical bond at site n.

While the T above does the job, it is not Hermitian. We can trade it for a Hermitian version,

$$T = V_3^{\frac{1}{2}} V_1 V_3^{\frac{1}{2}}. \tag{8.7}$$

Now, each $V^{\frac{1}{2}}$ will capture half the energies of the horizontal bonds in the two rows in question, the other half coming from the next and previous insertion of T in the computation of Z.

We will ignore the Hermiticity issue since it will disappear in the τ-continuum limit we will be focusing on. We will also drop the $(\cosh K_\tau^*)^N$ in the denominator of V_1 and the additive contribution of $\log\cosh K_\tau^*$ it makes to βf. The dedicated reader can keep track of these points.

8.2 The Jordan–Wigner Transformation and Majorana Fermions

One may ask why it is hard to diagonalize the transfer matrix given that in the exponent we have just a translationally invariant sum of terms linear or quadratic in the Pauli matrices. Why not resort to a Fourier transform? The answer is that the Pauli matrices anticommute at one site but commute at different sites. Thus they are neither bosonic nor fermionic. The Fourier transforms will not have canonical commutation or anticommutation properties. We now consider a way to trade the Pauli matrices for genuine Majorana fermions, which will have global canonical anticommutation rules before and after Fourier transformation.

In case Majorana fermions are new to you, let us begin with the usual fermion operators Ψ and $\Psi^\dagger$, which obey the anticommutation rules

$$\{\Psi, \Psi^\dagger\} = 1, \quad \{\Psi, \Psi\} = \{\Psi^\dagger, \Psi^\dagger\} = 0. \tag{8.8}$$

These will be referred to as Dirac fermions. With a pair of them we can associate a fermion number density $n_F = \Psi^\dagger \Psi$, which we have seen only has eigenvalues $n_F = 0, 1$.

Now consider Hermitian or "real" combinations

$$\psi_1 = \frac{\Psi + \Psi^\dagger}{\sqrt{2}}, \tag{8.9}$$

$$\psi_2 = \frac{\Psi - \Psi^\dagger}{\sqrt{2}i}, \tag{8.10}$$

which obey

$$\{\psi_i, \psi_j\} = \delta_{ij}. \tag{8.11}$$

More generally, N fermions ψ_i, $i = 1,\ldots,n$ obeying the above anticommutation rules are called *Majorana fermions*. The algebra they obey is that of Dirac's γ matrices in N dimensions (also called the *Clifford algebra*).

The inverse relation is

$$\Psi = \frac{\psi_1 + i\psi_2}{\sqrt{2}}, \tag{8.12}$$

$$\Psi^\dagger = \frac{\psi_1 - i\psi_2}{\sqrt{2}}. \tag{8.13}$$

Imagine that at each site we have a pair of Majorana fermions $\psi_1(n)$ and $\psi_2(n)$. These live in a two-dimensional space, as can be seen by considering the Dirac fermions $\Psi(n)$ and $\Psi^\dagger(n)$ we can make out of them. The two states correspond to $n_F = 0, 1$. On the full lattice the fermions will need a Hilbert space of dimension 2^N.

But this is also the dimensionality of the Pauli matrices. The relation between them, due to Jordan and Wigner, is, however, non-local:

$$\psi_1(n) = \left[\begin{array}{ll} \frac{1}{\sqrt{2}}\left[\prod_{l=1}^{n-1}\sigma_1(l)\right]\sigma_2(n) & \text{if } n > 1 \\ \frac{1}{\sqrt{2}}\sigma_2(1) & \text{for } n = 1; \end{array}\right. \tag{8.14}$$

$$\psi_2(n) = \left[\begin{array}{ll} \frac{1}{\sqrt{2}}\left[\prod_{l=1}^{n-1}\sigma_1(l)\right]\sigma_3(n) & \text{if } n > 1 \\ \frac{1}{\sqrt{2}}\sigma_3(1) & \text{for } n = 1. \end{array}\right. \tag{8.15}$$

Exercise 8.2.1 *Verify that*

$$\{\psi_i(n), \psi_j(n')\} = \delta_{ij}\delta_{nn'}. \tag{8.16}$$

If you carried out Exercise 8.2.1, you will see that the role of the string is to ensure that ψ's at different sites n and n' anticommute. Let us say $n' > n$. In the product of two fermions, the $\sigma_1(n)$ in the string of $\psi_{1,2}(n')$ will anticommute with both the $\sigma_2(n)$ and the $\sigma_3(n)$ of $\psi_{1,2}(n)$.

Although Shultz, Mattis, and Lieb used Dirac fermions in their Jordan–Wigner transformation, we will use the Majorana fermions, which are more natural here.

While the string is good for ensuring global anticommutation relations, simple operators involving a few spins will typically involve a product of a large number of fermions.

Amazingly, something very nice happens when we consider the two operators that appear in T:

$$\sigma_1(n) = -2i\psi_1(n)\psi_2(n), \tag{8.17}$$

$$\sigma_3(n)\sigma_3(n+1) = 2i\psi_1(n)\psi_2(n+1). \tag{8.18}$$

Things are not so nice in general. For example, if we wanted to couple the spins to a magnetic field we would need a sum over $\sigma_3(n)$, and each σ_3 would involve arbitrarily long strings of fermions:

$$\sigma_3(n) = \sqrt{2}\left[\prod_{l=1}^{n-1}(-2i\psi_1(l)\psi_2(l))\right]\psi_2(n). \tag{8.19}$$

This is why no one has been able to solve the two-dimensional Ising model in a magnetic field using fermions (or any other device).

Exercise 8.2.2 *Prove Eq. (8.19)*

Returning to the transfer matrix in Eq. (8.2), we have, upon dropping the prefactor $\left[\cosh K_\tau^*\right]^{-N}$,

$$\begin{aligned} T &= \exp\left[\sum_{n=1}^{N} -2iK_\tau^*\psi_1(n)\psi_2(n)\right]\exp\left[\sum_{n=1}^{N} 2iK_x\psi_1(n)\psi_2(n+1)\right] \\ &\equiv V_1V_3. \end{aligned} \tag{8.20}$$

Thus we see that V_1 and V_3 contain expressions quadratic in fermions, which we can diagonalize in momentum space. This was the key ingredient in Onsager's *tour de force*, communicated to mortals by Kauffman and then Schultz, Mattis, and Lieb.

How do we deal with the fact that there are *two* exponentials, each with its own quadratic form with $\mathcal{O}(N)$ terms in it? They need to be combined into a single exponential using the BCH formula:

$$e^Ae^B = e^{A+B+\frac{1}{2}[A,B]+\frac{1}{12}[A,[A,B]]-\frac{1}{12}[B,[A,B]]\cdots}. \tag{8.21}$$

Another worry is that the combination that emerges may not be in quadratic form. This concern can be dispensed with right away: the combined exponential will also be in quadratic form since the commutator of any two bilinears in fermions will either be a c-number or another bilinear, as explained in Exercise 8.2.3.

Exercise 8.2.3 *Consider the commutator $\left[f_1f_2,f_3f_4\right]$, where the f's are fermion creation or destruction operators with anticommutator $\{f_i,f_j\} = \eta_{ij}$ a c-number, possibly 0. Rearrange $f_1f_2f_3f_4$ into $f_3f_4f_1f_2$ by anticommuting f_3 and f_4 to the left, picking up some η's. You will see that the commutator is at most quadratic in the f's. This result also follows from the very useful formula that expresses commutators of fermion bilinears in terms of anticommutators of fermions:*

$$[AB,CD] = A\{B,C\}D - C\{A,D\}B - \{A,C\}BD + CA\{B,D\}. \tag{8.22}$$

As for combining the exponentials with so many terms, we will find a great simplification after we do the Fourier transform.

8.3 Boundary Conditions

The remarkable feature of the Jordan–Wigner transformation was that the terms linear and quadratic in the Pauli matrices mapped into fermion bilinears, as shown in Eqs. (8.17) and (8.18).

This would be the end of the story in an infinite chain, but we would like to analyze the model in a finite chain as well. To ensure translation invariance at finite N, we shall impose periodic boundary conditions for the spins by closing the line into a circle, with site N right next to site 1. (This is equivalent to identifying the site i with site $i+N$ on an infinite chain.) We may naively expect that the additional term $\sigma_3(N)\sigma_3(1)$ is given by Eq. (8.18) as

$$\sigma_3(N)\sigma_3(1) = 2i\psi_1(N)\psi_2(1). \tag{8.23}$$

This is not true, because $\psi_2(1)$ has no string of σ_1's to cancel the string that $\psi_1(N)$ has. You may verify that instead we get

$$\sigma_3(N)\sigma_3(1) = -\left[\prod_1^N \sigma_1(l)\right] 2i\psi_1(N)\psi_2(1) \equiv (-1)\cdot\mathcal{P}\cdot 2i\psi_1(N)\psi_2(1), \tag{8.24}$$

where

$$\mathcal{P} = \prod_1^N \sigma_1(l). \tag{8.25}$$

Thus, it seems that in demanding translation invariance we have introduced a minus sign and a very nasty non-local term $\mathcal{P}$ in relation to the naive expectation. Luckily, $\mathcal{P}$ commutes with the transfer matrix T, since it flips all spins, and that is a symmetry of T. Since $\mathcal{P}^2 = I$, its eigenvalues are ± 1. The simultaneous eigenstates of T and $\mathcal{P}$ can be divided into those where $\mathcal{P} = 1$ (the even sector) and those where $\mathcal{P} = -1$ (odd sector).

In the odd sector where $\mathcal{P} = -1$, we find that

$$\sigma_3(N)\sigma_3(1) = 2i\psi_1(N)\psi_2(1) = 2i\psi_1(N)\psi_2(N+1), \tag{8.26}$$

$$\psi_2(N+1) = \psi_2(1) \quad \text{by definition.} \tag{8.27}$$

Thus, the periodic spin–spin interaction becomes

$$\sum_{n=1}^N \sigma_3(n)\sigma_3(n+1) = 2i\sum_{n=1}^N \psi_1(n)\psi_2(n+1). \tag{8.28}$$

But how about the even sector where $\mathcal{P} = +1$? Now it looks like we should demand

$$\psi_2(N+1) = -\psi_2(1) \tag{8.29}$$

if we are to once again obtain

$$\sum_{n=1}^{N} \sigma_3(n)\sigma_3(n+1) = 2i \sum_{n=1}^{N} \psi_1(n)\psi_2(n+1). \tag{8.30}$$

Look at Eq. (8.29). Can the fermion return to minus itself when we go around the loop and come back to the same point? Yes it can, just like the spinors do after a rotation by 2π. The reason, in both cases, is that physical quantities (observables) depend only on bilinears in the spinors and bilinears in the fermion fields.

To summarize: *we must use antiperiodic (periodic) boundary conditions for the fermion in the even (odd) sectors with* $\mathcal{P} = 1\ (-1)$.

We now characterize the two sectors in the fermion language. At each site, let us form a Dirac fermion and its adjoint:

$$\Psi(n) = \frac{\psi_1(n) + i\psi_2(n)}{\sqrt{2}}, \quad \Psi^\dagger(n) = \frac{\psi_1(n) - i\psi_2(n)}{\sqrt{2}}. \tag{8.31}$$

The fermion number associated with site n is

$$N_\Psi(n) = \Psi^\dagger(n)\Psi(n) = \frac{1 + 2i\psi_1(n)\psi_2(n)}{2} = \frac{1 - \sigma_1(n)}{2}. \tag{8.32}$$

If we now use

$$\sigma_1 = -ie^{\frac{i\pi}{2}\sigma_1} = e^{\frac{i\pi}{2}(\sigma_1 - 1)}, \tag{8.33}$$

we find

$$\mathcal{P} = \prod_1^N \sigma_1(n) = e^{\frac{i\pi}{2}\sum_n(\sigma_1(n)-1)} = e^{-\frac{i\pi}{2}\sum_n(2\Psi^\dagger(n)\Psi(n))} = e^{-i\pi N_\Psi} = (-1)^{N_\Psi}, \tag{8.34}$$

where

$$N_\Psi = \sum_{n=1}^{N} \Psi^\dagger(n)\Psi(n) \tag{8.35}$$

is the total fermion number operator associated with the Dirac field Ψ. (Later we will run into similar numbers associated with other operators.)

Thus, $\mathcal{P} = (-1)^{N_\Psi}$ is the fermion parity of the state, and it depends on N_Ψ mod 2. We are to use periodic boundary conditions in the odd sector $\mathcal{P} = -1$ with an odd N_Ψ and antiperiodic boundary conditions in the even sector with $\mathcal{P} = +1$ and even N_Ψ. *Periodic solutions in the even sector and antiperiodic solutions in the odd sector are physically irrelevant and should be discarded.* This point is key to proving why, in the thermodynamic limit, the model has two degenerate ground states at low temperatures (polarized up or down) and only one in the high-temperature state (disordered).

8.4 Solution by Fourier Transform

We now introduce Fourier transforms, which will dramatically simplify the problem. For convenience we will assume N, the number of sites, is a multiple of four.

Let us introduce N momentum space operators $c_\alpha(m)$ (and their adjoints) obeying

$$\{c_\alpha(m), c_\beta^\dagger(m')\} = \delta_{\alpha\beta}\delta_{mm'}. \tag{8.36}$$

In the even sector the N allowed momenta that ensure antiperiodic boundary conditions (APBC) are:

$$k_m = \left[\frac{m\pi}{N},\ m = \pm 1, \pm 3, \pm 5, \ldots, \pm(N-1)\right]. \tag{8.37}$$

Thus, for example, if $N = 8$, $k_m = \left[\pm\frac{\pi}{8}, \pm\frac{3\pi}{8}, \pm\frac{5\pi}{8}, \pm\frac{7\pi}{8}\right]$.

The odd sector, with periodic boundary conditions (PBC), requires

$$k_m = \left[\frac{m\pi}{N},\ m = 0, \pm 2, \pm 4, \ldots, \pm(N-2), N\right]. \tag{8.38}$$

Thus, for example, if $N = 8$, $k_m = \left[0, \pm\frac{2\pi}{8}, \pm\frac{4\pi}{8}, \pm\frac{6\pi}{8}, \pi\right]$.

The momenta 0 and π, which are equal to minus themselves, require special treatment.

Caution: Unfortunately, the allowed values of m are even in the odd parity sector and vice versa. When I refer to even and odd sectors I shall always mean the parity of the fermion number and not the parity of the momentum index m.

8.4.1 The Fermionic Transfer Matrix in the Even Sector

In the even sector we expand the Majorana fermions as

$$\psi_\alpha(n) = \frac{1}{\sqrt{N}} \sum_{m=-(N-1)}^{N-1} c_\alpha(m) e^{ik_m n} \quad (m \text{ is odd for APBC}). \tag{8.39}$$

Demanding $\psi_\alpha(n) = \psi_\alpha^\dagger(n)$ gives the condition

$$c_\alpha(k_m) = c_\alpha^\dagger(-k_m). \tag{8.40}$$

This makes sense because we can only trade N Hermitian (or real) pairs for half as many non-Hermitian (complex) pairs. The expansion that makes all this explicit is

$$\psi_\alpha(n) = \frac{1}{\sqrt{N}} \sum_{m=1,3,N-1} \left[c_\alpha(m) e^{ik_m n} + c_\alpha^\dagger(m) e^{-ik_m n}\right]. \tag{8.41}$$

Observe that the momenta in the sum are restricted to be positive. The anticommutation rules for ψ_α follow from those of the c's and $c^\dagger$'s.

Exercise 8.4.1 *Verify that the anticommutation rules for ψ_α follow from those of the c's and $c^\dagger$'s.*

Now we evaluate the exponentials in T in terms of these Fourier modes and find, in the even sector,

$$T_{\mathrm{E}} = \exp\left[\sum_{m=1,3,\ldots}^{N-1} -2iK_\tau^* \left[c_1(m)c_2^\dagger(m) + c_1^\dagger(m)c_2(m)\right]\right]$$
$$\times \exp\left[\sum_{m=1,3,\ldots}^{N-1} 2iK_x \left[c_1(m)c_2^\dagger(m)e^{-ik_m} + c_1^\dagger(m)c_2(m)e^{ik_m}\right]\right]. \tag{8.42}$$

We do not get any cc or $c^\dagger c^\dagger$ terms. This is because the sum over spatial sites (which enforces momentum conservation) vanishes unless the momenta of the two operators add to zero, and this is not possible for terms of this type with momenta of the same sign [Eq. (8.41)].

Now for the good news on combining exponentials: Although the terms in the two exponentials do not all commute, the terms at any one m commute with terms at any other m' since fermion bilinears at different momenta commute. Thus we may write T as a tensor product over momenta as

$$T_{\mathrm{E}} = \prod_{\otimes m=1,3,\ldots}^{N-1} T(m), \tag{8.43}$$

where

$$T(m) = \exp\left[-2iK_\tau^*(c_1(m)c_2^\dagger(m) + c_1^\dagger(m)c_2(m))\right]$$
$$\times \exp\left[2iK_x(c_1(m)c_2^\dagger(m)e^{-ik_m} + c_1^\dagger(m)c_2(m)e^{ik_m})\right]. \tag{8.44}$$

The operators c_α and $c_\alpha^\dagger$ live in a four-dimensional space, spanned by states $|00\rangle, |01\rangle, |10\rangle, |11\rangle$ labeled by the occupation numbers for $c_1^\dagger c_1$ and $c_2^\dagger c_2$. So we just need to combine exponentials of 4×4 matrices. It gets even better. Since $|00\rangle$ and $|11\rangle$ can neither be raised nor lowered, they get annihilated by the fermion bilinears in the exponent, and $T(m)$ becomes the unit operator in this subspace. So we just need to combine non-trivial 2×2 matrices in the $|01\rangle, |10\rangle$ subspace. For an application of these ideas, go to Exercise 8.4.2.

8.4.2 The Fermionic Transfer Matrix in the Odd Sector

The Hermiticity condition Eq. (8.40),

$$c_\alpha(k_m) = c_\alpha^\dagger(-k_m), \tag{8.45}$$

when applied to momenta that equal minus themselves, takes the form

$$c_\alpha(0) = c_\alpha^\dagger(0) \equiv \eta_\alpha(0), \tag{8.46}$$

$$c_\alpha(\pi) = c_\alpha^\dagger(\pi) \equiv \eta_\alpha(\pi), \tag{8.47}$$

where $\eta_{0,\pi}$ are Majorana fermions themselves. The Fourier expansion then takes the form

$$\psi_\alpha(n) = \frac{1}{\sqrt{N}}\left[\sum_{m=2,4,\ldots}^{N-2}\left[c_\alpha(m)e^{ik_m n} + c_\alpha^\dagger(m)e^{-ik_m n}\right] + \eta_\alpha(0) + \eta_\alpha(\pi)e^{i\pi n}\right]. \tag{8.48}$$

Once again, you may verify that this momentum expansion reproduces the anticommutation rules in real space.

The transfer matrix in the odd sector takes the form

$$T_O = \left[\prod_{\otimes m=2}^{N-2} T(m)\right] \cdot e^{(K_\tau^* - K_x)[-2i\eta_1(0)\eta_2(0)]} \cdot e^{(K_\tau^* + K_x)[-2i\eta_1(\pi)\eta_2(\pi)]}, \tag{8.49}$$

where, just as in the even sector,

$$\begin{aligned} T(m) &= \exp\left[-2iK_\tau^*(c_1(m)c_2^\dagger(m) + c_1^\dagger(m)c_2(m))\right] \\ &\times \exp\left[2iK_x(c_1(m)c_2^\dagger(m)e^{-ik_m} + c_1^\dagger(m)c_2(m)e^{ik_m})\right]. \end{aligned} \tag{8.50}$$

Equations (8.43)/(8.44) and (8.49)/(8.50) form the starting point for the exact solution of the model. The two quadratic forms can be diagonalized using Fourier transformation and can be easily combined at each k. Schultz *et al.* found the eigenvalues of the 2×2 blocks and obtained all the eigenvalues of T, paying due attention to the correlation between the boundary conditions and the parity of the state. We will follow the easier route of doing all this in the τ-continuum limit.

Exercise 8.4.2 *One can interpret [3] the non-trivial 2×2 blocks of $T(m)$ as the transfer matrix for a $d = 1$ Ising model with k_m dependent parameters:*

$$Z = \sum_{s_i} \exp\left[\sum_i \left[J(k_m)(s_i s_{i+1} - 1) + \frac{1}{2}h(k_m)(s_1 + s_{i+1}) + f_0(k_m)\right]\right]. \tag{8.51}$$

Show (turning to the reference for help if needed) that, in the isotropic case,

$$f_0(m) = \frac{1}{2}\ln\left[1 + S^{*2}(k_m)\cos^2(k_m)\right], \tag{8.52}$$

$$e^{-2J(m)} = \frac{S^* \sin k_m}{\sqrt{1 + S^{*2}\sin^2 k_m}}, \tag{8.53}$$

$$h(m) = \frac{1}{2}\ln\left[\frac{C^* + S^*\cos k_m}{C^* - S^*\cos k_m}\right] - 2K, \quad \textit{where} \tag{8.54}$$

$$C^* = \cosh 2K^*, \qquad S^* = \sinh 2K^*. \tag{8.55}$$

The free energy of the $d=2$ Ising model is then the sum (and when the momenta become some continuous k, the integral over k) of the free energies of these $d=1$ models.

The main aim of [3] was to consider a variant of what is called the McCoy–Wu model in which the horizontal bonds are the same within a row but vary randomly from row to row between two possible values K_1 and K_2. Verify that we can still use translational invariance in each row to write

$$T(i) = \prod_{\otimes m} T(m, K_i), \quad \text{where } i = 1 \text{ or } 2, \tag{8.56}$$

and

$$Z = Tr \prod_{i=1}^{\text{rows}} T(i) = \prod_{i} Tr \left[\prod_{\otimes m} T(m, K_i) \right], \tag{8.57}$$

so that in the end the free energy of the $d=2$ model with correlated bond randomness is the integral over free energies of $d=1$ Ising models with random bonds and fields. See the references at the end of the chapter for further details.

8.5 Qualitative Analysis in the τ-Continuum Limit

Having shown that we can solve the Ising model exactly for general anisotropic couplings and on a finite periodic lattice by mapping to fermions, we will now relax and solve it in the τ-continuum limit since it is a lot easier way to achieve the limited goal of this book, which is to describe certain universal features of the phase transition. Figure 8.1 shows the phase diagram for general coupling.

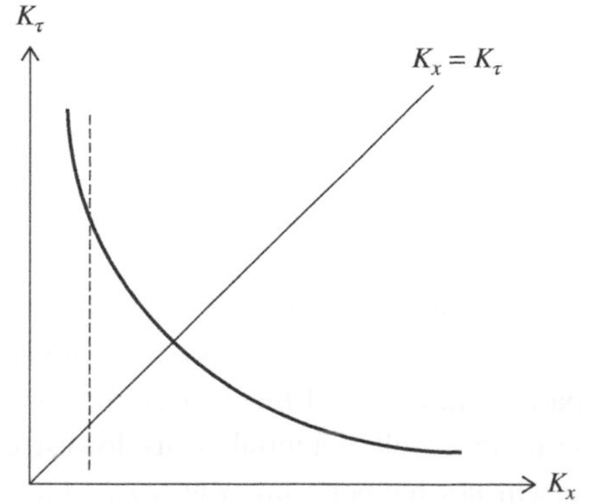

Figure 8.1 The parameter space for the anisotropic Ising model is the K_x–K_τ plane. The solid curve $\tanh K_\tau = e^{-2K_x}$ (shown schematically) separates the two phases. The region where both K's are small is disordered and the other is ordered. Each point in one phase has a dual in the other, and the phase boundary is the line of self-dual points. The isotropic model lies on the 45° line. The dotted vertical line at small (eventually infinitesimal) K_x is the range probed by the τ-continuum limit. Moving up and down this line we can still see the two phases and the transition.

In the limit

$$K_\tau^* = \lambda\tau, \quad K_x = \tau, \tag{8.58}$$

with $\tau \to 0$, we may freely combine the two exponentials in T (the errors being of order τ^2) to obtain, in terms of spins or fermions, the simplified version of T:

$$T = e^{-H\tau}, \quad \text{where} \tag{8.59}$$

$$H = \sum_1^N [-\lambda\sigma_1(n) - \sigma_3(n)\sigma_3(n+1)] \tag{8.60}$$

$$= \sum_1^N \left[2i\lambda\psi_1(n)\psi_2(n) - 2i\psi_1(n)\psi_2(n+1)\right], \tag{8.61}$$

where, in the fermionic version, the constraint $\psi(N+1) = \mp\psi(1)$ for $(-1)^{N_\Psi} = \pm 1$ is understood.

Before proceeding with the solution by Fourier transformation, let us understand some qualitative features of the τ-continuum Hamiltonian, most transparent in Eq. (8.60) written in terms of spins.

Our H contains non-commuting operators σ_1 and σ_3, just like the kinetic and potential terms in quantum mechanics made up of P's and X's that do not share any eigenstates.

When $\lambda = 0$, we can ignore the σ_1 terms and minimize the mutually commuting $-\sigma_3(n)\sigma_3(n+1)$ terms by choosing spins to be all up or all down. We denote by $|+++\cdots++\rangle \equiv |\Uparrow\rangle$ and $|----\cdots--\rangle \equiv |\Downarrow\rangle$ these two degenerate ground states or vacua. In the thermodynamic limit of the classical Ising model, the magnetization is the expectation value of $\sigma_3(0)$ in the ground state of H. (In a translationally invariant system we can obtain $\langle M\rangle$ from the average of σ_3 at any site, not necessarily the origin.) We find that

$$\langle M\rangle = \langle s\rangle = \langle 0|\sigma_3(0)|0\rangle = \pm 1. \tag{8.62}$$

These two degenerate ground states $|\Uparrow / \Downarrow\rangle$ clearly correspond to the fully ordered states of lowest classical energy that dominate Z at $T = 0$. Thus, $T = 0$ corresponds to $\lambda = 0$.

This is an example of spontaneous symmetry breaking: the Hamiltonian does not prefer one sign of σ_3 over another, and yet the system chooses one. It is, however, a trivial example, like the particle in classical mechanics choosing to sit still at one of two degenerate minima of the double-well potential in its lowest energy state. In quantum mechanics the same thing will not happen *unless we artificially turn off the kinetic term that does not commute with* X. But of course it is generally there, and causes tunneling between the two minima, producing a ground state that is a symmetric combination with $\langle X\rangle = 0$.

The Hamiltonian with $\lambda = 0$ is also classical in the sense that H has no non-commuting terms. So let us now turn on a small λ and treat the σ_1 terms perturbatively, starting with any one of the vacua, say $|\Uparrow\rangle$. The perturbation flips spins and the perturbed ground state is an admixture of states with down spins, which lower the magnetization $\langle M\rangle$ from $+1$. *Thus, increasing λ in H corresponds to increasing temperature in the classical Ising model.* At order $\mathcal{O}(N)$ the state $|\Downarrow\rangle$ with all spins down enters the mix, and in the end the ground state will end up having $\langle M\rangle = 0$. Thus, the finite system will not break the symmetry spontaneously for any non-zero λ. On the other hand, if we first let $N \to \infty$ and then do our perturbation theory, we will never connect the sectors built around $|\Uparrow\rangle$ and $|\Downarrow\rangle$, and we will end up with two degenerate ground states. Thus, spontaneous symmetry breaking requires an infinite system (assuming all other parameters are generic).

We turn to the other limit of $\lambda = \infty$. Now we rewrite Eq. (8.60) as

$$\frac{H}{\lambda} = -\sum_n \sigma_1(n) - \lambda^{-1}\sum \sigma_3(n)\sigma_3(n+1). \tag{8.63}$$

At large λ we just focus on the σ_1 term, which is minimized if at each site we choose the states with $\sigma_1 = +1$, which we denote by $|\Rightarrow\rangle$. In this *unique vacuum*, $\langle\sigma_3\rangle = 0$. This corresponds to the totally disordered state of the classical model at $T = \infty$, $K = 0$.

It makes sense that there are two ordered ground states at low T, there being two choices for the global spin, and just one disordered ground state at high T. Somewhere in between is the phase transition.

8.6 The Eigenvalue Problem of T in the τ-Continuum Limit

We know from our earlier analysis of the transfer matrix that there are actually two Hamiltonians that arise: one in the even sector,

$$\begin{aligned} H_{\mathrm{E}} = \sum_{m=1,3,\ldots}^{N-1} \Big[& 2i\lambda(c_1(m)c_2^\dagger(m) + c_1^\dagger(m)c_2(m)) \\ & -2i(c_1(m)c_2^\dagger(m)e^{-ik_m} + c_1^\dagger(m)c_2(m)e^{ik_m})\Big], \end{aligned} \tag{8.64}$$

with k_m restricted to the positive odd values for m,

$$k_m = \frac{m\pi}{N}, \qquad m = 1,3,\ldots,(N-1), \tag{8.65}$$

and one in the odd sector,

$$\begin{aligned} H_{\mathrm{O}} = \sum_{m=2,4,\ldots}^{N-2} \Big[& 2i\lambda(c_1(m)c_2^\dagger(m) + c_1^\dagger(m)c_2(m)) \\ & -2i(c_1(m)c_2^\dagger(m)e^{-ik_m} + c_1^\dagger(m)c_2(m)e^{ik_m})\Big] \\ & +2i(\lambda-1)\eta_1(0)\eta_2(0) + 2i(\lambda+1)\eta_1(\pi)\eta_2(\pi), \end{aligned} \tag{8.66}$$

with k_m restricted to positive even values for m,

$$k_m = \frac{m\pi}{N}, \quad m = 2, 4, \ldots, (N-2). \tag{8.67}$$

Our goal is to solve for the spectrum of each, with due attention to the restrictions imposed by the boundary conditions: the even (odd) sector should have an even (odd) number of particles N_Ψ:

$$N_\Psi = \sum_{n=1}^{N} \Psi^\dagger(n)\Psi(n). \tag{8.68}$$

The ground state of the Ising Hamiltonian will be the lowest (allowed) energy state, which could be from either sector. We will find that for $\lambda > 1$, the even sector provides the clear winner. For $\lambda < 1$, the even sector wins by an exponentially small amount in the finite system and the two solutions are tied *in the thermodynamic limit.*

The treatment can be tricky at times, and I ask you to work through it if you want to see how, once again, adherence to the mathematics gives us the correct physical answers in a most remarkable fashion.

8.6.1 Ground State Energy in the Even Sector

We may write in matrix notation:

$$H_{\mathrm{E}} = \sum_{k_m > 0} \left(c_1^\dagger(m), c_2^\dagger(m)\right) \begin{pmatrix} 0 & 2i(\lambda - e^{ik_m}) \\ -2i(\lambda - e^{-ik_m}) & 0 \end{pmatrix} \begin{pmatrix} c_1(m) \\ c_2(m) \end{pmatrix}. \tag{8.69}$$

Do not be fooled by the 2×2 matrix involving operators; at each m, we have a *four*-dimensional space labeled by the number operators $n_1(m) = c_1^\dagger(m)c_1(m)$ and $n_2(m) = c_2^\dagger(m)c_2(m)$. *Suppressing the label m when it is apparent*, we write:

$$|n_1, n_2\rangle = |00\rangle, \quad \text{where } c_1|00\rangle = c_2|00\rangle = 0, \tag{8.70}$$

$$|n_1, n_2\rangle = |10\rangle = c_1^\dagger|00\rangle, \tag{8.71}$$

$$|n_1, n_2\rangle = |01\rangle = c_2^\dagger|00\rangle, \tag{8.72}$$

$$|n_1, n_2\rangle = |11\rangle = c_1^\dagger c_2^\dagger|00\rangle. \tag{8.73}$$

The total of number of c-type particles (at each m),

$$N_c(m) = n_1 + n_2 = c_1^\dagger(m)c_1(m) + c_2^\dagger(m)c_2(m), \tag{8.74}$$

has values $(0, 1, 1, 2)$ in the four states above.

The Hamiltonian H_{E} may be diagonalized by the following unitary transformation:

$$c = U\eta, \quad \text{or, in more detail,} \tag{8.75}$$

$$\begin{pmatrix} c_1(m) \\ c_2(m) \end{pmatrix} = \frac{1}{\sqrt{2}} \begin{pmatrix} 1 & -e^{i\theta} \\ e^{-i\theta} & 1 \end{pmatrix} \begin{pmatrix} \eta_+(m) \\ \eta_-(m) \end{pmatrix}, \tag{8.76}$$

$$e^{i\theta} = \frac{i(\lambda - e^{ik_m})}{|(i(\lambda - e^{ik_m})|}. \tag{8.77}$$

The final result is

$$H_{\mathrm{E}} = \sum_{k_m > 0} \left[\eta_+^\dagger(m)\eta_+(m)\varepsilon(k_m) - \eta_-^\dagger(m)\eta_-(m)\varepsilon(k_m) \right], \tag{8.78}$$

$$\varepsilon(k) = 2\sqrt{1 - 2\lambda\cos k + \lambda^2}. \tag{8.79}$$

Observe that $\varepsilon(k)$ is always positive and that it is an even function of its argument:

$$\varepsilon(k) = \varepsilon(-k). \tag{8.80}$$

(Although so far we have invoked this function only for $k \geq 0$, we will soon bring in $k < 0$ and use this symmetry.)

Exercise 8.6.1 *Derive Eqs. (8.78) and (8.79), employing the unitary transformation Eq. (8.76).*

Typically, this is when we would declare victory since we have diagonalized the quadratic Hamiltonian. We can read off not just the ground state, but all excited states. But we are not done yet, because of the restriction that the states in the even sector should have even values of N_Ψ. To this end, we have to relate N_Ψ to N_c and N_η, the number of c and η particles.

First of all, since c and η are related by a unitary transformation, we have (at each m):

$$N_c = c_1^\dagger c_1 + c_2^\dagger c_2 = \eta_+^\dagger \eta_+ + \eta_-^\dagger \eta_- = N_\eta, \tag{8.81}$$

so we can use N_c or N_η to describe the parity. I will now show that

$$e^{i\pi N_\Psi} = (-1)^{N_\Psi} = (-1)^{N_c} (= (-1)^{N_\eta}). \tag{8.82}$$

Start with

$$N_\Psi = \sum_1^N \Psi^\dagger(n)\Psi(n) \tag{8.83}$$

$$= \sum_1^N \frac{1 + 2i\psi_1(n)\psi_2(n)}{2} \tag{8.84}$$

$$= \sum_1^N \frac{1}{2} + \sum_{m>0} \left[i(c_1^\dagger(m)c_2(m) - c_2^\dagger(m)c_1(m) \right]. \tag{8.85}$$

Let us assume that N is a multiple of 4 so that the $\frac{1}{2}$ can be dropped mod 2.

We see a problem here: the number N_Ψ is not simply a sum over the fermions in the momentum states. Indeed, the operator we sum over is not even diagonal in the indices 1 and 2. How are we to select the allowed states if we are not working with states of definite N_c or N_η? We are redeemed by the fact that what we really want is the parity $e^{i\pi N_\Psi}$ and not N_Ψ.

First, we see that on the empty and doubly occupied states (of c) on which $(-1)^{N_c} = +1$:

$$\exp\left[i\pi(i(c_1^\dagger(m)c_2(m) - c_2^\dagger(m)c_1(m))\right]|00\rangle = 1\cdot|00\rangle, \tag{8.86}$$

$$\exp\left[i\pi(i(c_1^\dagger(m)c_2(m) - c_2^\dagger(m)c_1(m))|11\rangle\right] = 1\cdot|11\rangle, \tag{8.87}$$

because when we expand out the exponential, only the 1 survives and the c or $c^\dagger$ annihilate the states to their right. So $e^{i\pi N_c} = e^{i\pi N_\Psi}$ in this sector.

Next, consider the states $|01\rangle$ and $|10\rangle$ on which $(-1)^{N_c} = -1$. On these states, $i(c_1^\dagger(m)c_2(m) - c_2^\dagger(m)c_1(m))$ acts like the second Pauli matrix, call it τ_y. Since the eigenvalues of τ_y are ± 1, on this sector

$$e^{i\pi\tau_y} = e^{\pm i\pi} = -1, \tag{8.88}$$

which is exactly the action of $e^{i\pi N_c}$.

Thus, in all four states at every m we find

$$e^{i\pi N_\Psi} = e^{i\pi N_c} = e^{i\pi N_\eta}. \tag{8.89}$$

In summary, the states in the even sector should contain an even number of η particles.

Before imposing the parity condition, we are going to make some notational changes in Eq. (8.78). Right now it uses only half the momenta ($k > 0$), and half the oscillators have negative excitation energies.

Let us trade the two Dirac operators $\eta_\pm$ (and their adjoints) at each $m > 0$ for just one η (with no subscript) but for every allowed m positive and negative. We make the following assignment, which preserves the anticommutation rules:

$$\eta(m) = \eta_+(m), \quad \eta^\dagger(m) = \eta_+^\dagger(m) \quad (m > 0), \tag{8.90}$$

$$\eta(-m) = \eta_-^\dagger(m), \quad \eta^\dagger(-m) = \eta_-(m) \quad (m > 0). \tag{8.91}$$

Under this relabeling,

$$N_\eta = \sum_{m>0}(\eta_+^\dagger(m)\eta_+(m) + \eta_-^\dagger(m)\eta_-(m)) \tag{8.92}$$

$$= \sum_{m>0}(\eta^\dagger(m)\eta(m) + \eta(-m)\eta^\dagger(-m)) \tag{8.93}$$

$$= \sum_{m>0}(\eta^\dagger(m)\eta(m) - \eta^\dagger(-m)\eta(-m) + 1) \tag{8.94}$$

$$= \sum_{m>0} (\eta^\dagger(m)\eta(m) + \eta^\dagger(-m)\eta(-m)) \tag{8.95}$$

$$= \sum_{m} \eta^\dagger(m)\eta(m), \tag{8.96}$$

where I have dropped the 1 in the sum over $m > 0$ (because it equals 0 mod 2, since we have assumed that N is a multiple of 4) and reversed the sign of the integer-valued operator $-\eta^\dagger(-m)\eta(-m)$, which is allowed mod 2.

The diagonalized Hamiltonian $H_{\rm E}$ responds as follows to this relabeling:

$$\begin{aligned} H_{\rm E} &= \sum_{k_m>0} \left[\eta_+^\dagger(m)\eta_+(m)\varepsilon(k_m) - \eta_-^\dagger(m)\eta_-(m)\varepsilon(k_m)\right] \\ &= \sum_{k_m>0} \left[\eta^\dagger(m)\eta(m)\varepsilon(k_m) - \eta(-m)\eta^\dagger(-m)\varepsilon(-k_m)\right] \\ &= \sum_{k_m>0} \left[\eta^\dagger(m)\eta(m)\varepsilon(k_m) + (\eta^\dagger(-m)\eta(-m) - 1)\varepsilon(-k_m)\right] \\ &= \sum_{k_m} \left[\eta^\dagger(m)\eta(m) - \frac{1}{2}\right]\varepsilon(k_m). \end{aligned} \tag{8.97}$$

Now all the excitation energies are positive and the allowed momenta are symmetric between positive and negative values.

This is as simple a form as $H_{\rm E}$ can assume. We are now ready to find the ground state in the even sector.

Since $\varepsilon > 0$, the lowest energy state is when $N_\eta(m) = \eta^\dagger(m)\eta(m) = 0 \; \forall \, m$. In the ground state, we keep all levels empty. This is an allowed state since 0 is even.

The lowest energy in the even sector is

$$E_{\rm E}^{\rm min} = -\frac{1}{2}\sum_{k_m} \varepsilon(m) = -\sum_{k_m} \sqrt{1 - 2\lambda \cos k_m + \lambda^2}. \tag{8.98}$$

The allowed excited states from this ground state must have an even number of fermions.

8.6.2 The Ground State Energy in the Odd Sector

Luckily, we have laid most of the groundwork in the even sector, but there are some inevitable changes. First, the number operator N_Ψ now has extra contributions at $k = 0$ and $k = \pi$ that go into themselves under sign change.

Starting with Eq. (8.66), and repeating what we did in the odd sector for all modes except the ones at 0 and π, we find:

$$\begin{aligned} H_{\rm O} = &\sum_{k_m \neq 0,\pi} \left[\eta^\dagger(m)\eta(m) - \frac{1}{2}\right]\varepsilon(k_m) \\ &+ 2i(\lambda - 1)\eta_1(0)\eta_2(0) + 2i(\lambda + 1)\eta_1(\pi)\eta_2(\pi). \end{aligned} \tag{8.99}$$

In terms of

$$\eta(0,\pi) = \frac{\eta_1(0,\pi) + i\eta_2(0,\pi)}{\sqrt{2}}, \tag{8.100}$$

we may finally write

$$H_{\rm O} = \sum_{k_m} \left[\eta^\dagger(m)\eta(m) - \frac{1}{2} \right] \varepsilon(k_m), \tag{8.101}$$

with the understanding that

$$\varepsilon(0) = 2(\lambda - 1), \quad \varepsilon(\pi) = 2(\lambda + 1). \tag{8.102}$$

Unlike other eigenvalues, $\varepsilon(0)$ is not necessarily positive. In particular,

$$\varepsilon(0) = 2(\lambda - 1) = \lim_{k\to 0} \varepsilon(k) \text{ for } \lambda > 1 \tag{8.103}$$

$$= 2(\lambda - 1) = -\lim_{k\to 0} \varepsilon(k) \text{ for } \lambda < 1. \tag{8.104}$$

If we compute N_Ψ, we find that the two Majorana modes contribute precisely $\eta_0^\dagger \eta_0 + \eta_\pi^\dagger \eta_\pi$, so that once again the parity of N_Ψ equals that of N_η. Therefore we have, once again,

$$e^{i\pi N_\Psi} = e^{i\pi N_\eta}, \tag{8.105}$$

and we want to have an odd number of η particles.

Consider the lowest energy state first. We clearly do not fill any of the states with $k_m \neq 0$, which allows these modes to contribute $-\frac{1}{2}\varepsilon(k_m)$ to the ground state energy. But we need an odd number, and the cheapest way is to fill the mode at $k = 0$. This makes a contribution $(\lambda - 1)$ to the energy. However, the implications are quite sensitive to whether $\lambda > 1$ or $\lambda < 1$. We will consider them in turn.

8.6.3 Asymptotic Degeneracy for $\lambda < 1$

First, consider $\lambda < 1$, when filling the $k_m = 0$ level is a good thing, for it adds a term $(\lambda - 1) = -|\lambda - 1| = -\frac{1}{2}\varepsilon(0^+)$ to the energy. The total ground state energy in the odd sector is then

$$E_{\rm O}^{\rm min} = -\frac{1}{2} \sum_{k_m} \varepsilon(m) = -\sum_{k_m} \sqrt{1 - 2\lambda \cos k_m + \lambda^2}, \tag{8.106}$$

where $k = 0$ is included.

When we compare this to the even sector, we find each ground state is a sum over N negative contributions, determined by the same function $-\sqrt{1 - 2\lambda \cos k_m + \lambda^2}$, but evaluated on slightly shifted interpenetrating sets of points, as illustrated in Figure 8.2 for the case $N = 8$. In the limit $N \to \infty$ it can be shown that the two sums are equal to within errors that vanish exponentially in N. We have *asymptotic degeneracy* of the ground state in the thermodynamic limit for $\lambda < 1$. These correspond to the two choices for magnetization.

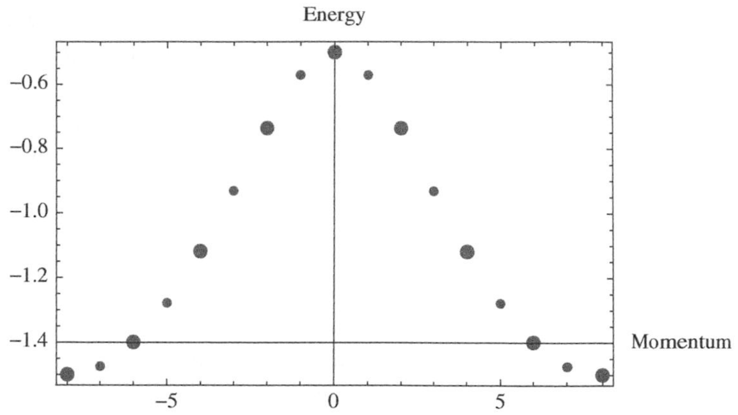

Figure 8.2 The points contributing to the ground state energy in the odd sector (big dots) and even sector (small dots) for $N = 8$ and $\lambda = 0.5$. The $k = 0$ mode is occupied and contributes a negative number $\lambda - 1$. The points $\pm\pi$ are both shown, though only one should be counted. At $N = 8$, the even sector has the lower energy – see Exercise 8.6.3. In the limit $N \to \infty$, sums in either sector approach the integral with differences exponentially small in N. However, if $\lambda > 1$ the point $k = 0$ that we are forced to occupy will contribute a positive amount $\lambda - 1$, the odd sector will clearly lose, and there will be a unique even ground state.

The even and odd ground states have even and odd parity under the flipping of all spins induced by the action of $\mathcal{P} = \prod \sigma_1(n)$. Let us write

$$|\text{even/odd}\rangle = \frac{|\text{“}\Uparrow\text{"}\rangle \pm |\text{“}\Downarrow\text{"}\rangle}{\sqrt{2}}, \tag{8.107}$$

$$\mathcal{P}|\text{even/odd}\rangle = \pm|\text{even/odd}\rangle, \tag{8.108}$$

where the quotes remind us that the states are polarized in the direction shown, but not fully polarized like the states $|\Uparrow / \Downarrow\rangle$.

At any finite N, the ground state from the even sector, symmetric under spin reflection, will be lower in energy, just as in the double well. (See Exercise 8.6.3.) The average $\langle M\rangle$ then vanishes in both even and odd ground states since an odd operator cannot have an average among parity eigenstates. This is like saying that in the double well with any tunneling allowed, $\langle X\rangle = 0$ in the symmetric and antisymmetric states of the lowest energy.

In the thermodynamic limit the states become degenerate, and mathematically any linear combination is a ground state, with a magnetization that ranges from positive to zero to negative depending on the admixture. But physically we must choose the ones with a definite sign of magnetization, for only these will obey clustering, a property expected from general locality arguments (recall the discussion in Chapter 1):

$$\lim_{|i-j|\to\infty} \langle \sigma_3(i)\sigma_3(j)\rangle \to \langle \sigma_3(i)\rangle\langle \sigma_3(j)\rangle. \tag{8.109}$$

In states of definite parity, clustering is absent since the left-hand side will approach the square of the magnetization while both factors on the right-hand side will vanish by symmetry.

To choose one or the other sector amounts in the classical Ising model to summing over only half the configurations in the Boltzmann sum. This is correct physically, since in the thermodynamic limit a system that starts out in one sector will not sample the other in any finite time – the time evolution is *non-ergodic*.

Exercise 8.6.2 *To understand clustering, consider the extreme case of $\lambda = 0$ and the ground states $|\Uparrow\rangle$ and $|\Downarrow\rangle$.*
(i) First show that they both obey clustering. (Remember that in these states every spin is up or down, so that $\sigma_3(i)\sigma_3(j)$ is trivially evaluated.)
(ii) Now consider states

$$|\theta\rangle = \cos\theta|\Uparrow\rangle + \sin\theta|\Downarrow\rangle. \tag{8.110}$$

Show that

$$\langle\theta|\sigma_3(i)\sigma_3(j)|\theta\rangle = 1, \tag{8.111}$$

$$\langle\theta|\sigma_3(i)|\theta\rangle = \langle\theta|\sigma_3(j)|\theta\rangle = \cos 2\theta, \tag{8.112}$$

and that clustering is obtained only for the so-called pure states *for which* $\cos 2\theta = \pm 1$, *which correspond to the states $|\Uparrow\rangle$ and $|\Downarrow\rangle$ up to a phase.*

Exercise 8.6.3 *Consider $N = 4$. List the allowed momenta in the even and odd sectors. Compute (numerically) the ground state energy in the even and odd sectors. These are not equal since N is only 4, but show that for very small λ they are very close. List all the allowed states, with eight coming from each sector with their energies, labeling them by the occupation numbers for η.*

8.6.4 The Unique Ground State of $\lambda > 1$

Now we pass to the high-temperature phase $\lambda > 1$. Things change when λ exceeds 1: in the odd sector the mandatory filling of the $k_m = 0$ mode (the cheapest way to obtain the desired odd parity) now costs a positive energy $\lambda - 1$, and the sum over mode energies is higher than the best we can do in the even sector. (Had we left the $k = 0$ state empty, its contribution $-(\lambda - 1)$ to the ground state energy would have made it asymptotically degenerate with the even sector. But that state is unphysical. In the physical sector we have to occupy this state, and that makes this sector higher by $2(\lambda - 1)$.) Thus there is a unique ground state in the high-temperature $\lambda > 1$ phase that is invariant under spin reversal.

The discontinuous behavior at $\lambda = 1$ tells us that it is the critical point separating the high and low temperature phases.

8.6.5 Summary of Ground States

The classification of various sectors and ground states was sufficiently complicated to warrant a short summary.

The way to handle the non-local parity operator was to divide the Hilbert space into its two eigen-sectors, even and odd under the reflection of all spins.

In the even sector the momenta that yielded the requisite antiperiodic boundary conditions came in equal and opposite pairs. In the odd sectors there were two special momenta, 0 and π, that were their own negatives.

The even (odd) sector had to have an even (odd) number of η particles in terms of which H was diagonal. There was one fermion associated with each k, positive, negative, or zero.

In the even sector, since every energy $\varepsilon(k)$ was positive, the ground state had 0 particles and the ground state energy was the sum of negative zero-point energies, $E_{\text{E}}^{\text{min}}$. This was true for all λ.

In the odd sector, since every energy $\varepsilon(k)$ was positive for $k > 0$, none of those states were filled. But the ground state in the odd sector had to have at least one particle, and this had to be at $k = 0$, which had the smallest energy $\frac{1}{2}\varepsilon(0) = \lambda - 1$.

For $\lambda < 1$, *filling* this state *lowered* the ground state energy and tied it with the ground state from the odd sector in the thermodynamic limit. In this limit one could add and subtract the even and odd states to form states obeying clustering and polarized in two possible directions.

For $\lambda > 1$, filling the $k = 0$ state raised the ground state energy above that of the even sector, leaving us with a unique spin-reversal-invariant (disordered) ground state in the high-temperature phase.

Figure 8.3 summarizes the situation.

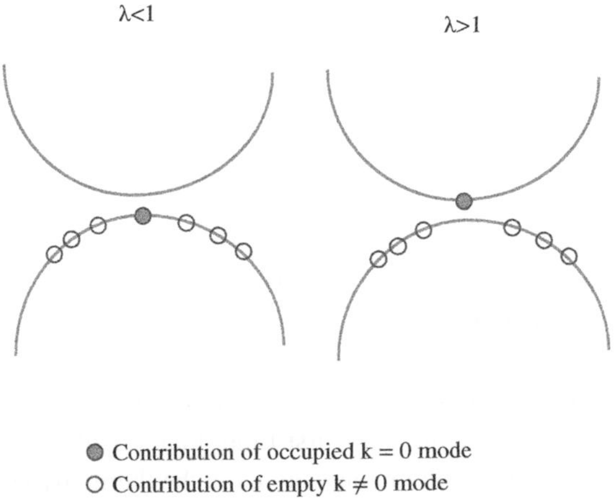

Figure 8.3 The role of the $k = 0$ mode, which has to be occupied in the odd sector. For $\lambda < 1$, the occupied state is flanked by empty $k \neq 0$ modes, making this sector asymptotically degenerate with the even sector. As λ increases, the two bands ($\pm\varepsilon(k)$) approach each other, touch at $\lambda = 1$, and then separate. The zero mode, which keeps moving up the whole time, joins the upper band for $\lambda > 1$, making the odd sector higher in energy.

8.7 Free Energy in the Thermodynamic Limit

At this point we shift our emphasis from the model at finite N to the limit $N \to \infty$, which is when we hope to see a singularity corresponding to the phase transition.

Let us recall from Chapter 1 that in a lattice with M rows and N columns the partition function is

$$Z = \lim_{M \to \infty} \mathrm{Tr}\, T^M \to \lambda_0^M = e^{-ME_0\tau}, \tag{8.113}$$

where λ_0 is the largest eigenvalue of T, E_0 is the smallest eigenvalue of H, and $T = e^{-H\tau}$. The free energy per site approaches

$$-\beta f \to \frac{1}{MN} \ln Z = -\frac{1}{N} E_0 \tau. \tag{8.114}$$

For $\lambda < 1$ there are two degenerate eigenvalues $E_{\mathrm{E}}^{\min} = E_{\mathrm{O}}^{\min} = E_0$ in this limit, so that $Z = 2e^{-ME_0\tau}$. The factor of 2 can be forgotten since it contributes $\ln 2/MN$ to $-\beta f$.

So we write, uniformly for all λ,

$$\beta f = \frac{1}{N} E_0 \tau = -\tau \frac{1}{N} \sum_{k_m} \frac{1}{2} \varepsilon(k_m). \tag{8.115}$$

We can forget the difference between the even and odd sectors since both sums over momenta become approximations to the integral of $\varepsilon(k)$ over k. Using the usual recipe (valid when the k points are spaced $2\pi/N$ apart),

$$\frac{1}{N} \sum_{k_m} \to \int_{-\pi}^{\pi} \frac{dk}{2\pi}, \tag{8.116}$$

we obtain, on recalling that $\tau = K_x$,

$$\beta f = -K_x \int_{-\pi}^{\pi} \sqrt{1 - 2\lambda \cos k + \lambda^2}\, \frac{dk}{2\pi}. \tag{8.117}$$

The most important goal before us is the demonstration that the free energy is singular at the critical point $\lambda = 1$ and to extract the nature of the singularity. In doing this we will pay no attention to overall prefactors or additive terms that are analytic, say due to the factors of $\left[\cosh K_\tau^*\right]^N$ that were unceremoniously jettisoned from T.

There are two reasons behind this choice of goal. The first is a matter of principle, to demonstrate, as Onsager did for the very first time, that singularities can arise from the seemingly innocuous sum over Boltzmann weights in the thermodynamic limit.

The second is to pave the way for the study of critical phenomena in general, where it is the nature of the singularity that matters, and not its prefactors. Thus, if f behaves as

$$f \simeq A_\pm (T - T_c)^{2/3} + 36T^2$$

near the critical temperature T_c (where $\pm$ refer to the two sides of the transition), it is the exponent $\frac{2}{3}$ that concerns us, not the value of T_c or the analytic term $36T^2$. The prefactors $A_\pm$ are not of interest either, but the ratio of such factors on the two sides of the transition is.

The widely used symbol to denote the deviation from criticality is t, which can be written in many equivalent ways:

$$t = \frac{T - T_c}{T_c} \simeq \lambda - \lambda_c. \tag{8.118}$$

With this in mind, we focus on the stripped-down expression for f_s, the singular part of the free energy:

$$f_s = -\int_0^\pi \sqrt{\lambda^2 - 2\lambda \cos k + 1}\, dk. \tag{8.119}$$

In addition to constant factors like π, I have also dropped temperature-dependent factors like K_x and β, which are regular at T_c.

Consider

$$f_s = -\int_0^\pi \sqrt{(\lambda - 1)^2 + 2\lambda(1 - \cos k)}dk \tag{8.120}$$

$$= -\int_0^\pi \sqrt{t^2 + k^2}dk, \tag{8.121}$$

where I have set $\lambda = 1$ where it is harmless to do so and recognized that since the singularity as $t \to 0$ resides at small k, we may approximate $\cos k \simeq 1 - \frac{1}{2}k^2$. Letting $k = t \sinh z$, we find, ignoring prefactors and analytic corrections (see Exercise 8.7.1),

$$f_s = t^2 \ln |t| + \text{regular terms}. \tag{8.122}$$

Exercise 8.7.1 *Provide the missing steps between Eqs. (8.121) and (8.122).*

The dominant term in the specific heat, as per Eq. (1.93), is

$$C_V = -T\frac{d^2 f}{dT^2} \simeq -\frac{d^2 f}{dt^2} \simeq -\ln |t|, \tag{8.123}$$

which is a landmark result.

The same singularity arises in the $\lambda > 1$ side, where there is a unique ground state.

8.8 Lattice Gas Model

When we solve the Ising model we solve a related problem for free. Consider a square lattice on the sites of which we can have zero or one particle: $n = 0, 1$. There is a chemical potential μ that controls the total number, and a coupling J that decides the energy cost of occupying neighboring sites $\langle ij \rangle$. The energy assigned to a configuration is

$$E = -\sum_{\langle ij \rangle} \left[4Jn_i n_j - \mu n_i \right]. \tag{8.124}$$

We can map this into an Ising model by defining an Ising spin

$$n_i = \frac{1+s_i}{2} = 0 \text{ or } 1, \tag{8.125}$$

and rewriting the energy (up to an irrelevant additive constant) as

$$E = -\sum_{\langle ij \rangle} \left[Js_i s_j + (2J - \frac{1}{2}\mu) s_i \right]. \tag{8.126}$$

The fully ordered phases $s_i = 1\ \forall i$ and $s_i = -1\ \forall i$ correspond to fully occupied and fully empty sites. If the "magnetic field" $h = (2J - \frac{1}{2}\mu) = 0$, the system can have a phase transition as a function of J or $K = \beta J$. I leave it to you to figure out the analogies with the Ising model observables like χ, the susceptibility, correlation length, and so on. The same exponents will appear, since it is just the Ising model in disguise.

8.9 Critical Properties of the Ising Model

Having encountered our first example of critical behavior in the form of the divergent specific heat, we are ready for a first pass at this subject, using the Ising model as a concrete example. There are excellent reviews available, the canonical ones being due to Kadanoff [5] and Fisher [6].

A critical point is one that separates distinct phases, the one under study being the transition from a magnetized state to an unmagnetized one. Of special interest to us are *second-order phase transitions* in which the *order parameter* (the magnetization $\langle M \rangle$ in our example) is non-zero in one phase and zero in the other, and vanishes continuously at the transition. In our case the order parameter also spontaneously breaks (up/down) symmetry in the ordered side. In addition, the correlation length diverges at the transition. By contrast, in a *first-order transition*, the order parameter drops to zero discontinuously at the critical point, and the correlation length remains finite. The divergent correlation length means what you think it does: a very large number of variables, eventually infinite, fluctuate in a correlated way.

The study of such transitions has revealed a feature called *universality*, which refers to the following. At and near the critical point one defines the following exponents, where h is the magnetic field:

$$C_V \simeq C_{\pm}|t|^{-\alpha}, \tag{8.127}$$

$$\langle M \rangle \simeq |t|^{\beta} \text{ as } t \to 0^-, \tag{8.128}$$

$$\chi = \left. \frac{\partial \langle M \rangle}{\partial h} \right|_{h=0} \simeq \chi_{\pm}|t|^{-\gamma}, \tag{8.129}$$

$$\xi \simeq \xi_{\pm}|t|^{-\nu}, \tag{8.130}$$

$$G(r) \to \frac{1}{r^{d-2+\eta}} \text{ at } t=0 \text{ in } d \text{ dimensions}, \tag{8.131}$$

$$\langle M \rangle \simeq h^{1/\delta} \text{at } t=0. \tag{8.132}$$

These have been written for the magnetic example. In general, $\langle M \rangle$ will be replaced by the order parameter and h by whatever field couples to the order parameter. For example, in the liquid–vapor transition, $(\rho-\rho_c)/\rho_c$ (where ρ is the density) plays the role of the order parameter $\langle M \rangle$, while $(P-P_c)/P_c$, (where P is the pressure) plays the role of h, the magnetic field. Occasionally other exponents arise, but these will do for us.

What was noticed experimentally was that these exponents were universal in the sense that they did not change as the microscopic interactions were modified: the exponents for the Ising model did not depend on the anisotropy, did not change with the addition of a reasonable amount of second-neighbor or multi-spin interactions, did not even depend on whether we were talking about spins or another problem with the same symmetry (under flipping the sign of the order parameter in the liquid–vapor transition).

There are, however, limits to universality: the exponents certainly varied with the number of spatial dimensions and the symmetries. For example, in problems with $O(N)$ symmetry they varied with N. (The Ising model corresponds to $O(1)=Z_2$.) Finally, it is the exponents that were universal, not T_c or the prefactors.

We will devote quite some time to understanding universality in later chapters. For now, let us see how many exponents we know for the Ising model.

We have just seen that $C_V \simeq |t|^{-\alpha}$ diverges logarithmically. So, one says

$$\alpha = 0^+ \tag{8.133}$$

to signify a divergence that is weaker than any power.

Consider next ν, which describes the divergence of the correlation length. Though the model has been solved completely, it is a non-trivial task to read off ν. The reason is that we want the correlation function of the spins, but what is simple is the correlation function of the Majorana fermion, and the two have a highly non-local relationship. Consider, for example, two spins separated by a distance n in the x direction. Then, recalling the Jordan–Wigner transformation Eq. (8.15),

$$G(n) = \langle \sigma_3(1)\sigma_3(n+1)\rangle \tag{8.134}$$

$$= \langle \sigma_3(1)\sigma_3(2)\sigma_3(2)\sigma_3(3)\sigma_3(3)\cdots\sigma_3(n+1)\rangle \tag{8.135}$$

$$= \left\langle \prod_1^n [2i\psi_1(i)\psi_2(i+1)] \right\rangle. \tag{8.136}$$

One can use Wick's theorem to evaluate this product, but the number of ψ's grows linearly with the separation n. The problem is, however, surmountable, and the behavior of $G(n \to \infty)$ has been calculated at and near criticality. You are directed to Exercise 9.3.3 for a taste of what is involved.

Unlike the case of σ_3, it is far easier to find correlators of $\sigma_1 = -2i\psi_1\psi_2$ since it is just a bilinear in the fermions. Choosing points separated by $n\tau$ in the τ-direction, we have

$$G(n_\tau) = \langle \mathcal{T}\sigma_1(n_\tau,0)\sigma_1(0,0)\rangle, \tag{8.137}$$

where the average is taken in the ground state $|\text{vac}\rangle$, the η vacuum. By inserting intermediate states j of energy E_j containing two fermions, we get a non-zero contribution

$$G(n_\tau) = \sum_j |\langle j|\sigma_1|\text{vac}\rangle|^2 e^{-E_j n_\tau \tau}. \tag{8.138}$$

The dominant term as $n \to \infty$ comes from a two-particle state of lowest energy, namely $4|\lambda - 1|$, and

$$G(n_\tau) \to Ae^{-4n_\tau\tau|\lambda-1|} = Ae^{-4n_\tau\tau|t|} \equiv Ae^{-n_\tau\tau/\xi(t)}, \tag{8.139}$$

which tells us that

$$\xi(t) \simeq |t|^{-1} \Rightarrow \nu = 1. \tag{8.140}$$

We have assumed that the correlation length for σ_1 diverges like that for σ_3. This happens to be true, but is not obvious given the very different dependencies of these operators on the fermions.

Consider the exponent β that describes how $\langle M\rangle$ vanishes as we approach $\lambda = 1$ from below. There are two routes: the hard one and the really hard one.

The first is to argue that the spin–spin correlation will approach $\langle M\rangle^2$ asymptotically as the separation grows to infinity. This is presumably how Onsager found it, though he never published his derivation, deciding instead to announce, at a meeting, the following exact expression for the magnetization in the general anisotropic case:

$$|\langle M\rangle| = \left[1 - \frac{1}{\sinh^2 2K_x \sinh^2 2K_\tau}\right]^{\frac{1}{8}}. \tag{8.141}$$

It is easy to show that, near the critical point in the τ-continuum limit,

$$|\langle M\rangle| \simeq |t|^{\frac{1}{8}}. \tag{8.142}$$

Thus,

$$\beta = \frac{1}{8}. \tag{8.143}$$

Exercise 8.9.1 *Let K_x and $K_\tau^* = \lambda K_x$ both go to zero. Show that $|\langle M\rangle|$ vanishes like $(1-\lambda)^{\frac{1}{8}}$ as $\lambda \to 1$.*

The second way, due to C. N. Yang [7], is to compute the free energy in the presence of a small magnetic field h and take the derivative with respect to h as $h \to 0$. The computation is very hard, even according to Yang.

We next turn to the Fisher exponent η that characterizes the decay of the spin–spin correlation at $\lambda = 1$. The direct way to evaluate this is to consider once again longer and longer strings of fermions in the critical theory. The result, quoted without proof, is

$$\eta = \frac{1}{4}. \tag{8.144}$$

I will provide a derivation of this later in the book. For now, let us assume it and move on.

Consider the divergence of susceptibility $\chi \simeq |t|^{-\gamma}$. In the classical two-dimensional language, on an $M \times N$ lattice,

$$\chi(t) = \frac{1}{MN}\frac{\partial^2 \ln Z}{\partial h^2} \tag{8.145}$$

$$= \frac{1}{MN}\left\langle \sum_{n_x,n_y} s(n_x,n_y) \sum_{n'_x,n'_y} s(n'_x,n'_y)\right\rangle_c \tag{8.146}$$

$$= \sum_{\boldsymbol{r}} G(\boldsymbol{r}) \tag{8.147}$$

$$\simeq \int_0^\infty r dr \frac{e^{-r/\xi(t)}}{r^{\frac{1}{4}}} \tag{8.148}$$

$$\simeq A\xi^{\frac{7}{4}} \simeq |t|^{-\frac{7}{4}}, \tag{8.149}$$

where: (i) in Eq. (8.147), $\boldsymbol{r}$ denotes the vector separation between (n'_x, n'_y) and (n_x, n_y) and translation invariance has been invoked; (ii) in Eq. (8.148), we have assumed that G falls like $r^{-1/4}$ for $r < \xi$ (i.e., ignored the falling exponential) and ignored the entire contribution for $r > \xi$, i.e., cut the integral off at $r = \xi$ due to the exponential fall-off; and finally (iii) scaled r by ξ and set the integral over $z = r/\xi$ equal to A in Eq.(8.149). The crude modeling in part (ii) gives the right exponent γ, which we know from a more careful analysis:

$$\gamma = \frac{7}{4}. \tag{8.150}$$

We will not try to derive the result $\delta = 15$. (Recall $\langle M\rangle \simeq h^{1/\delta}$ at $t = 0$.) We will later on discuss some relations between exponents, valid quite generally and not for just the $d = 2$ Ising model. One such example is

$$\alpha + 2\beta + \gamma = 2; \qquad \text{for Ising: } 0 + \tfrac{1}{4} + \tfrac{7}{4} = 2. \tag{8.151}$$

These relations allow us to express all the exponents in terms of two independent ones. They will let us compute δ in terms of the ones we have discussed so far.

Here is what we have for the Ising model:

$$\alpha = 0^+, \tag{8.152}$$

$$\beta = \frac{1}{8}, \tag{8.153}$$

$$\gamma = \frac{7}{4}, \tag{8.154}$$

$$\delta = 15, \tag{8.155}$$

$$\eta = \frac{1}{4}, \tag{8.156}$$

$$\nu = 1. \tag{8.157}$$

8.10 Duality in Operator Language

Recall that Kramers–Wannier duality, which we studied in Section 7.4, related the model of classical spins on a two-dimensional lattice at (K_x, K_τ) to one at the dual values of the parameters (K_τ^*, K_x^*) – see Eq. (7.25). We will now see how it emerges in the operator formalism in the τ-continuum limit, following Fradkin and Susskind [8]. Let us begin with

$$H = -\lambda \sum_n \sigma_1(n) - \sum_n \sigma_3(n)\sigma_3(n+1). \tag{8.158}$$

The problem is addressed in perturbation theory in λ starting with one of the fully ordered ground states. Any series will break down at $\lambda = 1$ when we encounter the singularity. Duality allows us to peek into the strong coupling side.

Define *dual variables*

$$\mu_1(n) = \sigma_3(n)\sigma_3(n+1), \tag{8.159}$$

$$\mu_3(n) = \prod_{-\infty}^{n} \sigma_1(l), \tag{8.160}$$

which have the same algebra as the σ's. Notice that the string runs all the way back to $-\infty$. We prefer these open boundary conditions to study the phenomena at hand.

The inverse formulas have the same form:

$$\sigma_1(n) = \mu_3(n-1)\mu_3(n), \tag{8.161}$$

$$\sigma_3(n) = \prod_{-\infty}^{n-1} \mu_1(l). \tag{8.162}$$

Exercise 8.10.1 *Verify that the μ's in Eqs. (8.159) and (8.160) obey the same algebra as the σ's: any two μ matrices at different sites commute, and the two at the same site anticommute. Derive the inverse relations Eqs. (8.161) and (8.162).*

Using dual variables, we may write

$$H(\lambda) = \sum [-\lambda\sigma_1(n) - \sigma_3(n)\sigma_3(n+1)] \tag{8.163}$$

$$= \sum [-\lambda\mu_3(n-1)\mu_3(n) - \mu_1(n)] \tag{8.164}$$

$$= \lambda H\left(\frac{1}{\lambda}\right), \tag{8.165}$$

where in the last step we have used the fact that H expressed in terms of the μ's is algebraically the same as H expressed in terms of the σ's: it will have the same spectrum, eigenstates (in the μ basis, of course), and so forth. After all, we can call the Pauli matrices by any name and they will have the same properties.

From Eq. (8.165) we see that the spectrum of H obeys

$$E(\lambda) = \lambda E\left(\frac{1}{\lambda}\right) \tag{8.166}$$

level by level. If the mass gap (the difference between the ground state and the first excited state) vanishes at some critical value λ_c, it must do so at $1/\lambda_c$ as well. If the transition is unique, it must occur at $\lambda_c = 1$.

Duality is seen to map the problem at λ *to a problem at* $\frac{1}{\lambda}$*, i.e., weak coupling to strong coupling and vice versa.* This is a general feature of duality. What we have is more: the model after the dual transformation is the same Ising model. This is called *self-duality* to distinguish it from simply *duality*, which maps one model at strong coupling to a different model at weak coupling. For example, the Ising model in $d = 3$ is dual to a different model, the gauge Ising model, which we will touch upon briefly.

We close with an analysis of the physics underlying the dual variables. Consider the zero-temperature case $\lambda = 0$, with two possible ground or vacuum states. Choose one, say with spins all up. When $\mu_3(n)$ acts on it, it flips all the spins from $-\infty$ to n, so that we now have the down vacuum from $-\infty$ to n and the up vacuum to the right:

$$\mu_3(n)|\cdots \uparrow\uparrow\uparrow\uparrow\uparrow\uparrow \cdots\rangle = |\cdots \downarrow\downarrow\downarrow\downarrow\uparrow\uparrow\uparrow\uparrow\uparrow \cdots\rangle. \tag{8.167}$$

Thus $\mu_3(n)$ creates a *kink* or *soliton* that interpolates between the two degenerate vacua, which too is a general feature of kinks and solitons. At any $0 < \lambda < 1$, the vacua are not simply all up or all down but predominantly so; the action of $\mu_3(n)$ is still to flip spins up to n.

The expectation value of μ_3 vanishes in the ordered ground state:

$$\langle \lambda < 1|\mu_3|\lambda < 1\rangle = 0, \tag{8.168}$$

since the state with an infinite number of spins flipped is orthogonal to states having any finite number of flipped spins.

If we impose periodic boundary conditions, only an even number of kinks can be created.

The presence of kinks reduces the average magnetization. For example, with two kinks at n_1 and n_2, the spins in between oppose the majority. Even though an infinite number of spins is flipped by one kink, the energy cost is finite, namely 2 in the $\lambda = 0$ limit because only one bond is broken. As λ increases, the cost of producing kinks goes down and eventually vanishes at $\lambda = 1$. From the Jordan–Wigner transformation we see that the kink does essentially what the Majorana fermions do with their string. It is to be expected that the kink mass vanishes when the fermion mass vanishes. As we approach the critical point, kinks are produced more and more easily, and the magnetization is correspondingly reduced.

What happens when we go beyond the point $\lambda = 1$ when the kink mass vanishes? The vacuum has a non-zero density of kinks, and we say that the kinks have *condensed*. More precisely, we say that

$$\langle \lambda > 1 | \mu_3 | \lambda > 1 \rangle \neq 0, \tag{8.169}$$

which follows because mathematically μ_3 behaves for large λ exactly the way σ_3 does for small λ. To say that the vacuum has a finite density of kinks is to say that the sign of the spin fluctuates wildly from site to site, as is expected in the disordered state.

Kadanoff (who studied this problem in the classical language) has dubbed the dual variables *disorder variables*. Indeed, while μ_3 is the disorder variable with respect to σ_3, we can equally well argue that it is the other way around, at least mathematically. Physically, only one of them, σ_3, describes physical moments that we can see and couple to a magnetic field.

The transformation to dual variables is independent of the spin Hamiltonian, but its utility is not. In general, a non-local change of variables will convert a local Hamiltonian to a non-local one. For example, if we add a magnetic field term $h \sum_n \sigma_3(n)$, it becomes a sum over infinitely long strings of μ_1's.

References and Further Reading

[1] L. Onsager, Physical Review, **65**, 117 (1944). A must-read, even if you cannot follow all the details, to understand how it inspires awe in all those who behold it.
[2] T. D Schultz, D. Mattis, and E. H. Lieb, Reviews of Modern Physics, **36**, 856 (1964). This is Onsager for the rest of us.
[3] R. Shankar and G. Murthy, Physical Review B, **36**, 536 (1987). For an extension and a minor correction to Eq. (2.16) of this paper, Eq. (8.53) in this book, see [4].
[4] M. E. Fisher and L. Mikheev, Physical Review Letters, **70** (1993).
[5] L. P. Kadanoff, Reviews of Modern Physics, **49**, 267 (1977).
[6] M. E. Fisher, in *Critical Phenomena*, ed. F. W. J. Hahne, Lecture Notes in Physics 186, Springer (1983).
[7] C. N. Yang Physical Review, **85**, 808 (1952).
[8] E. Fradkin and L. Susskind, Physical Review D, **17**, 2637 (1978).

9

Majorana Fermions

Majorana fermions have a life of their own, quite outside the solution to the Ising model. In this chapter we explore some topics connected to them.

9.1 Continuum Theory of the Majorana Fermion

The Hamiltonian

$$H = \sum_n \left[2i\lambda\psi_1(n)\psi_2(n) - 2i\psi_1(n)\psi_2(n+1)\right] \tag{9.1}$$

$$= \sum_n \left[2i(\lambda-1)\psi_1(n)\psi_2(n) - 2i\psi_1(n)(\psi_2(n+1) - \psi_2(n))\right] \tag{9.2}$$

describes a theory on a lattice of points. The lattice is assumed to be infinite, and boundary conditions will not enter the discussion. The lattice description is perfectly appropriate for the Ising model whose spin degrees of freedom do indeed live on a lattice made of atoms, say an Å apart. Correlations fall with a correlation length $\xi = 1/(\lambda - 1)$ in the x and τ directions. (To be precise, ξ is $1/4(\lambda - 1)$ for the two-fermion intermediate state. We will drop such factors like 4 in this discussion.) If, say, $\lambda = 0.9$, we know $\xi = 10$, which means 10 lattice units. Suppose you want a theory which has a correlation length $\xi = 10$ lattice spacings based on some experiment done on the system. You simply pick $\lambda = 0.9$. If you want $\xi = 20$, you pick, $\lambda = 0.95$ and you are done. You can predict the correlations for any separation having fitted the single parameter in the model to data.

But suppose you are a field theorist who strictly holds the view that space and time are continuous. This is certainly reasonable since we do not see any lattice structure in space or time. You measure distances in meters and masses in kilograms. You wish to formulate a field theory of Majorana fermions for this spacetime continuum using a lattice as an intermediate device, either to put things on a computer or, in more complicated cases, to deal with some infinities that arise in a frontal assault on the continuum.

Let us assume the asymptotic form of correlations,

$$G(|x|) = e^{-|x|/\xi_c}, \tag{9.3}$$

where ξ_c is the correlation length of the continuum theory measured in meters.

Suppose you measure the correlation between two points x_1 and x_2 separated by x meters and vary x until you find that the correlation is $1/e$. This value of x is then your ξ_c.

Now you formulate the problem on a lattice. To begin with you choose a lattice with $\mathcal{N} = 100$ points per ξ_c. Then the lattice spacing is $a = \xi_c/100\,\text{m}$. Since your points of interest, x_1 and x_2, are 100 lattice units apart, you want $\xi = 100$ in these lattice units. So you should pick the lattice model with $\lambda = 0.99$. This will ensure that the correlation between the points is $e^{-100/100} = 1/e$. Having fit the theory to the data, you can calculate the correlation between points at other distances, say two correlation lengths apart.

You can also get the correlations for smaller distances like $\xi/10$ lattice units apart. But you cannot ask for a separation less than $a = \xi_c/100$. Also, you suspect that by the time you come down to separations of the order of the lattice spacing a, the lattice approximation will not be very faithful to the continuum. If you want to do a good job down to $\xi/100$, you can choose a lattice with, say, $\mathcal{N} = 1000$ and $a = \xi_c/1000$. If you do this, 1000 points separate x_1 and x_2 but you want the correlation to still be $1/e$. This means you need $\xi = 1000$ in the lattice problem, or $\lambda = 0.999$.

Thus we need to choose λ to be a function of a such that in general we would want the dimensionless correlation length to obey

$$\xi(\lambda(a)) = \frac{\xi_c}{a}. \tag{9.4}$$

In our case we have a simple result, $\xi = 1/|\lambda - 1|$, which means

$$\lambda(a) = 1 \pm \frac{a}{\xi_c}, \tag{9.5}$$

the two solutions corresponding to critical models on either side of the transition.

Now, in particle physics one measures not correlations but particle masses. A correlation mediated by a particle of mass m decays as (dropping power-law prefactors and factors of c and $\hbar$, etc.)

$$G(x) = e^{-mx}, \quad \text{so that} \quad \xi_c = \frac{1}{m}. \tag{9.6}$$

Thus, the general condition for choosing $\lambda(a)$ is

$$\xi(\lambda(a)) = \frac{1}{ma}, \tag{9.7}$$

which for the Ising case becomes

$$\lambda(a) = 1 \pm ma. \tag{9.8}$$

No matter how small a is, we can find a $\xi(\lambda)$ such that $a\xi(\lambda) = \frac{1}{m}$ since ξ diverges at the transition. In other words, we can reproduce the desired correlation between two fixed points in the continuum even as the number of lattice points separating them goes to infinity because $\xi(\lambda)$ can be made as large as we want near a second-order critical point.

Although we have introduced a lattice, we have chosen the parameter $\lambda(a)$ such that the decay of the correlation depends on the observed m and not the artificially introduced spacing a, which is being driven to zero.

This process of choosing the coupling of the lattice model as a function of lattice spacing to give the same physics in laboratory units is called *renormalization*. But we generally need to do one more thing before all reference to the lattice is gone. To illustrate this, I use a more accurate description of the fermionic correlation near the critical point that displays the power-law prefactor:

$$\langle \psi(0)\psi(n)\rangle = \frac{e^{-n/\xi(\lambda)}}{n}. \tag{9.9}$$

If we want the correlators in the continuum theory between points a distance x apart, we must choose n such that $na = x$. Thus, we will find

$$\langle \psi(0)\psi(x/a)\rangle = \frac{e^{-x/a\xi(\lambda)}}{n} = a\frac{e^{-mx}}{x}. \tag{9.10}$$

The right-hand side is free of a except for the a up front. If we define *renormalized continuum fields* $\psi_{\mathrm{r}}(x)$,

$$\psi_{\mathrm{r}}(x) = \frac{1}{\sqrt{a}}\psi\left(\frac{x}{a}\right), \tag{9.11}$$

the correlation of renormalized fields is completely independent of a:

$$\langle \psi_{\mathrm{r}}(0)\psi_{\mathrm{r}}(x)\rangle = \frac{e^{-mx}}{x}, \tag{9.12}$$

and is expressed entirely in terms of the laboratory length x and mass m.

This process of rescaling the fields and varying the coupling λ with a to ensure that the physical quantities are left fixed and finite in laboratory units as we let $a \to 0$ is called *taking the continuum limit*.

It follows from Eq. (9.11) that the renormalized fields obey delta function anticommutation rules in this limit:

$$\{\psi_{\alpha\mathrm{r}}(x), \psi_{\beta\mathrm{r}}(x')\} = \delta_{\alpha\beta}(x - x'). \tag{9.13}$$

The Hamiltonian in Eq. (9.2) becomes

$$H = \sum_n \left[2i\lambda\psi_1(n)\psi_2(n) - 2i\psi_1(n)\psi_2(n+1)\right] \tag{9.14}$$

$$= (\lambda - 1)\sum_n (i\psi_1(n)\psi_2(n) - i\psi_2(n)\psi_1(n))$$
$$-i\sum_n (\psi_1(n)\psi_2(n+1) - \psi_2(n+1)\psi_1(n)) \tag{9.15}$$

$$= -a\sum_n \left(m(i\psi_1(n)\psi_2(n) - i\psi_2(n)\psi_1(n))\right.$$
$$+a\sum_n \psi_1 \frac{-i\Delta\psi_2}{a} + \psi_2 \frac{-i\Delta\psi_1}{a}, \tag{9.16}$$

where I have added and subtracted $i(\psi_1(n)\psi_2(n) - \psi_2(n)\psi_1(n))$, defined the difference operation Δ,

$$\Delta f(n) = f(n+1) - f(n), \tag{9.17}$$

and recalled that

$$1 - \lambda = ma. \tag{9.18}$$

If we now trade ψ for ψ_{r}, $a\sum_n$ for $\int dx$, $\Delta\psi/a$ for $d\psi/dx$ (because the correlation length is much larger than the lattice spacing), we find the renormalized Hamiltonian

$$H_{\mathrm{r}} = \frac{H}{2a} = \frac{1}{2}\int dx\, (\psi_{1\mathrm{r}}(x), \psi_{2\mathrm{r}}(x)) \begin{pmatrix} 0 & -im - i\partial \\ +im - i\partial & 0 \end{pmatrix} \begin{pmatrix} \psi_{1\mathrm{r}}(x) \\ \psi_{2\mathrm{r}}(x) \end{pmatrix}$$
$$\equiv \frac{1}{2}\int dx \psi^{\mathrm{T}}(\alpha P + \beta m)\psi, \tag{9.19}$$

where the subscript r has been dropped and

$$\alpha = \begin{pmatrix} 0 & 1 \\ 1 & 0 \end{pmatrix}, \quad \beta = \begin{pmatrix} 0 & -i \\ i & 0 \end{pmatrix}. \tag{9.20}$$

Duality takes a very simple form in the fermion language. Since $\lambda \to 1/\lambda$, the mass $m = 1 - \lambda$ changes sign under duality (near $\lambda = 1$) and $m = 0$ is the self-dual point. It is easily seen that duality corresponds to the simple transformation

$$\begin{pmatrix} \psi_{1\mathrm{r}}(x) \\ \psi_{2\mathrm{r}}(x) \end{pmatrix} \to \alpha \begin{pmatrix} \psi_{1\mathrm{r}}(x) \\ \psi_{2\mathrm{r}}(x) \end{pmatrix}, \tag{9.21}$$

for if we do this inside H_{r} above we find that it reverses m because α and β anticommute.

Thus, duality simply exchanges the components of ψ! The reason duality is local in terms of the ψ's while it was very non-local in terms of the σ's is that the ψ's themselves are non-local in terms of the σ's. I invite you to verify that the fermions that emerge from a Jordan–Wigner transformation on the μ's are again the same fermions except for the exchange $1 \leftrightarrow 2$ and a factor at the beginning of the chain.

By a Fourier transformation that respects the Hermiticity of ψ, this problem can be solved along the same lines as the one on the lattice, to yield

$$H_{\mathrm{r}} = \int_{-\infty}^{\infty} \psi^{\dagger}(p)\psi(p)\sqrt{m^2 + p^2}\,\frac{dp}{2\pi} - \frac{1}{2}\int_{-\infty}^{\infty} \sqrt{m^2 + p^2}\,\frac{dp}{2\pi}, \tag{9.22}$$

where p is now the momentum in laboratory units.

Notice that there are just particles in this theory, no antiparticles.

The Ising model gives the simplest example of the process called taking the continuum limit, which consists of taking a sequence of lattice Hamiltonians for smaller and smaller lattice spacings until the limit $a \to 0$ is reached, all the while keeping physical quantities like masses fixed in laboratory units.

The continuum limit is mandatory if the lattice is an artifact. But even if it is real, as in condensed matter physics, one may invoke the continuum limit to describe the physics at length scales much larger than the lattice spacing. In both cases we want the ratio of distances of interest to the lattice spacing to go to infinity.

Because the Ising model corresponded to a free-field theory, we could take the continuum limit so easily. When we see later how it is done for interacting theories, this concrete case of the Ising model will help you better follow those discussions.

9.2 Path Integrals for Majorana Fermions

Whereas the Dirac fermion Ψ has a distinct adjoint $\Psi^\dagger$, the Majorana fermion operator is Hermitian:

$$\boldsymbol{\psi}^\dagger = \boldsymbol{\psi}. \tag{9.23}$$

(In particle physics the Majorana fermion is its own antiparticle.)

The defining anticommutation rule is

$$\{\boldsymbol{\psi}_i, \boldsymbol{\psi}_j\} = \delta_{ij}. \tag{9.24}$$

It is natural to begin by looking for the usual coherent states for Majorana fermions with a Grassmann eigenvalue ψ,

$$\boldsymbol{\psi}|\psi\rangle = \psi|\psi\rangle. \tag{9.25}$$

But Eq. (9.24) implies

$$\boldsymbol{\psi}^2 = \frac{1}{2}, \tag{9.26}$$

whereas on applying $\boldsymbol{\psi}$ once again to Eq. (9.25), we find

$$\psi^2 = \frac{1}{2}, \tag{9.27}$$

which disagrees with $\psi^2 = 0$.

So we cannot find coherent states for Majorana fermions. We can, however, even in this case, define the Grassmann integral as usual:

$$\int \psi d\psi = 1. \tag{9.28}$$

Let us now see how we can write a path integral for Majorana fermions.

Suppose we are given a quadratic Hermitian Hamiltonian

$$H = \frac{i}{2}\sum_{ij} \boldsymbol{\psi}_i h_{ij} \boldsymbol{\psi}_j \equiv \frac{i}{2}\boldsymbol{\psi}^{\mathrm{T}} h \boldsymbol{\psi}, \tag{9.29}$$

$$h_{ij} = -h_{ji}. \tag{9.30}$$

We may safely assume that h is antisymmetric since different $\boldsymbol{\psi}_j$'s anticommute and $\boldsymbol{\psi}_i^2 = \frac{1}{2}$. It is necessarily even dimensional.

How do we find the corresponding path integral, given that coherent states of $\boldsymbol{\psi}$ do not exist?

One option is to introduce a second Majorana operator $\boldsymbol{\eta}$ with no term in the Hamiltonian:

$$H(\boldsymbol{\psi}, \boldsymbol{\eta}) = \frac{i}{2}\sum_{ij} \boldsymbol{\psi}_i h_{ij} \boldsymbol{\psi}_j. \tag{9.31}$$

Now we can form a canonical (Dirac) fermion and its adjoint:

$$\boldsymbol{\Psi}_i = \frac{\boldsymbol{\psi}_i + i\boldsymbol{\eta}_i}{\sqrt{2}}, \quad \boldsymbol{\Psi}_i^\dagger = \frac{\boldsymbol{\psi}_i - i\boldsymbol{\eta}_i}{\sqrt{2}}, \tag{9.32}$$

which obey

$$\{\boldsymbol{\Psi}_i^\dagger, \boldsymbol{\Psi}_j\} = \delta_{ij}. \tag{9.33}$$

The Hamiltonian becomes

$$H = \frac{i}{4}\sum_{ij} \left[(\boldsymbol{\Psi}_i + \boldsymbol{\Psi}_i^\dagger) h_{ij} (\boldsymbol{\Psi}_j + \boldsymbol{\Psi}_j^\dagger)\right]. \tag{9.34}$$

In terms of coherent states (labeled by capital $(\bar{\Psi}, \Psi)$ just in this section),

$$\boldsymbol{\Psi}|\Psi\rangle = \Psi|\Psi\rangle, \quad \langle\bar{\Psi}|\boldsymbol{\Psi}^\dagger = \langle\bar{\Psi}|\bar{\Psi}, \tag{9.35}$$

we have the usual action

$$S = \bar{\Psi}\left(-\frac{d}{d\tau}\right)\Psi - H(\boldsymbol{\Psi}_i^\dagger \to \bar{\Psi}, \boldsymbol{\Psi} \to \Psi). \tag{9.36}$$

(Careful readers will note that the Hamiltonian must be normal ordered with all $\boldsymbol{\Psi}_i^\dagger$ to the left if we are to make the substitution $\boldsymbol{\Psi}^\dagger \to \bar{\Psi}, \boldsymbol{\Psi} \to \Psi$. But our H has one term in which the order is opposite. We can reverse the order at the cost of a minus sign (remember $i \neq j$ everywhere and all fermion operators anticommute), get the action, and once again reverse the order of the corresponding Grassmann variables $(\psi, \bar{\psi})$ at the cost of a compensating minus sign. In short, you did not have to worry about normal ordering on this occasion.)

We now go back to new Grassmann variables ψ and η (the Jacobian is unity) defined by

$$\Psi_i = \frac{\psi_i + i\eta_i}{\sqrt{2}}, \quad \bar{\Psi}_i = \frac{\psi_i - i\eta_i}{\sqrt{2}}. \tag{9.37}$$

The action separates and the partition function factorizes as follows:

$$S(\psi,\eta) = \frac{1}{2}\psi^{\mathrm{T}}\left(-\frac{d}{d\tau}\right)\psi - H(\boldsymbol{\psi} \to \psi) \tag{9.38}$$

$$+\frac{1}{2}\eta^{\mathrm{T}}\left(-\frac{d}{d\tau}\right)\eta \tag{9.39}$$

$$Z = \int e^{S(\psi)}\,[\mathcal{D}\psi] \times \int e^{S(\eta)}\,[\mathcal{D}\eta]. \tag{9.40}$$

There are no cross terms between η and ψ in the time-derivatives. The putative terms vanish if we integrate by parts and invoke anticommutativity:

$$\frac{i}{2}\left[\psi^{\mathrm{T}}\left(-\frac{d}{d\tau}\right)\eta - \eta^{\mathrm{T}}\left(-\frac{d}{d\tau}\right)\psi\right] \tag{9.41}$$

$$= \frac{i}{2}\left[\psi^{\mathrm{T}}\left(-\frac{d}{d\tau}\right)\eta + \left(-\frac{d\eta^{\mathrm{T}}}{d\tau}\right)\psi\right] \tag{9.42}$$

$$= \frac{i}{2}\left[\psi^{\mathrm{T}}\left(-\frac{d}{d\tau}\right)\eta - \psi^{\mathrm{T}}\left(-\frac{d}{d\tau}\right)\eta\right] = 0. \tag{9.43}$$

It follows that as far as $\boldsymbol{\psi}$ goes, we can write down a path integral as we would for Dirac fermions with just two changes: the time-derivative has a $\frac{1}{2}$ in front and only ψ or ψ^{T} appear anywhere, there is no $\bar{\psi}$. The path integral is over just $[\mathcal{D}\psi]$. The part that depends on η is a spectator when we compute correlation functions of $\boldsymbol{\psi}$, and we can just forget about it. It is as if coherent states existed for Majorana fermions.

Another way to obtain the path integral, as long as there are an even number of Majorana operators, is to pair them into half as many Dirac operators and their adjoints, use coherent states of the latter pairs, get the path integral, and go back to Majorana Grassmann variables in the action. This is slightly messier to carry out, but does not involve the spectators $\boldsymbol{\eta}$. The end result is the same: as if coherent states existed for Majorana fermions.

In any case, we see that we will have to do integrals over just ψ as there is no $\bar{\psi}$. This leads to different rules for integration of Majorana fermions, compared to the usual Dirac fermions where we integrate over ψ and $\bar{\psi}$. We begin our study of this problem.

Exercise 9.2.1 *Write down the path integral for a problem with just two Majorana operators,*

$$H = i\omega\psi_1\psi_2, \tag{9.44}$$

by introducing one canonical Dirac pair and their coherent states.

9.3 Evaluation of Majorana Grassmann Integrals

As in the Dirac case, we will get used to integrals over a finite and countable number of Majorana Grassmann numbers, and then pass to the limit of the path integral.

We briefly discuss *Pfaffians*, which are essential to what follows. They are connected to determinants and play the role the determinant did for Dirac fermions in the Gaussian integrals.

For an *antisymmetric* $2N \times 2N$ matrix A we define the Pfaffian as

$$\mathrm{Pf}(A) = \frac{1}{2^N N!} \sum_p (-1)^p A_{1_p 2_p} A_{3_p 4_p} \cdots A_{2N-1_p, 2N_p}, \tag{9.45}$$

where p is a permutation of $\{1, 2, \ldots, 2N\}$, 3_p is the permuted version of 3, and so on, and $(-1)^p) = \pm 1$ is the sign of the permutation. Thus the first term, corresponding to no permutation, is $A_{12}A_{34} \cdots A_{2N-1,2N}$.

Using the antisymmetry we can abridge the sum to one where terms related by exchanging labels within a pair or different orderings of the pairs are not considered distinct. Then,

$$\mathrm{Pf}(A) = \sum_p (-1)^p A_{1_p 2_p} A_{3_p 4_p} \cdots A_{2N-1_p, 2N_p}. \tag{9.46}$$

Here is a simple example. If

$$A = \begin{pmatrix} 0 & A_{12} \\ -A_{12} & 0 \end{pmatrix}, \tag{9.47}$$

then

$$\mathrm{Pf}(A) = A_{12}. \tag{9.48}$$

We do not count both A_{12} and $-A_{21} = A_{12}$ in the abridged sum.

Observe something in this simple case:

$$\det A = \mathrm{Pf}(A)^2. \tag{9.49}$$

This happens to be true for all N.

Exercise 9.3.1 *Consider*

$$A = \begin{pmatrix} 0 & A_{12} & A_{13} & A_{14} \\ -A_{12} & 0 & A_{23} & A_{24} \\ -A_{13} & -A_{23} & 0 & A_{34} \\ -A_{14} & -A_{24} & -A_{34} & 0 \end{pmatrix}. \tag{9.50}$$

Verify (using the abridged sum) that

$$Pf(A) = A_{12}A_{34} - A_{13}A_{24} + A_{14}A_{23}, \tag{9.51}$$

$$\det A = Pf(A)^2. \tag{9.52}$$

In the unabridged sum, $A_{12}A_{34}$ would be counted eight times:

$$\begin{aligned} A_{12}A_{34} &= -A_{21}A_{34} = -A_{12}A_{43} = A_{21}A_{43} \\ &= A_{34}A_{12} = -A_{34}A_{21} = -A_{43}A_{12} = A_{43}A_{21}. \end{aligned} \tag{9.53}$$

We have seen that just as complex (Dirac) fermions can be described by a Grassmann integral, so can the real or Majorana fermions. We postulate the basic integrals

$$\int \psi d\psi = 1, \qquad \int 1 d\psi = 0. \tag{9.54}$$

Consider a Gaussian Grassmann integral $Z(A)$ associated with $2N$ anticommuting Grassmann numbers $[\psi_1, \psi_2, \ldots, \psi_{2N}]$ obeying

$$\{\psi_i, \psi_j\} = 0, \qquad \int \psi_i d\psi_j = \delta_{ij}, \tag{9.55}$$

and an antisymmetric matrix A:

$$Z(A) = \int \exp\left[-\frac{1}{2}\sum_{i,j,1}^{2N} \psi_i A_{ij} \psi_j\right] d\psi_1 \cdots d\psi_{2N}. \tag{9.56}$$

The reason we spent some time on the Pfaffian is the following result:

$$Z(A) = \mathrm{Pf}(A). \tag{9.57}$$

Exercise 9.3.2 *Starting with* $\det A = \frac{1}{\det A^{-1}}$, *argue that*

$$Pf(A^{-1}) = \pm\frac{1}{Pf(A)}. \tag{9.58}$$

For a 2×2 *matrix, verify that the minus sign applies. Pick an easy* 4×4 *case and check that the plus sign applies.*

Exercise 9.3.3 *Prove Eq. (9.57) for* $2N = 2$ *and* $2N = 4$ *by explicit evaluation.*

Show that

$$\langle \psi_1 \psi_2 \rangle \equiv \frac{\int \exp\left[-\frac{1}{2}\sum_{i,j,1}^{2N} \psi_i A_{ij} \psi_j\right] \psi_1 \psi_2 d\psi_1 \cdots d\psi_{2N}}{\int \exp\left[-\frac{1}{2}\sum_{i,j,1}^{2N} \psi_i A_{ij} \psi_j\right] d\psi_1 \cdots d\psi_{2N}} \tag{9.59}$$

$$= (A^{-1})_{12} \tag{9.60}$$

for the cases $2N = 2, 4$. *(You will need the fact that* $\det A = Pf(A)^2$ *and the definition of the inverse in terms of cofactors.)*

Show that for $2N = 4$ *we have Wick's theorem:*

$$\langle \psi_1 \psi_2 \psi_3 \psi_4 \rangle = \langle \psi_1 \psi_2 \rangle \langle \psi_3 \psi_4 \rangle - \langle \psi_1 \psi_3 \rangle \langle \psi_2 \psi_4 \rangle + \langle \psi_1 \psi_4 \rangle \langle \psi_2 \psi_3 \rangle, \tag{9.61}$$

where $\langle \psi_i \psi_j \rangle = (A^{-1})_{ij}$. *It works for larger values of 2N for the same reason.*

Finally, consider $\langle \psi_1 \psi_2 \psi_3 \psi_4 \rangle$ when $2N > 4$. It is still true that

$$\langle \psi_1 \psi_2 \psi_3 \psi_4 \rangle = A^{-1}_{12} A^{-1}_{34} - A^{-1}_{13} A^{-1}_{24} + A^{-1}_{14} A^{-1}_{23}. \tag{9.62}$$

This is just the Pfaffian of a submatrix of A^{-1} containing just the first four rows and columns.

Returning to the problem of computing correlations of Ising spins n sites apart in the space directions, we see that we need a Pfaffian of a matrix of size that grows as $\simeq 2n$.

Having explicitly worked out a few integrals we now consider a general action $S(\psi)$ and add source terms $J^{\mathrm{T}}\psi$ to get the generating function $Z(J)$ of correlation functions. The following result for Gaussian actions is very important:

$$\int e^{-\frac{1}{2}\eta^{\mathrm{T}}A\eta+J^{\mathrm{T}}\eta}[\mathcal{D}\eta]=\mathrm{Pf}(A)e^{-\frac{1}{2}J^{\mathrm{T}}A^{-1}J}, \tag{9.63}$$

where J is a column vector made of Grassmann numbers and

$$[\mathcal{D}\eta]=d\eta_1\cdots d\eta_{2N}. \tag{9.64}$$

You should verify Eq. (9.63) for the 2×2 case. You can show it in general by shifting variables,

$$\eta=\eta'-A^{-1}J, \qquad \eta^{\mathrm{T}}=(\eta')^{\mathrm{T}}+J^{\mathrm{T}}A^{-1}, \tag{9.65}$$

in the exponent. Correlation functions can be obtained by differentiation. For example,

$$\langle\psi_1\psi_2\rangle=\frac{1}{Z}\left.\frac{\partial^2 Z}{\partial J_1\partial J_2}\right|_{J=0}. \tag{9.66}$$

9.4 Path Integral for the Continuum Majorana Theory

In Exercises 9.3.1 and 9.3.3 you were introduced to the notion of Grassmann Gaussian integrals over Majorana fermions and the process of computing "averages" using Wick's theorem which yielded Pfaffians. Now we turn to path integrals for the Ising fermions in the continuum theory. We begin with Eq. (9.19). As explained earlier [Eq. (9.38)], the action is

$$S=\frac{1}{2}\int dx d\tau\, \psi^{\mathrm{T}}(x,\tau)\left[-\frac{\partial}{\partial\tau}-\alpha P-\beta m\right]\psi(x,\tau), \tag{9.67}$$

where ψ^{T} is the transpose of the two-component spinor ψ.

Exercise 9.4.1 *Derive Eq. (9.67) using Eq. (6.113).*

For this discussion let us choose

$$\beta=\sigma_2=\gamma_0, \tag{9.68}$$

$$\alpha=\sigma_3, \tag{9.69}$$

$$\gamma_1=\sigma_1. \tag{9.70}$$

We may write the action in a Lorentz-invariant form as

$$S = -\frac{1}{2}\int dx d\tau \bar{\psi}(\not{\partial} + m)\psi, \tag{9.71}$$

$$\bar{\psi} = \psi^{\mathrm{T}}(\sigma_2), \tag{9.72}$$

$$\not{\partial} = \gamma_0 \partial_\tau + \gamma_1 \partial_x, \tag{9.73}$$

$$\gamma_0 = \sigma_2, \quad \gamma_1 = \sigma_1. \tag{9.74}$$

Though I call it Lorentz invariant, it is actually rotationally invariant. In the Euclidean two-dimensional spacetime the rotation is generated by

$$\sigma_{01} = \frac{i}{2}\gamma_0\gamma_1. \tag{9.75}$$

The rotation operator is

$$R = e^{-\frac{1}{2}\gamma_0\gamma_1\theta} = e^{\frac{1}{2}i\sigma_3\theta} = R^{\mathrm{T}}, \tag{9.76}$$

under which

$$\psi \to R\psi, \tag{9.77}$$

$$\bar{\psi} \to \psi^{\mathrm{T}} R^{\mathrm{T}} \sigma_2 = \psi^{\mathrm{T}} R \sigma_2 = \psi^{\mathrm{T}} \sigma_2 R^{-1} = \bar{\psi} R^{-1}, \tag{9.78}$$

$$\bar{\psi}\psi \to \bar{\psi}\psi, \tag{9.79}$$

$$\bar{\psi}\not{\partial}\psi \to \bar{\psi}\not{\partial}\psi, \tag{9.80}$$

where the last equation is true because γ_μ and ∂_μ both get rotated.

Exercise 9.4.2 *Derive the Lorentz-invariant form of the action in Eq. (9.71).*

Note that $\bar{\psi}$ is not an independent Grassmann variable: it is made of just ψ_1 and ψ_2. This is made explicit when we write the measure for path integration:

$$Z = \int \prod_{x,\tau} [d\psi_1(x,\tau) d\psi_2(x,\tau)] e^S. \tag{9.81}$$

(The Jacobian for changing from $\int \prod d\bar{\psi} d\psi$ to $\int \prod d\psi_2 d\psi_1$ is unity.)

Although at finite temperatures we limit τ to the interval $[0, \beta]$, in the zero-temperature theory we shift the origin so that we integrate from $-\infty < \tau < \infty$ and denote by d^2x this range over the entire x–τ plane.

A useful result for later use: If we had a Dirac fermion Ψ, we could trade the action for two decoupled Majorana fermions χ and η as follows:

$$\int d^2x \bar{\Psi}(\not{\partial} + m)\Psi = \frac{1}{2}\int d^2x \left[\bar{\eta}(\not{\partial} + m)\eta + \bar{\chi}(\not{\partial} + m)\chi\right], \tag{9.82}$$

where

$$\Psi = \frac{\chi + i\eta}{\sqrt{2}}, \qquad \bar{\Psi} = \frac{\bar{\chi} - i\bar{\eta}}{\sqrt{2}}, \tag{9.83}$$

and $\bar{\chi} = \chi^{\mathrm{T}}\sigma_2$, $\bar{\eta} = \eta^{\mathrm{T}}\sigma_2$.

Exercise 9.4.3 *Prove Eq. (9.82) using integration by parts.*

Finally, with an eye on applications to conformal field theory, let us return to

$$S = \frac{1}{2}\int d^2x \psi^{\mathrm{T}}(-\partial_\tau + i\sigma_3\partial_x)\psi \tag{9.84}$$

for the massless case. In terms of $\psi_\pm$, the eigenstates of σ_3, we may separate S out as follows:

$$S = \frac{1}{2}\int d^2x \big[\psi_+(-\partial_\tau + i\partial_x)\psi_+ - \psi_-(\partial_\tau + i\partial_x)\psi_-\big]. \tag{9.85}$$

The equations of motion for $\psi_\pm$, obtained by varying S, tell us that

$$\frac{\partial\psi_+}{\partial z^*} = 0, \qquad \frac{\partial\psi_-}{\partial z} = 0, \tag{9.86}$$

which means $\psi_\pm$ are functions of z and z^* respectively. One refers to $\psi_\pm$ as right-/left-handed or analytic/antianalytic fields. A mass term mixes these two fields.

Suppose we want the correlation function of the analytic field. By familiar arguments we have, ignoring constants like π,

$$\langle\psi_+(z_1)\psi_+(z_2)\rangle = \left[\frac{1}{-\partial_\tau + i\partial_x}\right]_{z_1z_2} = \left[\frac{\partial_\tau + i\partial_x}{-\nabla^2}\right]_{z_1z_2} \tag{9.87}$$

$$= \frac{\partial}{\partial z_{12}}\ln z_{12}z_{12}^* = \frac{1}{z_{12}}, \tag{9.88}$$

where

$$z_{12} = z_1 - z_2. \tag{9.89}$$

Exercise 9.4.4 *Derive Eq. (9.88) using intermediate momentum states. It may be easier to work with just x and τ and corresponding derivatives rather than $z = x + i\tau$ and $z^* = x - i\tau$ and corresponding derivatives.*

Higher-order correlations will yield Pfaffians upon invoking Wick's theorem, as in Eq. (9.61). For example,

$$\langle\psi_+(z_1)\psi_+(z_2)\psi_+(z_3)\psi_+(z_4)\rangle = \frac{1}{z_{12}}\frac{1}{z_{34}} - \frac{1}{z_{13}}\frac{1}{z_{24}} + \frac{1}{z_{14}}\frac{1}{z_{23}}. \tag{9.90}$$

9.5 The Pfaffian in Superconductivity

We will now see how Pfaffians arise naturally in the many-body wavefunction for the BCS state.

Suppose we create an N-electron state from the Fock vacuum $|\emptyset\rangle$ as follows:

$$|N\rangle = \int dx_1 \cdots dx_N \, \phi(x_1,\ldots,x_N)\Psi^\dagger(x_N)\cdots\Psi^\dagger(x_1)|\emptyset\rangle, \tag{9.91}$$

where x could be a spatial coordinate in any number of dimensions. (The spins are assumed to be frozen in the symmetric all-up state.) Given the state $|N\rangle$, we could extract the wavefunction as follows:

$$\phi(x_1,\ldots,x_N) = \langle\emptyset|\Psi(x_1)\cdots\Psi(x_N)|N\rangle. \tag{9.92}$$

Exercise 9.5.1 *Verify Eq. (9.92) for $\phi(x_1,x_2)$.*

In what are called $p+ip$ superconductors [1], we are interested in a state

$$|\mathrm{BCS}\rangle = e^{\frac{1}{2}\int \Psi^\dagger(x)g(x-y)\Psi^\dagger(y)dxdy}|\emptyset\rangle \tag{9.93}$$

where the wavefunction of the *Cooper pair*, $g(x-y)$, is odd in $x-y$. Expanding the exponential in a series, we find states with all possible even numbers of particles. In this case it is natural to introduce a *generating function* of wavefunctions for any number of particles from which the wavefunctions can be obtained by differentiating with respect to the Grassmann source $J(x)$:

$$Z(J) = \langle\emptyset|e^{\int dxJ(x)\Psi(x)}|\mathrm{BCS}\rangle, \tag{9.94}$$

$$\phi(x_1,\ldots,x_N) = \left.\frac{\partial^N Z}{\partial J(x_1)\cdots\partial J(x_N)}\right|_{J=0}. \tag{9.95}$$

We now introduce the identity just once and proceed as follows:

$$\begin{aligned} Z(J) &= \langle\emptyset|e^{\int dxJ(x)\Psi(x)}|\mathrm{BCS}\rangle \\ &= \langle\emptyset|e^{\int dxJ(x)\Psi(x)}\cdot I\cdot e^{\frac{1}{2}\int \Psi^\dagger(x)g(x-y)\Psi^\dagger(y)dxdy}|\emptyset\rangle \\ &\equiv \int \left[d\bar{\psi}d\psi\right]e^{-\bar{\psi}\psi}\langle\emptyset|e^{J\Psi}|\psi\rangle\langle\bar{\psi}|e^{\frac{1}{2}\Psi^\dagger g\Psi^\dagger}|\emptyset\rangle; \end{aligned} \tag{9.96}$$

in the last step, we have resorted to a compact notation and inserted the following resolution of the identity in terms of Grassmann coherent states:

$$I = \int |\psi\rangle\langle\bar{\psi}|e^{-\bar{\psi}\psi}\left[d\bar{\psi}d\psi\right], \tag{9.97}$$

and it is understood, for example, that

$$|\psi\rangle = \prod_x |\psi(x)\rangle, \qquad \left[d\bar{\psi}d\psi\right] = \prod_x \left[d\bar{\psi}(x)d\psi(x)\right]. \tag{9.98}$$

It is important to remember that $\bar{\psi}$ and ψ are independent and dummy variables. Using the defining property of coherent states,

$$\mathbf{\Psi}|\psi\rangle = \psi|\psi\rangle, \quad \langle\bar{\psi}|\mathbf{\Psi}^{\dagger} = \langle\bar{\psi}|\bar{\psi}, \tag{9.99}$$

in Eq. (9.96), we find

$$Z(J) = \int \left[d\bar{\psi}d\psi\right] e^{-\bar{\psi}\psi} e^{J\psi} e^{\frac{1}{2}\bar{\psi}g\bar{\psi}}, \tag{9.100}$$

where we have used the fact that

$$\langle\emptyset|\psi\rangle\langle\bar{\psi}|\emptyset\rangle = 1 \tag{9.101}$$

since at each site

$$|\psi\rangle = |0\rangle - \psi|1\rangle, \quad \langle\bar{\psi}| = \langle 0| - \langle 1|\bar{\psi}, \tag{9.102}$$

and $|\emptyset\rangle = |0\rangle \otimes |0\rangle \otimes \cdots \otimes |0\rangle$.

There are two ways to do the Gaussian integral Eq. (9.100). One is to recall the Grassmann delta function from Eq. (6.96),

$$\int e^{(J-\bar{\psi})\psi} d\psi = \delta(J - \bar{\psi}), \tag{9.103}$$

in the ψ integral and then integrate over $\bar{\psi}$ to obtain

$$Z(J) = e^{\frac{1}{2}JgJ}. \tag{9.104}$$

The other is to do the Gaussian integrals over $\bar{\psi}$ with $e^{-\bar{\psi}\psi}$ as the source term and then over ψ with $e^{J\psi}$ as the source term, to find the same result, upon using $\mathrm{Pf}(A)\mathrm{Pf}(A^{-1}) = 1$.

The pair wavefunction is

$$\phi(x_1, x_2) = \left.\frac{\partial^2 Z}{\partial J(x_1)\partial J(x_2)}\right|_{J=0} = -g(x_1 - x_2). \tag{9.105}$$

Other wavefunctions follow from Wick's theorem. For example,

$$\begin{aligned}\phi(x_1, x_2, x_3, x_4) = g(x_1 - x_2)g(x_3 - x_4) \\ -g(x_1 - x_3)g(x_2 - x_4) \\ +g(x_1 - x_4)g(x_2 - x_3).\end{aligned} \tag{9.106}$$

References and Further Reading

[1] D. Green and N. Read, Phys. Rev. B **61**, 10267 (2000).

10

Gauge Theories

We now study a family of models that are bizarre in the following sense: all the correlation functions you would naively think of vanish identically. Yet these models have meaningful parameters like temperature, and even exhibit phase transitions. More importantly, the electromagnetic, weak, and strong interactions are described by such *gauge theories*. So we have to take them very seriously. This chapter is only a brief introduction to this vast subject and its novel features, aimed at preparing you for more advanced and exhaustive treatments. Although many references will be furnished along the way, the review article by J. Kogut [1] will come in handy everywhere.

Gauge theories can be constructed on the lattice or the continuum. We only consider the lattice version here. In the modern view, even continuum versions must first be defined on a lattice, whose spacing must be made to vanish in a certain limiting process. The continuum Majorana theory that came from the Ising model is an example. It is, however, a trivial example because the final continuum theory describes free fermions. Defining interacting theories in the continuum will require the identification and detailed description of a more complicated second-order phase transition. It will also require the use of the renormalization group, to be described later.

Why would one dream up a model with no order parameter in statistical mechanics? The motivation comes from the *XY model*, which one can show has no order parameter and yet at least two phases. Let us simply follow it far enough to understand the notion of phase transitions without an order parameter, a point made by Stanley and Kaplan [2]. It will teach us a way to classify phases without referring to the order parameter.

10.1 The *XY* Model

Let us recall how we know the Ising model (in $d > 1$) has two phases. At high temperatures, the $\tanh K$ expansion shows exponentially decaying correlation functions. The magnetization $M(T)$, given by the square root of the asymptotic two-point function, vanishes. The expansion has a finite, non-zero radius of convergence. At low T, the spins start out fully aligned $M(0) = 1$ and get steadily disordered as T increases (for $d > 1$). This expansion also has a non-zero radius. If $M(T)$ is non-zero in one range and identically zero in another, it must be singular somewhere. There must be at least one phase transition in T.

The non-zero radius of convergence at both ends is essential to this argument and is absent in $d = 1$, where $M(T > 0) = 0$, i.e., the order immediately disappears when we turn on T. So it is a single-phase system. The transition from order to disorder, if you like, is *at* $T = 0$.

The corresponding argument for a transition is more complicated in the XY model.

The XY model is a member of a family of *non-linear sigma models*, defined in all dimensions and involving an $O(N)$ isovector $\boldsymbol{n}$ of unit length with an $O(N)$-invariant interaction:

$$Z(K) = \int \prod_{\boldsymbol{r}} d\boldsymbol{n}(\boldsymbol{r}) \exp\left[K \sum_{\langle \boldsymbol{r},\boldsymbol{r}' \rangle} \boldsymbol{n}(\boldsymbol{r}) \cdot \boldsymbol{n}(\boldsymbol{r}') \right], \qquad |\boldsymbol{n}| = 1. \tag{10.1}$$

This is the simplest nearest-neighbor ferromagnetic interaction between the unit vectors. It is not a Gaussian theory, despite appearances, because of the constraint $|\boldsymbol{n}| = 1$. If $\boldsymbol{n}$ has just one component, it must be ± 1, which is just the Ising model.

We consider here $d = 2$ and a two-component unit vector $\boldsymbol{n}$, imagined to lie in an *internal* X–Y plane. (The equality of the internal and spatial dimensions in this instance is a coincidence.) At each site $\boldsymbol{r}$ of a square lattice there is a variable

$$\boldsymbol{n}(\boldsymbol{r}) = \boldsymbol{i}\cos\theta(\boldsymbol{r}) + \boldsymbol{j}\sin\theta(\boldsymbol{r}). \tag{10.2}$$

The partition function becomes, in terms of θ,

$$Z(K) = \int \prod_{\boldsymbol{r}} \frac{d\theta(\boldsymbol{r})}{2\pi} \exp\left[K \sum_{\langle \boldsymbol{r},\boldsymbol{r}' \rangle} \left[\cos(\theta(\boldsymbol{r}) - \theta(\boldsymbol{r}'))\right] \right], \tag{10.3}$$

where $\langle \boldsymbol{r},\boldsymbol{r}' \rangle$ means $\boldsymbol{r}$ and $\boldsymbol{r}'$ are nearest neighbors. Since θ is defined only modulo 2π, a suitable variable is $e^{i\theta(\boldsymbol{r})}$. One can easily deduce the correlators of $\cos\theta$ or $\sin\theta$ from those of $e^{i\theta}$.

A correlation function we will consider is

$$G_{\alpha\beta}(\boldsymbol{r} - \boldsymbol{r}') = \langle e^{i\alpha\theta(\boldsymbol{r})} e^{i\beta\theta(\boldsymbol{r}')} \rangle, \tag{10.4}$$

where α and β must be integers to ensure periodicity under $\theta \to \theta + 2\pi$.

The Boltzmann weight or action, and the integration measure, are invariant under a global rotation of all θ's by some θ_0:

$$\theta(\boldsymbol{r}) \to \mathcal{S}(\theta(\boldsymbol{r})) = \theta(\boldsymbol{r}) + \theta_0 \quad \forall\, \boldsymbol{r}. \tag{10.5}$$

We shall use the symbol $\mathcal{S}$(variable) to denote the variable when acted by the symmetry operation $\mathcal{S}$.

This symmetry implies that $G_{\alpha\beta}(\boldsymbol{r}-\boldsymbol{r}')$ is unaffected by $\mathcal{S}$:

$$\langle \mathcal{S}(e^{i\alpha\theta(\boldsymbol{r})})\mathcal{S}(e^{i\beta\theta(\boldsymbol{r}')})\rangle = \langle e^{i\alpha\theta(\boldsymbol{r})}e^{i\beta\theta(\boldsymbol{r}')}\rangle. \tag{10.6}$$

Let us understand this equality. In evaluating the average on the left-hand side, we can change variables in the action and the integration measure from θ to $\mathcal{S}(\theta(\boldsymbol{r}))$ without changing either. This makes the average equal to the one on the right-hand side, because the two differ only in the renaming of dummy variables.

Since

$$\mathcal{S}(e^{i\alpha\theta(\boldsymbol{r})}) = e^{i\alpha\theta(\boldsymbol{r})+i\alpha\theta_0} = e^{i\alpha\theta_0}\cdot e^{i\alpha\theta(\boldsymbol{r})}, \tag{10.7}$$

Eq. (10.6) implies

$$e^{i(\alpha+\beta)\theta_0}\cdot\langle e^{i\alpha\theta(\boldsymbol{r})}e^{i\beta\theta(\boldsymbol{r}')}\rangle = \langle e^{i\alpha\theta(\boldsymbol{r})}e^{i\beta\theta(\boldsymbol{r}')}\rangle. \tag{10.8}$$

Consequently,

$$G_{\alpha\beta}(\boldsymbol{r}-\boldsymbol{r}') = 0 \quad \text{unless } \alpha+\beta = 0. \tag{10.9}$$

This result has obvious generalizations:

- If we consider a string with more such variables, their product has to be invariant under the symmetry, i.e.,

$$\alpha+\beta+\cdots = 0 \tag{10.10}$$

  if it is to have a non-zero average.
- *In any theory, the average of any product of variables that is not invariant under the symmetries of the action and measure will vanish.* The reason is the same: we will be able to show in such cases that

$$G = (\text{factor} \neq 1)\times G. \tag{10.11}$$

Exercise 10.1.1 *For the Ising model without a magnetic field, show, by a similar argument, that correlations of an odd number of spins will vanish.*

Now consider the temperature dependence of

$$G(\boldsymbol{r}) = \langle e^{i\theta(\mathbf{O})}e^{-i\theta(\boldsymbol{r})}\rangle, \tag{10.12}$$

where one of the points has been chosen to be at the origin $\mathbf{O}$ using translation invariance.

10.1.1 High-Temperature Limit of $G(\mathbf{r})$

In the high-temperature limit $K = 0$, this correlator vanishes because both factors of $e^{\pm i\theta}$ individually integrate to zero thanks to

$$\int_0^{2\pi} \frac{d\theta}{2\pi} e^{i\theta} = 0, \tag{10.13}$$

analogously to what happens when we sum over an isolated Ising spin:

$$\sum_i s_i = 0. \tag{10.14}$$

There, say in $d = 1$, the first non-zero contribution to $\langle s_0 s_n \rangle$ in the high-temperature expansion appeared at order $(\tanh K)^n$, when the spins at the ends of the n bonds that linked 0 to n wiped out all unpaired spins:

$$s_0 \left[(s_0 s_1)(s_1 s_2)(s_2 s_3) \cdots (s_{n-1} s_n) \right] s_n. \tag{10.15}$$

Of course, in $d > 1$, longer paths could be included as corrections, but these will not alter the exponential decay, only the decay rate and prefactor.

We similarly want to neutralize the $e^{\pm i\theta}$ factors in G, so we first expand the factor $e^{K\cos(\theta(\mathbf{r})-\theta(\mathbf{r}'))}$ in each of the bonds as follows

$$e^{K\cos(\theta(\mathbf{r})-\theta(\mathbf{r}'))} = 1 + \frac{K}{2}\left(e^{i(\theta(\mathbf{r})-\theta(\mathbf{r}'))} + e^{-i(\theta(\mathbf{r})-\theta(\mathbf{r}'))} \right) + \cdots \tag{10.16}$$

In the Ising case, there were just two terms (in $\tanh K$); here, however, there are an infinite number in the K-expansion. The leading terms in the high-temperature expansion come from the term linear in K. The cheapest way to get rid of the two unpaired factors of $e^{\pm i\theta}$ in G is if, as in the Ising model, we connect the origin to $\mathbf{r}$ by the shortest path in a K expansion. For example, if the two points lie at $x = 0$ and $x = n$, we would cancel $e^{i\theta(0)} e^{-i\theta(n)}$ by generating the term

$$\left[e^{i(\theta(1)-\theta(0))} e^{i(\theta(2)-\theta(1))} \cdots e^{i(\theta(n)-\theta(n-1))} \right] \tag{10.17}$$

at order K^n. To this we can add terms from longer paths that occur at higher orders in K and terms with higher powers of K in each bond, so that we have, as $K \to 0$,

$$G(\mathbf{r}) \lim_{K\to 0} C(n) K^n [1 + \cdots] = e^{-n/\xi(K)} [1 + \cdots], \tag{10.18}$$

$$\xi(K)^{-1} = (\ln K^{-1}) + \cdots \tag{10.19}$$

In the above, n is the Manhattan distance between the end points if they are not collinear and $C(n)$ comes from the multiplicity of the shortest paths when $\mathbf{r}$ is not along the axes. The n-dependence of G will be dominated by the falling exponential.

All we need to remember going forward is that in the high-temperature phase, spin–spin correlations fall exponentially and the order parameter is zero, as in the Ising model.

10.1.2 Low-Temperature Limit of $G(r)$

The low-temperature behavior is very different from the Ising model because θ, unlike s, is a continuous variable. It can fluctuate by arbitrarily small amounts. Since $K \to \infty$, we need to consider $\cos(\theta(\boldsymbol{r}) - \theta(\boldsymbol{r}'))$ only near its maximum and expand the action as follows:

$$\exp\left[\sum_{\boldsymbol{r}} K\cos(\theta(\boldsymbol{r}) - \theta(\boldsymbol{r}'))\right] \tag{10.20}$$

$$\simeq \exp\sum_{\boldsymbol{r}}\left[K\left(1 - \frac{1}{2}(\theta(\boldsymbol{r}) - \theta(\boldsymbol{r}'))^2 + \cdots\right)\right] \tag{10.21}$$

$$\simeq \exp\left[\sum_{\boldsymbol{r}} -\frac{a^2K}{2}\left(\frac{\theta(\boldsymbol{r}) - \theta(\boldsymbol{r}')}{a}\right)^2\right] \tag{10.22}$$

$$= e^{-\int \frac{K}{2}(\nabla\theta)^2 d^2\boldsymbol{r}}, \tag{10.23}$$

where I have introduced a lattice spacing a and dropped a constant K at each site. Once again, since $K \to \infty$, we can assume θ varies smoothly from site to site, and also safely extend the range of θ to $\pm\infty$. We find ourselves with the familiar Gaussian action.

We may interpret this action as follows. We know that $\theta \to \theta + \theta_0$ is a symmetry. So all states with some fixed θ have the same action, namely zero. In the Ising model we had two such states at $K = \infty$, namely all up or all down, which could be perturbed in powers of e^{-2K}, in the number of broken bonds. In the XY model, all points on the unit circle ($|\boldsymbol{n}| = 1$) parametrized by θ are possible sectors. Let us pick one of these, say $\theta = \theta^*$, as our least action state and allow fluctuations at $T > 0$. Unlike with Ising spins, we do not "flip" θ away from θ^* to create excitations, we vary it smoothly in space. The energy density of such excitations is precisely $\frac{K}{2}(\nabla\theta)^2$ in all dimensions.

Imagine such a *spin wave* in which θ changes by 2π in one of the spatial directions. The presence of such an excitation kills the average θ within its extent. If the excitation has a size L in all d dimensions, and θ varies along one of them by 2π, the cost of this excitation is just the integral of the $(\nabla\theta)^2$ energy density:

$$\delta E \simeq \left[\frac{2\pi}{L}\right]^2 L^d \simeq L^{d-2}. \tag{10.24}$$

For $d > 2$, these excitations, whose proliferation destroys order, are limited in size and the order in θ survives for some non-zero range of $1/K$. In $d < 2$, the excitations grow to large sizes for essentially no cost and order is impossible. The case $d = 2$ is borderline and needs more careful analysis.

These arguments have been made rigorous by the Hohenberg–Mermin–Wagner theorem [3, 4], which says that *no continuous symmetry can be broken in* $d \leq 2$. The result was established in the context of relativistic field theories by Coleman [5].

While we are at it, let us show that for discrete symmetries, order is stable against fluctuations for $d > 1$. The reason is that an island of spins opposed to the majority spins is separated by broken bonds (the analog of the gradient here) all over a domain "wall" of volume L^{d-1}. In the case of continuous symmetries we cannot concentrate all the change in θ at some wall, for this would cause the $(\nabla\theta)^2$ energy density to diverge.

Returning to the *XY* model in $d = 2$, we can settle the issue of spontaneous magnetization by looking at $G(r \to \infty)$.

If we add a source J, the generating function of correlation functions in this Gaussian model is

$$Z(J) = Z(0) \exp\left[\int d^2\boldsymbol{r} \int d^2\boldsymbol{r}' J(\boldsymbol{r}) c(K) \left[\frac{1}{\nabla^2}\right]_{\boldsymbol{r}\boldsymbol{r}'} J(\boldsymbol{r}')\right], \tag{10.25}$$

where $c(K)$ is a constant depending on K, π, and so on.

The two-point function we are interested in,

$$G(\boldsymbol{r}) = \langle e^{-i\theta(\boldsymbol{r})} e^{i\theta(\mathbf{O})} \rangle, \tag{10.26}$$

corresponds to adding a source J with delta functions of area ∓ 1 at $\boldsymbol{r}$ and the origin respectively. The cross-term between the point-sources in Eq. (10.25) gives the distance dependence. Now,

$$\left[\frac{1}{\nabla^2}\right]_{\boldsymbol{r}\boldsymbol{r}'} = -\frac{1}{2\pi} \ln|\boldsymbol{r} - \boldsymbol{r}'|, \tag{10.27}$$

which means that

$$G(\boldsymbol{r}) \simeq e^{-\eta(K)\ln r} = \frac{1}{r^{\eta(K)}}, \tag{10.28}$$

where the exponent $\eta(K)$ *varies continuously with* K. The correlations fall off much slower than an exponential, but still vanish at infinite separation, as does the order parameter.

As advertised, the *XY* model exemplifies a system with no order parameter but necessarily at least one phase transition between small and large K because $G(\boldsymbol{r})$ *cannot go analytically from being a power to an exponential.*

The *XY* model teaches us a valuable lesson for gauge theories: one way to distinguish between different phases is to study the decay of correlation functions.

However, the *XY* model is itself not a gauge theory. Indeed, the low-temperature order parameter is non-zero in $d > 2$ and the phase transition is a garden variety order–disorder transition. For example, in $d = 3$,

$$\left[\frac{1}{\nabla^2}\right]_{\boldsymbol{r}\boldsymbol{r}'} = -\frac{1}{4\pi|\boldsymbol{r} - \boldsymbol{r}'|}, \tag{10.29}$$

$$G(r) \lim_{r\to\infty} \simeq \text{constant} \simeq M^2. \tag{10.30}$$

By contrast, the lattice gauge theories do not have an order parameter in *any* number of dimensions. Let us see how this comes about.

10.2 The Z_2 Gauge Theory in $d = 2$

Let us begin with the $d = 2$ model, even though it does not exhibit the full richness of gauge theories.

Consider the square lattice in Figure 10.1. The Ising variables are located on the *bonds*, and

$$Z(K) = \sum_{s} \exp\left[K \sum_{\square} s_i s_j s_k s_l\right], \tag{10.31}$$

where the $\square$ in the summation denotes the plaquettes or squares in the lattice and the product $s_i s_j s_k s_l$ is taken around every one of them.

The variables on each bond, $s = \pm 1$, can be viewed as elements of the two-element group Z_2, which is why this is called the Z_2 gauge theory. All gauge models have group elements on the links. This gauge theory was invented by Wegner [6] (and independently by Polyakov). The non-Abelian version was invented by Wilson [7] to describe quantum chromodynamics, and in particular to explain confinement.

This model is obviously invariant under the reversal of *all* spins,

$$s_i \to -s_i \;\; \forall \, i, \tag{10.32}$$

which is the familiar *global symmetry* we have seen in the Ising model. It ensures that products of odd numbers of spins have no average.

But the model has a much bigger *gauge* symmetry:

$$s_i \to -s_i, \;\; i \in B(\boldsymbol{r}), \tag{10.33}$$

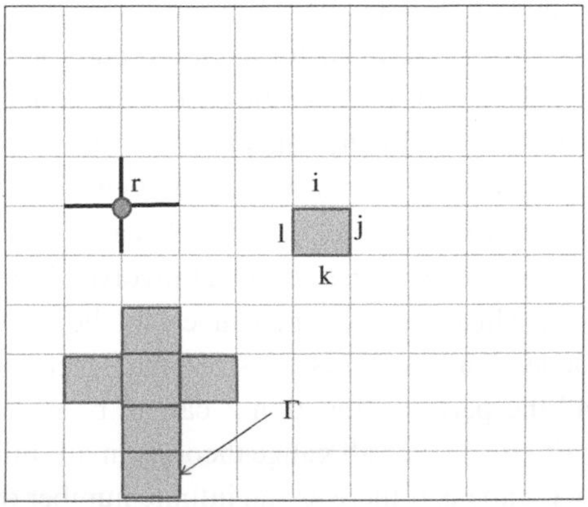

Figure 10.1 The Z_2 model in $d = 2$. The action contains the product of spins around every plaquette (whose edges are labeled i, j, k, l in one example). The local gauge transformation at site $\boldsymbol{r}$ (shown by a dot) flips the (four) spins on the bonds connected to that site. The T-shaped region with six plaquettes is bounded by its perimeter Γ.

where $\boldsymbol{r}$ is *any* site of the lattice, and $B(\boldsymbol{r})$ are the bonds incident on site $\boldsymbol{r}$. In other words, we can arbitrarily pick some sites $\boldsymbol{r}$, flip only the spins on bonds connected to those sites, and leave the rest alone. An example of the four bonds connected to a site is shown in Figure 10.1. This symmetry of the plaquette action follows because either no bonds from a plaquette are flipped (if that site is not part of a plaquette) or two are flipped (if it is).

The remarkable thing about this symmetry is that it is *local*: only the spins incident at a point are flipped. While we had just one symmetry operator $\mathcal{S}$ that flipped all the spins in the Ising model, now we have a transformation $\mathcal{S}_{\boldsymbol{r}}$ for each site. We can display this by rewriting Eq. (10.33) as follows:

$$\mathcal{S}_{\boldsymbol{r}}(s_i) = -s_i, \quad \text{for } i \in B(\boldsymbol{r}). \tag{10.34}$$

We have seen that non-zero correlations exist only for operators that are invariant under the symmetry. Thus $s_i s_j$ will not survive averaging, even though it is invariant under global spin flip, since there are operations $\mathcal{S}_{\boldsymbol{r}}$ that will flip just one of them.

So what kind of objects will survive averaging? It is objects like the plaquette which had two or zero bonds flipped by the action of any $\mathcal{S}_{\boldsymbol{r}}$. *Thus, O_Γ, the products of spins around a closed path G composed of bonds, will survive.* These *gauge-invariant* (GI) loop operators are named after Wegner or after Wilson, who extended the study of lattice gauge theories to continuous groups. One such T-shaped loop Γ is shown in Figure 10.1.

In the Ising model, in the absence of a magnetic field, we can argue that $\langle s \rangle = 0$ by symmetry. However, we still considered phases in which $\langle s \rangle = M$, the spontaneous magnetization, was non-zero. This was due to non-ergodicity: if the system begins in the sector with $M > 0$, it cannot evolve into the sector with $M < 0$ in any finite time, or to any finite order in perturbation theory in e^{-2K}, *in the thermodynamic limit*. Even though the initial and final states are degenerate by symmetry, the intermediate configurations pose an infinite energy barrier. A large domain wall of infinite cost and infinite perimeter in $d = 2$ or infinite area in $d = 3$ will first have to form. In the end, order is destroyed (and the low-temperature series breaks down) because the cost of the domain walls is overwhelmed by their multiplicity or entropy, i.e., by the vanishing free energy of large domain walls.

Such infinite barriers between symmetry-related configurations do not exist in a gauge theory, even in the thermodynamic limit. This is because every configuration has a degenerate one related by a symmetry operation that involves flipping just a finite number (four in $d = 2$) of spins. Thus, the system can easily hop over these finite barriers (whose magnitude is independent of the system size) and never get locked into any sector. Evolution is ergodic, and the partition function Z cannot be restricted to any one sector. This was established by Elitzur [8] for all gauge theories in all finite dimensions and finite group volumes. The only escape is if there are an infinite number of variables in each bond or if we are in infinite spatial dimensions. In such cases, two configurations separated by a local gauge transformation may be separated by an infinite barrier.

In a gauge theory, one correctly takes the view that the only physical quantities are the GI ones. After all, the others have identically vanishing averages by themselves or when multiplied by GI observables. Gauge theories therefore constitute a redundant description in which physical (GI) objects (plaquette variables, for example) appear along with unphysical non-invariant objects (spins on the bonds). When we compute an average of a product of GI observables, many, many configurations, all related by gauge transformations, contribute with equal weight. In every one of them the GI observables have the same value and change only if we move to a configuration not related by a gauge transformation. Consider, for example, the case where Γ is an elementary plaquette. In the configuration with all $s_i=+1$, it has a value $+1$. Starting with this, if we apply some $\mathcal{S}_{\boldsymbol{r}}$, we get a different configuration of spins but with no change in any of the GI variables, including the plaquette variable we are averaging, even if $\boldsymbol{r}$ is part of the plaquette. All configurations related by gauge transformations form a class. It can be shown that every *gauge class* has the same number of representatives, call it $\mathcal{V}$, the volume of the group of gauge transformations. If we want, we can choose any one representative member from each class in evaluating the average of the GI observables O_{GI}. This will greatly reduce the number of terms kept in the numerator and denominator of such an average:

$$\langle O_{\text{GI}}\rangle = \frac{\sum_s O_{\text{GI}} e^{S(s)}}{\sum_s e^{S(s)}} \tag{10.35}$$

$$= \frac{\mathcal{V}\sum_{\text{GI}} O_{\text{GI}} e^{S(s)}}{\mathcal{V}\sum_{\text{GI}} e^{S}(s)}, \tag{10.36}$$

where $\sum_{\text{GI}}$ is the sum over one representative element from each gauge class. The precise manner in which a representative element is chosen is called *gauge fixing*. We will resort to this shortly.

Bear in mind that in a gauge-fixed evaluation, non-gauge-invariant objects will no longer have zero averages. These averages will generally have different values in different gauges, namely different recipes for choosing the representative elements. These non-zero averages have no significance, unless they correspond to some gauge-invariant observable written in that gauge, in which case they will be unaffected by a change of gauge.

Exercise 10.2.1 *The bonds that are flipped within a plaquette may not be adjacent, but opposite. Consider the product of two gauge transformations, one at the origin* $(0,0)$ *and the other at* $(0,1)$. *The bond connecting these points is then not flipped. Consider all six plaquettes affected by this transformation and identify the six flipped bonds.*

10.2.1 Computing Some Averages: High-Temperature Limit

As a warm-up, let us compute the partition function

$$Z = \sum_s \exp\left[K\sum_{\square} s_i s_j s_k s_l\right] \tag{10.37}$$

$$= \sum_s \prod_{\square} \cosh K\left[1+\tanh K s_i s_j s_k s_l\right]. \tag{10.38}$$

This is just as in the usual Ising model, because $s_i s_j s_k s_l$ is also an Ising variable that squares to $+1$, but now the product over bonds is replaced by a product over plaquettes. To zeroth order in the tanh expansion, we pick up a 1 from every plaquette and the sum over spins gives

$$Z = (2\cosh K)^{\mathcal{N}}, \tag{10.39}$$

where $\mathcal{N}$ is the number of plaquettes. At higher orders, the average of the products of the $\tanh K s_i s_j s_k s_l$'s from any number of plaquettes will all vanish, as there will always be some unpaired spins that will sum to zero. For example, if the plaquettes fill the T-shaped region in Figure 10.1, the spins at overlapping edges will square to $+1$, while those at the perimeter will average to zero. This is true even if we fill the entire lattice in Figure 10.1, because there will still be unpaired spins at the edges. If, however, we impose *periodic or toroidal boundary conditions*, so that the top edge is glued to the bottom and the left edge to the right, and the plane is wrapped into a torus, every spin occurs twice, and we have

$$Z = (2\cosh K)^{\mathcal{N}} \left[1 + (\tanh K)^{\mathcal{N}}\right]. \tag{10.40}$$

This result is also exactly as in the $d = 1$ Ising model, when the second term, $(\tanh K)^{\mathcal{N}}$, exists only under periodic boundary conditions, which allow us to connect all $\mathcal{N}$ bonds to form a closed ring.

Next, we consider $\langle O_\Gamma \rangle$, the average of the product of spins around a closed loop Γ. In the high-temperature, small-K region, *we need to neutralize every spin on the loop by tiling the interior with plaquettes*. This is done to order A, where A is the area of the loop. For example, if the loop Γ is the perimeter of the T-shaped region in Figure 10.1, then $\mathcal{A} = 6$. This result is clearly true for loops of any size:

$$\langle O_\Gamma \rangle = \left\langle \prod_{i \in \Gamma} s_i \right\rangle = (\tanh K)^{\mathcal{A}} = e^{\mathcal{A} \ln \tanh K}. \tag{10.41}$$

Since at any finite K, $\tanh K < 1$, we say that the Wegner loop decays with an *area law*.

I will now show that this behavior, deduced in the small-K expansion, is actually good for all finite K. The Z_2 model in $d = 2$ has no phase transitions at any finite K.

10.2.2 Area Law in d = 2 without Gauge Fixing

It is desirable, but rarely possible, to work with a set of complete and independent GI observables. While the action is gauge invariant, the spins being summed over are not and furnish a redundant description.

The Z_2 theory in $d = 2$ is a rare exception. A complete set of GI variables is the product of spins around each plaquette:

$$p(\square) = \prod_{s \in \square} s = \pm 1. \tag{10.42}$$

It is clearly GI, but is it complete? We have seen that the only gauge-invariant observables are products of spins around any loop Γ. But these can be written as the product of spins around the plaquettes that tile the loop: the spins on the interior bonds get squared, while those on the perimeter remain to form the product of spins in Γ. See, for example, the T-shaped loop Γ in Figure 10.1.

Let us now find $\langle O_\Gamma \rangle$ with only GI objects appearing in the action or in the sum over configurations. Suppressing the degeneracy factor $\mathcal{V}$, which cancels out, we have

$$\langle O_\Gamma \rangle = \frac{\sum_{p(\Box)} \left[\prod_{\Box \text{inside } \Gamma} p(\Box)\right] \prod_\Box e^{Kp(\Box)}}{\sum_{p(\Box)} \prod_\Box e^{Kp(\Box)}} \tag{10.43}$$

$$= (\tanh K)^{A(\Gamma)}, \tag{10.44}$$

where $A(\Gamma)$ is the area of the loop Γ. The computation of the average for a single $p(\Box)$, which behaves like an independent Ising spin in a "magnetic field" K, is left as an exercise.

Exercise 10.2.2 *Prove Eq. (10.44). Treat the spin sums over $p(\Box)$ inside Γ separately from the ones outside.*

This answer is good for all $\tanh K$: there is no question of any singular behavior in the average over one plaquette. The result is independent of system size and valid in the thermodynamic limit as well. To be absolutely clear: if we were dealing with an infinite series in $\tanh K$, we could certainly worry that it would be invalid beyond its radius of convergence. But here, each plaquette has its own series containing just two terms, namely e^K and e^{-K} for $p(\Box) = \pm 1$.

In summary, the Z_2 theory in $d = 2$ is a single-phase system with an area law for the decay of O_Γ. It just factors into independent plaquettes, leaving no room for any phase transitions. The original goal, of devising a system with no order parameter but phase transitions, has eluded us in $d = 2$.

Things change in $d = 3$. Why should the preceding argument fail? *It fails because in $d = 3$ or higher, the plaquette variables are not independent: the product of six plaquettes that form a cube is identically $+1$ and hence all six cannot be chosen independently.* Here is an analogy: In $d = 1$, (with open boundary conditions) we used Ising variables $t_i = s_1 s_{i+1}$ in terms of which we could construct any product of s's. The partition function factorized in t. We cannot do this in $d = 2$ since the t's are not independent: the product of t's around a plaquette is identically $+1$. Coming back to the Z_2 theory, in $d = 3$ and 4 there are two phases. What can be the difference between them without an order parameter?

We want to address this, but first a rederivation of the same area law in $d = 2$ in the low-temperature phase to illustrate calculations with gauge fixing.

10.2.3 Low-Temperature Limit in d = 2 with Gauge Fixing

As $K \to \infty$, the dominant configurations should have $p(\Box) = 1$. Plaquettes with $p(\Box) = -1$ are said to be *frustrated*. One configuration with all $p(\Box) = 1$ is one where every $s_i = 1$. Its gauge-equivalent configurations also have this property, and we want to find a way to pick just one member from this class as well as other classes. For this we employ the *temporal gauge*, defined as follows:

$$\text{Temporal gauge:} \quad s_i = +1 \ \forall \text{ vertical bonds.} \tag{10.45}$$

The name derives from the fact that the vertical direction is usually the Euclidean time direction.

Given a general configuration on a finite square lattice, we can transform it to the temporal gauge as follows. Suppose the lowest vertical bond in the left-most column is $+1$. Then leave it alone. If not, apply $\mathcal{S}$ to the site at its upper end. Now go to the second vertical bond in this column and see what is there *after* the previous transformation. If it is $+1$, leave it alone; if not, apply $\mathcal{S}$ to the upper end of this second vertical bond. Keep doing this until you get to the top-most time-like bond of the left-most column. Now do this for the second column. Convince yourself that this will not affect the vertical spins in the first column. Go all the way to the right-most column. The final configuration cannot be subject to any further gauge transformations without violating the gauge condition. The gauge is totally fixed. (Sometimes, we may fix the gauge partially, in that some gauge condition is met, but each gauge class still has many representatives. We will see an example in the $d = 3$ model.)

This transformation to the temporal gauge fails if we have periodic boundary conditions in the time direction. When we fix the top-most bond in any column to make it $+1$, we could end up flipping the bottom-most bond in that column, which now shares the top-most site by periodicity. This is as it should be, since the product of all vertical bonds now corresponds to a closed loop Γ which is GI and called the *Polyakov loop*. For example, if all but one vertical bonds were $+1$ in a column, $\Gamma = -1$ and will stay that way. If we could have made all vertical bonds $+1$, we would have managed to change this GI observable, which is impossible. We will not bother with this point and work with open or free boundary conditions. However, these closed loops in the periodic direction are not always ignorable or uninteresting. In some higher-dimensional theories, these are sometimes the only residual variables in a low-energy description.

In the temporal gauge, the spins on the horizontal bonds have physical meaning since they cannot be altered by a gauge transformation. The action now takes the form

$$S = K \sum_{i=1}^{M} \sum_{j=1}^{N-1} s_{i,j} s_{i,j+1} \tag{10.46}$$

for a lattice with M columns and N rows. Notice that every plaquette variable has collapsed to the product of the spins on its two horizontal bonds because the vertical bonds are fixed at $+1$. In addition, there is no communication between columns. What we have is a collection of M one-dimensional Ising chains that are N units long. This shows that we have a single-phase system.

The correlation function O_Γ can be evaluated in this gauge. Suppose Γ is a rectangle of width 1 and height h. Since all vertical bonds are $+1$, O_Γ becomes just the product of the top-most and bottom-most horizontal bonds, all from the same columnar Ising chain:

$$\langle O_\Gamma \rangle = \langle s^{\text{top}} s_{\text{bot}} \rangle = (\tanh K)^h. \tag{10.47}$$

If we now glue together many such rectangles, we can form any Γ we want, with an area equal to the sum of the areas and an average obeying the area law:

$$\langle O_\Gamma \rangle = (\tanh K)^A. \tag{10.48}$$

10.3 The Z_2 Theory in $d=3$

The partition function has the same form in $d=3$:

$$Z = \sum_s \exp\left[K \sum_\square s_i s_j s_k s_l\right] \tag{10.49}$$

$$= \sum_s \prod_\square \cosh K \left[1 + \tanh K s_i s_j s_k s_l\right], \tag{10.50}$$

except that plaquettes can now lie in all three planes and the gauge transformations $\mathcal{S}_r$ will flip all *six* spins incident on site $\boldsymbol{r}$. But now, the $\tanh K$ expansion has non-zero terms. At sixth order we can assemble six plaquettes to form a cube. Every edge occurs twice and every spin is squared. At higher orders, bigger and more elaborate closed surfaces will come in the expansion with a factor $(\tanh K)^A$.

Consider such closed surfaces. They will arise naturally in the *low-temperature expansion* of a $d=3$ Ising model if we start with the all-up state and surround flipped spins with closed surfaces on the dual lattice. In $d=2$, these "closed surfaces" were, in fact, closed loops, and they described the *same* Ising model at the dual temperature.

Here, the high-temperature Z_2 theory is dual to the low-temperature non-gauge Ising model in $d=3$. Since the latter has a phase transition as e^{-2K^*} grows, so must the Z_2 theory as $\tanh K$ grows or the temperature is lowered.

How do we understand this Curie transition of the Ising model in Z_2 terms? It is clearly not an order–disorder transition since there is no order parameter. *It is reflected in $\langle O_\Gamma \rangle$, which goes from decaying as the exponential of the area at small K to the exponential of the perimeter at large K as we consider very large loops.*

There are many ways to demonstrate this, and here is one that teaches many other useful concepts along the way.

10.3.1 *Hamiltonian Treatment of the Z_2 Model in $d=3$*

First, let us pick the temporal gauge and choose couplings K_τ and K for the temporal and spatial plaquettes with $e^{-2K_\tau} = \lambda\tau$ and $K \to \tau$. This leads to a τ-continuum Hamiltonian of the form

$$H = -\sum_{\square} \sigma_3(i)\sigma_3(j)\sigma_3(k)\sigma_3(l) - \lambda \sum_{\text{bonds}} \sigma_1(\text{bonds}), \tag{10.51}$$

where the plaquettes in the first sum lie in the x–y plane.

Exercise 10.3.1 *Derive Eq. (10.51). Hint: The derivation is very similar to the usual Ising model. In the spatial plaquettes we simply use $s \to \sigma_3$ in going from the classical model to the quantum operators. In the temporal plaquettes the spins in the vertical sides equal $+1$ and those in the horizontal sides (at adjacent times) are multiplied as in the Ising model, leading to the σ_1 terms.*

We need the temporal gauge to be able to obtain the Hamiltonian. This is because a Hamiltonian is supposed to predict the unique future state given the initial state. This is incompatible with the freedom to change the spins in the future without affecting the initial value data using a time-dependent gauge transformation. However, the temporal gauge is not completely fixed. We can still make time-independent, space-dependent gauge transformations implemented in the operator language by

$$\mathcal{S}_r = \prod \sigma_1, \tag{10.52}$$

where the product is over the bonds incident at $\boldsymbol{r}$, where $\boldsymbol{r}$ labels lattice points in space. Clearly, this operator commutes with the σ_1 part of H and flips two or zero σ_3's in the plaquette terms. It is generally true of gauge theories that in the temporal gauge, the Hamiltonian is invariant under space-dependent transformations implemented by unitary operators. Since gauge symmetry cannot be spontaneously broken, the ground state must be gauge invariant. In such a state, only gauge-invariant operators will have an expectation value. (These can be Heisenberg operators at equal or unequal times since the time-evolution operator is also gauge invariant.)

The limit $\lambda \to \infty$ is deep in the high-temperature phase, when every bond is in the eigenstate of $\sigma_1 = +1$ and σ_3 has zero average, while as $\lambda \to 0$, the plaquette term dominates and every plaquette likes the product around it to equal $+1$. We now study the behavior of $\langle O_\Gamma \rangle$ in the two regimes.

10.3.2 High-Temperature Phase in $d = 3$

The Wegner loop we use to study the phase need not be planar. Let us, however, use one lying in the spatial plane for simplicity. The area law can be established in the $\tanh K$ expansion as within the classical isotropic model in $d = 2$, except that now the tiling of the loop need not be in the spatial plane – it can be any surface with the loop Γ as its boundary. However, the leading term, with the fewest powers of $\tanh K$, will still come from the planar tiling and we can still say

$$\langle O_\Gamma \rangle \lim_{K \to 0} \simeq e^{-c(K)A}, \tag{10.53}$$

where $c(K) \simeq \tanh K + \cdots$.

Let us now rederive this in operator language.

In the operator language we want to evaluate $\langle O_\Gamma \rangle$ as an operator average in the ground state $|\emptyset\rangle$ (assumed normalized),

$$\langle O_\Gamma \rangle = \langle \emptyset | \prod_{i \in \Gamma} \sigma_3(i) | \emptyset \rangle. \tag{10.54}$$

Since $\lambda \to \infty$, let us pull it out of H and work with H':

$$H' = \frac{H}{\lambda} = -\frac{1}{\lambda} \sum_{\square} \sigma_3(i)\sigma_3(j)\sigma_3(k)\sigma_3(l) - \sum_{\text{bonds}} \sigma_1(\text{bonds}), \tag{10.55}$$

which has the same set of states and the same physics. At $\frac{1}{\lambda} = 0$, the unperturbed ground state $|\emptyset\rangle = |0\rangle$, the $+1$ eigenstate of every σ_1. So the expectation value of any $\sigma_3 = 0$ and $\langle O_\Gamma \rangle$ vanishes. Evidently, we must perturb the ground state to order $(1/\lambda)^A$ in the plaquette term to introduce enough factors of σ_3 to neutralize all the σ_3's in Γ. In other words, we must tile the loop with plaquettes.

So the average will begin as

$$\langle \emptyset | \prod_{i \in \Gamma} \sigma_3(i) | \emptyset \rangle \simeq \left(\frac{1}{\lambda} \right)^A \simeq e^{-A \ln \lambda} \tag{10.56}$$

in the limit $\frac{1}{\lambda} \to 0$.

10.3.3 Low-Temperature Phase in $d = 3$

Consider, as $\lambda \to 0$, the decay of the average of a large loop of perimeter L in the spatial plane. Then $\langle O_\Gamma \rangle$ becomes the expectation value of the corresponding operator in the normalized ground state $|\emptyset\rangle$ of H:

$$\left\langle \prod_{i \in \Gamma} s_i \right\rangle = \langle \emptyset | \prod_{i \in \Gamma} \sigma_3(i) | \emptyset \rangle. \tag{10.57}$$

At $\lambda = 0$, $|\emptyset\rangle$ is just $|0\rangle$, the unperturbed ground state with every plaquette equal to $+1$: it is the normalized uniform sum of the state with all spins up and its gauge transforms. It is the sum that makes it GI, since a gauge transformation simply rearranges the sum. In this class the loop will have an average of $+1$. To order λ, the perturbation will flip σ_3 on a bond and make two plaquettes unhappy at an energy cost 4. It will flip the loop to -1 only if the flipped bond lies on it.

To order λ, the perturbed, *unnormalized* ground state $|\emptyset\rangle$ is

$$
\begin{aligned}
|\emptyset\rangle &= |0\rangle + \frac{\lambda}{4}\sum_{\text{bonds}} \sigma_1(n)|0\rangle \\
&= |0\rangle + \frac{\lambda}{4}\sum_{n \notin \Gamma} |\text{bond n is flipped}\rangle \\
&\quad + \frac{\lambda}{4}\sum_{n \in \Gamma} |\text{bond n is flipped}\rangle; \\
\prod_{i\in\Gamma}\sigma_3(i)|\emptyset\rangle &= |0\rangle + \frac{\lambda}{4}\sum_{n \notin \Gamma} |\text{bond n is flipped}\rangle \\
&\quad - \frac{\lambda}{4}\sum_{n \in \Gamma} |\text{bond n is flipped}\rangle; \\
\langle\emptyset|\emptyset\rangle &= 1 + \frac{\lambda^2\mathcal{N}}{16}; \\
\frac{\langle\emptyset|\prod_{i\in\Gamma}\sigma_3(i)|\emptyset\rangle}{\langle\emptyset|\emptyset\rangle} &= \frac{1+\frac{\lambda^2}{16}(\mathcal{N}-2L)}{1+\frac{\lambda^2}{16}\mathcal{N}} = 1 - \frac{\lambda^2 L}{8} + \cdots \\
&\simeq \exp\left[-\frac{\lambda^2 L}{8}\right].
\end{aligned}
\tag{10.58}
$$

When I say "bond n is flipped," I mean in gauge-invariant terms that the two plaquettes attached to it have become frustrated. I am also using the fact that all states with one bond flipped (one plaquette frustrated) are orthogonal to the one with no bonds flipped (no frustrated plaquettes) and that states with frustration in different locations are orthogonal. The exponentiation in the last line, done too glibly, is correct if we sum over an arbitrary number of bonds being flipped on the lattice, treating them as independent. But they do become dependent if they get very close, though this will only produce subdominant corrections.

In summary, the Z_2 gauge theory in $d = 3$ has two phases: a high-temperature small-K phase with an area law and a low-temperature phase with a perimeter law for the decay of large loops. The small-K gauge theory is dual to an ordinary Ising model at K^*.

Exercise 10.3.2 *As in the path integral, Eqs. (10.35) and (10.36), you may choose one representative state from a gauge-equivalent family to do the preceding calculation. Suppose $|i\rangle$ and $|j\rangle$ are two representative states from their families. Their gauge-invariant versions are*

$$|I\rangle = \frac{1}{\sqrt{\mathcal{V}}}\sum_\alpha g_\alpha|i\rangle, \tag{10.59}$$

$$|J\rangle = \frac{1}{\sqrt{\mathcal{V}}}\sum_\beta g_\beta|j\rangle, \tag{10.60}$$

where α and β run over all the independent gauge transformations g_α and g_β, and $\mathcal{V}$ is the number of states in the sum. Show that if $\mathcal{O}$ is a GI operator then

$$\langle I|\mathcal{O}|J\rangle = \langle i|\mathcal{O}|j\rangle. \tag{10.61}$$

Hints: $\sum_\alpha g_\alpha \sum_\beta g_\beta = \mathcal{V}\sum_\beta g_\beta$. *No g_α can take a state out of its family.*

10.4 Matter Fields

We are now going to introduce matter fields that live at the sites and are included in a gauge-invariant way. We switch to the different but widely used notation in lattice gauge theories depicted in Figure 10.2.

- The sites will be labeled by a vector $\boldsymbol{r}$. The field at the site will be called $\phi(\boldsymbol{r})$.
- The bonds will be denoted by a pair of labels $(\boldsymbol{r},\boldsymbol{e}_\mu)$, where $\boldsymbol{r}$ is the starting point and $\boldsymbol{r}+\boldsymbol{e}_\mu$ the end point and where $\boldsymbol{e}_\mu$ is the unit vector in the μ direction of the cubic lattice. The bond variables will be called $U_\mu(\boldsymbol{r})$ to remind us that they are group elements and that they live on the bond emanating from $\boldsymbol{r}$ and terminating at $\boldsymbol{r}+\boldsymbol{e}_\mu$. The bond variables are *oriented*: if the bond is traversed in the direction of decreasing coordinate, we should use the inverse $U^\dagger$. That is why the bonds pointing into $\boldsymbol{r}+\boldsymbol{j}$ and $\boldsymbol{r}$ carry $U^\dagger$. This distinction is absent in Z_2, where each element $U=\pm 1$ is its own inverse.

The simplest gauge-invariant action for the Z_2 problem is

$$S=\beta\sum_{\boldsymbol{r}\mu}\phi(\boldsymbol{r})U_\mu(\boldsymbol{r})\phi(\boldsymbol{r}+\boldsymbol{e}_\mu)+K\sum_{\boldsymbol{r}\mu\nu}\left[U_\mu(\boldsymbol{r})U_\nu(\boldsymbol{r}+\boldsymbol{e}_\mu)U^\dagger_\mu(\boldsymbol{r}+\boldsymbol{e}_\nu)U^\dagger_\nu(\boldsymbol{r})\right], \tag{10.62}$$

where $\boldsymbol{r}\mu\nu$ labels a plaquette with one corner at $\boldsymbol{r}$ and the edges meeting there aligned with the μ and ν axes. If the group were not Z_2, we would add the complex conjugate of this S and divide by 2.

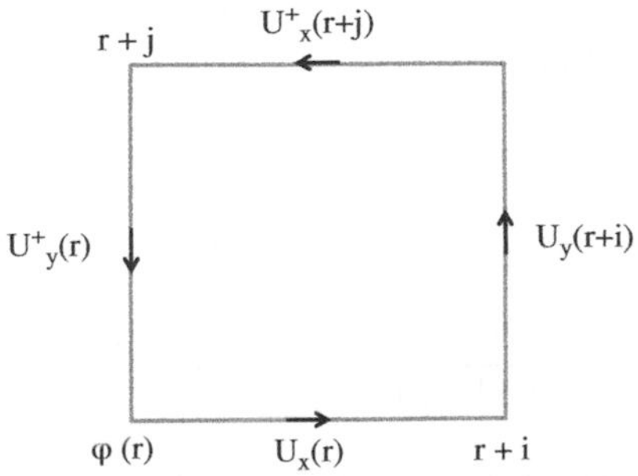

Figure 10.2 The definition of the gauge element $U_\mu(\boldsymbol{r})$ which resides in the bond emanating from $\boldsymbol{r}$ and terminating at $\boldsymbol{r}+\boldsymbol{e}_\mu$. The bonds traversed against the unit vectors (such as the one going *into* $\boldsymbol{r}$ against $\boldsymbol{j}$) will be the adjoints $U^\dagger$. The matter fields $\phi(\boldsymbol{r})$ reside at the four corners. If U were a matrix, a trace of the product around the loop should be taken to ensure gauge invariance.

The action is invariant under the gauge transformation:

$$\phi(\boldsymbol{r}) \to t(\boldsymbol{r})\phi(\boldsymbol{r}), \tag{10.63}$$

$$U_\mu(\boldsymbol{r}) \to t(\boldsymbol{r})U_\mu(\boldsymbol{r})t(\boldsymbol{r}+\boldsymbol{e}_\mu), \tag{10.64}$$

$$t(\boldsymbol{r}) = \pm 1. \tag{10.65}$$

In other words, the action is invariant if we multiply each $\phi(\boldsymbol{r})$ by a site-dependent Z_2 element $t(\boldsymbol{r}) = \pm 1$, provided we also multiply the two ends of every U by the t's at the ends. This means that variables like $\phi(\boldsymbol{r})$ or $\phi(\boldsymbol{r})\phi(\boldsymbol{r}')$ are unphysical, as is U. But variables like $\phi(\boldsymbol{r})U_\mu(\boldsymbol{r})\phi(\boldsymbol{r}+\boldsymbol{e}_\mu)$ are physical because they are gauge invariant.

What are the phases of this theory? We can say one thing for sure right away: large loops will never decay with an area law. Recall that in a pure gauge theory, the entire interior of the loop Γ had to be tiled with plaquettes to cancel the spins in Γ, and this occurred at order $(\tanh K)^A$. But, with matter fields we can neutralize the spins in Γ using the $\beta\phi(\boldsymbol{r})U_\mu(\boldsymbol{r})\phi(\boldsymbol{r}+\boldsymbol{e}_\mu)$ terms end-to-end all along Γ just L times: the ϕ from the end of one link will cancel that at the beginning of the next, while the U's in the L links will cancel the U's in the contour Γ. Thus, the decay of the loop will go as the perimeter L:

$$\langle O_\Gamma \rangle \simeq \beta^L \simeq e^{-f(\beta)L}. \tag{10.66}$$

The distinction between the area and perimeter laws has profound significance in quantum chromodynamics (QCD), the theory of quarks and gluons, where the quarks are described by ϕ and the gluons by U. Consider the pure gauge theory with a rectangular Wilson loop of spatial width R and height T in time. We saw in the Z_2 case that O_Γ can be obtained by stringing together hopping terms $\phi(\boldsymbol{r})U_\mu(\boldsymbol{r})\phi(\boldsymbol{r}+\boldsymbol{e}_\mu)$. The closed loop then describes a ϕ particle going in a loop in spacetime. At any time, if we slice it we find a particle going forward in time and a particle going backward, i.e, an antiparticle. With the trivial deformation of the loop at the initial and final times, we can make it look like a particle–antiparticle pair was created at $\tau = 0$ and annihilated at $\tau = T$. Now, we have seen that as $\tau \to \infty$, the partition function is just

$$Z \simeq e^{-E_0 \tau}, \tag{10.67}$$

where E_0 is the ground state energy. If we include the loop, the result becomes

$$\frac{Z(\text{loop})}{Z(\text{no loop})} = \langle O_\Gamma \rangle = e^{-V(R)T}, \tag{10.68}$$

where $V(R)$ is the additional (potential) energy of the static quark–antiquark pair as a function of the separation R. If the area law is obeyed we find, for a loop of width R and height T,

$$V(R)T = RT \quad \longrightarrow V(R) \propto R, \tag{10.69}$$

i.e., the potential between the quark–antiquark pair grows linearly, or that the force between them approaches a constant as $R \to \infty$, a result established by Wilson [7]. In contrast to electrons and positrons, the forces between which vanish as they are separated infinitely far from each other, the quark–antiquark pair is *confined*. If, on the other hand, the perimeter law is obeyed, the pair can be separated arbitrarily far with a finite cost of energy. Now, in the real world, if you pull apart the quark and antiquark in a meson far enough, another quark–antiquark pair will eventually come out of the vacuum (because their price of $2mc^2$ is cheaper than the increasing $V(R)$), partner with the original pair, and form two mesons. It is like trying to break a bar magnet in half, to get just a south or north pole in each hand, and finding instead two bar magnets, each with two opposite poles. That screening of the force at long distance does not happen in a *pure* gauge theory coupled to the loop O_Γ since the only pair that exists is the one we put in by hand in the loop. This is why they are called non-dynamical quarks. They stick to the contour we assign to them, and there is no sum over paths for them or any question of creating pairs from the vacuum. If, however, we add dynamical quarks to the gauge theory, i.e., with the terms $\phi(\boldsymbol{r})U_\mu(\boldsymbol{r})\phi(\boldsymbol{r}+\boldsymbol{e}_\mu)$ in the action, these will screen the external source O_Γ as described.

In summary, a pure gauge theory can have an area or perimeter law (corresponding to confined or deconfined phases) depending on whether the external non-dynamical pair experiences a linear potential or not. In a theory with dynamical charges, the test charges in the Wilson loop could get screened. I say "could" because it may not happen in general. In the Z_2 theory, there is only kind of matter field (± 1) and it will always screen the test charge (± 1), of which there is also just one kind. In a theory like electrodynamics or QCD, the dynamical charges could be in a representation that cannot combine to shield the test charges, say when the dynamical charges are in the adjoint representation and the test charges are in the fundamental (quark) representation. In this case the area law can be used as a diagnostic for confinement.

Let us be clear about this: quarks can be (and are believed in our world to be) in the confining phase, in that we will never see them isolated from their antiparticles, but we cannot use the area law of the Wilson loop as a diagnostic for confinement.

We now turn to a more systematic treatment of what can and cannot happen in general in gauge theories with matter and what will decide the outcome. To this end we must first acquaint ourselves with one more topic that will enter this discussion: Z_q gauge theories.

10.4.1 Z_q Gauge Theories

The elements of the Z_q group generalize the Ising variables $s = \pm 1$, which may be viewed as the two roots of unity. They form a group under multiplication. If we write

$$+1 = e^{i\pi \cdot 0}, \quad -1 = e^{i\pi \cdot 1}, \tag{10.70}$$

the exponents add modulo 2 under group multiplication.

The elements of the group Z_q (often called Z_N) are the q roots of unity:

$$z_n = \exp\left[\frac{2\pi i n}{q}\right], \quad n = 0, 1, \ldots, q-1. \tag{10.71}$$

The Ising model corresponds to Z_2.

Under multiplication, the exponents add modulo q:

$$z_n z_m = z_{[m+n \text{ mod } q]}. \tag{10.72}$$

For $q > 2$, the elements are distinct from their complex conjugates.

The element living on the bond starting at $\boldsymbol{r}$ and pointing in the μ direction is denoted as

$$U_\mu(\boldsymbol{r}) = \exp\left[\frac{2\pi i n_\mu(\boldsymbol{r})}{q}\right], \quad n_\mu = 0, 1, \ldots, (q-1), \tag{10.73}$$

and gauge transforms as follows:

$$U_\mu(\boldsymbol{r}) \to z(\boldsymbol{r}) U_\mu(\boldsymbol{r}) z^*(\boldsymbol{r}+\mu), \tag{10.74}$$

where $z(\boldsymbol{r})$ is any element of Z_q at site $\boldsymbol{r}$. The gauge-invariant action is

$$S = \frac{K}{2} \sum_{\boldsymbol{r}\mu\nu} \left[U_\mu(\boldsymbol{r}) U_\nu(\boldsymbol{r}+\boldsymbol{e}_\mu) U_\mu^\dagger(\boldsymbol{r}+\boldsymbol{e}_\nu) U_\nu^\dagger(\boldsymbol{r}) + cc \right], \tag{10.75}$$

where $U^\dagger = U^*$ for Z_q.

Exercise 10.4.1 *Verify the invariance of S in Eq. (10.75) under the gauge transformation in Eq. (10.74).*

If we include matter fields $\phi(\boldsymbol{r})$, also elements of Z_q at the sites $\boldsymbol{r}$, the action

$$\begin{aligned} S = {} & \frac{\beta}{2} \sum_{\boldsymbol{r}\mu} \left(\phi^*(\boldsymbol{r}) U_\mu(\boldsymbol{r}) \phi(\boldsymbol{r}+\boldsymbol{e}_\mu) + cc \right) \\ & + \frac{K}{2} \sum_{\boldsymbol{r}\mu\nu} \left[U_\mu(\boldsymbol{r}) U_\nu(\boldsymbol{r}+\boldsymbol{e}_\mu) U_\mu^\dagger(\boldsymbol{r}+\boldsymbol{e}_\nu) U_\nu^\dagger(\boldsymbol{r}) + cc \right] \end{aligned} \tag{10.76}$$

would be invariant under

$$U_\mu(\boldsymbol{r}) \to z(\boldsymbol{r}) U_\mu(\boldsymbol{r}) z^*(\boldsymbol{r}+\mu), \tag{10.77}$$

$$\phi(\boldsymbol{r}) \to \phi(\boldsymbol{r}) z(\boldsymbol{r}). \tag{10.78}$$

10.5 Fradkin–Shenker Analysis

A revealing treatment of gauge fields coupled to matter was given by Fradkin and Shenker [9]. I give only the highlights, directing you to the very readable original for more details.

10.5.1 Matter in the Fundamental Representation

Consider first the Z_2 theory with action

$$S=\beta\sum_{r\mu}\phi(\boldsymbol{r})U_\mu(\boldsymbol{r})\phi(\boldsymbol{r}+\boldsymbol{e}_\mu)+K\sum_{r\mu\nu}\left[U_\mu(\boldsymbol{r})U_\nu(\boldsymbol{r}+\boldsymbol{e}_\mu)U_\mu(\boldsymbol{r}+\boldsymbol{e}_\nu)U_\nu(\boldsymbol{r})\right]. \quad (10.79)$$

Let us use this opportunity to learn about the *unitary gauge*. In this gauge the matter fields are fully eliminated as follows. In schematic form we begin with

$$S=\beta\sum_{\text{bonds}}\phi_i U_{ij}\phi_j+K\sum_{\square}U_{ij}U_{jk}U_{kl}U_{li}. \quad (10.80)$$

Now, the following gauge transformations will preserve the action and the measure for sum (or, later, integration) over configurations:

$$\phi_i\to\phi'_i=t_i\phi_i,\qquad t_i=\pm1, \quad (10.81)$$
$$\phi_j\to\phi'_j=t_j\phi_j, \quad (10.82)$$
$$U_{ij}\to U'_{ij}=t_iU_{ij}t_j. \quad (10.83)$$

The resulting action will be

$$S'=\beta\sum_{\text{bonds}}\phi'_iU'_{ij}\phi'_j+K\sum_{\square}U'_{ij}U'_{jk}U'_{kl}U'_{li}. \quad (10.84)$$

Let us now choose, for any given configuration of ϕ's, the gauge transformation $t_i=\phi_i$. Then

$$\phi'_i=+1\ \forall\, i, \quad (10.85)$$
$$S'=\beta\sum_{\text{bonds}}U'_{ij}+K\sum_{\square}U'_{ij}U'_{jk}U'_{kl}U'_{li}. \quad (10.86)$$

The phase diagram is shown in Figure 10.3.

On the edge $K=0$, the action reduces to the sum over independent U's of $e^{\beta U}$, the partition function of decoupled Ising spins in some magnetic field. There is no room for any phase transition as a function of β.

On the edge $\beta=0$, we have a pure Ising gauge theory with its perimeter (large-K) and area (small-K) laws, assuming $d>2$. However as soon as $\beta>0$, the perimeter law sets in because the matter fields screen the area law.

On the edge $\beta=\infty$, all U' must equal $+1$ and nothing interesting is possible as a function of K.

On the right edge, where $K\to\infty$, every plaquette must have a value $+1$. This can be realized as an identity by introducing site variables $s=\pm1$ in terms of which

$$U'_{ij}=s_is_j. \quad (10.87)$$

The action describes an Ising model with coupling β with its order–disorder transition as a function of β. This phase where the symmetry is broken is called the Higgs phase.

But weren't gauge symmetries always unbroken? Yes, that is correct, and this breaking of symmetry is an illusion due to gauge fixing.

We can understand this better if we go back to the original action. When $K = \infty$ we want every plaquette to equal $+1$. Now the configuration with all $U_{ij} = 1$ (where i and j are neighbors connected by the link U_{ioj} lives on) certainly does the job, but so do all its gauge transforms. Let us pick this one with $U_{ij} \equiv 1 \; \forall \; ij$ as the representative element. Consider the correlator $\langle \phi_1 \phi_N \rangle$, where sites 1 and $N \to \infty$ are very far apart along the x axis. Suppose this average tends to a non-zero constant, signaling symmetry breaking and a non-zero average of what seems to be a non-gauge-invariant object. How can that be? The answer is that this object is actually the gauge-invariant entity $\langle \phi_1 U_{1,2} U_{2,3} \cdots U_{N-1,N} \phi_N \rangle$ *in the gauge* $U = 1$. When you hear particle physicists say that the Higgs field has acquired a vacuum expectation value, this is what they mean. Once you pick a gauge, a seemingly non-gauge-invariant object can have a non-zero average and be physically significant.

Fradkin and Shenker showed that the critical points $(0, K_c)$ and (β_c, ∞) extend into the square perturbatively. Figure 10.3 shows the simplest continuation inside. Both the Higgs and confinement phases obey the perimeter law in the case of Z_2 with matter fields.

In the electroweak theory, ϕ is called the Higgs field and the phase in the top right-hand corner, connected to the ordered phase, is called the *Higgs phase*. In a strip to the left of $K = K_c$ is the *confining phase*. When $\beta = 0$, the pure gauge theory indeed goes from area to perimeter law at this point. When the matter coupling is turned on, this phase merges smoothly with the Higgs phase since no phase transitions separate them. All this seems very reasonable in the Z_2 theory, but was far from obvious in particle physics. In the Higgs

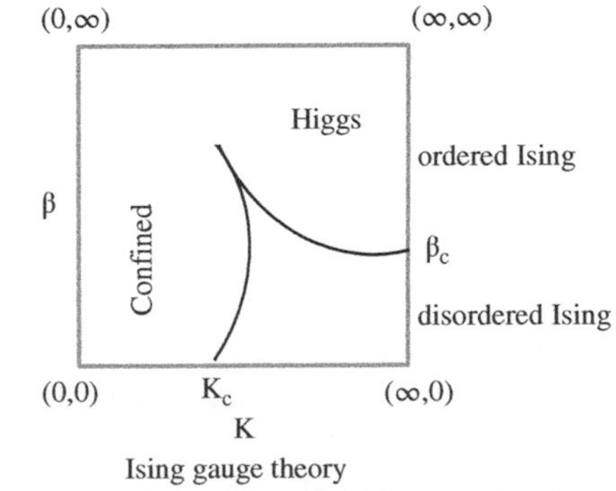

Figure 10.3 The topology of phase diagram of the Z_2 lattice gauge theory. When $\beta = 0$ we have the pure Ising gauge theory with area $(K < K_c)$ and perimeter $(K > K_c)$ phases. It is assumed the transitions at K_c and β_c exist in the dimensions considered. For any $\beta > 0$ the area law is replaced by the perimeter law due to matter field coupling. When $K = \infty$ we have a non-gauge Ising model with its ordered (or Higgs) and disordered phases as a function of β. Nothing happens on the other two sides.

phase an order parameter in ϕ has appeared and the gauge bosons have become massive and appear in the spectrum: these are the W and Z bosons found at CERN. In the confinement phase the gauge bosons are trapped and cannot appear in the physical spectrum. Susskind (unpublished) and 't Hooft had realized and argued that these were still the same phase with the same kind of particle spectrum. The work of Fradkin and Shenker, along with others, fleshed out this idea.

10.5.2 Matter Field in a Higher Representation

It was pointed out that if the dynamical matter fields were in a representation of higher charge they could not screen a loop of the lowest charge and the area and perimeter phases would be distinct. Unfortunately, the Z_2 theory cannot be used to illustrate this fact since it has only one representation. So we consider the least complicated model that can illustrate this result: an Abelian $U(1)$ model analyzed by Banks and Rabinovici [10–12].

The action is

$$S = \frac{\beta}{2}\sum_{r\mu}\left[\phi^*(\boldsymbol{r}+\boldsymbol{r}_\mu)U^q_\mu(\boldsymbol{r})\phi(\boldsymbol{r}) + cc\right]$$
$$+\frac{K}{2}\sum_{r\mu\nu}\left[U_\mu(\boldsymbol{r})U_\nu(\boldsymbol{r}+\boldsymbol{e}_\mu)U^\dagger_\mu(\boldsymbol{r}+\boldsymbol{e}_\nu)U^\dagger_\nu(\boldsymbol{r}) + cc\right]. \tag{10.88}$$

The $U(1)$ variables are

$$\phi(\boldsymbol{r}) = e^{i\theta(r)}, \tag{10.89}$$
$$U_\mu(\boldsymbol{r}) = \exp\left[iA_\mu(\boldsymbol{r})\right], \tag{10.90}$$
$$U^q_\mu(\boldsymbol{r}) = \exp\left[iqA_\mu(\boldsymbol{r})\right], \tag{10.91}$$

where θ and A_μ lie in the interval $[0, 2\pi]$, $U^\dagger = U^*$, and q is a parameter restricted to be an integer to keep U^q_μ single-valued, and will be shortly identified as the charge of the matter field. Finally, and in schematic form,

$$Z = \prod_{r\mu}\int_0^{2\pi} d\theta(\boldsymbol{r})\int_0^{2\pi} dA_\mu(\boldsymbol{r})e^S \equiv \left[\int \mathcal{D}\theta\mathcal{D}A_\mu\right]e^S. \tag{10.92}$$

In terms of θ and A_μ, the action becomes

$$S = \beta\sum_{r\mu}\cos\left[\Delta_\mu\theta(\boldsymbol{r}) - qA_\mu(\boldsymbol{r})\right] + K\sum_{\square\mu\nu}\cos F_{\mu\nu}, \tag{10.93}$$

where

$$\Delta_\mu f(\boldsymbol{r}) = f(\boldsymbol{r}+\boldsymbol{e}_\mu) - f(\boldsymbol{r}), \tag{10.94}$$
$$F_{\mu\nu} = \Delta_\mu A_\nu - \Delta_\nu A_\mu. \tag{10.95}$$

Exercise 10.5.1 *Prove Eq. (10.93) using the definitions Eqs. (10.94) and (10.95).*

The gauge transformations are

$$\theta(\boldsymbol{r}) \rightarrow \theta'(\boldsymbol{r}) = \theta(\boldsymbol{r}) + q\lambda(\boldsymbol{r}), \tag{10.96}$$

$$A_\mu(\boldsymbol{r}) \rightarrow A'_\mu(\boldsymbol{r}) = A_\mu(\boldsymbol{r}) + \Delta_\mu \lambda(\boldsymbol{r}). \tag{10.97}$$

By comparison to electrodynamics in the continuum, we see that q corresponds to the charge of the matter field. But there are differences. In the continuum, q is also continuous because the description is entirely in terms of $-\infty < A < \infty$ and not U, while on a lattice we can introduce link variables $U = e^{iA}$, with a compact range $0 \leq A_\mu < 2\pi$, which is why q has to be an integer. If U is non-Abelian, the charge gets quantized by the representation of the group that is chosen for ϕ.

We will use the unitary gauge where θ is banished from the picture, i.e. set equal to 0 modulo 2π. This is done by choosing

$$q\lambda(\boldsymbol{r}) = -\theta(\boldsymbol{r}) + 2\pi m, \quad m = 0, 1, \ldots, q-1 \tag{10.98}$$

or

$$\lambda = -\frac{\theta}{q} + \frac{2\pi m}{q}, \qquad m = 0, 1, \ldots, q-1. \tag{10.99}$$

The net result is that any of these choices will lead to

$$\theta'(\boldsymbol{r}) = \theta(\boldsymbol{r}) + q\lambda(\boldsymbol{r}) = 2\pi m \simeq 0, \tag{10.100}$$

$$A'_\mu(\boldsymbol{r}) = A_\mu(\boldsymbol{r}) + \Delta_\mu \lambda(\boldsymbol{r}), \tag{10.101}$$

$$S(\text{unitary}) = \beta \sum_{\boldsymbol{r}\mu} \cos\left[qA'_\mu(\boldsymbol{r})\right] + K \sum_{\square\mu\nu} \cos F'_{\mu\nu}. \tag{10.102}$$

Hereafter, the prime on A' will be dropped and the unitary gauge assumed.

Consider the edges of the phase diagram in Figure 10.4. When $K = 0$, the variables $A_\mu(\boldsymbol{r})$ are unrestricted and completely decoupled in space, and nothing can happen as a function of β.

When $\beta = 0$, we have a pure gauge theory as a function of K with its area and perimeter laws.

When $K = \infty$ we want F to vanish, i.e., A to be a pure gauge,

$$A_\mu = \Delta_\mu \theta, \tag{10.103}$$

which leads to an XY model, which will have its order–disorder transition as a function of β. The ordered phase is the Higgs phase. In this phase the gauge boson is massive and the perimeter law applies.

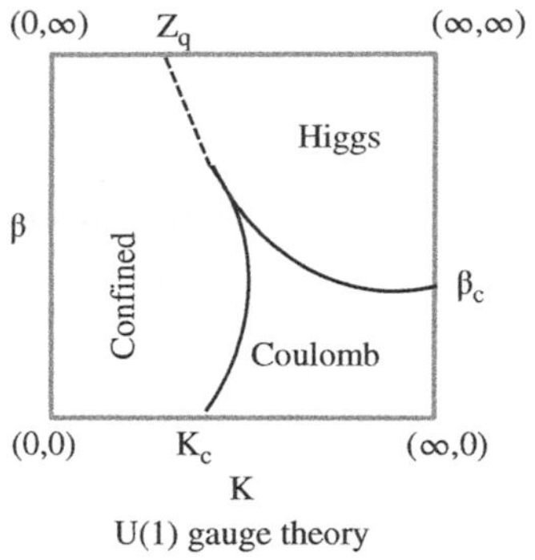

Figure 10.4 Phase diagram of the $U(1)$ lattice gauge theory. If the charge is the smallest ($q = 1$), ignore the dotted line terminating at the top edge at Z_q, where the pure Z_q gauge theory has a phase transition. It is assumed that the transitions at K_c and β_c exist in the dimensions considered.

None of the three edges considered depends sensitively on q, the representation of the matter field. Now consider the edge $\beta = \infty$. We need $\cos qA_\mu = 1$, or

$$A_\mu = \frac{2\pi j}{q}, \quad j = 0, 1, \ldots, q-1, \tag{10.104}$$

which makes $U = e^{iA}$ the elements of Z_q, the q roots of unity.

For $q = 1$, this makes $A_\mu = 0$ and nothing can happen as we vary K. This means that the Higgs and confinement phases are smoothly connected, i.e., one and the same. As soon as $\beta > 0$, the perimeter law sets in if $q = 1$ (matter charge has lowest value).

For $q > 1$, the K terms describe a pure Z_q gauge theory that will have at least two phases (assuming $d > 2$), corresponding to area and perimeter laws. Now a phase transition will separate the Higgs and confinement phases, as shown by the dashed line in Figure 10.4. The area law will hold in the confining region for lowest-charge loops that cannot be screened by the dynamical charges.

Finally, consider the bottom-right corner. Since θ is a continuous variable, this will be a *Coulomb phase*, with a gapless photon and charges that can be arbitrarily widely separated (deconfined). Focus on the K term for large K, when $K \cos F_{\mu\nu}$ will restrict $F_{\mu\nu}$ to be small. (We do not have this option in the Z_2 theory because the corresponding $UUUU$ terms can only be ± 1.) In the $U(1)$ case we will obtain

$$S = -\frac{K}{2} \sum_{\square \mu\nu} F_{\mu\nu}^2, \tag{10.105}$$

which describes the familiar Maxwell theory. By bringing in a lattice spacing a, we may define the continuum photon field by $A_\mu \to aA_\mu$ and replace the sum over $\square$ by an integral over Euclidean spacetime.

Such a Coulomb phase may not always exist: for example, in non-Abelian theories the Coulomb phase is absent for $d \leq 4$. The Abelian case does have a Coulomb phase in $d = 4$, and corresponds to the familiar photon, which when coupled to matter yields quantum electrodynamics. It is absent in $d = 3$, as shown by Polyakov [13].

References and Further Reading

[1] J. B. Kogut, Reviews of Modern Physics, **51**, 659 (1979).
[2] E. Stanley and T. Kaplan, Physical Review Letters, **17**, 913 (1966).
[3] D. Mermin and H. Wagner, Physical Review Letters, **17**, 1133 (1966).
[4] P. Hohenberg, Physical Review, **158**, 383 (1967).
[5] S. Coleman, Communications in Mathematical Physics, **31**, 259 (1973).
[6] F. Wegner, Journal of Mathematical Physics, **12**, 2259 (1971).
[7] K. G. Wilson, Physical Review D, **10**, 2445 (1974).
[8] S. Elitzur, Physical Review D, **12**, 3978 (1975).
[9] E. Fradkin and S. Shenker, Physical Review D, **19**, 3682 (1979).
[10] T. Banks and E. Rabinovici, Nuclear Physics B, **160**, 349 (1979).
[11] G. 't Hooft, in *Recent Developments in Field Theory*, ed. G. 't Hooft *et al.*, Plenum (1980).
[12] M. Einhorn and R. Savit, Physical Review D, **19**, 1198 (1979).
[13] A. M. Polyakov, *Gauge Fields and Strings*, CRC Press (1987).

11

The Renormalization Group

The renormalization group or RG was originally introduced to deal with the unwanted infinities that arose in relativistic field theory. In this incarnation it was rather hard to follow, as I know from experience as a budding field theorist. Just when I finally got accustomed to divergent answers and subtracting infinities, I moved to condensed matter physics, and was promptly traumatized when my first integral converged on my face. How was I to extract the physics? Would nature be so cruel as to bury her secrets inside convergent integrals? I was then rescued, not a moment too soon, by the modern version of the RG due to Kadanoff and Wilson. It was love at first sight. It is this approach I will discuss here, though at a later stage I will discuss its relation to the renormalization program of field theory.

11.1 The Renormalization Group: First Pass

The need for the RG arises in myriad problems that have nothing to do with infinities. I will focus on its application to problems that can be cast in the form of a partition function. While this does not cover all the possibilities, it covers a lot. Besides the partition functions of traditional classical and quantum statistical mechanics, all of Euclidean quantum mechanics (and this includes quantum field theory) can be formulated in terms of partition functions.

Consider the very simple partition function

$$Z(a,b) = \int dx \int dy e^{-a(x^2+y^2)} e^{-b(x+y)^4}, \tag{11.1}$$

where a, b are the parameters and x and y are two variables that fluctuate with the Boltzmann weight represented by the exponential. The average of any quantity $f(x,y)$ is

$$\langle f \rangle = \frac{\int dx \int dy f(x,y) e^{-a(x^2+y^2)} e^{-b(x+y)^4}}{\int dx \int dy e^{-a(x^2+y^2)} e^{-b(x+y)^4}}. \tag{11.2}$$

First consider $b = 0$, *the Gaussian model.* Suppose that we are just interested in the *statistical properties* of functions of x. Then we have the option of integrating out y and working with the new partition function as follows:

$$\langle f \rangle = \frac{\int dx f(x) e^{-ax^2} \int dy e^{-ay^2}}{\int dx e^{-ax^2} \int dy e^{-ay^2}} \tag{11.3}$$

$$= \frac{\int dx f(x) e^{-ax^2}}{Z(a)}, \quad \text{where} \tag{11.4}$$

$$Z(a) = \int dx e^{-ax^2}. \tag{11.5}$$

In other words, if we are only interested in x and never in y, we may use $Z(a)$ to perform all averages. In this simple case, if we do not care about y, what we have done amounts to simply setting $y = 0$ everywhere. But this will not work in the non-Gaussian case $b \neq 0$ when there are cross terms between x and y. Now we have to derive the partition function $Z(a', b', c', \ldots)$ that describes x by *integrating out* y as follows.

Let us begin with

$$\langle f \rangle = \frac{\int dx f(x) \int dy e^{-a(x^2+y^2)} e^{-b(x+y)^4}}{\int dx \left[\int dy e^{-a(x^2+y^2)} e^{-b(x+y)^4} \right]}. \tag{11.6}$$

Look at the y-integral. It must be positive, because the integrand is, and hence can be written as the exponential of a function of x with some real coefficients. The function has to be even, because we can change x to $-x$ and get the same answer if we simultaneously change y to $-y$ in the integral. In other words, we may assert that

$$\int dy e^{-a(x^2+y^2)} e^{-b(x+y)^4} = e^{-a'x^2 - b'x^4 - c'x^6 + \cdots}, \tag{11.7}$$

where $a', b', c', \ldots$ are defined by the above equation. Continuing,

$$\langle f \rangle = \frac{\int dx f(x) e^{-a'x^2 - b'x^4 - c'x^6 + \cdots}}{\int dx e^{-a'x^2 - b'x^4 - c'x^6 + \cdots}} \tag{11.8}$$

$$\equiv \frac{\int dx f(x) e^{-a'x^2 - b'x^4 - c'x^6 + \cdots}}{Z(a', b', \ldots)}, \quad \text{where} \tag{11.9}$$

$$Z(a', b', \ldots) = \int dx \left[\int dy e^{-a(x^2+y^2)} e^{-b(x+y)^4} \right]. \tag{11.10}$$

If we write in the notation of Euclidean path integrals,

$$Z(a', b', c', \ldots) = \int dx e^{-S'(a', b', c', \ldots)}, \tag{11.11}$$

where S' is the *action*, we have the central result

$$e^{-S'(a', b', c', \ldots; x)} = \int dy e^{-a(x^2+y^2)} e^{-b(x+y)^4} \equiv \int dy e^{-S(a, b, \ldots; x, y)}. \tag{11.12}$$

The primed coefficients a', b', and so on define the parameters of the *effective theory* for x, and $S'(a',b',c',\ldots)$ is the *effective action*. These primed parameters will reproduce exactly the same averages for x as the original ones, but with no reference to y. However, we did not eliminate y by simply setting it equal to zero; we included its full impact on the fate of x by *integrating it out*, to get the same answers as in the presence of y. In other words, the fate of x was tied to that of y due to the coupling between them in the action. What we have in the end is an effective theory or action S' which *modifies the coupling of x to itself* in such a way that the same answers are obtained for any questions involving x. It is not obvious *a priori* that such an effective theory exists, but we know it does, having just outlined its construction.

A more advanced example is where x stands for the charged particle degrees of freedom and y for those of the electromagnetic field. The interaction between the two will be encoded in cross terms involving both x and y. Suppose we only care about particles. If we simply set $y=0$, we will end up describing non-interacting particles. On the other hand, if we integrate out the electromagnetic field, we will get an effective theory of interacting electrons with predictions identical to the one at the outset.

The evolution of parameters $(a,b,c,\ldots) \to (a',b',c',\ldots)$ upon the elimination of uninteresting degrees of freedom is what we mean these days by renormalization. As such, it has nothing to do with infinities. We just saw it happen in a problem with just two variables and no infinities in sight.

The parameters $(a,b,c,\ldots)$ or $(a',b',c',\ldots)$ are called *couplings*, and the terms they multiply ($(x+y)^4$, for example) are called *interactions*. The x^2 term is called the *kinetic* or *free-field* term and its coefficient a or a' can be chosen to be a constant like $\frac{1}{2}$ by rescaling x. Of course, once we have rescaled a' in this manner, we are stuck with the values of the other couplings.

Notice that to get the effective theory we generally need to do a non-Gaussian integral. This can only be done perturbatively. At the simplest *tree level*, we simply drop y and find $b' = b$. At higher orders, we expand the non-quadratic exponential in powers of y and integrate monomials of the form $x^m y^n e^{-ay^2}$ with respect to y (term by term), re-exponentiate the answer, and generate effective interactions for x. For those who know them, this procedure can be represented by Feynman diagrams in which *integrals in the loops are over the variables being eliminated or integrated*. We will discuss this in some detail later on.

So this is how we renormalize. But why do we bother to do that?

One common reason is that certain tendencies of x are not so apparent when y is around, but come to the surface as we zero in on x. For example, we are going to consider a problem in which x stands for low-energy variables and y for high-energy variables. Suppose our focus is on the ground state and its low-lying excitations. Upon integrating out high-energy variables to obtain the effective Hamiltonian for the low-energy sector, a coupling that used to be numerically small in the full theory can grow in size and dominate (or an initially impressive one diminish into oblivion). By focusing our attention on the dominant couplings we can often guess the nature of the ground state.

Renormalization is invoked in understanding problems that have very different Boltzmann weights or actions and yet exhibit identical singularities near a second-order phase transition. This confounding feature is now demystified: the initially different Boltzmann weights renormalize to the same Boltzmann weight for the surviving variables, which alone control the singularities at the phase transition.

These notions can be made more precise as follows. Consider the Gaussian model in which only $a \neq 0$. We have seen that this value does not change as y is eliminated because x and y do not talk to each other. This is called a *fixed point of the RG*. Now turn on new couplings or "interactions" (corresponding to higher powers of x, y, etc.) with coefficients b, c, and so on. Let a', b', ... be the new couplings after y is eliminated. The mere fact that $b' > b$ does not mean b is more important for the physics of x. This is because a' could also be bigger than a. So we rescale x so that the kinetic part, x^2, has the same coefficient as before. If the quartic term still has a bigger coefficient (still called b'), we say it is a *relevant coupling*. If $b' < b$ we say it is an *irrelevant coupling*. This is because in reality y stands for many variables, and as they are eliminated one by one, the relevant couplings will keep growing and the irrelevant ones will shrink to zero. If a coupling neither grows nor shrinks it is called a *marginal coupling*. This classification holds even for non-Gaussian fixed points at which many couplings (not just a) could be non-zero. Deviations from this fixed point can again be classified as relevant, irrelevant, or marginal depending on whether they grow, shrink, or remain the same under renormalization.

Two problems that flow to the same fixed point will have the same behavior with respect to the surviving variables. This, as I mentioned above, explains why different initial Boltzmann weights have the same singular critical behavior.

Kadanoff [1] was the first to employ this strategy of trading the given problem for a new one with fewer degrees of freedom and different couplings. It was then elevated to dizzying heights by the *tour de force* of Wilson [2]. Wilson's paper is straight from the source, and very easy to read. It describes his language of fixed points and relevant, irrelevant, and marginal operators. A careful reading of it will prevent many common misunderstandings in the application of the RG.

11.2 Renormalization Group by Decimation

The term *decimation* is generally used to describe the elimination of a subset of degrees of freedom labeled by points in space. (It was used in ancient Rome as a punishment in which a tenth of an army unit, chosen by lots, was sentenced to death, the sentence often to be carried out by the other nine-tenths.) I will now describe the simplest non-trivial example first discussed by Nelson and Fisher [3].

Consider the $d = 1$ Ising model with

$$Z(K) = \sum_{s_i=\pm 1} \exp\left[K \sum_{i=-\infty}^{\infty} s_i s_{i+1} \right]. \tag{11.13}$$

Let us say that our goal is to find the correlation function between two spins a distance n apart:

$$G(n,K) = \frac{\sum_{s_i=\pm 1} s_0 s_n \exp\left[K \sum_{i=-\infty}^{\infty} s_i s_{i+1}\right]}{Z(K)}. \tag{11.14}$$

Let us assume $n=32$. It follows that we can get an effective theory for computing $G(32,K)$ even if we eliminate all the odd sites. Thus, the spins at $(\ldots, -5, -3, -1, 1, 3, 5, \ldots)$ are to be summed over. This will take some work because the even sites talk to the odd sites. Look at what happens when we eliminate s_1. Begin by asking where it makes its appearance:

$$Z(K) = \sum_{s_i=\pm 1} \cdots e^{Ks_{-3}s_{-2}} \cdot e^{Ks_{-2}s_{-1}} \cdot e^{Ks_{-1}s_0} \cdot e^{Ks_0 s_1} \cdot e^{Ks_1 s_2} \cdots \tag{11.15}$$

The sum over s_1 requires us to evaluate just the part involving s_1 that talks to its neighbors s_0 and s_2:

$$\sum_{s_1=\pm 1} e^{Ks_0 s_1} \cdot e^{Ks_1 s_2} = \sum_{s_1=\pm 1} e^{Ks_1(s_0+s_2)}. \tag{11.16}$$

This sum over s_1 is positive, expressible as a real exponential, and it depends only on s_0 and s_2. It must be unaffected under $s_0 \to -s_0$, $s_2 \to -s_2$, for this change can be absorbed by redefining $s_1 \to -s_1$ in Eq. (11.16). It can only contain the zeroth and first powers of these variables because they square to $+1$. Thus, it must be that

$$\sum_{s_1=\pm 1} e^{Ks_1(s_0+s_2)} = e^{K' s_0 s_2 + F}, \tag{11.17}$$

where K' is the renormalized coupling constant that describes the surviving even spins and F is the contribution to the free energy from the sum over s_1. To evaluate K' and F we need two independent equations, and these follow from choosing $(s_0, s_2 = +1, +1)$ and $(s_0, s_2 = +1, -1)$ on both sides:

$$\sum_{s_1=\pm 1} e^{Ks_1(1+1)} = e^{K' 1\cdot 1 + F}, \tag{11.18}$$

$$\sum_{s_1=\pm 1} e^{Ks_1(1-1)} = e^{K' 1\cdot(-1) + F}. \tag{11.19}$$

More explicitly, we have

$$\sum_{s_1=\pm 1} e^{2Ks_1} = e^{K'+F}, \tag{11.20}$$

$$\sum_{s_1=\pm 1} e^{0\cdot Ks_1} = e^{-K'+F}, \tag{11.21}$$

or

$$2\cosh 2K = e^{K'+F}, \tag{11.22}$$
$$2 = e^{-K'+F}. \tag{11.23}$$

(The other two spin choices will give the same results.) These equations can be solved to give

$$e^{2K'} = \cosh 2K, \tag{11.24}$$
$$e^{2F} = 4\cosh 2K. \tag{11.25}$$

(We will ignore F in the rest of this discussion.) The first equation may be rewritten as

$$\tanh K' = \tanh^2 K. \tag{11.26}$$

Exercise 11.2.1 *Verify Eq. (11.26) starting with Eq. (11.24).*

After this decimation of the odd sites, the separation between the old points s_0 and s_{32} is now 16 in the new lattice units. The correlation between them is, of course, the same as before since we have not touched them in any way. But we now see it as the correlation of spins 16 sites apart in a theory with coupling

$$K' = \tanh^{-1}(\tanh^2 K). \tag{11.27}$$

In other words, we assert

$$G(32,K) = G(16,K'). \tag{11.28}$$

This agrees with the familiar result

$$G(n,K) = (\tanh K)^n, \tag{11.29}$$

according to which

$$G(32,K) = (\tanh K)^{32} = (\tanh^2 K)^{16} = G(16,K'). \tag{11.30}$$

Having decimated once, we can do it again and again. At every stage the spins will come closer by a factor of two and the running value of $\tanh K$ will get squared. For $n = 32$, the original spins will become nearest neighbors after four such renormalizations or decimations. Had the second spin been at $n = 2^{r+1}$, the two would become neighbors after r renormalizations. If we denote by $K^{(r)}$ the coupling after r iterations we have the sequence

$$K^{(0)} \to K^{(1)} \to K^{(2)} \to K^{(3)} \dots \tag{11.31}$$
$$n \to \frac{1}{2}n \to \frac{1}{2^2}n \to \frac{1}{2^3}n \cdots \tag{11.32}$$

The origin of the name renormalization *group* is evident above. One decimation followed by another is equivalent to a third one in which $K^{(r)} \to K^{(r+2)}$ and the lattice spacing is quadrupled or the distance between sites (in new lattice units) reduced by a factor of four. The same goes for combining any two decimations. The RG is, however, a *semigroup* since the process has no inverse: once some variable is summed over, all information about it is lost. In other words, a renormalized or effective interaction can have many ancestors that led to it on elimination of variables.

One is often interested in the behavior of $G(n,K)$ as $n \to \infty$ for some given K. The RG allows us to map this into the G for a fixed separation (say nearest neighbor) but at a coupling renormalized r times, assuming $n = 2^{r+1}$ for convenience.

In terms of the dimensionless correlation length $\xi(K)$ defined by

$$G(n,K) = e^{-n/\xi(K)} \tag{11.33}$$

and measured in the current lattice units, the result

$$G(n,K) = G\left(\frac{n}{2}, K'\right) \tag{11.34}$$

becomes

$$\xi(K') = \frac{1}{2}\xi(K). \tag{11.35}$$

This is a very profound result.

Recall that a fixed point is one that does not flow under renormalization. In our example, this means

$$K'(K) = K = K^*, \text{ say.} \tag{11.36}$$

Then Eq. (11.35) implies

$$\xi(K^*) = \frac{1}{2}\xi(K^*). \tag{11.37}$$

Alert! In this context K^* denotes the fixed-point value of K and not its dual, related by $\tanh K^* = e^{-2K}$. In what follows I will denote all fixed-point quantities with asterisks.

This has two solutions: $\xi(K^*) = 0$ or ∞. Or, going back to Eq. (11.26), Eq. (11.36) implies

$$\tanh K^* = \tanh^2 K^*, \tag{11.38}$$

which corresponds to two different fixed points:

$$\xi(K^*) = 0, \quad \tanh K^* = 0, \quad K^* = 0, \quad T^* = \infty; \tag{11.39}$$

$$\xi(K^*) = \infty, \quad \tanh K^* = 1, \quad K^* = \infty, \quad T^* = 0, \tag{11.40}$$

where T is the temeprature.

The fixed point $T = \infty$ or $K = 0$ corresponds to spins not coupled to their neighbors, which explains the correlation length of 0. This is called a *trivial fixed point* and is not very interesting.

More important and interesting is the fixed point at $T = 0$ or $K = \infty$, which has infinite correlation length: $G(n) = (\tanh K^*)^n = 1^n$ does not fall with distance at all.

Exercise 11.2.2 *Perform the decimation with a magnetic field starting with*

$$\sum_{s_1} e^{Ks_1(s_0+s_2)+hs_1} = e^{K's_0s_2+\frac{1}{2}\delta h(s_0+s_2)+F}. \tag{11.41}$$

Observe that each surviving site s_0 and s_2 receives only half *the contribution to δh from the elimination of s_1. The other half for s_0 and s_2 comes from the elimination of s_{-1} and s_3 respectively. You should find:*

$$e^{2h'} = e^{2h+2\delta h} = e^{2h}\frac{\cosh(2K+h)}{\cosh(2K-h)}, \tag{11.42}$$

$$e^{4K'} = \frac{\cosh(2K-h)\cosh(2K+h)}{\cosh^2 h}. \tag{11.43}$$

11.2.1 Decimation in $d = 2$

The success we had in $d = 1$ is atypical. The following example in $d = 2$ illustrates the difficulty encountered in general. Consider the square lattice shown in Figure 11.1. The analogs of the even and odd sites of $d = 1$ are the sites on the even and odd sublattices for which $n_x + n_y$, the sum of the integer coordinates n_x and n_y of the lattice point, is even or odd. The spin s_0 in one sublattice, say sublattice A, is surrounded by the four spins $s_1, \dots, s_4$ in the B sublattice. We are going to eliminate the A sublattice. As in $d = 1$, we write

$$\sum_{s_0=\pm 1} e^{Ks_0(s_1+s_2+s_3+s_4)} = e^{\frac{1}{2}K'(s_0s_1+s_2s_3+s_3s_4+s_4s_1)} \times e^{K'_2(s_1s_3+s_2s_4)+K'_4s_1s_2s_3s_4+F}. \tag{11.44}$$

There are some differences compared to $d = 1$.

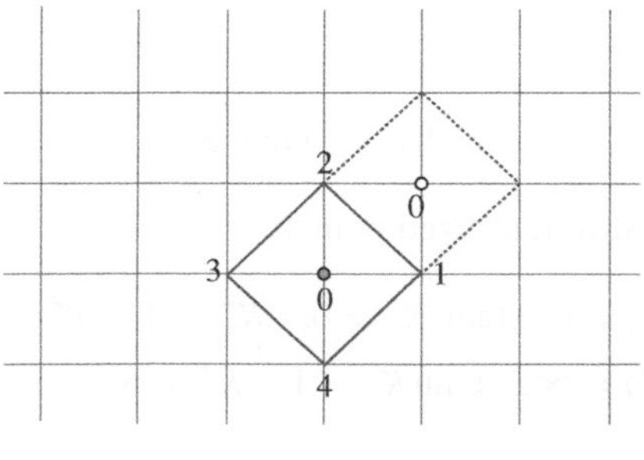

Figure 11.1 The central spin s_0 is eliminated and the result is a lattice rotated by $\frac{\pi}{4}$ and a spacing $\sqrt{2}$ in the old lattice units. Besides a nearest-neighbor interaction $K's_1s_2$, there are second-neighbor terms like $K_2s_1s_3$ and a four-spin term $K_4s_1s_2s_2s_4$. The dotted diamond centered at $0'$ is the adjacent one which provides an equal contribution to K' when the spin there is eliminated.

First, the coefficient in front of the nearest-neighbor terms is $K'/2$ and not K', because these nearest-neighbor couplings will receive equal contributions from the decimation of neighboring spins. For example, the coupling between s_1 and s_2 will be generated once again when we sum over s'_0, the counterpart of s_0 in the adjoining diamond shown by dotted lines.

Next, besides the renormalized nearest-neighbor coupling K', two more couplings K'_2 and K'_4 (with initial values $K_2 = K_4 = 0$) are generated by the decimation. If we decimate again, starting with these three couplings, even more complicated couplings will arise. If we compute K', K'_2, and K'_4 ignoring this complication for now, we find

$$K' = \frac{1}{4}\ln\cosh 4K, \tag{11.45}$$

$$K'_2 = \frac{1}{8}\ln\cosh 4K, \tag{11.46}$$

$$K'_4 = \frac{1}{8}\ln\cosh 4K - \frac{1}{2}\ln\cosh 2K. \tag{11.47}$$

Exercise 11.2.3 *Prove Eqs. (11.45)–(11.47) using any three independent configurations (not related by overall spin reversal).*

How are we to describe the flow of couplings given that two new ones have arisen? The simplest option is to simply ignore K'_2 and K'_4 and set

$$K' = \frac{1}{4}\ln\cosh 4K. \tag{11.48}$$

Unfortunately, this choice admits no fixed point because the equation

$$K^* = \frac{1}{4}\ln\cosh 4K^* \tag{11.49}$$

cannot be satisfied except at $K^* = 0$, which describes decoupled spins. For small $K^* > 0$, the right-hand side is quadratic in K^* and lies below the left-hand side, which is linear. As $K^* \to \infty$, the right-hand side approaches $K^* - \frac{\ln 2}{4}$ and still falls short of K^*.

We know the $d = 2$ Ising model has a critical point with $\xi = \infty$, which must correspond to a fixed point. So we try to include the coupling K_2 in some fashion. One reasonable way is to augment K' by an amount $\delta K'$ in a way that incorporates the ordering tendency produced by a non-zero positive K'_2. There is no way to do this exactly. One approximate way is to choose $K' + \delta K'$ such that in the fully ordered state the energy due to nearest-neighbor coupling $K' + \delta K'$ equals what we would get with K' and K'_2. Since there are as many nearest-neighbor bonds as there are second-neighbor bonds, this leads to $\delta K' = K'_2$ and a final value of

$$K' + \delta K' = K' + K'_2 = \frac{3}{8}\ln\cosh 4K. \tag{11.50}$$

Now there is a fixed point, a solution to

$$K^* = \frac{3}{8}\ln\cosh 4K^*, \tag{11.51}$$

because the right-hand side of Eq. (11.50) now approaches $\frac{3}{2}K$ as $K \to \infty$ and will cross the graph of K, which has unit slope.

The fixed point at $K^* = 0.506$ is close to Onsager's $K^* = 0.440$. For decimation on a triangular lattice, see [4].

11.3 Stable and Unstable Fixed Points

We can learn more about the RG program if we ask what happens if we start slightly away from the fixed point K^* and run the flow. Figure 11.2 shows that the deviation in either direction gets amplified. For example, if we begin with the point 1, the RG map $K \to K'(K)$ takes us straight up to point 2. We now take this output $K'(K)$ and feed it in as the input value by moving horizontally to point 3 and letting the flow $(K \to K'(K))$ take us up to point 4, and so on. Likewise, if we start below K^*, the flow takes us further down along $1' \to 2' \to 3' \ldots$, ending up at $K = 0$, which describes decoupled spins with no correlations. This fixed point at $K^* = 0$ is called the *trivial fixed point* for this reason.

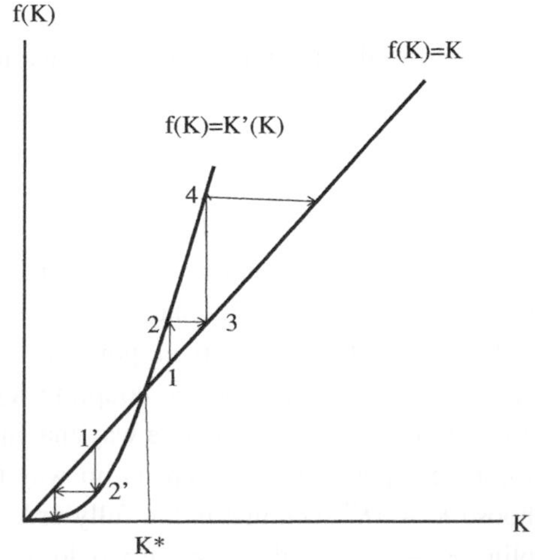

Figure 11.2 If we begin slightly away from the fixed point K^* we move away even more. For example, if we start at 1, we renormalize to 2, and if we feed that in the flow takes us from 3 to 4, which is even further away. The same thing happens if we start below K^*, ending up at $K^* = 0$. One says the fixed point at $K = 0$ is a *stable fixed point*, while the one at the K^* value is an *unstable fixed point*.

The interesting physics of the Ising model is to be found at and near the other fixed point, which we found was at $K^* = 0.506$. In the decimation in $d = 1$ we said

$$\xi(K') = \frac{1}{2}\xi(K) \tag{11.52}$$

because in the new lattice with a spacing twice as big as the old one, the same physical correlation length appears to be half as big.

In $d = 2$ the lattice spacing goes up by a factor $\sqrt{2}$ (and gets rotated by $\frac{\pi}{4}$), and we may assert

$$\xi(K') = \frac{1}{\sqrt{2}}\xi(K), \tag{11.53}$$

where the ξ's on the two sides are measured in units of the lattices they are defined on. We know that $\xi(K^*) = \infty$, but how does it diverge as a function of

$$\Delta K = K - K^*? \tag{11.54}$$

In particular, assuming that on both sides of the fixed point

$$\xi(\Delta K) \simeq |\Delta K|^{-\nu}, \tag{11.55}$$

what is ν, the correlation length exponent?

We need to find out two things and combine them to get to the result:

- What happens to ξ after one RG iteration?
- What happens to the deviation from the fixed point, $\Delta K = K - K^*$, after one iteration?

If ΔK is the deviation to begin with and it becomes $\Delta K'$ after one round of the RG, we have the kinematical result (from the relative sizes of the old and new lattice units in terms of which ξ is measured)

$$\xi(\Delta K') = \frac{1}{\sqrt{2}}\xi(\Delta K). \tag{11.56}$$

Let us see what happens to the deviation ΔK after one iteration. We know by definition of the fixed point that K^* goes into itself:

$$K'(K^*) = K^*. \tag{11.57}$$

Suppose we begin at point $K^* + \Delta K$. It ends up at

$$K'(K^* + \Delta K) = K^* + \left.\frac{dK'(K)}{dK}\right|_{K^*} \Delta K \tag{11.58}$$

$$\equiv K^* + D'(K^*)\Delta K, \tag{11.59}$$

where I have defined

$$D'(K^*) = \left.\frac{dK'(K)}{dK}\right|_{K^*}. \tag{11.60}$$

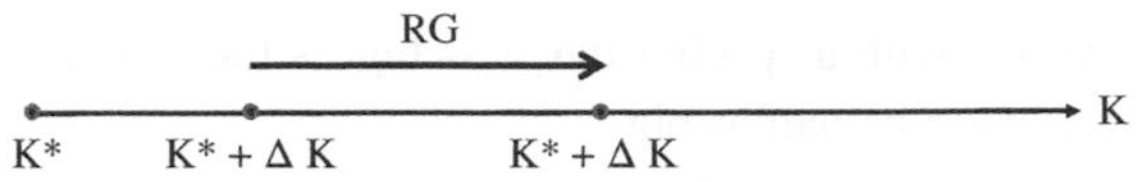

Figure 11.3 A point at $K^* + \Delta K$ flows under the RG to $K^* + \Delta K'$. The deviation from the fixed point gets amplified by $D'(K^*)$ and ξ gets reduced by $1/\sqrt{2}$ (in this example).

So, the new deviation is

$$\Delta K' \equiv K'(K^* + \Delta K) - K^* = D'(K^*)\Delta K. \tag{11.61}$$

The last equation makes precise how ΔK, the initial deviation from the fixed point, is related to the deviation $\Delta K'$ upon one RG iteration: it gets amplified by $D'(K^*)$. So the derivative of the flow at the fixed point is very important. Figure 11.3 describes the situation.

Consequently,

$$\xi(\Delta K') = \frac{1}{\sqrt{2}}\xi(\Delta K), \tag{11.62}$$

$$|\Delta K'|^{-\nu} = |D'(K^*)\Delta K|^{-\nu} \tag{11.63}$$

$$= \frac{1}{\sqrt{2}}|\Delta K|^{-\nu} \text{ from Eq. (11.62)}, \tag{11.64}$$

which means

$$|D'(K^*)|^{-\nu} = \frac{1}{\sqrt{2}}, \text{ or} \tag{11.65}$$

$$\nu = \frac{\ln\sqrt{2}}{\ln|D'(K^*)|} = 0.94, \tag{11.66}$$

which compares very well with Onsager's $\nu = 1$. However, repeating this with K_4' included in a similar manner makes things worse! This is because adding K_2' and K_4' as we did is not a systematic scheme for improvement. Ken Wilson was of the view that one should add all interactions of a given range, and when probed told me unhesitatingly that he would add around 1000 interactions for the next range.

11.4 A Review of Wilson's Strategy

The previous example, despite some caveats to be stated later, is an excellent way to illustrate Wilson's strategy. Let us focus on the $d = 2$ Ising transition, and in particular the computation of ν, the correlation length exponent. The most reliable answer $\nu = 1$ is due to Onsager's legendary calculation. Great though it is, it cannot be applied if we throw in a small second-neighbor or four-spin term. A method that can handle this variation uses the high- and low-temperature series, either coming down from infinite temperature or up from zero temperature. Given enough terms in the series and fancy resummation techniques, one can indeed compute ν to great accuracy after considerable work. For example, we can obtain $\nu = 0.629971(4)$ in $d = 3$. (These indices were derived in the past using the RG or series expansions. The figure quoted here is the best to date, from the *conformal bootstrap approach*. See [5].) A frontal attack on the problem runs into the difficulty that we are trying to use a power series precisely where it is known to diverge. Approximations do not work where a function is singular, and that is exactly what happens to the thermodynamic functions and correlation function at criticality.

Yet in Wilson's RG approach we were able to derive the approximate flow equation for K near K^* and then *even take its derivative right at* K^*! How is it possible to do this and yet get singular answers in the end? Here I provide the analogy from Wilson's paper which says it best.

Consider a particle obeying the equation of motion (not Newton's law)

$$\frac{dx}{dt} = -\frac{dV}{dx} \equiv -V'(x), \tag{11.67}$$

where $V(x)$ is a function of the form shown in Figure 11.4. Suppose we release a particle near the maximum at x^* and ask for the time $T_{\frac{1}{2}}$ it takes to roll halfway down hill to $x_{\frac{1}{2}}$ as a function of the point of release x_0. This time is infinite if we begin at the fixed point $x_0 = x^*$. For nearby points this time is given by

$$T_{\frac{1}{2}}(x_0) = -\int_{x_0}^{x_{\frac{1}{2}}} \frac{dx}{V'(x)}. \tag{11.68}$$

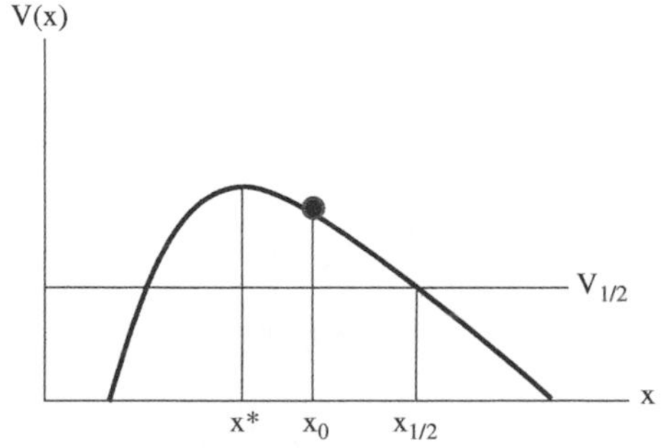

Figure 11.4 The time $T_{\frac{1}{2}}$ taken by a particle released at a point x_0 near the maximum at x^* to reach the halfway point $V_{\frac{1}{2}}$ diverges as $x_0 \to x^*$. The divergence is controlled by the way the slope $V'(x)$ vanishes at x_0.

We see that it diverges as x_0 approaches x^*, where the velocity V' vanishes. The divergence is controlled by how the velocity vanishes. Expanding it near its zero, we find

$$V'(x) = 0 + V''(x^*)(x - x^*) + \cdots \tag{11.69}$$

We will keep just the leading term because all the action is near x^*. The details of $V'(x)$ away from x^* will not be important. Thus,

$$T_{\frac{1}{2}}(x_0) = -\int_{x_0}^{x_{\frac{1}{2}}} \frac{dx}{V''(x^*)(x - x^*)}. \tag{11.70}$$

The integral clearly diverges if the initial point $x_0 \to x^*$. Notice how we have managed to find a singular quantity, namely $T(x_0 \to x_{\frac{1}{2}})$, by analyzing the flow near the fixed point where the velocity vanishes. We were able to use a Taylor expansion of $V'(x)$ near x^* *because there was nothing singular about* V' *itself: it merely had a zero there.*

Note also that the nature of the singularity is controlled by the zero of V' at x^* and does not change if we make the destination $x_{1/4}$ or $x_{3/4}$ in obvious notation.

The relation of this to the Ising model analysis we just performed is as follows. The continuous flow of this example is replaced by the discrete map $K \to K'(K)$ depicted in Figure 11.3. The fixed point of this flow analogous to x^* is K^*, and the "velocity" near K^* corresponds to $D'(K^*)$. The initial deviation from K^* is $\Delta K = K - K^*$, and it gets amplified into

$$\Delta K' = D'(K^*)\Delta K \tag{11.71}$$

during one round of RG. During every iteration the correlation length drops by a factor $\sqrt{2}$. Let us iterate until we come to a point $K - K^* = 1$, which is far from the critical region and where $\xi(1)$ is some fixed and finite number. (Here I assume $\Delta K > 0$.) If we get there after N iterations, we can say two things:

$$\xi(\Delta K) = (\sqrt{2})^N \xi(1), \tag{11.72}$$

$$\Delta K \cdot |D'(K^*)|^N = 1, \text{ which means} \tag{11.73}$$

$$N = \frac{\ln(1/\Delta K)}{\ln|D'(K^*)|}, \tag{11.74}$$

which in turn means

$$\frac{\xi(\Delta K)}{\xi(1)} = \left[\sqrt{2}\right]^{\frac{\ln(1/\Delta K)}{\ln|D'(K^*)|}} \tag{11.75}$$

$$\equiv \Delta K^{-\nu}, \quad \text{i.e.,} \tag{11.76}$$

$$\nu = \frac{\ln\sqrt{2}}{\ln|D'(K^*)|}, \tag{11.77}$$

as in Eq. (11.66).

Let me recap. In the RG one translates the large correlation length to the large number of RG steps it takes to go from the tiny $\Delta K = K - K^*$ to $K - K^* = 1$. This number, N

in Eq. (11.74), diverges as $\ln(1/\Delta K)$. (The choice $K - K^* = 1$ is not written in stone; it merely controls the prefactor in front of $\xi(\Delta K)$ and not its singular structure.)

You might object that we used the rescaling factor $\Delta K' = D'(K^*)\Delta K$ per iteration even away from K^* although it was derived very close to K^*. This is inconsequential, because most of the iterations are near the fixed point, in the slow escape from smaller and smaller initial ΔK's to a value of order unity. (In the case of the ball rolling downhill, more and more time is spent near the maximum as we begin closer and closer to it, which is why the derivative of the slope, $V''(x^*)$, controls the divergence of $T(x_0 \to x_{\frac{1}{2}})$.)

Now for the caveats. As we delve deeper into the subject we will run into the following new features:

- Instead of a scale change by 2 or $\sqrt{2}$ we will encounter scale changes of the form $1 + \varepsilon$, with $\varepsilon \to 0$. The flow, given by a discrete map $K \to K'$ like we saw in the examples of decimation, will be replaced by a continuous flow of coupling with a continuous change of length scale. Such continuous scale change is not natural in decimations in spatial lattices of the type we studied, but arises naturally either in problems posed in the spatial continuum or in momentum space. This difference, of discrete versus continuous flow, is not a profound one.
- There will generally be an unlimited number of couplings generated, even if we begin with just one. We saw one coupling beget three in $d = 2$. If these three are fed into the RG many more will appear, and so on. I will denote them all by a vector $\boldsymbol{K}$ in the space of all couplings. All its components have to be included in the flow, which in turn will determine their fate under the RG. The fixed point $\boldsymbol{K}^*$ will reside in this infinite-dimensional space. The derivative $D'(K^*)$ of the Ising example will be replaced by a matrix of derivatives that controls the flow near the fixed point. It is clear that some scheme for organizing the generated terms into a manageable hierarchy will be needed to make any progress. We will see one such scheme devised by Wilson and Fisher [6].
- It will be found that at the fixed point only a handful of components (usually just two) of the vector deviation $\Delta\boldsymbol{K} = \boldsymbol{K} - \boldsymbol{K}^*$ will get amplified, while the rest will shrink back to the fixed point, the way K did near the trivial fixed point $K^* = 0$. (See Figure 11.2.) The few couplings that flow away will describe exponents like ν. The significance of the others will be detailed later. This is an essentially new feature.
- No matter what, *the flow will be analytic and singularity free even at the fixed point.* I have emphasized how important this is to the success of Wilson's program. In our two examples, $K'(K)$ was obtained by summing an entire function of K (the Boltzmann weight) over a single spin. Such a process, involving a finite number of degrees of freedom, cannot lead to any singularities. When correctly done, the RG flow will always have this feature. Wilson's idea is that singularities in the correlation functions or free energy and its derivatives should arise from running the RG flow for more and more iterations as we start closer and closer to the fixed point, and not from singularities in the flow equations or the couplings $(a, b, c, \ldots)$.

We now turn to critical phenomena, where this machinery finds one of its finest applications.

References and Further Reading

[1] L. P. Kadanoff, Physics, **2**, 263 (1966).
[2] K. G. Wilson, Physical Review B, **4**, 3174 (1971).
[3] D. R. Nelson and M. E. Fisher, Annals of Physics, **91**, 226 (1975).
[4] T. Niemeijer and J. M. J. van Leeuven, Physical Review, **31**, 1411 (1973).
[5] F. Kos, D. Poland, D. Simmons-Duffin, and A. Vichi, Journal of High Energy Physics, **1608**, 036 (2016).
[6] K. G. Wilson and M. E. Fisher, Physical Review Letters, **28**, 240 (1972).

12

Critical Phenomena: The Puzzle and Resolution

A critical point separates distinct phases, say a magnetized state and an unmagnetized one, at a phase transition. Of special interest to us are *second-order phase transitions*, in which the *order parameter* (the magnetization $\langle M \rangle$ in the magnetic example), which is non-zero in one phase and zero in the other, vanishes continuously at the transition and where, in addition, the correlation length diverges. In the magnetic example the order parameter also spontaneously breaks (up/down) symmetry in the ordered side. In a *first-order transition*, the order parameter drops to zero discontinuously at the critical point, and the correlation length remains finite.

We have already encountered a concrete example of critical phenomena in Section 8.9 in our study of the $d = 2$ Ising model. Being fully solvable, the Ising model is a marvelous pedagogical tool, like the harmonic oscillator is in quantum mechanics. I will briefly recall some key ideas in critical phenomena using the magnetic case to illustrate them. My treatment will be tailor-made to expose the underlying ideas and is not meant to provide an exhaustive review of the subject matter or its history. There are entire books [1–5] and reviews [6–10] dedicated to the subject.

At and near the critical point one defines the following exponents in terms of t, a dimensionless measure of the deviation from the critical temperature T_c,

$$t = \frac{T - T_c}{T_c}, \tag{12.1}$$

and h, a similar dimensionless measure of the magnetic field:

$$\begin{aligned}
C_V &\simeq |t|^{-\alpha}, && (12.2)\\
\langle M \rangle &\simeq |t|^{\beta} \text{as } t \to 0^-, && (12.3)\\
\chi &= \left.\frac{\partial \langle M \rangle}{\partial h}\right|_{h=0} \simeq |t|^{-\gamma}, && (12.4)\\
\xi &\simeq |t|^{-\nu}, && (12.5)\\
G(r) &\to \frac{1}{r^{d-2+\eta}} \quad \text{at } t = 0 \text{ in } d \text{ dimensions}, && (12.6)\\
\langle M \rangle &\simeq h^{1/\delta} \text{at } t = 0. && (12.7)
\end{aligned}$$

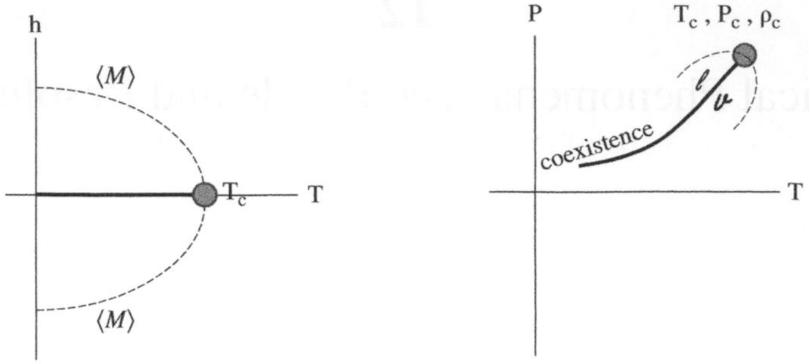

Figure 12.1 Left: The magnetic system. The critical point is at $h = 0$, $T = T_c$ and the coexistence line is $h = 0$, $T < T_c$. The dotted lines represent $\langle M \rangle$, which vanishes continuously with exponent β as we approach T_c from the left. Right: The liquid–vapor critical line in the (P,T) plane. The dotted lines represent $\rho_l - \rho_c$ and $(\rho_c - \rho_v)$. The order parameter $\rho_l - \rho_v$ vanishes as we approach (T_c, P_c), also with the same β.

The left half of Figure 12.1 describes a magnetic system in the (h,T) plane. For $T > T_c$ and $h = 0$, the magnetization $\langle M \rangle = 0$ and its value responds continuously to any applied field h. Below T_c, the average magnetization $\langle M \rangle$ has two non-zero choices of opposite sign $\pm|\langle M \rangle|$ when $h \to 0^{\pm}$. These are shown by dotted lines and labeled $\langle M \rangle$. Crossing the $h = 0$ line yields a first-order transition that can be brought about by an infinitesimal h. Since the magnetic interaction in the absence of h has up/down symmetry, the non-zero $\langle M \rangle$ below T_c for $h = 0^{\pm}$ constitutes *spontaneous* symmetry breaking. As $T \to T_c$ or $t \to 0^-$, the spontaneous magnetization vanishes continuously: $|\langle M \rangle| \simeq |t|^{\beta}$.

In what follows I will often denote the magnetization by m:

$$\langle M \rangle \equiv m. \tag{12.8}$$

What was noticed experimentally and analytically (using series, say) was that these exponents exhibited the remarkable property of *universality*: they did not change as the microscopic interactions were modified. For example, the exponents for the Ising model did not depend on the anisotropy, did not vary with the addition of a reasonable amount of second-neighbor or multi-spin interactions. The exponents did not even depend on whether we were talking about spins or another problem in the same *universality class*. An example of another member of the Ising universality class is the liquid–vapor phase transition. Look at the right half of Figure 12.1. In the (P,T) plane there is a line of coexistence between liquid and vapor called the *coexistence curve*. (This is the counterpart of the line at $h = 0$ and $T \le T_c$.) Just above and below the line are liquid with density ρ_l and vapor with density ρ_v. There is a discontinuous jump in density as we cross the line by an infinitesimal amount, say by changing P, analogous to the jump in $\langle M \rangle$ when we reverse the sign of the infinitesimal magnetic field below T_c. The order parameter is $\rho_l - \rho_v$ and it vanishes as $|t|^{\beta}$, with the same β as in the magnetic case. However, there is no spontaneous symmetry breaking involved here because neither the liquid nor the vapor breaks any symmetry; they are both structureless and just happen to have different densities. Even though there is no microscopic symmetry that relates the densities on the two sides, such a symmetry emerges near the critical point (see the review by Fisher [9]). One finds $\rho_l - \rho_c = -(\rho_v - \rho_c)$ (where ρ_c is the density at (T_c, P_c)) just the way $\langle M \rangle$ was equal and opposite on two sides of the first-order line. The figure tries to suggest this.

There are, however, limits to universality: the exponents certainly vary with the number of spatial dimensions and the symmetries. For example, they vary with N in problems with $O(N)$ symmetry. (The Ising model corresponds to $O(1) = Z_2$.) Finally, it is the exponents that are universal, not T_c or the prefactors. For this reason in my analysis of critical behavior I will pay no attention to such prefactors.

The central problem is to understand this universality. *Even if we could solve every problem in a universality class, we still would not understand* why *the same exponents appeared at all their critical points.*

The RG gives a natural explanation for universality. Before revealing it, I will give you a brief history of the earlier contributions to critical phenomena by Landau, Widom, and Kadanoff. This culminated in Wilson's grand scheme for the RG with its language of flows, fixed points, relevant and irrelevant operators, and so forth. Wilson and Fisher then showed how the theory could turn out critical exponents as a series in a small parameter $\varepsilon = 4 - d$, and the flood gates opened. I recommend Fisher's article [9] for a personal and thorough description of this final conquest.

12.1 Landau Theory

Here is one way to motivate Landau theory. Let us start with Ising spins s_i with nearest-neighbor coupling J, in a magnetic field H, on a lattice in d dimensions. Let us form a field $m(\boldsymbol{r})$ in the continuum as an average over some volume $\mathcal{V}$, say a sphere of radius R centered at $\boldsymbol{r}$:

$$m(\boldsymbol{r}) = \frac{1}{\text{sites in } \mathcal{V}} \sum_{i \in \mathcal{V}} s_i. \tag{12.9}$$

We then write the partition function, originally a sum over s_i, as a functional integral over $m(\boldsymbol{r})$:

$$Z(J,T,H) = \sum_{s_i} \exp\left[\frac{1}{kT}\left(\sum_{\langle ij \rangle} J s_i s_j + \sum_i H s_i\right)\right] \tag{12.10}$$

$$= \sum_{s_i} \int \prod_{\boldsymbol{r}} dm(\boldsymbol{r}) \delta\left(m(\boldsymbol{r}) - \frac{1}{\text{sites in } \mathcal{V}} \sum_{i \in \mathcal{V}} s_i\right)$$

$$\times \exp\left[\frac{1}{kT}\left(\sum_{\langle ij \rangle} J s_i s_j + \sum_i H s_i\right)\right] \tag{12.11}$$

$$\equiv \int \prod_r \int dm(\boldsymbol{r}) e^{-f[m(\boldsymbol{r})]} \tag{12.12}$$

$$\equiv \int [\mathcal{D}m]\, e^{-f[m(\boldsymbol{r})]}, \tag{12.13}$$

which defines the functional $f[m(\boldsymbol{r})] \equiv f[m(\boldsymbol{r}), T, H]$ that enters the Boltzmann weight for $m(\boldsymbol{r})$, and where

$$\int [\mathcal{D}m] = \prod_r \int dm(\boldsymbol{r}). \tag{12.14}$$

These steps need some qualifications. At any fixed $\boldsymbol{r}$, the factor $e^{-f(m(\boldsymbol{r}))}$ as defined above is a sum of δ-functions on the m-axis. The location of these δ-functions is decided by what the spins in $\mathcal{V}$ sum up to, and their height is proportional to the multiplicity of ways of getting that sum (peaked at $m = 0$), as well as a smoothly varying factor from the Ising Boltzmann weight. However, since we are going to integrate over m, we can replace these spikes by a smooth function that makes the same contribution over some small but fixed interval in m. It is this smooth integrand I mean by $e^{-f(m(\boldsymbol{r}))}$.

Often, many authors [11] begin with Eq. (12.13) and assume $m(\boldsymbol{r})$ stands for some coarse-grained representation of the underlying spins and $f(m)$ is a smooth function of this m. In doing this switch, are we not doing violence to the Ising model? Have we not gone from spins with values ± 1 on a lattice to a smooth function defined in the continuum? Yes, but we may still hope that since the essential features of the model, such as symmetry under $m \to -m$ and its dimensionality, have been captured, quantities that do not depend too sensitively on short-distance physics, in particular critical behavior, will be preserved.

Consider the free energy per unit volume $f(t,h)$ (not to be confused with $f[m(\boldsymbol{r})]$, which determines the Boltzmann weight), defined as usual:

$$Z(t,h) = e^{-Vf(t,h)} = \int [\mathcal{D}m]\, e^{-f[m(\boldsymbol{r})]}. \tag{12.15}$$

In Landau theory, one approximates the functional integral by its value at m^*, the minimum of $f[m(\boldsymbol{r})]$, and ignores fluctuations:

$$f(t,h) \simeq f(m^*). \tag{12.16}$$

For this reason it is referred to as a *mean-field theory*. (The variable m^* has implicit dependence on t and h, which f will inherit.)

But even this simplification is not enough: we do not really know the function f defined by Eqs. (12.11) and (12.12) because the integrals involved are unmanageable. Landau finessed this problem as follows. He began with the following facts:

- The function $f(m)$ is analytic in m and may be expanded in a power series in m since it was generated by an averaging procedure that could not produce singularities.
- If $h = 0$, $f(m)$ must be an even function of m.

- Near the transition, m will be small and an expansion in powers of m makes sense. At each stage one should try to get away with the smallest powers of m.
- The coefficients of the expansion will be analytic functions of T for the same reason as in the first item.
- If the correlation length is large (as it will be in the critical region) we may choose a large R for averaging and expect only small momenta $k \propto 1/R$ in the Fourier expansion of m. Thus, an $\boldsymbol{r}$-dependent m can be expanded in powers of its gradient.

First, consider only spatially uniform m and set $h = 0$. The functional $f(m)$ has an expansion

$$f(m) = r_0(T)m^2, \tag{12.17}$$

where $r_0(T)$ is determined in principle by Eq. (12.12) but not in practice. If $r_0(T) > 0$, the minimum of $f(m)$ is at $m^* = 0$. If $r_0(T) < 0$ the Gaussian integral over m diverges. To obtain a phase transition to the ordered phase, we are forced to add the next even power:

$$f(m) = r_0(T)m^2 + u_0(T)m^4. \tag{12.18}$$

The stationary points are the solutions to

$$0 = \left.\frac{df}{dm}\right|_{m^*} = 2r_0(T)m^* + 4u_0(T)m^{*3}. \tag{12.19}$$

If $r_0 > 0$, there is only one solution and it is at

$$m^* = 0, \tag{12.20}$$

and $f(m)$ has an absolute minimum at $m = 0$ as shown in the left half of Figure 12.2. On the other hand, if $r_0(T) < 0$, $f(m)$ initially curves down, is then turned around by the quartic term, and assumes the form shown in the right half of Figure 12.2. There are two stationary points:

$$m^* = 0 \quad \text{(a local maximum)}, \tag{12.21}$$

$$m^*(T) = \pm\sqrt{\frac{2|r_0(T)|}{4u_0(T)}} \quad \text{(degenerate minima)}. \tag{12.22}$$

The two degenerate minima are the symmetry-broken solutions.

Now Landau reveals his trump card. He argues that while we do not know the function $r_0(T)$ (which comes from a complicated multidimensional integral that expresses the original theory in terms of s_i into one in terms of $m(\boldsymbol{r})$), *we do know that the transition to the ordered state occurs when $r_0(T)$ changes sign.* That is, $r_0(T)$ vanishes at the critical point. Expanding $r_0(T)$ near T_c (given its analytic nature),

$$r_0(T) = a(T - T_c) \simeq t, \tag{12.23}$$

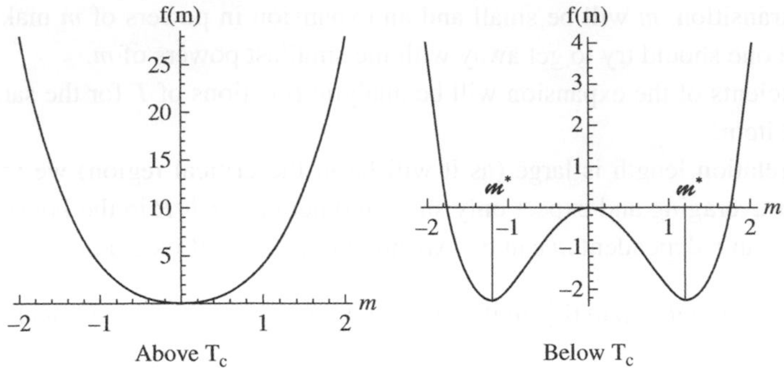

Figure 12.2 Left: The free energy in Landau theory above T_c, when $r_0(T) > 0$. The origin is the only minimum. Right: The free energy below T_c, when $r_0(T) < 0$. There are now two minima with equal and opposite m.

we learn from Eqs. (12.22) and (12.23) that

$$m^*(t) \simeq |t|^{\frac{1}{2}}, \tag{12.24}$$

which gives us our first exponent:

$$\beta = \frac{1}{2}. \tag{12.25}$$

We have no knowledge of the numerical factor in front of m^*. We only know the exponent. This is fine because only the exponent is universal. The exponents of Landau theory are also called *mean-field exponents*.

Landau's answer is independent of dimensionality and would be the same even if m had many components. How would we ever get any other answer? One proposal may be to assume that $r_0(T)$ itself vanished like some power $r_0(T) \simeq (T - T_c)^\theta$ so that $\beta = \frac{\theta}{2}$. This is no good because $r_0(T)$ has to be analytic by its very construction, resulting as it does from summing over Ising spins over a finite volume. The way to get non-Landau or *non-classical exponents* is to confront the fluctuations around the saddle point integral over $m(\boldsymbol{r})$. This is easier said than done because these fluctuations grow as we approach criticality. One slays this dragon piece by piece using the RG.

Next we turn to specific heat, which is the second derivative of the free energy:

$$C_V \simeq \frac{d^2 f(t)}{dt^2}. \tag{12.26}$$

The equilibrium value of f for $T < T_c$ is, given $m^* \simeq \sqrt{r_0} \simeq t^{\frac{1}{2}}$,

$$f(t,h) = f(m^*) \simeq r_0(t)(m^*)^2 + u_0(t)\ (m^*)^4 \propto r_0^2 \simeq t^2, \tag{12.27}$$

and so

$$C_V \simeq \text{constant for } T < T_c. \tag{12.28}$$

On the other hand, for $T > T_c$ we have seen that $m^* = 0$, so that $f = 0$ as well and

$$C_V = 0 \quad \text{for } T > T_c. \tag{12.29}$$

Thus in Landau theory C_V has a *jump discontinuity* at T_c.

To find χ and δ we turn on an infinitesimal h, beginning with

$$f(m) = r_0(T)m^2 - hm, \tag{12.30}$$

and locate m^* for $T > T_c$ or $r_0(T) > 0$:

$$0 = \left.\frac{df}{dm}\right|_{m^*} = 2r_0(T)m^* - h \tag{12.31}$$

$$m^* \simeq \frac{h}{r_0(T)} \tag{12.32}$$

$$\chi = \frac{dm^*}{dh} \simeq \frac{1}{r_0(T)} \simeq \frac{1}{t} \stackrel{\text{def}}{=} t^{-\gamma}, \quad \text{which means} \tag{12.33}$$

$$\gamma = 1. \tag{12.34}$$

To find δ we need, by its very definition, to relate m to h *at criticality*. Since $r_0(T_c) = 0$, we begin with the quartic term, whose coefficient $u_0(T_c)$ is generically non-zero, and proceed as follows:

$$f(m) = u_0(T_c)m^4 - hm \tag{12.35}$$

$$0 = \left.\frac{df}{dm}\right|_{m^*} = 3u_0(T_c)m^{*3} - h \tag{12.36}$$

$$m^* \simeq h^{\frac{1}{3}} \stackrel{\text{def}}{=} h^{\frac{1}{\delta}}, \quad \text{which means} \tag{12.37}$$

$$\delta = 3. \tag{12.38}$$

To find exponents related to correlation functions we need to introduce the derivative term in f in the absence of which the degrees of freedom at different points are decoupled. The simplest modification is

$$f = \int d^d r \left[\frac{1}{2}(\nabla m)^2 + \frac{1}{2} r_0(T) m^2\right], \tag{12.39}$$

where I have inserted a $\frac{1}{2}$ in front of r_0 because this is the traditional form of the "mass term" in field theory. This trivial change allows us to read off standard results derived in this convention without affecting critical exponents.

Now we have, when $h = 0$,

$$Z(T) = \int [\mathcal{D}m] e^{-[f(m,T)]} \tag{12.40}$$

$$= \int [\mathcal{D}m] \exp\left[-\int d^d r \left[\frac{1}{2} (\nabla m)^2 + \frac{1}{2} r_0(T) m^2 \right] \right] \tag{12.41}$$

$$= \int [\mathcal{D}m] \exp\left[-\int d^d r \left[\frac{1}{2} m(\boldsymbol{r})(-\nabla^2 + r_0(T)) m(\boldsymbol{r}) \right] \right]. \tag{12.42}$$

From Section 6.11, and in particular Eq. (6.203), we know that the correlation function is

$$\langle m(\boldsymbol{r}) m(0) \rangle = G(\boldsymbol{r}) = \int \frac{d^d k}{(2\pi)^d} \frac{e^{i\boldsymbol{k}\cdot\boldsymbol{r}}}{k^2 + r_0(T)}. \tag{12.43}$$

When $r_0 = 0$, we find, by dimensional analysis,

$$G(r) \simeq \frac{1}{r^{d-2}} \stackrel{\text{def}}{=} \frac{1}{r^{d-2+\eta}}, \quad \text{that is,} \tag{12.44}$$

$$\eta = 0. \tag{12.45}$$

To find the correlation length exponent we need to find the exponential fall-off of $G(r)$. We can examine the behavior of $\xi(T)$ $T > T_c$. Let us first consider $d = 3$, even keeping track of factors of π just to show we can do it in a crunch:

$$G(r) = \frac{1}{(2\pi)^3} \int_0^\infty k^2 dk \int_{-1}^{1} d\cos\theta \int_0^{2\pi} d\phi \frac{e^{ikr\cos\theta}}{k^2 + r_0(T)} \tag{12.46}$$

$$= \frac{1}{2\pi^2 r} \int_0^\infty k dk \frac{\sin kr}{(k + i\sqrt{r_0(T)})(k - i\sqrt{r_0(T)})} \tag{12.47}$$

$$= \frac{1}{4\pi^2 r} \int_{-\infty}^{\infty} k dk \operatorname{Im} \left[\frac{e^{ikr}}{(k + i\sqrt{r_0(T)})(k - i\sqrt{r_0(T)})} \right]. \tag{12.48}$$

We changed the limits to $(-\infty, \infty)$ and multiplied by $\frac{1}{2}$ because $k \sin kr$ is even. This allows us to perform the k-integration using Cauchy's theorem.

Closing the contour in the upper half-plane where e^{ikr} converges and the pole at $k = i\sqrt{r_0(T)}$ contributes, we arrive at the answer

$$G(r) = \frac{1}{4\pi r} e^{-r\sqrt{r_0}} \simeq \frac{e^{-r\sqrt{t}}}{r}. \tag{12.49}$$

The correlation length is determined by the exponential part of G (the power of r in front makes subdominant contributions). Comparing to the standard form

$$e^{-r\sqrt{t}} = \exp\left[-\frac{r}{\xi(t)} \right], \quad \text{we infer} \tag{12.50}$$

$$\xi(t) = t^{-\frac{1}{2}}, \tag{12.51}$$

$$\nu = \frac{1}{2}. \tag{12.52}$$

Table 12.1 *Critical exponents: comparison.*

	Landau	$d=2$	$d=3$	$d=4$
α	jump	0^+	0.110	0^+
β	$\frac{1}{2}$	$\frac{1}{8}$	0.326	$\frac{1}{2}$
γ	1	$\frac{7}{4}$	1.237	1
δ	3	15	5.1	3
ν	$\frac{1}{2}$	1	0.630	$\frac{1}{2}$
η	0	$\frac{1}{4}$	0.036	0

Table 12.1 compares the exponents in Landau theory to known answers in $d=2,3,4$ (from Onsager or numerical work). We see that it does better with increasing dimensionality.

Exercise 12.1.1 *Show that $\xi(t) \simeq t^{-\frac{1}{2}}$ in $d=2$. Unlike in $d=3$, the integral over k, which goes from 0 to ∞, cannot be extended to the range $(-\infty,\infty)$ because the integrand is odd. The θ integral will give you a Bessel function $J_0(kr)$ and the integral over the positive k-axis can be looked up in a table of integrals to give*

$$G(r) \simeq K_0(r\sqrt{r_0}) \lim_{r\to\infty} \simeq \frac{e^{-r\sqrt{t}}}{\sqrt{r}}. \tag{12.53}$$

12.2 Widom Scaling

Consider the function $f(x) = \sin x$. It does not respond in a simple way to the rescaling $x \to \lambda x$, for constant λ. By this I mean we cannot relate $\sin \lambda x$ to $\sin x$ in any simple way. Consider, on the other hand,

$$f(x) = x^a, \tag{12.54}$$

where a is some exponent. It obeys

$$f(\lambda x) = (\lambda x)^a = \lambda^a f(x). \tag{12.55}$$

Conversely, imagine we were told that

$$f(\lambda x) = \lambda^a f(x) \tag{12.56}$$

is true for any λ. Choosing

$$\lambda = \frac{1}{x}, \tag{12.57}$$

we discover

$$f(1) = \lambda^a f(x) = x^{-a} f(x), \tag{12.58}$$

$$f(x) \simeq x^a. \tag{12.59}$$

The scaling properties of functions can give us clues to the underlying physics. For example, it was the observation by Bjorken [12] that the scattering amplitudes of virtual photons of energy ν and momentum q on nucleons depended only on the ratio ν/q^2 that suggested to Feynman the idea of partons as constituents of nucleons.

What Widom did [13] was to demonstrate from the data that the *singular part of the free energy per site* or unit volume $f_s(t,h)$, the part that gives rise to divergences when differentiated with respect to t or h, exhibited *scaling*. I will drop the subscript on f_s from now on.

Widom observed that

$$f(t\lambda^a, h\lambda^b) = \lambda f(t,h). \tag{12.60}$$

Let us choose λ so that

$$t\lambda^a = 1. \tag{12.61}$$

It follows that

$$f(t,h) = t^{\frac{1}{a}} f(1, ht^{-\frac{b}{a}}). \tag{12.62}$$

We can see why Widom's arguments (or any RG arguments I espouse later) apply only to the *singular* part of the free energy. If we set $t = 0$ above we find that $f = 0$, which is not true of the total free energy at criticality. There are analytic terms that are non-zero, as you can check with the $d = 2$ Ising model. The singular part, however, does vanish (like $t^2 \ln|t|$ in the $d = 2$ Ising case), and its various t-derivatives (like $C_V(t)$) can diverge at criticality.

Equation 12.62 means that the function of two variables (t,h) can be reduced to a function of one variable $ht^{-b/a}$ if we rescale f at the same time:

$$\frac{f(t,h)}{t^{\frac{1}{a}}} = f(1, ht^{-\frac{b}{a}}). \tag{12.63}$$

This equation has the following implication for data analysis. Suppose you have a plot of f as a function of h for two temperatures t_1 and t_2. Now take each point (f,h) at t_1 and assign it to the point $(ft_1^{-1/a}, ht_1^{-b/a})$ to get a new curve. Do this for t_2 and you will find the two graphs collapse into one. Indeed, any number of f versus h graphs at various temperatures will collapse into this one graph if the axes are scaled in this manner.

So what? Widom showed that if skillfully exploited, scaling laws could be used to express the critical exponents a, β, γ, δ (associated with thermodynamic quantities) in terms of just a and b. This also meant that between any three of them there had to be a relation.

First, upon taking two t-derivatives of Eq. (12.62) at $h = 0$, we find

$$C_V \simeq \frac{\partial^2 f(t,0)}{\partial t^2} \simeq t^{\frac{1}{a}-2} C_V(1,0) \stackrel{\text{def}}{=} t^{-\alpha}, \tag{12.64}$$

$$\alpha = 2 - \frac{1}{a}. \tag{12.65}$$

(I should really be using $|t|$, but I will leave that implicit here and elsewhere.) Next we take an h-derivative at $h = 0$:

$$m(t) \simeq \left.\frac{\partial f(t,h)}{\partial h}\right|_{h=0} \simeq t^{\frac{1-b}{a}} m(1,0) \simeq t^{\beta}, \tag{12.66}$$

$$\beta = \frac{1-b}{a}. \tag{12.67}$$

Taking another h-derivative yields the susceptibility:

$$\chi(t) \simeq \left.\frac{\partial^2 f(t,h)}{\partial h^2}\right|_{h=0} \simeq t^{\frac{1-2b}{a}} \chi(1,0) \stackrel{\text{def}}{=} t^{-\gamma}, \tag{12.68}$$

$$\gamma = \frac{2b-1}{a}. \tag{12.69}$$

We can combine the exponent formulas above to get the relation

$$\alpha + 2\beta + \gamma = 2, \tag{12.70}$$

first noted by Essam and Fisher [14]. Subsequently Rushbooke (1963, unpublished) proved a rigorous inequality:

$$\alpha + 2\beta + \gamma \geq 2. \tag{12.71}$$

To find the δ that relates m to h at criticality, we cannot scale out t as before because it vanishes. Let us go back one step and begin with Eq. (12.60) with $t = 0$:

$$f(0,h) = \lambda^{-1} f(0, h\lambda^b) \tag{12.72}$$

$$m = \frac{\partial f(0,h)}{\partial h} = \lambda^{b-1} m(0, h\lambda^b), \tag{12.73}$$

$$\simeq h^{\frac{1-b}{b}} \quad \text{(upon setting } h\lambda^b = 1 \text{ above)} \tag{12.74}$$

$$\stackrel{\text{def}}{=} h^{1/\delta}, \quad \text{which means} \tag{12.75}$$

$$\delta = \frac{b}{1-b}. \tag{12.76}$$

Let us apply these results to the $d = 2$ Ising model starting with the known α and β to find a and b and thence δ, which is not so easily found:

$$\alpha = 2 - \frac{1}{a} = 0 \to a = \frac{1}{2}, \tag{12.77}$$

$$\beta = \frac{1-b}{a} = \frac{1}{8} \to b = \frac{15}{16}, \tag{12.78}$$

$$\delta = \frac{b}{1-b} = 15. \tag{12.79}$$

We cannot get η and ν from Widom scaling because it does not address spatial correlations. But it gave a marvelous fillip to Kadanoff, who gave a physical basis for it and used that to lay the path to the modern theory of critical phenomena. His approach also yielded expressions for ν and η in terms of a and b.

12.3 Kadanoff's Block Spins

In Widom scaling,

$$f(t,h) = \lambda^{-1} f(t\lambda^a, h\lambda^b), \tag{12.80}$$

λ was a scale parameter with no obvious physical interpretation. Kadanoff provided one which proved to be seminal [15].

Suppose we take a system of Ising spins s_i with parameters (K,h) near criticality and group them into blocks labeled by an index I and of size L in each of the d directions. Figure 12.3 shows 3×3 blocks in $d = 2$. Assume that due to the large correlation length near criticality, all the spins in one block $s_{i \in I}$ will be in the same state, which we will call the *block spin* S_I. Thus if $S_I = +$, all spins in the block I, namely $s_{i \in I}$, are also $+$. What will be the effective parameters K_L and h_L for the block spins?

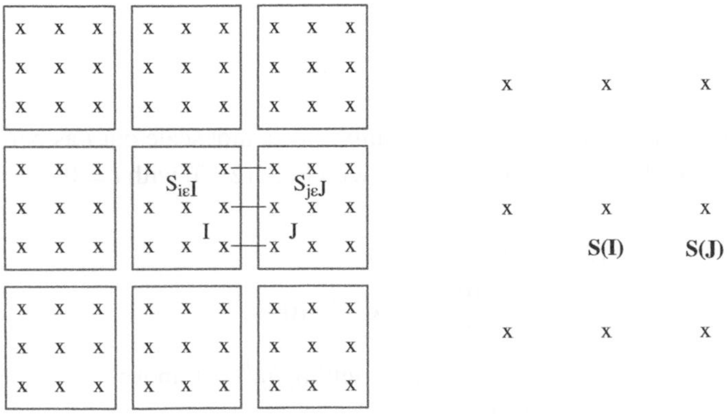

Figure 12.3 3×3 blocks in $d = 2$. Spins in block I (J) are called $s_{i \in I}$ ($s_{j \in J}$). The block spins I and J are coupled by the three adjacent pairs by bonds K that straddle the blocks.

The block field h_L that couples to S_I must reproduce the coupling to h of the nine spins in the block (all of which equal S_I):

$$h_L S_I = \sum_{i\in I} h s_i = \sum_{i\in I} h S_I = 9hS_I \tag{12.81}$$

$$h_L = 9h, \quad \text{or more generally} \tag{12.82}$$

$$h_L = L^d h. \tag{12.83}$$

Likewise, the coupling of neighboring blocks I and J is mediated by three couplings of strength K between the spins $s_{i\in I}$ and $s_{j\in J}$ at the interface. Thus, we demand that the block spins couple to their neighbors with strength

$$K_L = 3K, \quad \text{or, more generally,} \tag{12.84}$$

$$K_L = L^{d-1}K. \tag{12.85}$$

A fixed point (K^*, H^*) has to obey

$$h^* = L^d h^*, \tag{12.86}$$

$$K^* = L^{d-1}K^*. \tag{12.87}$$

The only solutions are trivial: 0 or ∞ in both cases.

This failure is to be expected since there will always be fluctuations within a block. There are fluctuations at all length scales even when $\xi \to \infty$. This is why Kadanoff defined the block spin as an average followed by a rescaling factor $\zeta^{-1}(L)$:

$$S_I = \zeta^{-1}(L)\frac{1}{L^d}\sum_{i\in I} s_i. \tag{12.88}$$

Notice that even if s_i were discrete to begin with, S_I will be nearly continuous if L is sufficiently large or if the blocking has been done many times. (So S_I should be integrated rather than summed over.) The coupling $h_L S_I$ is chosen to replicate that of the underlying spins that S_I represents:

$$h_L S_I = h_L \zeta^{-1}(L)\frac{1}{L^d}\sum_{i\in I} s_i = h\sum_{i\in I} s_i, \quad \text{which implies} \tag{12.89}$$

$$h_L = L^d \zeta(L) h \equiv L^{bd} h, \tag{12.90}$$

which defines the index b. Likewise, the effective t_L, which couples neighboring block spins, is assumed to be of the form

$$t_L = tL^{ad}. \tag{12.91}$$

These power laws ensure that blocking is a semigroup. For example, a blocking by L_1 followed by L_2 equals a single blocking by $L_1 L_2$:

$$t \xrightarrow{L_1} t_{L_1} = tL_1^{ad} \xrightarrow{L_2} tL_1^{ad}L_2^{ad} = t(L_1L_2)^{ad}. \tag{12.92}$$

A single site of the block spins packs in the free energy per site of L^d old sites. Thus,

$$f(t_L, h_L) = f(tL^{ad}, hL^{bd}) = L^d f(t,h), \tag{12.93}$$

which is just Widom's scaling with

$$\lambda = L^d. \tag{12.94}$$

But now we can interpret L as the size of the blocks and t_L and h_L as the renormalized couplings of the block spins to each other and to the external field. The exponents a and b have the same significance as before, because

$$L^{ad} = (L^d)^a = \lambda^a, \quad \text{and likewise for } b. \tag{12.95}$$

Consequently, Eqs. (12.65)–(12.69) expressing α, β, and γ in terms of a and b continue to hold.

12.3.1 Expressions for ν and η

Kadanoff's approach allows us to go beyond Widom's on another front: we can obtain the exponents ν and η involving correlation functions in terms of a and b.

Consider ν. Since the distance between block spins is L times smaller than the distance between the corresponding original spins (in their respective lattice units), the dimensionless correlation lengths (i.e., measured in corresponding lattice units) will be related as follows:

$$\xi(t) = L\xi(t_L) = L\xi(tL^{ad}) \tag{12.96}$$

$$= t^{-\frac{1}{ad}}\xi(1) \simeq t^{-\nu}, \tag{12.97}$$

$$\nu = \frac{1}{ad}. \tag{12.98}$$

Next, consider the correlation function G. Let R be the distance in block lattice units between two block spins I and J. They are separated by $r = LR$ units of the original lattice. Let us express the correlation between block spins in terms of the correlations between the original spins in each block:

$$G(R, t_L, h_L) = \langle S_I S_J \rangle \tag{12.99}$$

$$= \zeta^{-2}(L)\frac{1}{L^{2d}}\sum_{i\in I}\sum_{j\in J}\langle s_i s_j\rangle. \tag{12.100}$$

Since $R \gg L$, the distance between any spin in I and any spin in J may be set equal to the same constant $r = LR$. So the L^{2d} correlators in the double sum are all equal to $G(r = RL, t, h)$. This multiplicity exactly neutralizes the L^{-2d} factor in front. Continuing,

$$G(R, t_L, h_L) = \zeta^{-2}(L)G(r = RL, t, h). \tag{12.101}$$

Let us rewrite the above as

$$G(r,t,h) = \zeta^2(L)G\left(\frac{r}{L}, t_L, h_L\right). \tag{12.102}$$

Why did we rescale the block spin with ζ but not the Ising spin during decimation in $d=1$? Consider Eq. (12.102) at $t=h=0$ and consequently $t_L = h_L = 0$. We find that

$$G(r,0,0) = \zeta^2(L)G\left(\frac{r}{L}, 0, 0\right). \tag{12.103}$$

Without the ζ^2 we would conclude, based on $G(R) = G(RL)$, that G was R-independent. This was actually correct: in the exact solution of the $d=1$ Ising model at its critical point, $T=0$ or $K=\infty$ and $(\tanh K)^L = 1^L$, but there the *surviving spin was just one of the original spins* and there was no reason to rescale it. But the block spin is an average of the original spins and of indefinite length. To get a non-trivial fixed point we have to leave room for a possible rescaling by ζ. Assume that

$$\zeta(L) = L^{-\omega}, \tag{12.104}$$

where the exponent ω will be nailed down shortly. If we choose $r = L$, we find, from Eq. (12.103),

$$G(r,0,0) = \frac{1}{r^{2\omega}}G(1,0,0) \stackrel{\text{def}}{=} \frac{1}{r^{d-2+\eta}}, \quad \text{which means} \tag{12.105}$$

$$\omega = \frac{1}{2}(d-2+\eta). \tag{12.106}$$

It looks like η is an independent exponent because it is related to the correlation function and not the free energy. However, it too may be related to (a,b) as follows.

The susceptibility χ is the second derivative of the free energy (per site or per unit volume) with respect to a constant h. We have seen in Section 8.9 that χ is the spatial integral (or sum) of G:

$$\chi(t) = \int d^d r G(r,t). \tag{12.107}$$

To extract γ, let us model G as

$$G(r,t) \simeq \frac{e^{-r/\xi(t)}}{r^{d-2+\eta}}, \tag{12.108}$$

and integrate it over all of space. For $r \ll \xi$ we can ignore the exponential, and for $r \gg \xi$ the exponential kills G off. So let us replace this smooth transition by an abrupt one that takes place at $r = \xi$ to obtain:

$$\chi(t) \simeq \int_0^{\xi} \frac{r^{d-1}}{r^{d-2+\eta}} dr \tag{12.109}$$

$$\simeq \xi(t)^{2-\eta} \quad \text{by dimensional analysis} \tag{12.110}$$

$$\simeq t^{-\nu(2-\eta)} \stackrel{\text{def}}{=} t^{-\gamma} \quad (12.111)$$
$$\gamma = (2-\eta)\nu. \quad (12.112)$$

This now allows us to express η in terms of (a,b) if we want.

Finally, we begin with Eq. (12.93) rewritten as follows and manipulate:

$$f(t) = L^{-d} f(t_L) = L^{-d} f(tL^{ad}) = L^{-d} f(tL^{\frac{1}{\nu}}) \quad (12.113)$$
$$C_V \simeq \frac{\partial^2 f}{\partial t^2} \simeq L^{\frac{2}{\nu}-d} C_V(tL^{\frac{1}{\nu}}). \quad \text{Setting } L = t^{-\nu}, \quad (12.114)$$
$$C_V \simeq t^{-(2-\nu d)} = t^{-\alpha}, \quad \text{which means} \quad (12.115)$$
$$\alpha = 2 - \nu d, \quad (12.116)$$

which is called the *hyperscaling relation*. It depends on d, the dimensionality. It relates the exponent ν defined by correlation functions to α, defined by the specific heat, a thermodynamic quantity. It is found to work for $d \leq 4$, but not above.

Here are some useful exponent relations:

$$\alpha + 2\beta + \gamma = 2, \quad (12.117)$$
$$\gamma = (2-\eta)\nu, \quad (12.118)$$
$$\alpha = 2 - \nu d. \quad (12.119)$$

We have derived a lot of mileage from Kadanoff's block spin idea of trading a given problem for a new one with fewer degrees of freedom and modified parameters. It is a crucial step in the modern theory of the RG. But it still leaves some questions. Why are the exponents universal? Is there some approximation in which they may be computed reliably? For this we turn to Wilson's program, which addresses these.

12.4 Wilson's RG Program

The general idea behind Wilson's program [11] will now be illustrated by a magnetic example like the Ising model.

Suppose we begin with a theory described by some couplings referred to collectively by the vector $\boldsymbol{K}$. Each one of its components K_α corresponds to an interaction: K_1 and K_2 could correspond to the nearest-neighbor and second-neighbor interactions for Ising spins. (We used to refer to K_1 as simply K when it was the only coupling in town. But now the subscript reminds us it is part of an infinite family.)

In the pre-RG days, we would pick a simple interaction, say with just $K_1 \neq 0$, and proceed to solve the problem exactly (Onsager) or approximately (mortals). But the RG instructs us to trade the given problem for another, with fewer degrees of freedom, different couplings, but the same asymptotic behavior near criticality. This process forces us to move to an infinite-dimensional space of all possible couplings where $\boldsymbol{K}$ resides. (We saw how in the first round of the RG with Ising spins the nearest-neighbor coupling K_1 morphed into K_1', K_2', and K_4', the renormalized nearest-neighbor coupling and the newly generated second-neighbor and four-spin couplings.)

The only way to keep some couplings out is by symmetry; for example, in the absence of a magnetic field, only couplings invariant under spin reversal will be generated. Couplings involving odd powers of spin will not arise. Apart from that, anything goes. For now, let us assume there is no magnetic field and the interactions are symmetric under sign reversal of the fluctuating variable. It is also assumed the interactions are short ranged, meaning they decay exponentially with separation. This ensures that there is no room for singularities to creep into the RG flow equations.

Suppose we now eliminate some degrees of freedom and end up with new couplings

$$\boldsymbol{K}_L' = \boldsymbol{K}'(\boldsymbol{K}), \tag{12.120}$$

where L is the size of the new lattice in the old lattice units, or, in general, the factor that relates dimensionless lengths before and after one round of the RG. Equivalently, the degrees of freedom have been thinned out by a factor L^{-d} during one round of the RG. The subscript L on $\boldsymbol{K}_L'$ will be implicit if it is not shown.

The dimensionless correlation length obeys:

$$\xi(\boldsymbol{K}') = L^{-1}\xi(\boldsymbol{K}). \tag{12.121}$$

The whole RG enterprise is predicated on there being a fixed point $\boldsymbol{K}^*$ of the transformation that obeys

$$\boldsymbol{K}'(\boldsymbol{K}^*) = \boldsymbol{K}^*. \tag{12.122}$$

It follows from Eq. (12.121) that $\xi(\boldsymbol{K}^*) = \infty$ (I ignore the uninteresting option $\xi = 0$).

If $\boldsymbol{K}^* \to \boldsymbol{K}^*$ under the RG, what about points nearby? Their fate is determined by analyzing the flow near $\boldsymbol{K}^*$. Let us ask how deviations from the fixed point behave after one iteration, repeating the old argument but with many couplings. Suppose we begin with a point

$$\boldsymbol{K} = \boldsymbol{K}^* + \Delta\boldsymbol{K}, \tag{12.123}$$

or, in terms of components,

$$K_\alpha = K_\alpha^* + \Delta K_\alpha. \tag{12.124}$$

The fate of the initial point is

$$K_\alpha'(\boldsymbol{K}^* + \Delta\boldsymbol{K}) = K_\alpha^* + \left.\frac{\partial K_\alpha'}{\partial K_\beta}\right|_{\boldsymbol{K}^*} \Delta K_\beta \tag{12.125}$$

$$\Delta K_\alpha' = \left.\frac{\partial K_\alpha'}{\partial K_\beta}\right|_{\boldsymbol{K}^*} \Delta K_\beta \tag{12.126}$$

$$= \sum_\beta T_{\alpha\beta}\Delta K_\beta, \tag{12.127}$$

where T is the *derivative matrix* that linearizes the flow near the fixed point. The matrix T is generally not symmetric but is assumed to have a basis of eigenvectors, $|\alpha\rangle$, not necessarily orthonormal but complete.

I will illustrate the general picture with just three eigenvectors:

$$T(L)|a\rangle = L^{ad}|a\rangle \qquad a > 0, \text{ relevant}; \tag{12.128}$$

$$T(L)|\omega_1\rangle = L^{-|\omega_1|d}|\omega_1\rangle \qquad \omega_1 < 0, \text{ irrelevant}; \tag{12.129}$$

$$T(L)|\omega_2\rangle = L^{-|\omega_2|d}|\omega_1\rangle \qquad \omega_2 < 0, \text{ irrelevant}. \tag{12.130}$$

The eigenvalues must be of the form $L^{\alpha d}$ to satisfy the group property:

$$T(L_1)T(L_2) = T(L_1L_2). \tag{12.131}$$

Any initial point $\boldsymbol{K}^* + \Delta\boldsymbol{K}$ near $\boldsymbol{K}^*$ may be expressed in terms of the eigenvectors as

$$|\Delta\boldsymbol{K}\rangle = t|a\rangle + g_1|\omega_1\rangle + g_2|\omega_2\rangle. \tag{12.132}$$

After the action of T the point moves to

$$T(L)|\Delta\boldsymbol{K}'\rangle = tL^{ad}|a\rangle + g_1L^{-|\omega_1|d}|\omega_1\rangle + g_2L^{-|\omega_2|d}|\omega_2\rangle \tag{12.133}$$

$$\equiv t_L|a\rangle + g_{1L}|\omega_1\rangle + g_{2L}|\omega_2\rangle. \tag{12.134}$$

If initially only $t > 0$ we see that the point moves away under the action of T:

$$t \to t_L = tL^{ad}, \tag{12.135}$$

which is why $|a\rangle$ is a relevant eigenvector. By contrast, the other two are irrelevant eigenvectors,

$$g_1 \to g_{1L} = g_1L^{-|\omega_1|d}, \qquad g_2 \to g_{2L} = g_2L^{-|\omega_2|d}. \tag{12.136}$$

If we acted with $T(L)$ a total of N times, this tendency would continue; eventually, only t_L would survive and the representative point would lie on the eigendirection $|a\rangle$.

Rather than vary N, the number of times T acts to produce a scale change L^N, let us view L itself as a variable, allowed to assume any size. (For the blocking example we considered, the allowed values of L^N will be 3^N, but we can ignore this restriction without any real consequence. Anyway, we will soon be focusing on cases where L is continuous.)

Let us now consider several cases for the parameters (t, g_1, g_2).

12.4.1 The Critical Surface $t = 0$

Consider the points with $t = 0$,

$$|\Delta\boldsymbol{K}\rangle = g_1|\omega_1\rangle + g_2|\omega_2\rangle, \tag{12.137}$$

which flow into $\boldsymbol{K}^*$ under the RG. These define a two-dimensional planar surface. By continuity, this *critical surface* will retain its identity as we move further away even if it is

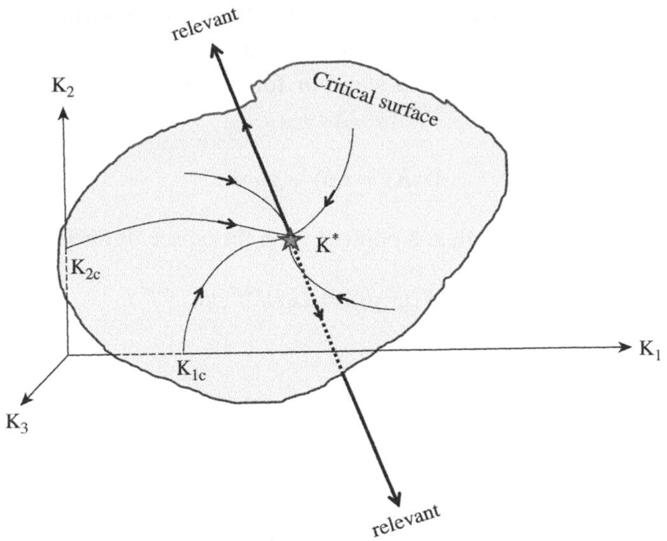

Figure 12.4 The critical surface, all the points in which flow to the fixed point $\boldsymbol{K}^*$. Points off the surface run away from $\boldsymbol{K}^*$ along the relevant direction (pointing up or down in the figure). In models of Ising spins, the K_1 axis would correspond to Onsager's nearest-neighbor coupling, which we called K.

no longer planar, as shown in Figure 12.4. The surface is so named because *every point on it is critical and has* $\xi = \infty$. This has to be so because the RG reduces ξ by a factor L and if *after this reduction* the original system flows to $\boldsymbol{K}^*$ (which has $\xi = \infty$), it must have had $\xi = \infty$ to begin with. The figure shows the familiar Onsager point at $K_{1c} = 0.440688$ lying on this surface. Also on this surface is the model with *only* the second-neighbor coupling, also at $K_{2c} = 0.440688$. (With only second-neighbor couplings, the lattice breaks into two interpenetrating sublattices with a nearest-neighbor coupling within each sublattice.) Other points on the critical surface are not familiar.

The presence of a whole surface of critical states is surprising. When we study the $d = 2$ Ising model, we find just one isolated critical point at $K_1 = K_{1c}$. It seems a rarity, and yet the three-dimensional coupling space has a two-dimensional surface full of critical states.

If we go to infinite dimensions, things get even more strange: there is still just one relevant direction and the critical surface occupies the rest; one says the surface has *codimension* 1. With so many couplings at play, do we not have to find a needle in a haystack in searching for a critical point? How can it be that we need to fix just one number, the coefficient in the relevant direction, to achieve criticality? But this is what we do all the time. Given a chunk of magnetic material, with its myriad couplings, of which we have little knowledge, we reach criticality simply by heating or cooling it! This is why the relevant coordinate is called t.

To describe what happens when we start slightly off the critical surface, I will strip down the analysis to a minimum and keep just one relevant eigenvector $|a\rangle$ and one irrelevant eigenvector $|\omega\rangle$, where the latter is a stand-in for an infinite number of its irrelevant siblings. That is, I will assume

$$|\Delta \boldsymbol{K}\rangle = t|a\rangle + g|\omega\rangle. \tag{12.138}$$

Consider the correlation length at a point (t,g) in this space. It obeys

$$\xi(t,g) = L\xi(t_L, g_L) = L\xi(tL^{ad}, gL^{-|\omega|d}). \tag{12.139}$$

Since $ad = 1/\nu$ let us write it as

$$\xi(t,g) = L\xi(tL^{\frac{1}{\nu}}, gL^{-|\omega|d}). \tag{12.140}$$

Let us choose

$$L = t^{-\nu} \tag{12.141}$$

above, to find

$$\xi(t,g) = t^{-\nu}\xi(1, gt^{|\omega|\nu d}). \tag{12.142}$$

We find that ξ does not immediately scale as $t^{-\nu}$ – there are *corrections to scaling* due to the irrelevant coupling(s) g. But these do vanish as $t \to 0$ because g_L scales as a positive power of t. This is the new input from Wilson's picture: the familiar scaling laws are only asymptotic and receive corrections that vanish only as $t \to 0$.

Equation (12.141) is very revealing. It says that L, the length scale by which we must renormalize to get to a finite distance from the fixed point, where ξ is some finite number (marked $t_L = 1$ in Figure 12.5), diverges exactly the way the correlation length diverges. *The divergence of ξ thus manifests itself in the RG analysis as the divergent amount of blocking needed to get to a fixed distance from $\boldsymbol{K}^*$.* That this singularity is found by linearizing the smooth flow equations near $\boldsymbol{K}^*$ is to be expected based on Wilson's analogy with the particle rolling down the hill.

12.4.2 Enter h

Let us introduce a magnetic field and the corresponding infinite number of couplings with odd powers of the spin. However, there will be just one more relevant coupling h and a relevant eigenvalue L^{bd}:

$$|\Delta \boldsymbol{K}\rangle = t|a\rangle + h|b\rangle + g|\omega\rangle \tag{12.143}$$

$$T(L)|\Delta \boldsymbol{K}\rangle = tL^{ad}|a\rangle + hL^{bd}|b\rangle + gL^{-|\omega|d}|\omega\rangle \tag{12.144}$$

$$= tL^{\frac{1}{\nu}}|a\rangle + hL^{d-\frac{\beta}{\nu}}|b\rangle + gL^{-|\omega|d}|\omega\rangle. \tag{12.145}$$

To evade the relevant flow in the t and h directions, to retain or attain criticality, we just have to tune to $t = h = 0$.

Next, let us consider the scaling corrections to the free energy. By familiar reasoning,

$$f(t,h,g) = L^{-d} f(tL^{\frac{1}{\nu}}, hL^{d-\frac{\beta}{\nu}}, gL^{-|\omega|d}). \tag{12.146}$$

Once again, we choose $L = t^{-\nu}$ to find

$$f(t,h,g) = t^{\nu d} f(1, ht^{\beta-\nu d}, gt^{|\omega|\nu d}). \tag{12.147}$$

The term proportional to h will actually grow as some negative power of t, so it has to be tuned to 0 if we are to be in the sway of this critical point. Assume that this is arranged. The coupling g, which scales as a positive power of t, represents vanishing corrections to Widom scaling by irrelevant interactions. Such corrections to scaling will percolate to the derivatives of the free energy like the magnetization or susceptibility.

So we see that irrelevant operators are not entirely irrelevant: in analyzing data we must include them in the fit before reaching asymptopia.

12.4.3 Practical Considerations

Our analysis so far has been for points in the vicinity of $\boldsymbol{K}^*$ with deviations t, h, g along $|a\rangle, |b\rangle, |\omega\rangle$, and so on. Now, $\boldsymbol{K}^*$ and its neighbors separated by t, h, or g stand for very complicated interactions *in the canonical basis* of nearest-neighbor, second-neighbor, four-spin interactions, and so forth. In practice we do not work with such interactions, we work with simple couplings, say, represented by points along the nearest-neighbor axis with coordinate K_1. How does the preceding analysis of points near $\boldsymbol{K}^*$ apply to the description of critical phenomena on the K_1 axis near K_{1c}? What, for example, is the t we should use in Eq. (12.143) if we begin slightly away from K_{1c}? In other words, how does the deviation from the *critical point* translate to the deviation from the *fixed point*, both of which I have been calling t? Here is the answer, where I choose $h = 0$ for simplicity. The situation is depicted in Figure 12.5. I do not show half the flow lines (above the relevant curve in the figure) to avoid clutter.

We know that if we start *exactly* at K_{1c} we will flow to $\boldsymbol{K}^*$, i.e. to $t = 0$, because it lies on the attractive RG trajectory. It follows that if we start *near* K_{1c} we will flow to a point with $t \propto K - K_{1c}$. Since the proportionality constant really does not affect the critical exponent, that is, since $t^{-\nu}$ and $(ct)^{-\nu}$ describe the same singularity in t, we might as well equate the standard $t = (T - T_c)/T_c$ to the t in the flow equations above. In other words, the critical behavior of a system that is at $T_c + t$ on the K_1 axis is the same as that of a system that starts out at $\boldsymbol{K}^* + t|a\rangle$. Of course, there will also be an irrelevant component g, whose overall scale is not obvious, but luckily it will eventually shrink to zero.

How will the RG flow manifest itself to a person confined to the K axis, in possession of the exact solution? The RG flow to large L will be replaced by the limit $r \to \infty$, where r stands collectively for all the differences in coordinates in the correlation functions. For

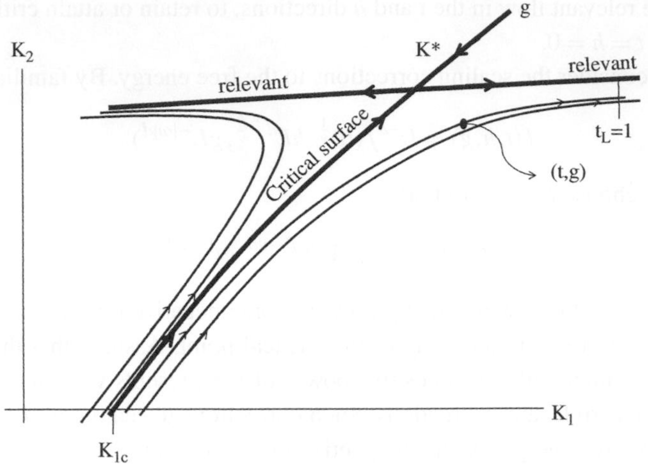

Figure 12.5 The flow with one relevant and one irrelevant direction at the fixed point. The nearest-neighbor Ising model at K_{1c} flows into $\boldsymbol{K}^*$. Points to the right or left of K_{1c} approach K^* initially. Soon, the difference from $\boldsymbol{K}^*$ in the relevant direction gets amplified and the trajectories veer away and flow along the relevant direction. Usually we follow the flow until we reach $t_L = 1$, where ξ is some fixed and finite number. I show only half the flow lines to avoid clutter.

example, at criticality the two-point function $G(\boldsymbol{n} = \boldsymbol{n}_2 - \boldsymbol{n}_1)$ will start out as a complicated function of the lattice separation $\boldsymbol{n}$, endowed with just fourfold symmetry (for the isotropic model with $K_x = K_y$), but will become rotationally invariant and assume the simple form $G(r) \simeq r^{-1/4}$ as $r \to \infty$. Irrelevant contributions will drop off not under $L \to \infty$ but under $r \to \infty$. Likewise, the free energy and its derivatives will assume the scaling form as $T \to T_c$ or $K_1 \to K_{1c}$ and $h \to 0$.

12.4.4 Marginal Operators

Eigenvalues that do not grow or shrink under RG are called *marginal*, and the corresponding interactions are called *marginal interactions*. Suppose there was a marginal direction at $\boldsymbol{K}^*$. This means that if we start out at a point that differs slightly from $\boldsymbol{K}^*$ in that direction, that deviation will not change under the RG. It means the displaced starting point is also a fixed point. Typically the fixed points will form a *fixed line*. Generally, each point on the fixed line can have its own set of eigenvalues and associated exponents. Fixed lines with varying exponents are rare in $d = 3$, but we have already encountered one in $d = 2$: the *XY* model, which has a critical line along which the exponent η varies continuously with T, as shown in Eq. (10.28).

12.5 The β-Function

I want to describe briefly how things change if the blocking size L is continuous. It is enough to consider just one coupling, K. First consider blocking by some generic L, say

$L = 2$, under which the coupling K goes to $K'(K,L)$, where I explicitly indicate that the new coupling depends on the old coupling and the block size L.

At the fixed point we have

$$K'(K^*,L) = K^*, \tag{12.148}$$

and near the fixed point

$$K'(K^* + \Delta K, L) - K^* \tag{12.149}$$

$$= \left.\frac{\partial K'(K,L)}{\partial K}\right|_{K^*} \Delta K \equiv D'(K^*,L)\Delta K \tag{12.150}$$

$$\Delta K' = D'(K^*,L)\Delta K, \tag{12.151}$$

where I make it explicit that D' depends on L. Given

$$\xi(\Delta K) = L\xi(\Delta K'), \quad \text{that is,} \tag{12.152}$$

$$(\Delta K)^{-\nu} = L\left[D'(K^*,L)\Delta K\right]^{-\nu}, \quad \text{we conclude} \tag{12.153}$$

$$\frac{1}{\nu} = \frac{\ln D'(K^*,L)}{\ln L} \quad (\text{I assume } \Delta K, D' > 0). \tag{12.154}$$

I rewrite this in a form that will be recalled shortly:

$$D' = \frac{\Delta K'}{\Delta K} = L^{\frac{1}{\nu}}. \tag{12.155}$$

Given the boundary condition

$$\ln 1 = 0, \tag{12.156}$$

$$D'(K^*,1) = 1 \quad (\text{no RG done, } K' = K), \tag{12.157}$$

we may write

$$\frac{1}{\nu} = \frac{\ln D'(K^*,L)}{\ln L} \tag{12.158}$$

$$= \frac{\ln D'(K^*,L) - \ln D'(K^*,1)}{\ln L - \ln 1}. \tag{12.159}$$

Now choose $L = 1 + \varepsilon$ or $\ln L = 0 + d\ln L$ to obtain

$$\frac{1}{\nu} \underset{d\ln L \to 0}{=} \left.\frac{\partial \ln D'(K^*,L)}{\partial \ln L}\right|_{L=1} \tag{12.160}$$

$$= \left.\frac{\partial^2 K'(K,L)}{\partial \ln L \partial K}\right|_{K=K^*,L=1}. \tag{12.161}$$

In a scheme where L varies continuously, the same mixed derivative occurs but in the opposite order. One first defines a *β-function*, which measures the rate of change of K under an infinitesimal scale change:

$$\left.\frac{dK'(L)}{d\ln L}\right|_{L=1} = \beta(K). \tag{12.162}$$

Notice that β depends on K but not L: the relation between the old and new K depends on the ratio of the old block size to the new block size, but not on the absolute size of the block (recall decimation in $d = 2$). The functions relating the new couplings to the old do not depend on whether this is the first or the fiftieth time we are blocking.

By definition,

$$\beta(K^*) = 0. \tag{12.163}$$

The fixed point is the zero of the β-function.

Near this zero we expand

$$\beta(K) = \beta'(K^*)(K - K^*) \tag{12.164}$$

and integrate the flow equation

$$\frac{dK'(L)}{d\ln L} = \beta'(K^*)(K - K^*) \tag{12.165}$$

$$\frac{d(K'(L) - K^*)}{(K - K^*)} = \beta'(K^*)d\ln L \tag{12.166}$$

$$\frac{K'(L) - K^*}{K - K^*} = L^{\beta'(K^*)} \tag{12.167}$$

$$\Delta K' = \Delta K \cdot L^{\beta'(K^*)} \stackrel{\text{def}}{=} \Delta K \cdot L^{\frac{1}{\nu}} \tag{12.168}$$

using Eq. (12.155). This gives us a result worth memorizing:

$$\frac{1}{\nu} = \beta'(K^*). \tag{12.169}$$

The derivative of the β-function (with respect to K) at the fixed point equals $\frac{1}{\nu}$.

Since β itself is the derivative of K' with respect to $\ln L$,

$$\frac{1}{\nu} = \beta'(K^*) \tag{12.170}$$

$$= \left.\frac{\partial \beta(K)}{\partial K}\right|_{K^*} \tag{12.171}$$

$$= \left.\frac{\partial^2 K'(K,L)}{\partial K \partial \ln L}\right|_{L=1,K^*}, \tag{12.172}$$

in agreement with Eq. (12.161).

Let us understand why the second mixed derivative enters the formula for ν, or more generally the linearized flow matrix. We need a derivative with respect to the scale factor L to find the flow. The flow is zero *at* the fixed point. To find the fate of points *near* the fixed point we need the derivative of the flow with respect to the coupling.

If there are many couplings, β and β' become matrices whose eigenvectors and eigenvalues control the flow and the exponents. The inverse of the relevant eigenvalue of β' will yield ν, and the inverses of the others the irrelevant ones.

In the language of groups, D refers to the group element of the RG while β refers to the generator of infinitesimal scale change.

References and Further Reading

[1] S.-K. Ma, *Modern Theory of Critical Phenomena* (Frontiers in Physics), W. A. Benjamin, 3rd edition (1976).
[2] J. Cardy, *Scaling and Renormalization in Statistical Physics*, Cambridge University Press (1996).
[3] L. Reichl, *A Modern Course in Statistical Physics*, Wiley (2009).
[4] N. Goldenfeld, *Lectures on Phase Transitions and the Renormalization Group*, Addison-Wesley (1992).
[5] M. Le Bellac, *Quantum and Statistical Field Theory*, Oxford University Press (1991).
[6] L. P. Kadanoff, in *Phase Transitions and Critical Phenomena*, vol. 5a, eds. C. Domb and M. S. Green, Academic Press (1976). See the entire series for many review articles on critical phenomena.
[7] F. J. Wegner, Physical Review B, **5**, 4529 (1972).
[8] F. J. Wegner, in *Phase Transitions and Critical Phenomena*, vol. 6, eds. C. Domb and M. S. Green, Academic Press (1976).
[9] M. E. Fisher, Reviews of Modern Physics, **70**, 653 (1998) (for a historical account).
[10] K. G. Wilson and J. B. Kogut, Physics Reports, **12**, 75 (1974).
[11] K. G. Wilson, Physical Review B, **4**, 3174 (1971).
[12] J. D. Bjorken, Physical Review **179**, 1547 (1969).
[13] B. Widom, Journal of Chemical Physics, **43**, 3898 (1965).
[14] W. Essam and M. E. Fisher, Journal of Chemical Physics, **38**, 802 (1963).
[15] L. P. Kadanoff, Physics, **2**, 263 (1966).

13

Renormalization Group for the ϕ^4 Model

We will now see the full power of the RG as applied to critical phenomena. The treatment, here and elsewhere, will emphasize the key ideas and eschew long and detailed calculations. In this and the next chapter I will focus on issues I found confusing rather than complicated. For example, a five-loop Feynman diagram is complicated but not confusing; I know what is going on. On the other hand, the relationship between renormalization of continuum field theories with one or two couplings and Wilson's program with an infinite number of couplings used to confuse me.

Because of universality, we can choose any member to study the whole class. For the Ising model the action has to have Z_2 symmetry, or invariance under sign reversal of the order parameter:

$$S(\phi) = S(-\phi). \tag{13.1}$$

(An infinitesimal symmetry-breaking term of the form $h\phi$ will be introduced to find exponents related to magnetization. Having done that, we will set $h = 0$ in the rest of the analysis.)

I will, however, discuss an action that enjoys a larger $U(1)$ symmetry:

$$S(\phi, \phi^*) = S(\phi e^{i\theta}, \phi^* e^{-i\theta}), \tag{13.2}$$

where θ is arbitrary. (We can also see this as $O(2)$ symmetry of S under rotations of the real and imaginary parts of ϕ.) I discuss $U(1)$ because the computations are very similar to the upcoming discussion of non-relativistic fermions, which also have $U(1)$ symmetry. I will show you how a minor modification of the $U(1)$ results yields the exponents of the Z_2 (Ising) models.

13.1 Gaussian Fixed Point

The Gaussian fixed point is completely solvable: we can find all the eigenvectors and eigenvalues of the linearized flow matrix T. It also sets the stage for a non-trivial model of magnetic transitions amenable to approximate calculations.

The partition function for the Gaussian model is

$$Z = \int [\mathcal{D}\phi^*(k)][\mathcal{D}\phi(k)]e^{-S^*(\phi,\phi^*)}, \tag{13.3}$$

where we have an asterisk on S^* because it will prove to be a fixed point. (The asterisk on ϕ^*, of course, denotes the conjugate.) The action is

$$S^*(\phi,\phi^*) = \int_0^\Lambda \phi^*(\mathbf{k})k^2\phi(\mathbf{k})\frac{d^dk}{(2\pi)^d}, \tag{13.4}$$

where the limits on the integral refer to the magnitude of the momentum k. This action typically describes some problem on a lattice. There are no ultraviolet singularities in this problem with a natural cut-off in momentum $k \simeq \frac{\pi}{a}$, where a is the lattice constant. We want to invoke the RG to handle possible infrared singularities at and near criticality. In that case we may focus on modes near the origin. We will begin with a ball of radius $\Lambda \ll \frac{1}{a}$ centered at the origin and ignore the shape of the Brillouin zone for $k \simeq \frac{1}{a}$.

The number of dimensions d is usually an integer, but we must be prepared to work with continuous d. In general, we will need to make sense of various quantities like vectors, dot products, and integrals in non-integer dimensions. For our discussions, we just need the integration measure for rotationally invariant integrands:

$$d^dk = k^{d-1}dkS_d, \tag{13.5}$$

where

$$S_d = \frac{2\pi^{\frac{d}{2}}}{\Gamma(\frac{d}{2})} \tag{13.6}$$

is the "area" of the unit sphere in d dimensions. We will rarely need this precise expression.

As a first step we divide the existing modes into slow and fast ones based on k:

$$\phi_{\mathrm{s}} = \phi(\boldsymbol{k}) \;\; \text{for } 0 \le k \le \Lambda/s \;\; \text{(slow modes)}, \tag{13.7}$$

$$\phi_{\mathrm{f}} = \phi(\boldsymbol{k}) \;\; \text{for } \Lambda/s \le k \le \Lambda \;\; \text{(fast modes)}, \tag{13.8}$$

where $s > 1$ is the scale parameter that decides how much we want to eliminate. We are going to eliminate the modes between the new cut-off Λ/s and the old one Λ.

The action itself separates into slow and fast pieces in momentum space:

$$S^*(\phi,\phi^*) = \int_0^{\Lambda/s} \phi^*(\mathbf{k})k^2\phi(\mathbf{k})\frac{d^dk}{(2\pi)^d} + \int_{\Lambda/s}^{\Lambda} \phi^*(\mathbf{k})k^2\phi(\mathbf{k})\frac{d^dk}{(2\pi)^d} \tag{13.9}$$

$$= S^*(\text{slow}) + S^*(\text{fast}), \tag{13.10}$$

and the Boltzmann weight factorizes over slow and fast modes. Thus, integrating over the fast modes just gives an overall constant $Z(\text{fast})$ multiplying the Z for the slow modes:

$$Z = \int [\mathcal{D}\phi_s^*(k)][\mathcal{D}\phi_s(k)]e^{-S^*(\phi_s)} \int [\mathcal{D}\phi_f^*(k)][\mathcal{D}\phi_f(k)]e^{-S_0^*(\phi_f)} \tag{13.11}$$

$$\equiv \int [\mathcal{D}\phi_s^*(k)][\mathcal{D}\phi_s(k)]e^{-S^*(\phi_s)}Z(\text{fast}) \tag{13.12}$$

$$= \int [\mathcal{D}\phi_s^*(k)][\mathcal{D}\phi_s(k)]e^{-S^*(\phi_s)+\ln Z(\text{fast})}. \tag{13.13}$$

We can ignore $\ln Z(\text{fast})$ going forward, because it is independent of ϕ_s and will drop out of all ϕ_s correlation functions.

The action after mode elimination,

$$S'^*(\phi,\phi^*) = \int_0^{\Lambda/s} \phi^*(\mathbf{k})k^2\phi(\mathbf{k})\frac{d^dk}{(2\pi)^d}, \tag{13.14}$$

is Gaussian, but not quite the same as the action we started with,

$$S^*(\phi,\phi^*) = \int_0^{\Lambda} \phi^*(\mathbf{k})k^2\phi,(\mathbf{k})\frac{d^dk}{(2\pi)^d} \tag{13.15}$$

because the allowed region for k is different.

We remedy this by defining a new momentum,

$$k' = sk, \tag{13.16}$$

which runs over the same range as k did before elimination. The action now becomes

$$S'^*(\phi,\phi^*) = \int_0^{\Lambda} \phi^*\left(\frac{\mathbf{k}'}{\mathbf{s}}\right)\left[\frac{k'^2}{s^2}\right]\phi\left(\frac{\mathbf{k}'}{\mathbf{s}}\right)\frac{d^dk'}{s^d(2\pi)^d} \tag{13.17}$$

$$= s^{-(d+2)}\int_0^{\Lambda} \phi^*\left(\frac{\mathbf{k}'}{\mathbf{s}}\right)k'^2\phi\left(\frac{\mathbf{k}'}{\mathbf{s}}\right)\frac{d^dk'}{(2\pi)^d}. \tag{13.18}$$

Due to this constant rescaling of units, the cut-off remains fixed and we may set $\Lambda = 1$ at every stage. (Of course, the cut-off decreases in fixed laboratory units. What we are doing is analogous to using the lattice size of the block spins as the unit of length as we eliminate degrees of freedom.)

To take care of the factor $s^{-(d+2)}$ we introduce a rescaled field:

$$\phi'(k') = s^{-(\frac{d}{2}+1)}\phi\left(\frac{k'}{s}\right) \equiv \zeta^{-1}(s)\phi\left(\frac{k'}{s}\right). \tag{13.19}$$

Notice that for every $\phi'(k')$ for $0 \le k' \le \Lambda$ there is a corresponding original field that survives elimination and is defined on a smaller sphere ($0 \le k \le \Lambda/s$).

In terms of ϕ', the new action coincides with the original one in every respect:

$$S'^*(\phi',\phi'^*) = \int_0^\Lambda \phi'^*(\mathbf{k}')k'^2\phi'(\mathbf{k}')\frac{d^dk'}{(2\pi)^d}. \tag{13.20}$$

Thus, S^* is a *fixed point* of the RG with the following three steps:

- Eliminate fast modes, i.e., reduce the cut-off from Λ to Λ/s.
- Introduce rescaled momenta, $k' = sk$, which now go all the way to Λ.
- Introduce rescaled fields $\phi'(k') = \zeta^{-1}\phi(k'/s)$ and express the effective action in terms of them. This action should have the same coefficient for the quadratic term. (In general, the field rescaling factor ζ^{-1} could be different from the $s^{-(1+d/2)}$ that was employed above.)

With this definition of the RG transformation, we have a mapping from actions defined in a certain k-space (a ball of radius Λ) to actions in the *same* space. Thus, if we represent the initial action as a point in a coupling constant space, this point will flow under the RG transformation to another point in the same space.

As the Gaussian action is a fixed point of this RG transformation, it must correspond to $\xi = \infty$, and indeed it does:

$$G(k) = \frac{1}{k^2} \leftrightarrow G(r) = \frac{1}{r^{d-2}}. \tag{13.21}$$

Here is our first critical exponent associated with the Gaussian fixed point:

$$\eta = 0. \tag{13.22}$$

This result is independent of d.

We now want to determine the flow of couplings near this fixed point. We will do this by adding various perturbations (also referred to as operators) and see how they respond to the three-step RG mentioned above. The perturbations will be linear, quadratic, and quartic in ϕ. (Operators with more powers of ϕ or more derivatives will prove highly irrelevant near $d = 4$, the region of interest to us.) We will then find the eigenvectors and eigenvalues of the linearized flow matrix T and classify them as relevant, irrelevant, or marginal. In general, the operators we add will mix under this flow and we must form linear combinations that go into multiples of themselves, i.e. the eigenvectors of T. The critical exponents and asymptotic behavior of correlation functions will follow from these.

13.1.1 Linear and Quadratic Perturbations

The perturbations can be an even or odd power of ϕ or ϕ^* (which I may collectively refer to as ϕ).

The uniform magnetic field couples linearly to ϕ and corresponds to an odd term $h\phi(0)$, where the argument (0) refers to the momentum. (Again, we must use the perturbation

$h\phi + h^*\phi^*$. We do not bother because the scaling is the same for both, and denote by h the coupling linear in the field.)

We have, from Eq. (13.19), which defines the rescaled field,

$$h\phi(0) = h\zeta\phi'(0\cdot s) = hs^{1+\frac{1}{2}d}\phi'(0) \stackrel{\text{def}}{=} h_s\phi'(0), \tag{13.23}$$

which means the renormalized h is

$$h_s = hs^{1+\frac{1}{2}d}. \tag{13.24}$$

Having found how an infinitesimal h gets amplified by RG, we will set $h = 0$ as we continue our analysis. We will not consider ϕ^3 because it is a *redundant operator* [1]. What this means is that if we began with

$$S = \int \left[h\phi + r_0\phi^2 + v\phi^3 + u\phi^4\right] d^dx, \tag{13.25}$$

the ϕ^3 term can be eliminated by a shift $\phi \to \phi - \frac{v}{4u}$.

Higher odd powers of ϕ or terms with an even number of extra derivatives will prove extremely irrelevant near $d = 4$, which will be our focus.

Now we turn to quadratic and quartic perturbations. First, consider the addition of the term

$$S_r = \int_0^\Lambda \phi^*(\mathbf{k}) r_0 \phi(\mathbf{k}) \frac{d^dk}{(2\pi)^d}, \tag{13.26}$$

which separates nicely into slow and fast pieces. Mode elimination just gets rid of the fast part, leaving behind the *tree-level term*

$$S_r^{\text{tree}} = \int_0^{\Lambda/s} \phi^*(\mathbf{k}) r_0 \phi(\mathbf{k}) \frac{d^dk}{(2\pi)^d}. \tag{13.27}$$

The adjective *tree-level* in general refers to terms that remain of an interaction upon setting all fast fields to 0. Keeping only this term amounts to eliminating fast modes by simply setting them to zero. Of course, the tree-level term will be viewed by us as just the first step.

If we express S_r^{tree} in terms of the new fields and new momenta, we find

$$S'_r = s^2 \int_0^\Lambda \phi'^*(\mathbf{k}') r_0 \phi'(\mathbf{k}') \frac{d^dk'}{(2\pi)^d} \tag{13.28}$$

$$\stackrel{\text{def}}{=} \int_0^\Lambda \phi'^*(\mathbf{k}') r_{0s} \phi'(\mathbf{k}') \frac{d^dk'}{(2\pi)^d}, \quad \text{which means} \tag{13.29}$$

$$r_{0s} = r_0 s^2. \tag{13.30}$$

In other words, after an RG action by a factor s, the coupling r_0 evolves into $r_{0s} = r_0 s^2$. Since S_r lacks the two powers of k that S_0 has, it is to be expected that r_0 will get amplified by s^2.

We have identified another relevant eigenvector in $r_0\phi^2$.

We may identify r_0 with t, the dimensionless temperature that takes us off criticality. Since the correlation length drops by $1/s$ under this rescaling of momentum by s,

$$\xi(r_0) = s^1\xi(r_{0s}) \tag{13.31}$$

$$= s^1\xi(r_0 s^2) \tag{13.32}$$

$$= (r_0)^{-\frac{1}{2}}\xi(1), \tag{13.33}$$

which means that

$$\nu = \frac{1}{2} \tag{13.34}$$

for all values of d. We do not need the RG to tell us this because we can find the propagator with a non-zero r_0 exactly:

$$G(k) \simeq \frac{1}{k^2 + r_0} \leftrightarrow G(r) \simeq \frac{e^{-\sqrt{r_0}r}}{r^{d-2}} \equiv \frac{e^{-r/\xi}}{r^{d-2}}, \quad \text{i.e.,} \tag{13.35}$$

$$\xi = r_0^{-\frac{1}{2}}. \tag{13.36}$$

In general, the quadratic perturbation could be with coupling $r(k)$ that varies with k. Given the analyticity of all the couplings in the RG actions (no singularities in and no singularities out), we may expand

$$r(k) = r_0 + r_2 k^2 + r_4 k^4 + \cdots \tag{13.37}$$

and show that

$$r_{2s} = r_2, \tag{13.38}$$

$$r_{4s} = s^{-2} r_4, \tag{13.39}$$

and so on. Thus, r_2 is marginal, and adding it simply modifies the k^2 term already present in S_0. If you want, you could say that varying the coefficient of k^2 in S_0 gives us a line of fixed points, but this line has the same exponents everywhere because the coefficient of k^2 may be scaled back to unity by field rescaling. (The Jacobian in the functional integral will be some constant.) The other coefficients like r_4 and beyond are irrelevant.

Exercise 13.1.1 *Derive Eqs. (13.38) and (13.39).*

13.1.2 Quartic Perturbations

When we consider quartic perturbations of the fixed point, we run into a new complication: the term couples slow and fast modes and we have to do more than just rewrite the old perturbation in terms of new fields and new momenta. In addition, we will find that mode elimination generates corrections to the flow of r_0, the quadratic term.

Consider the action

$$S = S^* + S_r + S_u \tag{13.40}$$

$$= \int_0^\Lambda \phi^*(\mathbf{k})k^2\phi(\mathbf{k})\frac{d^dk}{(2\pi)^d} + r_0\int_0^\Lambda \phi^*(\mathbf{k})\phi(\mathbf{k})\frac{d^dk}{(2\pi)^d}$$

$$+\frac{u_0}{2!2!}\int_{|k|<\Lambda} \phi^*(\mathbf{k}_4)\phi^*(\mathbf{k}_3)\phi(\mathbf{k}_2)\phi(\mathbf{k}_1)\prod_{i=1}^{3}\frac{d^dk_i}{(2\pi)^d} \tag{13.41}$$

$$\equiv S_0 + S_\mathrm{I}, \quad \text{where} \tag{13.42}$$

$$S_0 = S^* + S_r = \int_0^\Lambda \phi^*(\mathbf{k})(k^2+r_0)\phi(\mathbf{k})\frac{d^dk}{(2\pi)^d}, \tag{13.43}$$

$$S_\mathrm{I} = S_u \equiv \int_\Lambda \phi^*(4)\phi^*(3)\phi(2)\phi(1)u_0, \tag{13.44}$$

$$\boldsymbol{k}_4 = \boldsymbol{k}_1 + \boldsymbol{k}_2 - \boldsymbol{k}_3. \tag{13.45}$$

Notice the compact notation used for the quartic interaction in Eq. (13.44). The subscript in S_I stands for *interaction*, which is what S_u is here.

Remember that S_0 *is not the fixed point action*, it is the *quadratic* part of the action and includes the r_0 term. With respect to the Gaussian fixed point S^*, it is true that S_r is a perturbation, but in field theories, S_r is part of the non-interacting action S_0 and only $S_u \equiv S_\mathrm{I}$ is viewed as a perturbation. In other words, S_r perturbs the Gaussian fixed point action, while S_I perturbs the non-interacting action. For what follows it is more expedient to use the decomposition $S = S_0 + S_\mathrm{I}$, where $S_\mathrm{I} = S_u$ is quartic.

13.1.3 Mode Elimination Strategy

I will now describe the strategy for mode elimination for the case

$$S(\phi_\mathrm{s},\phi_\mathrm{f}) = S_0(\phi_\mathrm{s}) + S_0(\phi_\mathrm{f}) + S_\mathrm{I}(\phi_\mathrm{s},\phi_\mathrm{f}), \tag{13.46}$$

where S_0 has been separated into slow and fast pieces whereas S_I cannot be separated in that manner.

Let us do the integration over fast modes:

$$Z = \int [\mathcal{D}\phi_\mathrm{s}^*][\mathcal{D}\phi_\mathrm{s}]e^{-S_0(\phi_\mathrm{s})}\int [\mathcal{D}\phi_\mathrm{f}^*][\mathcal{D}\phi_\mathrm{f}]e^{-S_0(\phi_\mathrm{f})}e^{-S_\mathrm{I}(\phi_\mathrm{s},\phi_\mathrm{f})} \tag{13.47}$$

$$\stackrel{\text{def}}{=} \int [\mathcal{D}\phi_\mathrm{s}^*][\mathcal{D}\phi_\mathrm{s}]e^{-S_\mathrm{eff}(\phi_\mathrm{s})}, \tag{13.48}$$

which defines the *effective action* $S_{\mathrm{eff}}(\phi_s)$. Let us manipulate its definition a little:

$$e^{-S_{\mathrm{eff}}(\phi_s)} = e^{-S_0(\phi_s)} \int [\mathcal{D}\phi_f^*][\mathcal{D}\phi_f] e^{-S_0(\phi_f)} e^{-S_I(\phi_s,\phi_f)}$$
$$= e^{-S_0(\phi_s)} \frac{\int [\mathcal{D}\phi_f^*][\mathcal{D}\phi_f] e^{-S_0(\phi_f)} e^{-S_I(\phi_s,\phi_f)}}{\int [\mathcal{D}\phi_f^*][\mathcal{D}\phi_f] e^{-S_0(\phi_f)}} \underbrace{\int [\mathcal{D}\phi_f][\mathcal{D}\phi_f^*] e^{-S_0(\phi_f)}}_{Z_{0f}}. \tag{13.49}$$

Dropping Z_{0f}, which does not affect averages of the slow modes,

$$e^{-S_{\mathrm{eff}}(\phi_s)} = e^{-S_0(\phi_s)} \left\langle e^{-S_I(\phi_s,\phi_f)} \right\rangle_{0\rangle} \stackrel{\mathrm{def}}{=} e^{-S_0 - \delta S'}, \tag{13.50}$$

where $\langle\ \rangle_{0\rangle}$ denotes averages with respect to the fast modes with action $S_0(\phi_f)$.

Combining Eq. (13.50) with the *cumulant expansion*, which relates the *mean of the exponential to the exponential of the means*,

$$\left\langle e^{\Omega} \right\rangle = e^{\left[\langle\Omega\rangle + \frac{1}{2}(\langle\Omega^2\rangle - \langle\Omega\rangle^2) + \cdots\right]}, \tag{13.51}$$

we find

$$S_{\mathrm{eff}} = S_0 + \langle S_I \rangle - \frac{1}{2}(\langle S_I^2 \rangle - \langle S_I \rangle^2) + \cdots \tag{13.52}$$

It is understood that this expression has to be re-expressed in terms of the rescaled fields and momenta to get the final contribution to the action. We will do this eventually.

Exercise 13.1.2 *Verify the correctness of the cumulant expansion Eq. (13.51) to the order shown. (Expand e^{Ω} in a series, average, and re-exponentiate.)*

Since S_I is linear in u, Eq. (13.52) is a weak coupling expansion. It is now clear what has to be done. Each term in the series contains some monomials in fast and slow modes. The former have to be averaged with respect to the quadratic action $S_0(\phi_f)$ by the use of Wick's theorem. The result of each integration will yield monomials in the slow modes. When re-expressed in terms of the rescaled fields and momenta, each will renormalize the corresponding coupling. In principle, you have been given all the information to carry out this process. There is, however, no need to reinvent the wheel. There is a procedure involving Feynman diagrams that automates this process. These rules will not be discussed here since they may be found, for example, in Sections 3–5 of [2], or in many field theory books. Instead, we will go over just the first term in the series in some detail and comment on some aspects of the second term. Readers familiar with Feynman diagrams should note that while these diagrams have the same multiplicity and topology as the field theory diagrams, the momenta being integrated out are limited to the shell being eliminated, i.e., $\Lambda/s \le k \le \Lambda$.

The leading term in the cumulant expansion in Eq. (13.52) has the form

$$\langle S_{\mathrm{I}}\rangle = \frac{1}{2!2!}\left\langle \int_{|k|<\Lambda} (\phi_{\mathrm{f}}+\phi_{\mathrm{s}})_4^*(\phi_{\mathrm{f}}+\phi_{\mathrm{s}})_3^*(\phi_{\mathrm{f}}+\phi_{\mathrm{s}})_2(\phi_{\mathrm{f}}+\phi_{\mathrm{s}})_1 u_0)\right\rangle_{0\rangle}, \quad (13.53)$$

where the subscript $0\rangle$ stands for the average with respect to the quadratic action of the fast modes. The 16 possible monomials fall into four groups:

- Eight terms with an odd number of fast modes.
- One term with all fast modes.
- One term with all slow modes.
- Six terms with two slow and two fast modes.

We have no interest in the first two items: the first because they vanish by symmetry and the second because it makes a constant contribution, independent of ϕ_{s}, to the effective action.

The one term with all slow fields makes a tree-level contribution

$$S_u^{\mathrm{tree}} = \frac{1}{2!2!}u_0 \int_{k<\Lambda/s} \phi^*(\mathbf{k}_4)\phi^*(\mathbf{k}_3)\phi(\mathbf{k}_2)\phi(\mathbf{k}_1)\prod_{i=1}^{3}\frac{d^dk_i}{(2\pi)^d} \quad (13.54)$$

to the action. The momentum and field have not been rescaled yet, as is evident from the cut-off.

That leaves us with six terms which have two fast and two slow fields. Of these, two are no good because both the fast fields are ϕ_{f}'s or ϕ_{f}^*'s, and these have zero average by $U(1)$ symmetry. This leaves us with four terms which schematically look like

$$\phi_{\mathrm{f}}^*\phi_{\mathrm{s}}^*\phi_{\mathrm{f}}\phi_{\mathrm{s}},\ \phi_{\mathrm{s}}^*\phi_{\mathrm{f}}^*\phi_{\mathrm{f}}\phi_{\mathrm{s}},\ \phi_{\mathrm{s}}^*\phi_{\mathrm{f}}^*\phi_{\mathrm{s}}\phi_{\mathrm{f}},\ \phi_{\mathrm{f}}^*\phi_{\mathrm{s}}^*\phi_{\mathrm{s}}\phi_{\mathrm{f}}. \quad (13.55)$$

All four terms make the same contribution to S_r (modulo dummy labels, which takes care of the $\frac{1}{2!2!}$ up front):

$$\delta S_r = u_0 \int \phi_{\mathrm{s}}^*(4)\phi_{\mathrm{s}}(2)dk_4dk_2\langle\phi_{\mathrm{f}}^*(3)\phi_{\mathrm{f}}(1)\rangle dk_3dk_1\delta(4+3-2-1), \quad (13.56)$$

where I am using a compact notation:

$$\delta(4+3-2-1) \equiv (2\pi)^d\delta(k_4+k_3-k_2-k_1), \quad (13.57)$$

$$dk \equiv \frac{d^dk}{(2\pi)^d}. \quad (13.58)$$

Using

$$\langle\phi_{\mathrm{f}}^*(3)\phi_{\mathrm{f}}(1)\rangle = \frac{\delta(3-1)}{k_3^2+r_0}, \quad (13.59)$$

we find (on changing some dummy variables and returning to the more explicit notation):

$$\delta S_r = \int_0^{\Lambda/s} \phi^*(\boldsymbol{k})\phi(\boldsymbol{k})\frac{d^d k}{(2\pi)^d}\left(u_0 \int_{\Lambda/s}^{\Lambda} \frac{d^d k_3}{(2\pi)^d}\frac{1}{k_3^2 + r_0}\right). \tag{13.60}$$

Let us see briefly how the above results follow in the diagrammatic approach. First, we associate with each of the 16 monomials contained in Eq. (13.53) a four-pronged X, as in Figure 13.1(a). The incoming arrows correspond to ϕ and the outgoing ones to ϕ^*. Each prong can stand for a ϕ_{s} or a ϕ_{f}, or its conjugate. Figure 13.1(a) shows the case where all four are slow. It contributes to u_0 at tree level. It merely has to be re-expressed in terms of new fields and momenta. Figure 13.1(b) is an example of the eight terms with an odd number of fast lines. These average to zero. Figure 13.1(c) describes the case with two fast modes (labels 1 and 3, with average $G(3)$) and two slow lines (labels 2 and 4), with both sets coming in complex conjugate pairs. The two fast lines are joined by the averaging, and the average is represented by the line joining them, the propagator $G(k_3)$. This is called the *tadpole diagram*. We are left with two slow fields, 2 and 4. This renormalizes the quadratic term S_r as per Eq. (13.60). Finally, Figure 13.1(d) describes the case where all lines are fast, come in complex-conjugate pairs, and average to two propagators which are integrated over. We do not consider this term since it contributes a constant independent of ϕ_{s}.

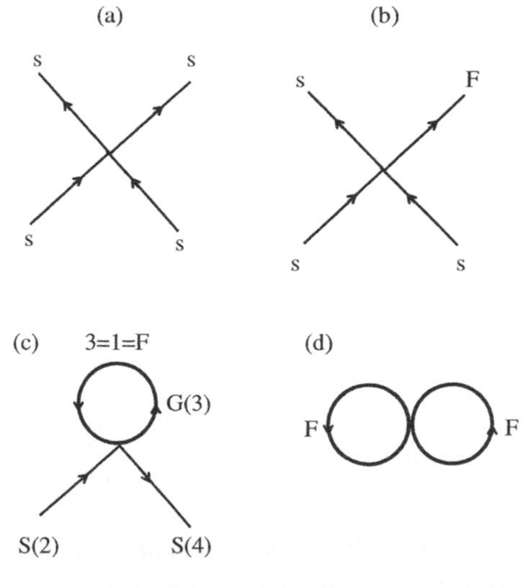

Figure 13.1 Diagrammatic description of the 16 monomials contained in Eq. (13.53). (a) The term with four slow fields, in complex-conjugate pairs. Contributes to u_0 at tree level. (b) One of the eight terms that have no average due to having odd powers of the fast field. (c) The tadpole graph that comes from averaging two fast fields (labeled 1 and 3, and with average $G(3)$) leaving behind a quadratic piece for the slow modes (labeled 2 and 4). It renormalizes r_0. (d) An ignorable ϕ_{s}-independent contribution with two fast pairs averaged.

Let us take stock. We began with

$$S = S^* + S_r + S_u \tag{13.61}$$

$$= \int_0^\Lambda \phi^*(\mathbf{k}) k^2 \phi(\mathbf{k}) \frac{d^d k}{(2\pi)^d} + r_0 \int_0^\Lambda \phi^*(\mathbf{k})\phi(\mathbf{k}) \frac{d^d k}{(2\pi)^d}$$
$$+ \frac{u_0}{2!2!} \int_{|k|<\Lambda} \phi^*(\mathbf{k}_4)\phi^*(\mathbf{k}_3)\phi(\mathbf{k}_2)\phi(\mathbf{k}_1) \prod_{i=1}^{3} \frac{d^d k_i}{(2\pi)^d} \tag{13.62}$$

$$\equiv S_0 + S_\mathrm{I}. \tag{13.63}$$

We ended up with

$$S_{\text{eff}} = \int_0^{\Lambda/s} \phi^*(\mathbf{k}) k^2 \phi(\mathbf{k}) \frac{d^d k}{(2\pi)^d}$$
$$+ r_0 \int_0^{\Lambda/s} \phi^*(\mathbf{k})\phi(\mathbf{k}) \frac{d^d k}{(2\pi)^d} + \int_0^{\Lambda/s} \phi^*(\boldsymbol{k})\phi(\boldsymbol{k}) \frac{d^d k}{(2\pi)^d} \left(u_0 \int_{\Lambda/s}^{\Lambda} \frac{d^d k_3}{(2\pi)^d} \frac{1}{k_3^2 + r_0} \right)$$
$$+ \frac{u_0}{2!2!} \int_{k<\Lambda/s} \phi^*(\mathbf{k}_4)\phi^*(\mathbf{k}_3)\phi(\mathbf{k}_2)\phi(\mathbf{k}_1) \prod_{i=1}^{3} \frac{d^d k_i}{(2\pi)^d}. \tag{13.64}$$

Now for the long awaited rescaling to switch to new momenta k and new fields ϕ':

$$k' = sk, \tag{13.65}$$

$$\phi'(k') = s^{-(\frac{d}{2}+1)} \phi\left(\frac{k'}{s}\right) \equiv \zeta^{-1}(s) \phi\left(\frac{k'}{s}\right). \tag{13.66}$$

I invite you to show that

$$S_{\text{eff}} = \int_0^{\Lambda} \phi^*(\mathbf{k}) k^2 \phi(\mathbf{k}) \frac{d^d k}{(2\pi)^d}$$
$$+ s^2 r_0 \int_0^{\Lambda} \phi^*(\mathbf{k})\phi(\mathbf{k}) \frac{d^d k}{(2\pi)^d} + s^2 \int_0^{\Lambda} \phi^*(\boldsymbol{k})\phi(\boldsymbol{k}) \frac{d^d k}{(2\pi)^d} \left(u_0 \int_{\Lambda/s}^{\Lambda} \frac{d^d k_3}{(2\pi)^d} \frac{1}{k_3^2 + r_0} \right)$$
$$+ s^{4-d} \frac{u_0}{2!2!} \int_{k<\Lambda} \phi^*(\mathbf{k}_4)\phi^*(\mathbf{k}_3)\phi(\mathbf{k}_2)\phi(\mathbf{k}_1) \prod_{i=1}^{3} \frac{d^d k_i}{(2\pi)^d}. \tag{13.67}$$

Exercise 13.1.3 *Carry out the rescaling of moment and fields and arrive at the preceding equation. (Only the momenta that are arguments of ϕ need rescaling, not k_3.)*

Exercise 13.1.4 *Suppose we begin with a quartic coupling $u(\boldsymbol{k}_1, \ldots, \boldsymbol{k}_4)$ instead of a constant u_0. Expand it in powers of k_i^2 and show that the coefficients of the higher powers are highly irrelevant near $d = 4$. This is analogous to what happened when we considered $r(k)$ instead of r_0 in Exercise 13.1.1.*

It is common to parametrize s as

$$s = e^l. \tag{13.68}$$

(Sometimes one sets $s = e^t$, and we will too, in a later chapter not connected to critical phenomena. But here it would be inviting trouble since t is associated with deviation from criticality.) In particular, for infinitesimal scale change we write

$$s = e^{dl} \simeq 1 + dl. \tag{13.69}$$

Look at the k_3 integral in Eq. (13.67):

$$\begin{aligned} u_0 \int_{\Lambda/s}^{\Lambda} \frac{d^d k_3}{(2\pi)^d} \frac{1}{k_3^2 + r_0} &= u_0 \int_{\Lambda(1-dl)}^{\Lambda} \frac{k_3^{d-3} dk_3 S_d}{(2\pi)^d} \frac{1}{k_3^2 + r_0} \\ &= u_0 \frac{\Lambda^{d-3} S_d \Lambda dl}{(2\pi)^d} \frac{1}{\Lambda^2 + r_0} \qquad (13.70) \\ &\stackrel{\text{def}}{=} u_0 \frac{A}{1 + r_0} dl, \qquad (13.71) \end{aligned}$$

which defines the constant A whose precise expression will not matter. As explained in the discussion following Eq. (13.18), we may set $\Lambda = 1$. (It is being used as the unit as we renormalize, just the way the new lattice spacing was used as the unit of distance after a block spin operation or decimation.)

Since we want to go to first order in r_0 and u_0 we may neglect the r_0 in Eq. (13.71) and approximate:

$$u_0 \frac{A}{1 + r_0} \simeq u_0 A. \tag{13.72}$$

Adding this induced quadratic term to the one from tree level, rescaling the momenta and fields, we find the following quadratic term:

$$(1 + 2dl) \int_0^{\Lambda} \phi^*(\mathbf{k}) \phi(\mathbf{k}) (r_0 + A u_0 dl) \frac{d^d k}{(2\pi)^d} \stackrel{\text{def}}{=} \int_0^{\Lambda} r_0(dl) \phi^*(\mathbf{k}) \phi(\mathbf{k}) \frac{d^d k}{(2\pi)^d}, \tag{13.73}$$

from which we deduce that

$$r_0(dl) = (1 + 2dl)(r_0 + A u_0 dl) \tag{13.74}$$

$$\frac{dr_0}{dl} = 2r_0 + A u_0. \tag{13.75}$$

Now for the u_0 term in Eq. (13.66). We are working to first order in r_0 and u_0. Since the tree-level term for u_0 is already first order, we stop with

$$u_0(dl) = u_0 s^{(4-d)} = u_0(1 + (4 - d)dl), \tag{13.76}$$

and conclude that

$$\frac{du_0}{dl} = (4-d)u_0. \tag{13.77}$$

Here are our final flow equations and β-functions:

$$\beta_r = \frac{dr_0}{dl} = 2r_0 + Au_0, \tag{13.78}$$

$$\beta_u = \frac{du_0}{dl} = (4-d)u_0. \tag{13.79}$$

These flow equations have only one fixed point, the Gaussian fixed point at the origin:

$$K^* = (r_0^* = 0, u_0^* = 0). \tag{13.80}$$

Near the fixed point a coupling $K_\alpha = K_\alpha^* + \delta K_\alpha$ flows as follows:

$$\frac{dK_\alpha}{dl} = \beta_\alpha(K^* + \delta K) = 0 + \left.\frac{\partial \beta_a}{\partial K_\beta}\right|^* (K_\beta - K_\beta^*), \tag{13.81}$$

$$\frac{d\delta K_\alpha}{dl} = \left.\frac{\partial \beta_a}{\partial K_\beta}\right|^* \delta K_\beta, \quad \text{where} \tag{13.82}$$

$$\delta K_a = K_\alpha - K_\alpha^*. \tag{13.83}$$

In our problem where $K^* = (0,0)$, $\delta K_a = K_\alpha$. That is, $\delta r_0 = r_0$ and $\delta u_0 = u_0$. Starting with Eqs. (13.78) and (13.79), and taking partial derivatives with respect to r_0 and u_0, we arrive at

$$\begin{pmatrix} \frac{dr_0}{dl} \\ \frac{du_0}{dl} \end{pmatrix} = \begin{pmatrix} 2 & A \\ 0 & (4-d) \end{pmatrix} \begin{pmatrix} r_0 \\ u_0 \end{pmatrix}. \tag{13.84}$$

The 2×2 matrix is not Hermitian and for good reason. The 0 in the lower left reflects the fact that r_0 does not generate any u_0, while the non-zero element A in the upper right says that u_0 generates some r_0.

Because the lower-left element vanishes, the relevant and irrelevant eigenvalues are the diagonal entries themselves:

$$ad = 2 \text{ (or } \nu = \tfrac{1}{2}), \qquad \omega d = 4 - d, \tag{13.85}$$

in the notation of Eqs. (12.143)–(12.145).

Even though the flow got more complicated by the introduction of the irrelevant term, the relevant exponent did not get modified. The asymmetric matrix has distinct left and right eigenvectors. The right eigenvectors, which we have been using all along, are given, in the notation of Section 12.4, by

$$|a\rangle = \begin{pmatrix} 1 \\ 0 \end{pmatrix}, \qquad |\omega\rangle = \begin{pmatrix} -\frac{A}{d-2} \\ 1 \end{pmatrix}. \tag{13.86}$$

In terms of canonical operators, the eigenvectors correspond to

$$|a\rangle = 1 \cdot \phi^2 + 0 \cdot \phi^4, \tag{13.87}$$

$$|\omega\rangle = -\frac{A}{d-2}\phi^2 + 1 \cdot \phi^4. \tag{13.88}$$

Under the action of T:

$$T|a\rangle = s^2|a\rangle, \tag{13.89}$$

$$T|\omega\rangle = s^{\varepsilon}|\omega\rangle, \tag{13.90}$$

$$\varepsilon = 4 - d, \tag{13.91}$$

where I have introduced the all-important parameter $\varepsilon = 4 - d$.

If we bring in the magnetic field, we have another eigenvector:

$$T|b\rangle = s^{1+\frac{d}{2}}|b\rangle. \tag{13.92}$$

This result is deduced as follows:

$$h\phi(0) = h\phi'(0/s)s^{1+\frac{d}{2}} \quad \text{(from Eq. (13.66))} \tag{13.93}$$

$$\equiv h_s\phi'(0). \tag{13.94}$$

So the Gaussian fixed point always has two relevant eigenvalues associated with temperature and magnetic field. The third eigenvalue $\omega d = 4 - d = |\varepsilon|$ is irrelevant for $d > 4$ and relevant for $d < 4$. It follows that for $d < 4$, the Gaussian fixed point does not describe the Ising class which can be driven to criticality by tuning just two parameters: $h = 0$, $t = 0$. It does describe critical phenomena with three relevant directions, but we will not go there. So, with one brief exception, we will study the Gaussian fixed point only for $d > 4$.

13.2 Gaussian Model Exponents for $d > 4$, $\varepsilon = 4 - d = -|\varepsilon|$

Figure 13.2 depicts the situation for $d > 4$ in the (r_0, u_0) plane, where r_0 is the coefficient of ϕ^2 and u_0 that of ϕ^4. The magnetic field is associated with a coordinate h and an eigenvector coming out of the page.

To attain criticality, we must first set $h = 0$. Next, in the (r_0, u_0) plane we need to tune just one parameter to hit the critical surface, the line

$$r_0 + \frac{Au_0}{d-2} = 0, \tag{13.95}$$

which just follows the irrelevant eigenvector flowing into the fixed point $K^* = (0,0)$. Equation 13.95 says that if we start with a non-zero u_0 we must tune r_0 to be $-\frac{Au_0}{d-2}$ to be critical.

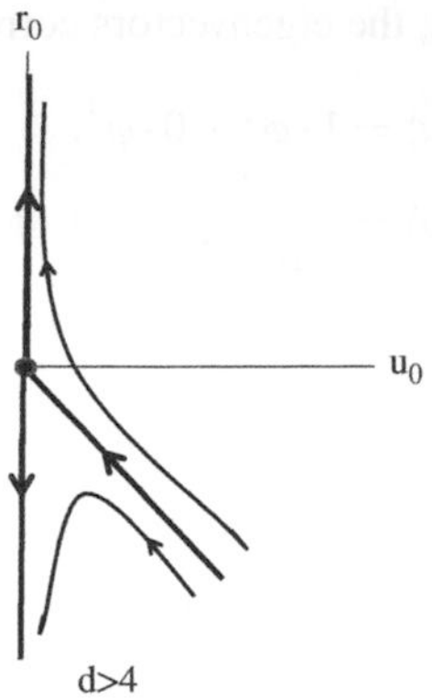

Figure 13.2 Flow in the Gaussian model for $d > 4$. The one-dimensional critical surface is shown by the attractive flow line into $K^* = (0,0)$.

In Section 12.4, Eqs. (12.143) and (12.145), we performed an abstract study of a flow with three parameters, two relevant and one irrelevant. I repeat those equations below with one minor change in notation: I replace L, the factor by which we change the spatial lattice size, with the factor s which produces the equivalent reduction in Λ:

$$|\Delta \boldsymbol{K}\rangle = t|a\rangle + h|b\rangle + g|\omega\rangle, \tag{13.96}$$

$$T(s)|\Delta \boldsymbol{K}\rangle = ts^{ad}|a\rangle + hs^{bd}|b\rangle + gs^{-|\omega|d}|\omega\rangle \tag{13.97}$$

$$= ts^{\frac{1}{\nu}}|a\rangle + hs^{d-\frac{\beta}{\nu}}|b\rangle + gs^{-|\omega|d}|\omega\rangle. \tag{13.98}$$

To evade relevant flow in the t and h directions, to retain or attain criticality, we just have to tune to $t = h = 0$.

What are h, t, and g in terms of h, r_0, and u_0?

Now h is just h. It gets rescaled as

$$h \to h_s = hs^{1+\frac{d}{2}}. \tag{13.99}$$

In the absence of u_0, the role of t is played by r_0, the coefficient of ϕ^2, since a non-zero r_0 takes us off the Gaussian fixed point. However, in the presence of u_0, t should be measured vertically up (in the r_0 direction) *from* the critical line $r_0 + \frac{Au_0}{d-2}u_0 = 0$. To understand this systematically we do the following:

- Express the initial position vector $r_0|1\rangle + u_0|2\rangle$, associated with the action

$$S = r_0\phi^2 + u_0\phi^4, \tag{13.100}$$

in terms of the eigenvectors of T:

$$|a\rangle = |1\rangle, \tag{13.101}$$

$$|\omega\rangle = -\frac{A}{d-2}|1\rangle + |2\rangle. \tag{13.102}$$

- Find the effect of an RG by scale factor s:

$$|a\rangle \to T|a\rangle = s^2|a\rangle, \tag{13.103}$$

$$|\omega\rangle \to T|\omega\rangle = s^{-|\varepsilon|}|\omega\rangle. \tag{13.104}$$

- Find the renormalized action in the canonical basis of ϕ^2 and ϕ^4.

The result of this exercise is that the action evolves as follows:

$$r_0\phi^2 + u_0\phi^4 \to \left[\left(r_0 + \frac{Au_0}{d-2}\right)s^2 - \frac{Au_0}{d-2}s^{-|\varepsilon|}\right]\phi^2 + u_0 s^{-\varepsilon}\phi^4. \tag{13.105}$$

Exercise 13.2.1 *Derive Eq. (13.105).*

Exercise 13.2.2 *Show that the initial point* $(r_0 = 0,\ u_0 = 1) = |2\rangle$ *does not flow to the fixed point even though* u_0 *is termed irrelevant. Do this by writing the initial state in terms of* $|a\rangle$ *and* $|\omega\rangle$.

As $s \to \infty$, we may drop the $s^{-|\varepsilon|}$ part compared to the s^2 part in the first term and identify

$$t = \left(r_0 + \frac{Au_0}{d-2}\right). \tag{13.106}$$

As expected, when $t = 0$ we lie on the critical line. The second term has the scaling form already and u_0 plays the role of g, the irrelevant coupling. But remember this: u_0 being irrelevant does not mean that if we add a tiny bit of it, the action will flow to the fixed point. Instead, u_0 will generate some r_0 and the final point will run off along the $|a\rangle$ axis, as discussed in Exercise 13.2.2.

To find the other exponents we need to begin with f, the free energy per unit volume,

$$f = -\frac{\ln Z}{\text{Volume}}, \tag{13.107}$$

and take various derivatives. Now the RG, in getting rid of fast variables, does not keep track of their contribution to f. These were the $\ln Z_0(\text{fast})$ factors which were dropped along the way as unimportant for the averages of the slow modes. Fortunately, these contributions were analytic in all the parameters, coming as they did from fast modes. What we want is f_s, the *singular part of the free energy*, which is controlled by the yet to be integrated soft modes near $k = 0$. This remains unaffected as we eliminate modes with one trivial modification: due to the change in scale that accompanies the RG, unit volume after RG corresponds to volume s^d before RG. Consequently, the free energy *per unit volume* behaves as follows in d dimensions:

$$f_s(t, h, u_0) = s^{-d} f_s(ts^2, hs^{1+\frac{1}{2}d}, u_0 s^{4-d} + \cdots), \tag{13.108}$$

where the ellipsis refers to even more irrelevant couplings like $w_0(\phi^*\phi)^3$, which can be safely set to zero and ignored hereafter.

Following the familiar route,

$$m(-|t|,h,u_0) \simeq \left.\frac{\partial f}{\partial h}\right|_{h=0} \tag{13.109}$$

$$= s^{1-\frac{1}{2}d} m(-|t|s^2, 0, u_0 s^{4-d}) \tag{13.110}$$

$$= |t|^{(-\frac{1}{2})(1-\frac{1}{2}d)} m(-1,0,u_0 t^{(d-4)/2}) \tag{13.111}$$

$$= |t|^{(-\frac{1}{2})(1-\frac{1}{2}d)} m(-1,0,0) \text{ when } t \to 0, \tag{13.112}$$

$$\beta = \frac{d-2}{4}. \tag{13.113}$$

Consider the arguments of m in Eq. (13.112). Starting with a small *negative* $t = -|t|$ (required for non-zero m) I have renormalized to a point where it has grown to a robust value of -1. The middle argument $h = 0$ once the h derivative has been taken. Finally, I have set $u_{0s} = u_0|t|^{(d-4)/2} = 0$ in the limit $t \to 0$.

Taking another h derivative,

$$\chi(t,0,u_0) = t^{-1}\chi(1,0,u_0 t^{(d-4)/2}) \simeq t^{-1}\chi(1,0,0), \text{ i.e.,} \tag{13.114}$$

$$\gamma = 1. \tag{13.115}$$

(Unlike m, which exists only for $t < 0$, χ and hence γ can be computed in the $t > 0$ region. Thus we can let t grow under RG to $+1$.)

To find C_V, we take two derivatives of f with respect to t and find, as usual,

$$\alpha = 2 - \frac{1}{2}d. \tag{13.116}$$

Finally, to find δ we begin with Eq. (13.108), take an h-derivative, and then set $t = 0$ to obtain

$$m(0,h,u_0) \simeq \left.\frac{\partial f_s}{\partial h}\right|_{t=0} = s^{1-\frac{1}{2}d} m(0, hs^{1+\frac{1}{2}d}, u_0 s^{4-d}) \tag{13.117}$$

$$= h^{\frac{d-2}{d+2}} m(0,1,u_0 h^{(d-4)/(1+\frac{1}{2}d)}) \tag{13.118}$$

$$= h^{\frac{d-2}{d+2}} m(0,1,0) \text{ when } h \to 0, \tag{13.119}$$

$$\delta = \frac{d+2}{d-2}. \tag{13.120}$$

Table 13.1 compares the preceding exponents of the Gaussian model for $d > 4$ to Landau theory.

The exponents β and δ agree only at $d = 4$ but not above. So which one is right? It turns out it is Landau's. Let me show you what was wrong with the way these two Gaussian exponents were derived.

Look at the passage from Eq. (13.111) to Eq. (13.112) for m. Using the RG scaling, we arrived at

$$m(-|t|, h=0, u_0) = |t|^{\frac{d-2}{4}} m(-1,0,u_0|t|^{(d-4)/2}), \tag{13.121}$$

Table 13.1 *Gaussian model for $d > 4$ versus Landau theory.*

Exponent	Landau	Gaussian $d > 4$
α	jump	$\frac{4-d}{2} < 0$
β	$\frac{1}{2}$	$\frac{d-2}{4}$
γ	1	1
δ	3	$\frac{d+2}{d-2}$
ν	$\frac{1}{2}$	$\frac{1}{2}$
η	0	0

which relates the magnetization in the critical region with a tiny negative t to its value at a point far from criticality, where t had been renormalized to -1, where fluctuations are negligible, and where we can use Landau's derivation with impunity. Landau's analysis gives, in the magnetized phase,

$$m(-|r|, 0, u) \simeq \sqrt{\frac{|r|}{u}}. \tag{13.122}$$

Applying this general result to our case,

$$m(-1, 0, u_0|t|^{(d-4)/2}) \simeq \sqrt{\frac{1}{u_0|t|^{(d-4)/2}}} \simeq |t|^{-\frac{d-4}{4}}, \tag{13.123}$$

and this means we cannot simply set $u_0|t|^{(d-4)/2}$ to zero as $t \to 0$, because it enters the denominator in the expression for m computed far from the critical point. Thus, $m(-1,0,0)$ is not some ignorable constant prefactor, but a factor with a divergent t dependence. Incorporating this singularity from Eq. (13.123) into Eq. (13.121), we are led to

$$m(-|t|, h = 0, u_0) = |t|^{\frac{d-2}{4}} |t|^{-\frac{d-4}{4}} \simeq |t|^{\frac{1}{2}}, \tag{13.124}$$

which is Landau's result.

So the error was in setting the irrelevant variable to 0 when it appeared in the *denominator* of the formula for m in the region far from criticality. For this reason, u_0 is called a *dangerous irrelevant variable*. A dangerous irrelevant variable is one which cannot be blindly set to zero even if it renormalizes to 0 far from the critical region. The singularity it produces must be incorporated with care in ascertaining the true critical behavior.

Likewise, given Landau's answer far from criticality,

$$m(0,h,u) \simeq \left[\frac{h}{u}\right]^{1/3}, \tag{13.125}$$

the derivation of δ must be modified as follows:

$$m(0,h,u_0) \simeq \left.\frac{\partial f_s}{\partial h}\right|_{t=0} \tag{13.126}$$

$$= s^{1-\frac{1}{2}d} m(0, hs^{1+\frac{1}{2}d}, u_0 s^{4-d}) \tag{13.127}$$

$$= h^{\frac{d-2}{d+2}} m(0, 1, u_0 h^{(d-4)/(1+\frac{1}{2}d)}) \tag{13.128}$$

$$= h^{\frac{d-2}{d+2}} \left(\frac{1}{u_0 h^{(d-4)/(1+\frac{1}{2}d)}}\right)^{1/3}, \tag{13.129}$$

$$= h^{\frac{1}{3}} \tag{13.130}$$

$$\delta = 3. \tag{13.131}$$

Now for why Landau theory works for $d > 4$ even though it ignores fluctuations about the minimum of the action. These fluctuations are computed perturbatively in u_0. For $d > 4$ these correction terms are given by convergent integrals. They do not modify the singularities of the theory at $u_0 = 0$.

Conversely, perturbation theory in u_0 fails in $d < 4$ near criticality *no matter how small u_0 is*. The true expansion parameter that characterizes the Feynman graph expansion ends up being $u_0 r_0^{\frac{1}{2}(d-4)}$ and not u_0. Thus, no matter how small u_0 is, the corrections due to it will blow up as we approach criticality.

One can anticipate this on dimensional grounds: since r_0 always has dimension 2 and u_0 has dimension $4-d$ in momentum units, the dimensionless combination that describes interaction strength is $u_0 r_0^{\frac{1}{2}(d-4)}$.

13.3 Wilson–Fisher Fixed Point $d < 4$

If $d < 4$, both directions become relevant at the Gaussian fixed point. (We have set $h = 0$.) It is totally unstable. *Two* parameters, namely r_0 and u_0, have to be tuned to hit criticality. This does not correspond to any Ising-like transition. This fixed point will be of interest later on, when we consider renormalization of quantum field theories, but for now let us move on to a fixed point in $d < 4$ that has only one relevant eigenvalue and describes Ising and Ising-like transitions.

Here is the trick due to Wilson and Fisher [3] for finding and describing it perturbatively in the small parameter

$$\varepsilon = 4 - d. \tag{13.132}$$

Their logic is that if mean-field theory works at $d = 4$, it should work near $d = 4$ with small controllable fluctuations. But this requires giving a meaning to the calculation for continuous d. As mentioned earlier, we just need to deal with the measure $d^d k$. The idea is to compute the exponents as series in ε and then set $\varepsilon = 1$, hoping to get reliable results for $d = 3$. Here we consider just the terms to order ε.

First for the renormalization of r_0. We already know that

$$\frac{dr_0}{dl} = 2r_0 + \frac{u_0 A}{1+r_0}. \tag{13.133}$$

The denominator is just $k^2 + r_0$ when $k = \Lambda = 1$.

We already have part of the flow for u_0 from Eq. (13.79):

$$\frac{du_0}{dl} = (d-4)u_0 + \mathcal{O}(u_0^2) = \varepsilon u_0 + \mathcal{O}(u_0^2). \tag{13.134}$$

We need the $\mathcal{O}(u_0^2)$ term to find the fixed point u to order ε. This term describes how u_0 renormalizes itself to order u_0^2.

We have to take two powers of the quartic interaction, each with 16 possible monomials, and perform the averages we need to get what we want: a term of the form

$$-\int \phi_s^*(k_4)\phi_s^*(k_3)\phi_s(k_2)\phi_s(k_1)u(4321), \tag{13.135}$$

which can be added on to the existing quartic term to renormalize it.

Look at Figure 13.3. Part (a) shows a contribution in which the two slow fields at the left vertex are part of the quartic term generated, and the two fast ones are averaged with their counterparts in the right vertex. Part (b) is identical except for the way the external momenta are attached to the vertices. Part (c) has a factor of $\frac{1}{2}$ to compensate for the fact that the vertical internal lines are both particles whose exchange does not produce a new contribution, in contrast to the internal lines in (a) and (b) which describe particle–antiparticle pairs. Part (d) describes a disconnected diagram which does not contribute to the flow and in fact cancels in the cumulant expansion.

Since we are looking only for the change in the marginal coupling $u_0 = u(0,0,0,0)$, we may assume the slow fields are all at $k = 0$. With no momentum flowing into the loops, all propagators have the same momentum $k = \Lambda$. All diagrams make the same contribution except for the $\frac{1}{2}$ in Figure 13.3(c).

You have three choices here. You can accept what I say next, or go through the 256 terms and collect the relevant pieces, or use Feynman diagrams which automate the process. For those of you who are interested, I show at the end of this section how the following result,

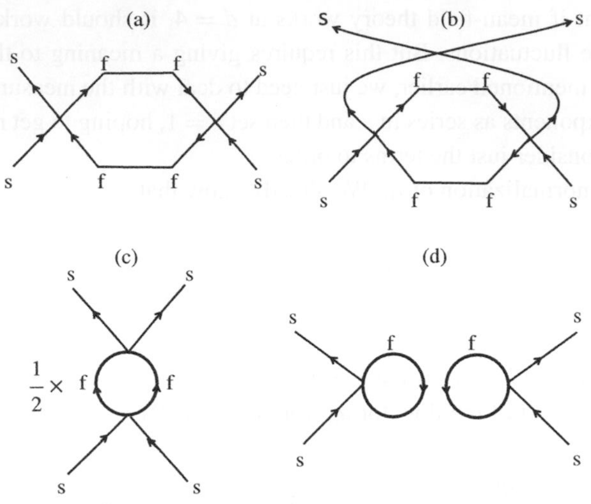

Figure 13.3 (a), (b), and (c) describe the three diagrams that contribute to the flow of u_0. Diagram (c) has a factor of $\frac{1}{2}$ due to the identity of two particles in the loop. Diagram (d) is a disconnected diagram. All external momenta vanish and all loop momenta are infinitesimally close to Λ.

Eq. (13.136), is obtained from Feynman diagrams:

$$u'_0 = (1+\varepsilon dl)\left[u_0 - \frac{5}{2}\frac{u_0^2 A dl}{(1+r_0)^2}\right], \tag{13.136}$$

$$\frac{du_0}{dl} = \varepsilon u_0 - \frac{5}{2}\frac{u_0^2 A}{(1+r_0)^2}, \tag{13.137}$$

where A is the same constant as in the tadpole graph that renormalized r_0 because it involves the same loop integral over one momentum.

Here, then, are our flow equations [Eqs. (13.133) and (13.137)]:

$$\beta_r = \frac{dr_0}{dl} = 2r_0 + \frac{u_0 A}{1+r_0} \simeq 2r_0 + u_0 A(1-r_0), \tag{13.138}$$

$$\beta_u = \frac{du_0}{dl} = \varepsilon u_0 - \frac{5}{2}\frac{u_0^2 A}{(1+r_0)^2} = \varepsilon u_0 - \frac{5}{2}u_0^2 A, \tag{13.139}$$

where in the last step I have set $r_0 = 0$ in the denominator with errors of order $u_0^2 r_0$.

From Eq. (13.139), we learn that to order ε the fixed point values u^* are

$$u_0^* = 0 \quad \text{Gaussian}, \tag{13.140}$$

$$u_0^* = \frac{2\varepsilon}{5A} \quad \text{Wilson–Fisher (WF)}. \tag{13.141}$$

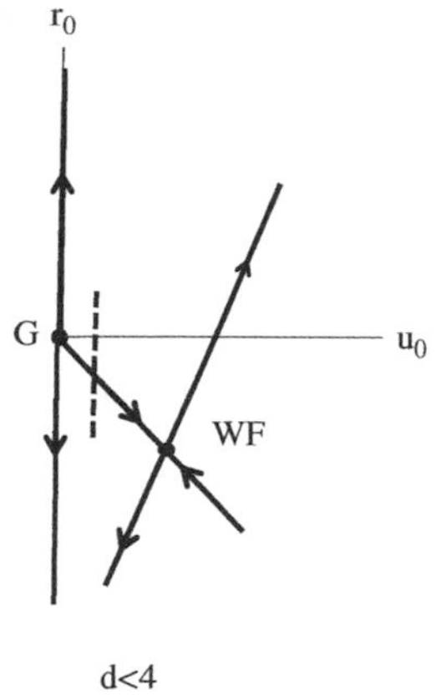

Figure 13.4 The Gaussian fixed point (with two relevant directions) and the Wilson–Fisher fixed point (with one relevant direction). Ignore the vertical dotted line for now.

Applying this to Eq. (13.138), we learn that the fixed point values r_0^* are, to order ε,

$$r_0^* = 0 \quad \text{Gaussian,} \tag{13.142}$$

$$r_0^* = -\frac{\varepsilon}{5} \quad \text{WF.} \tag{13.143}$$

The situation for $d < 4$ is shown in Figure 13.4. I will not consider the Gaussian fixed point at the origin, other than to note what we already know: that it is repulsive in both directions. The linearized flow near the WF fixed points is

$$\begin{pmatrix} \frac{d\delta r_0}{dl} \\ \frac{d\delta u_0}{dl} \end{pmatrix} = \begin{pmatrix} \frac{\partial \beta_r}{\partial r_0} & \frac{\partial \beta_r}{\partial u_0} \\ \frac{\partial \beta_u}{\partial r_0} & \frac{\partial \beta_u}{\partial u_0} \end{pmatrix}_{r_0^*, u_0^*} \begin{pmatrix} \delta r_0 \\ \delta u_0 \end{pmatrix} \tag{13.144}$$

$$= \begin{pmatrix} 2 - \frac{2}{5}\varepsilon & A(1 + \frac{\varepsilon}{5}) \\ 0 & -\varepsilon \end{pmatrix} \begin{pmatrix} \delta r_0 \\ \delta u_0 \end{pmatrix}, \tag{13.145}$$

where the matrix is evaluated at the fixed point and terms of higher order than ε have been dropped. The most interesting consequence is

$$\nu = \frac{1}{2 - \frac{2}{5}\varepsilon} = \frac{1}{2} + \frac{\varepsilon}{10} = 0.6 \text{ for } d = 3. \tag{13.146}$$

Observe that the answer does not depend on A.

To get the Ising result from this $U(1) = O(2)$ result we replace the $\frac{5}{2}$ in Eq. (13.139) and above by $\frac{6}{2}$ above because:

- the factor $\frac{1}{2!2}$ which is replaced by $\frac{1}{4!}$ is exactly canceled by the multiplicities;
- for real scalars there is no distinction between particles and antiparticles and all three graphs contribute equally.

Table 13.2 *The ε expansion versus others (# denotes numerical results).*

Exponent	Landau	Ising ($\mathcal{O}(\varepsilon)$)	Ising (#)	U(1)
α	jump	$\frac{\varepsilon}{6} = 0.17$	0.110	$\frac{\varepsilon}{10}$
β	$\frac{1}{2}$	$\frac{1}{2} - \frac{\varepsilon}{6} = 0.33$	0.326	$\frac{1}{2} - \frac{3\varepsilon}{20}$
γ	1	$1 + \frac{\varepsilon}{6} = 1.17$	1.24	$1 + \frac{\varepsilon}{5}$
δ	3	$3 + \varepsilon = 4$	4.79	$3 + \varepsilon$
ν	$\frac{1}{2}$	$\frac{1}{2} + \frac{\varepsilon}{12} = 0.58$	0.630	$\frac{1}{2} + \frac{\varepsilon}{5}$
η	0	$\mathcal{O}(\varepsilon^2)$	0.036	$\mathcal{O}(\varepsilon^2)$

The result is:

$$\nu = \frac{1}{2 - \frac{2}{6}\varepsilon} = \frac{1}{2} + \frac{\varepsilon}{12} = 0.58 \text{ for } d = 3. \tag{13.147}$$

Now for the other exponents at $d = 4 - \varepsilon$. The scaling of h is still

$$h_s = hs^{1+\frac{1}{2}d} \tag{13.148}$$

for real and complex fields because the field rescaling needed to obtain a fixed point action has not been altered to order ε: the coefficient of the k^2 term receives no corrections to order ε. The other exponents then follow from the usual analysis. Only the *deviation* δu_0 from the fixed point is irrelevant and flows to 0, but the fixed point itself is not at $u_0 = 0$. So u_0 is not dangerous and the extraction of exponents has no pitfalls. Table 13.2 compares the Ising exponents to order ε with Landau theory and the best numerically known answers in $d = 3$. I also show the $U(1)$ answers to $\mathcal{O}(\varepsilon)$.

Observe from the table that every correction to the Landau exponents is in the right direction, toward the numerical answers. The exponents obey $\alpha + 2\beta + \gamma = 2$, as is assured by the scaling arguments used in their derivation.

At higher orders in ε one faces two problems. The first is that the ε expansion is asymptotic: it does not represent a convergent series; beyond some point, agreement will worsen and fancy resummation techniques will have to be invoked. The second is that the Wilson approach gets very complicated to the next order. We need to include more operators (like $(\phi^*\phi)^3$). The kinematics can get messy. Consider, for example, the one-loop diagrams in Figure 13.3. The momenta in the loop have to be limited to a narrow sliver of width $d\Lambda$ near the cut-off. Since the external momenta were chosen to be zero, if one of the momenta was at Λ then so was the other one automatically, by momentum conservation. Had we been interested in irrelevant corrections to $u(k)$, we would have had to consider non-zero external momenta. In this case, if one of the momenta k were within the shell at Λ, restricting the other to the shell would have been very cumbersome. For these reasons one employs what is called the "field theory approach," which is less intuitive but more efficient. This will be described later.

Another feature we see at higher orders in ε is that the factor for field rescaling changes from

$$\zeta = s^{1+\frac{1}{2}d}, \tag{13.149}$$

which was derived in the non-interacting theory to keep the kinetic term with coefficient $\phi^* k^2 \phi$. At higher orders, the $\phi^* k^2 \phi$ term can get corrections as modes are eliminated. To bring the coefficient back to unity, a different ζ will be needed.

13.3.1 Digression on Feynman Diagrams

The diagrams of Figure 13.3 may be deduced from the Feynman diagrams shown in Figure 13.5. I show them in $d = 4$ for convenience. In quantum field theory, they describe

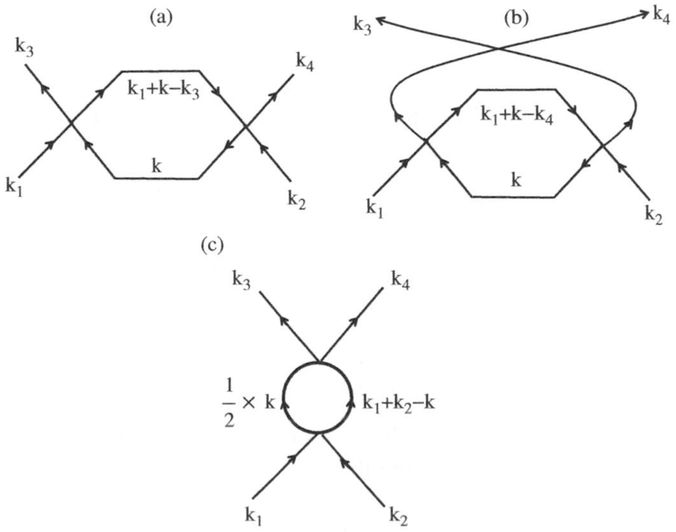

Figure 13.5 The Feynman diagrams that renormalize the quartic interaction at one loop. To get Eq. (13.136), all external momenta are set to 0 and loop integrals are restricted to lie between $k = \Lambda = 1 - dl$ and $k = \Lambda$.

the one-loop correction to the coupling u and are given by

$$u_0' = u_0 - u_0^2 \left[\int_0^\Lambda \frac{d^4k}{(2\pi)^4} \frac{1}{(k^2+r_0)(|\mathbf{k}+\mathbf{k}_1-\mathbf{k}_3|^2+r_0)} + \int_0^\Lambda \frac{d^4k}{(2\pi)^4} \frac{1}{(k^2+r_0)(|\mathbf{k}+\mathbf{k}_1-\mathbf{k}_4|^2+r_0)} + \frac{1}{2} \int_0^\Lambda \frac{d^4k}{(2\pi)^4} \frac{1}{(k^2+r_0)(|-\mathbf{k}+\mathbf{k}_1+\mathbf{k}_2|^2+r_0)} \right]. \tag{13.150}$$

The field theory diagrams agree on multiplicity and topology with the Wilsonian ones, but differ as follows:

- The loop momenta go from 0 to Λ.
- The external momenta have some general values $\boldsymbol{k}_1, \ldots, \boldsymbol{k}_4$.

We can borrow the corresponding integrals for the WF calculation if we set all external momenta to 0, set $k = \Lambda = 1$ in every loop, and multiply by $dk = dl$ to represent the integration over an infinitesimal shell of thickness dl at the cut-off. Finally, in $d = 4 - \varepsilon$ we must rescale the coupling by $1 + \varepsilon dl$, as in Eq. (13.136), because it is dimensionful. With these values, and $S_4 = 2\pi^2$, all three loops contribute an equal amount

$$-\frac{u_0^2 dl}{8\pi^2(1+r_0)^2}, \tag{13.151}$$

which we then multiply by $\frac{5}{2}$ to account for the three diagrams, the last of which contributes with a relative size of $\frac{1}{2}$. As an aside, we note that the constant $A = \frac{1}{8\pi^2}$.

13.4 Renormalization Group at $d = 4$

The case of $d = 4$ is very instructive. (Remember that d is the number of *spatial dimensions* and usually $d = 3$. However, as described in [4–6], certain systems in $d = 3$ have low-energy propagators that resemble those from $d = 4$ in the infrared. The following analysis applies to them.)

$$u_0' = u_0 - \frac{5u_0^2}{2} \int_{d\Lambda} \frac{k^3 dk d\Omega}{(2\pi)^4 k^4}, \tag{13.152}$$

$$\frac{du_0}{dl} = -\frac{5u_0^2}{16\pi^2}, \tag{13.153}$$

where I have again used the fact that the area of a unit sphere in $d = 4$ is $S_4 = 2\pi^2$. To one-loop accuracy we have the following flow:

$$\frac{dr_0}{dl} = 2r_0 + au_0, \tag{13.154}$$

$$\frac{du_0}{dl} = -bu_0^2, \tag{13.155}$$

where a and b are positive constants whose precise values will not matter.

We shall now analyze these equations. First, besides the Gaussian fixed point at the origin, there are no other points where both derivatives vanish. Next, the equation for u_0 is readily integrated to give

$$u_0(l) = \frac{u_0(0)}{1+bu_0(0)l}. \tag{13.156}$$

This means that if we start with a positive coupling $u_0(0)$ at $\Lambda = \Lambda_0$ and renormalize to $\Lambda = \Lambda_0 e^{-l}$, the effective coupling renormalizes to zero as $l \to \infty$. One says that u_0 *is marginally irrelevant*. In the present case it vanishes as follows as $l \to \infty$:

$$u(t) \lim_{t\to\infty} \frac{1}{bl}. \tag{13.157}$$

This statement needs to be understood properly. *In particular, it does not mean that if we add a small positive u_0 to the Gaussian fixed point, we will renormalize back to the Gaussian fixed point.* This is because the small u_0 will generate an r_0, and that will quickly grow under renormalization. What is true is that ultimately u_0 will decrease to zero, but r_0 can be large. To flow to the Gaussian fixed point, we must start with a particular combination of r_0 and u_0 which describes the critical surface. All this comes out of Eq. (13.154) for r_0, which is integrated to give

$$r_0(l) = e^{2l}\left[r_0(0) + \int_0^l e^{-2l'} \frac{au_0(0)}{1+bu_0(0)l'} dl' \right]. \tag{13.158}$$

Let us consider large l. Typically, r_0 will flow to infinity exponentially fast due to the exponential prefactor, unless we choose r_0 such that the object in brackets vanishes as $l \to \infty$:

$$r_0(0) + \int_0^\infty e^{-2l'} \frac{au_0(0)}{1+bu_0(0)l'} dl' = 0. \tag{13.159}$$

If we introduce this relation into Eq. (13.158), we find that

$$r_0(l) = e^{2l}\left[- \int_l^\infty e^{-2l'} \frac{au_0(0)}{1+bu_0(0)l'} dl' \right] \simeq -\frac{a}{2bl}. \tag{13.160}$$

Combined with the earlier result Eq. (13.157), we find, for large l,

$$r_0(l) \lim_{l\to\infty} -\frac{a}{2bl}, \tag{13.161}$$

$$u_0(l) \lim_{l\to\infty} \frac{1}{bl}, \tag{13.162}$$

which defines the critical surface (a line) in the r_0–u_0 plane:

$$r_0(l) = -\frac{au_0(l)}{2}. \tag{13.163}$$

This flow into the fixed point at $d=4$ resembles the flow in $d>4$ depicted in Figure 13.2, except that the approach to the fixed point is logarithmic and not power law.

Look at Eq. (13.156). It tells us that in the deep infrared, the coupling actually vanishes as $1/l$, and that in the large-l region we can make reliable weak coupling calculations. This is, however, thanks to the RG. In simple perturbation theory, we would have found the series

$$u_0(l) = u_0(0) - u_0^2 bl + \mathcal{O}(l^2) \tag{13.164}$$

with ever increasing terms as $l \to \infty$. The RG sums up the series for us [Eq. (13.156)] and displays how the coupling flows to 0 in the infrared.

References and Further Reading

[1] F. J. Wegner, Journal of Physics C, **7**, 2098 (1974).
[2] K. G. Wilson and J. R. Kogut, Physics Reports, **12**, 74 (1974).
[3] K. G. Wilson and M. E. Fisher, Physical Review Letters, **28**, 240 (1972).
[4] G. Ahlers, A. Kornbilt, and H. Guggenheim, *Proceedings of the International Conference on Low Temperature Physics*, Finland, 176 (1975). Describes experimental work showing $d = 4$ behavior in $d = 3$.
[5] A. Aharony, Physical Review B, **8**, 3363 (1973).
[6] A. I. Larkin and D. E. Khmelnitskii, Zh. Eksp. Teor. Fiz., **56**, 2087 (1969) [Journal of Experimental and Theoretical Physics, **29**, 1123 (1969)]. Contains earlier theoretical work.

14
Two Views of Renormalization

Here I discuss the relationship between two approaches to renormalization: the older one based on removing infinities in the quest for field theories in the continuum, and the more modern one due to Wilson based on obtaining effective theories. My focus will be on a few central questions. No elaborate calculations will be done.

14.1 Review of RG in Critical Phenomena

Let us recall the problem of critical phenomena and its resolution by the RG. Suppose we have some model on a lattice with some parameters, like K_1, K_2, ... of the Ising model. At very low and very high temperatures ($K \to \infty$ or $K \to 0$) we can employ perturbative methods like the low-temperature or high-temperature expansions to compute correlation functions. These series are predicated on the smooth change of physics as we move away from these extreme end points. By definition, these methods will fail at the critical point (and show signs of failing as we approach it) because there is a singular change of phase. One signature of trouble is the diverging correlation length ξ. The RG beats the problem by trading the original system near the critical point for one that is comfortably away from it (and where the series work) and things like ξ can be computed. The RG then provides a dictionary for translating quantities of original interest in terms of new ones. For example,

$$\xi(r_0) = 2^N \xi(r_N), \tag{14.1}$$

where $r_0 \simeq t$ is the deviation from criticality, N is the number of factor-of-2 RG transformations performed, and r_N the coupling that r_0 evolves into. At every step,

$$r_0 \to r_0 2^{ad} = r_0 2^{1/\nu}. \tag{14.2}$$

We keep renormalizing until r_N has grown to a safe value far from criticality, say

$$r_N = r_0 2^{N/\nu} = 1, \quad \text{that is} \tag{14.3}$$
$$2^N = r_0^{-\nu}. \tag{14.4}$$

Then, from Eq. (14.1),

$$\xi(r_0) = r_0^{-\nu}\xi(1). \tag{14.5}$$

The divergence in ξ is translated into the divergence in N, the number of steps needed to go from r_0 to $r_N = 1$ as r_0 approaches the critical value of 0.

In terms of the continuous scale s (which replaces 2^N), these relations take the form

$$\xi(r_0) = s\xi(r_s), \tag{14.6}$$

$$r_s = r_0 s^{1/\nu}. \tag{14.7}$$

Typically one finds some *approximate* flow equations, their fixed points $\boldsymbol{K}^*$, the linearized flow near $\boldsymbol{K}^*$, and, eventually, the exponents.

14.2 The Problem of Quantum Field Theory

Consider the field theory with action (with $c = 1 = \hbar$)

$$S = \int \left[\frac{1}{2}(\nabla\phi(x))^2 + \frac{1}{2}m_0^2\phi^2(x) + \frac{\lambda_0}{4!}\phi^4(x)\right] d^4x \tag{14.8}$$

$$= S_0 + S_{\mathrm{I}}, \tag{14.9}$$

where S_0 is the quadratic part. I have chosen $d = 4$, which is relevant to particle physics and serves to illustrate the main points, and a real scalar field to simplify the discussion. The parameters m_0 and λ_0 are to be determined by computing some measurable quantities and comparing to experiment.

To this end, we ask what is typically computed, how it is computed, and what information it contains.

Consider the two-point correlation function

$$G(x) = \langle\phi(x)\phi(0)\rangle \tag{14.10}$$

$$= \frac{\int[\mathcal{D}\phi]\phi(x)\phi(0)e^{-S}}{\int[\mathcal{D}\phi]e^{-S}}. \tag{14.11}$$

First, let us assume that $\lambda_0 = 0$. Doing the Gaussian functional integral we readily find

$$G(r) \simeq \frac{e^{-m_0 r}}{r^2}. \tag{14.12}$$

How do we determine m_0 from experiment? In the context of particle physics, m_0 would be the particle mass, measured the way masses are measured. If, instead, the ϕ^4 theory were being used to describe spins on a lattice of spacing a, we would first measure the dimensionless correlation length ξ (in lattice units) from the exponential decay of correlations and relate it to m_0 by the equation

$$m_0 = \frac{1}{a\cdot\xi}. \tag{14.13}$$

Sometimes I will discuss correlations of four ϕ's. They will also be called G, but will be shown with four arguments. If not, assume we are discussing the two-point function.

In momentum space we would consider the Fourier transform

$$\langle \phi(\boldsymbol{k}_1)\phi(\boldsymbol{k}_2)\rangle = (2\pi)^4\delta^4(\boldsymbol{k}_1-\boldsymbol{k}_2)G(k), \text{ where} \tag{14.14}$$

$$G(k) = G_0(k) = \frac{1}{k^2+m_0^2}; \tag{14.15}$$

the subscript on G_0 reminds us that we are working with a free-field theory. Correlations with more fields can be computed as products of two-point functions $G_0(k)$ using Wick's theorem. If this explains the data, we are done.

14.3 Perturbation Series in λ_0: Mass Divergence

Let us say the $\lambda_0 = 0$ theory does not explain the data. For example, the particles could be found to scatter. The $\lambda_0 = 0$ theory cannot describe that. So we toss in a λ_0 and proceed to calculate correlation functions, and fit the results to the data to determine m_0 and λ_0.

When $\lambda_0 \neq 0$, we resort to perturbation theory. We bring the $\lambda_0\phi^4$ term in S downstairs as a power series in λ_0 and do the averages term-by-term using Wick's theorem. To order λ_0, we find

$$G(x) = \langle \phi(x)\phi(0)\rangle \tag{14.16}$$

$$= \frac{\int [\mathcal{D}\phi]\phi(x)\phi(0)e^{-S_0(\phi)}\left[1-\frac{\lambda_0}{4!}\int\phi^4(y)d^4y\right]}{\int [\mathcal{D}\phi]e^{-S_0(\phi)}\left[1-\frac{\lambda_0}{4!}\int\phi^4(y)d^4y\right]}. \tag{14.17}$$

In the denominator, we pair the four $\phi(y)$'s two-by-two to obtain

$$\text{denominator} = 1-\frac{\lambda_0}{8}\int G_0^2(0)d^4y. \tag{14.18}$$

In the numerator, one option is to pair $\phi(x)$ and $\phi(0)$, which are being averaged, and pair the fields inside the y integral with each other. This will give $G_0(x)\cdot\left(1-\frac{\lambda_0}{8}\int G_0^2(0)d^4y\right)$. The factor in parentheses will get canceled by the normalizing partition function in the denominator. This happens in general: any contribution in which the fields being averaged do not mingle with the ones in the interaction, the so-called *disconnected terms*, may be dropped.

This leaves us with contributions where $\phi(x)$ and $\phi(0)$ are paired with the $\phi(y)$'s. The result is, *to order* λ_0,

$$G(x) = G_0(x) - \frac{1}{2}\lambda_0\int G_0(x-y)G_0(y-y)G_0(y-0)d^4y. \tag{14.19}$$

Since the second term is of order λ_0, we may set the normalizing denominator $1-\frac{\lambda_0}{8}\int G_0^2(0)d^4y$ to 1.

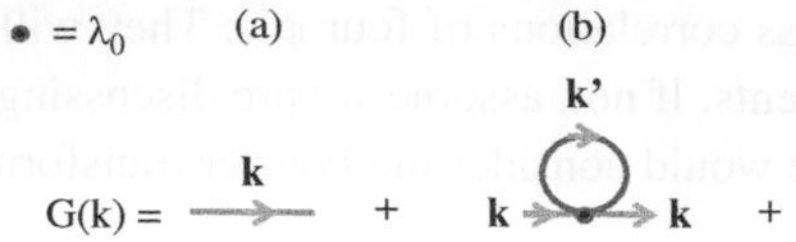

Figure 14.1 (a) $G_0(k) = G(k)$ in free-field theory. (b) One-loop correction to m_0^2. The lines with arrows denote the free propagator $G_0(k) = \frac{1}{k^2+m_0^2}$.

In momentum space,

$$G(k) = \frac{1}{k^2+m_0^2} - \frac{1}{k^2+m_0^2} \underbrace{\left[\frac{1}{2} \int_0^{\infty} \frac{\lambda_0}{k'^2+m_0^2} \frac{d^4k'}{(2\pi)^4} \right]}_{\delta m_0^2} \frac{1}{k^2+m_0^2} \cdots \tag{14.20}$$

This series is represented in Figure 14.1.

To this order in λ_0 we may rewrite this as

$$G(k) = \frac{1}{k^2+m_0^2+\delta m_0^2}. \tag{14.21}$$

We conclude that the mass squared in the interacting theory is

$$m^2 = m_0^2 + \delta m_0^2. \tag{14.22}$$

The next natural thing to do is compare the measured m^2 to this result and find a relation constraining m_0^2 and λ_0.

It is here we encounter the serious trouble with continuum field theory: *δm_0^2 is quadratically divergent in the ultraviolet*:

$$\delta m_0^2 = \frac{1}{2} \int_0^{\infty} \frac{\lambda_0}{k'^2+m_0^2} \frac{d^4k'}{(2\pi)^4}. \tag{14.23}$$

So no matter how small λ_0 is, the change in mass δm_0^2 is infinite. The infinity comes from working in the continuum with no limit on the momenta in Fourier expansions. The theory seems incapable of describing the experiment with a finite m, assuming m_0 and λ_0 are finite.

Let us set this aside and compute the scattering amplitude, to compare it with experiment to constrain m_0^2 and λ_0.

14.4 Scattering Amplitude and the Γ's

We must clearly begin with the correlation of four fields, two each for the incoming and outgoing particles. The momenta are positive flowing inwards and there is no difference between particles and antiparticles.The correlation function $G(\boldsymbol{k}_1,\ldots,\boldsymbol{k}_4)$ is depicted in

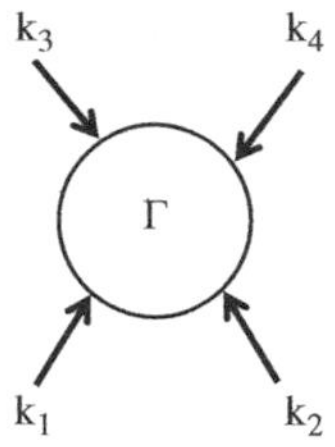

Figure 14.2 The scattering amplitude Γ is a function of the particle momenta, all chosen to point inwards. Their vector sum is zero.

Figure 14.2 and is defined as follows after pulling out the momentum-conserving δ function:

$$\langle \phi(\boldsymbol{k}_1)\phi(\boldsymbol{k}_2)\phi(\boldsymbol{k}_3)\phi(\boldsymbol{k}_4)\rangle = (2\pi)^4\delta^4(\boldsymbol{k}_1+\boldsymbol{k}_2+\boldsymbol{k}_3+\boldsymbol{k}_4)G(\boldsymbol{k}_1,\ldots,\boldsymbol{k}_4). \tag{14.24}$$

To lowest order in λ_0, we get, upon pairing the four external ϕ's with the four ϕ's in the $\lambda_0\phi^4$ interaction,

$$G(\boldsymbol{k}_1,\ldots,\boldsymbol{k}_4) = G(\boldsymbol{k}_1)G(\boldsymbol{k}_2)G(\boldsymbol{k}_3)G(\boldsymbol{k}_4)\lambda_0. \tag{14.25}$$

However, $G(\boldsymbol{k}_1,\ldots,\boldsymbol{k}_4)$ is not the scattering amplitude which we should square to get the cross section. The four external propagators do not belong there. (In Minkowski space, the propagators will diverge because $k^2 = m^2$.) The scattering amplitude $\Gamma(\boldsymbol{k}_1,\ldots,\boldsymbol{k}_4)$ is defined as follows:

$$G(\boldsymbol{k}_1,\ldots,\boldsymbol{k}_4) = G(\boldsymbol{k}_1)G(\boldsymbol{k}_2)G(\boldsymbol{k}_3)G(\boldsymbol{k}_4)\Gamma(\boldsymbol{k}_1,\ldots,\boldsymbol{k}_4). \tag{14.26}$$

That is,

$$\Gamma(\boldsymbol{k}_1,\ldots,\boldsymbol{k}_4) = G^{-1}(\boldsymbol{k}_1)G^{-1}(\boldsymbol{k}_2)G^{-1}(\boldsymbol{k}_3)G^{-1}(\boldsymbol{k}_4)G(\boldsymbol{k}_1,\ldots,\boldsymbol{k}_4). \tag{14.27}$$

To lowest order,

$$\Gamma(\boldsymbol{k}_1,\ldots,\boldsymbol{k}_4) = \lambda_0. \tag{14.28}$$

Do we really need to bring in another function $\Gamma(\boldsymbol{k}_1,\ldots,\boldsymbol{k}_4)$ if it is just $G(\boldsymbol{k}_1,\ldots,\boldsymbol{k}_4)$ with the four external legs chopped off? Actually, we could get by with just the $G(\boldsymbol{k}_1,\ldots,\boldsymbol{k}_4)$'s, but in doing so would miss some important part of quantum field theory (QFT). First, $\Gamma(\boldsymbol{k}_1,\ldots,\boldsymbol{k}_4)$ is not alone, it is part of a family of functions, as numerous as the G's. That is, there are entities $\Gamma(\boldsymbol{k}_1,\ldots,\boldsymbol{k}_n)$ for all n. They provide an alternate, equally complete, description of the theory to the G's, just like the Hamiltonian formalism is an alternative to the Lagrangian formalism. They are better suited than the $G(\boldsymbol{k}_1,\ldots,\boldsymbol{k}_n)$'s for discussing renormalization. And they are not just $G(\boldsymbol{k}_1,\ldots,\boldsymbol{k}_n)$'s with the external legs amputated.

In view of time and space considerations, I will digress briefly to answer just two questions:

- Where do the Γ's come from?
- What are the Feynman diagrams that contribute to them?

Consider the partition function $Z(J)$ with a source:

$$Z(J) = \int [\mathcal{D}\phi] e^{-S} e^{\int J(x)\phi(x)dx} \equiv e^{-\mathbf{W}(J)}. \tag{14.29}$$

The *generating functional* $\mathbf{W}(J)$ yields $G_{\rm c}(x_1,\ldots,x_n)$ upon repeated differentiation by $J(x)$, where the subscript c stands for *connected*:

$$\mathbf{W}(J) = -\int \frac{dx_1\cdots dx_n}{n!} G_{\rm c}(x_1,\ldots,x_n)J(x_1)\cdots J(x_n). \tag{14.30}$$

In particular,

$$\bar{\phi}(x) \equiv \langle \phi(x) \rangle = -\frac{\partial \mathbf{W}}{\partial J(x)}. \tag{14.31}$$

(It is understood here and elsewhere that the derivatives are taken at $J = 0$.) Taking one more derivative gives

$$\langle \phi(x)\phi(y) \rangle_{\rm c} = -\frac{\partial^2 \mathbf{W}}{\partial J(x)\partial J(y)} = G_{\rm c}(x,y). \tag{14.32}$$

Given this formalism, in which $\mathbf{W}(J)$ is a functional of J and $\bar{\phi}$ is its derivative, it is natural to consider a Legendre transform to a functional $\mathbf{\Gamma}(\bar{\phi})$ with J as its derivative. By the familiar route one follows to go from the Lagrangian to the Hamiltonian or from the energy to the free energy, we are led to

$$\mathbf{\Gamma}(\bar{\phi}) = \int J(y)\bar{\phi}(y)dy + \mathbf{W}(J). \tag{14.33}$$

By the usual arguments,

$$\frac{\partial \mathbf{\Gamma}(\bar{\phi})}{\partial \bar{\phi}(y)} = J(y). \tag{14.34}$$

The Taylor expansion

$$\mathbf{\Gamma}(\bar{\phi}) \stackrel{\text{def}}{=} \int \frac{dx_1\cdots dx_n}{n!} \Gamma(x_1,\ldots,x_n)\bar{\phi}(x_1)\cdots\bar{\phi}(x_n) \tag{14.35}$$

defines the Γ's with n arguments. A similar expansion in terms of $\bar{\phi}(\boldsymbol{k})$ defines $\Gamma(\boldsymbol{k}_1,\ldots,\boldsymbol{k}_n)$.

Given this definition, and a lot of work, one can show that $\Gamma(\boldsymbol{k}_1,\ldots,\boldsymbol{k}_n)$ will have the following diagrammatic expansion:

- Draw the connected diagrams that contribute to $G(\boldsymbol{k}_1,\ldots,\boldsymbol{k}_n)$ with the same incoming lines, except for those diagrams that can be split into two disjoint parts by cutting just one internal line. For this reason the Γ's are called *1PI* or *one-particle irreducible* correlation functions.
- Append a factor $G^{-1}(\boldsymbol{k})$ for every incoming particle of momentum $\boldsymbol{k}$.

To get acquainted with this formalism, let us derive the relation between $\Gamma(k)$ and $G(k)$ that it implies. Given that $J(x)$ and $J(y)$ are independent, it follows that

$$\delta(x-y) = \frac{\partial J(x)}{\partial J(y)} \tag{14.36}$$

$$= \frac{\partial^2 \mathbf{\Gamma}}{\partial J(y)\partial\bar{\phi}(x)} \tag{14.37}$$

$$= \frac{\partial^2 \mathbf{\Gamma}}{\partial\bar{\phi}(x)\partial J(y)} \tag{14.38}$$

$$= \int dz \frac{\partial^2 \mathbf{\Gamma}}{\partial\bar{\phi}(x)\partial\bar{\phi}(z)} \frac{\partial\bar{\phi}(z)}{\partial J(y)} \tag{14.39}$$

$$= -\int dz \frac{\partial^2 \mathbf{\Gamma}}{\partial\bar{\phi}(x)\partial\bar{\phi}(z)} \frac{\partial^2 \mathbf{W}}{\partial J(z)\partial J(y)} \tag{14.40}$$

$$= \int dy \Gamma(x,z) G(z,y), \tag{14.41}$$

which leads to the very interesting result that the matrices Γ and G with elements $\Gamma(x,z)$ and $G(z,y)$ are inverses:

$$\Gamma = G^{-1}. \tag{14.42}$$

This agrees with the rules given above for computing the two-point function $\Gamma(k)$ from $G(k)$: If we take the two-point function $G(k)$ and multiply by two inverse powers of $G(k)$ (one for each incoming line) we get $\Gamma(k) = G^{-1}(k)$.

Upon further differentiation with respect to $\bar{\phi}(\boldsymbol{k})$, one can deduce the relation between the G's and Γ's and the Feynman rules stated above.

14.4.1 Back to Coupling Constant Renormalization

Let us now return to the scattering amplitude $\Gamma(\boldsymbol{k}_1,\ldots,\boldsymbol{k}_4)$. To lowest order in λ_0,

$$\Gamma(\boldsymbol{k}_1,\ldots,\boldsymbol{k}_4) = \lambda_0. \tag{14.43}$$

It is $|\lambda_0|^2$ you must use to compute cross sections.

In general, $\Gamma(\boldsymbol{k}_1,\ldots,\boldsymbol{k}_4)$ will depend on the external momenta. However, to this order in λ_0, we find Γ does not have any momentum dependence and coincides with the coupling λ_0 in the action.

As we go to higher orders, $\Gamma(0,0,0,0)$ will be represented by a power series in λ_0. *We will then define* $\Gamma(0,0,0,0)$ *as the coupling* λ, *not the* λ_0 *in the action.* This fixes the interaction strength completely. I am not saying that the external momenta vanish in every scattering event, but that in any one theory, given $\Gamma(0,0,0,0)$, a unique $\Gamma(\boldsymbol{k}_1,\ldots,\boldsymbol{k}_4)$ is given by Feynman diagrams.

The trick of comparing the observed scattering rate to the one calculated from Eq. (14.43) to extract λ_0 will work only if λ_0 is small and higher-order corrections are negligible. Let us assume that λ_0 is very small, just like in electrodynamics where the analog of $\lambda_0 \simeq \frac{1}{137}$.

We will now consider scattering to order λ_0^2, even though it is one order higher than the correction to m_0^2. The reason is that it is also given by a one-loop graph, as shown in Figure 14.3, and *the systematic way to organize perturbation theory is in the number of loops*. If we restore the $\frac{1}{\hbar}$ in front of the action, we will find (Exercise 14.4.1) that the tree-level diagram, which is zeroth order in the loop expansion, is of order $\frac{1}{\hbar}$ and that each additional loop brings in one more positive power of $\hbar$. The loop expansion is therefore an $\hbar$ expansion. (During Christmas, we have a tree in our house but no wreath on the door, making us Christians at tree level but not one-loop level.)

Exercise 14.4.1 *Introduce* $\hbar^{-1}$ *in front of the action and see how this modifies* G_0 *and* λ_0. *Look at the diagrams for* G *and* Γ *to one loop and see how the loop brings in an extra* $\hbar$.

The one-loop corrections to scattering are depicted in Figure 14.3. They correspond to the following expression:

$$\Gamma(0,0,0,0) = \lambda_0 - 3\lambda_0^2 \int_0^\infty \frac{1}{(k^2+m_0^2)^2} \frac{d^4k}{(2\pi)^4} \equiv \lambda_0 + \delta\lambda_0 \equiv \lambda. \tag{14.44}$$

This defines the coupling λ to next order.

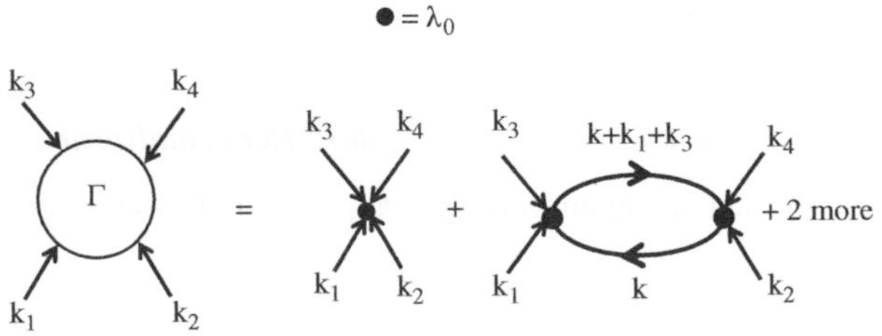

Figure 14.3 One-loop correction to Γ and $\lambda = \Gamma(0,0,0,0)$. Two more diagrams with external momenta connected to the vertices differently are not shown. They make the same contributions when external momenta vanish. The incoming arrows denote momenta and not propagators of that momentum (which have been amputated).

The factor of 3 comes from three loops with different routing of external momenta to the interaction vertices. Since all external momenta vanish, the graphs make identical contributions. *Unfortunately,* $\delta\lambda_0$ *is logarithmically divergent.*

14.5 Perturbative Renormalization

How do we reconcile these infinities in mass and coupling with the fact that actual masses and cross sections are finite? We employ the notion of *renormalization.*

First, we introduce a large momentum cut-off Λ in the loop integrals so that everything is finite but Λ-*dependent*:

$$\delta m_0^2(\Lambda) = \frac{\lambda_0}{2}\int_0^\Lambda \frac{1}{k^2+m_0^2}\frac{d^4k}{(2\pi)^4}, \tag{14.45}$$

$$\delta\lambda_0(\Lambda) = -3\lambda_0^2\int_0^\Lambda \frac{1}{(k^2+m_0^2)^2}\frac{d^4k}{(2\pi)^4}. \tag{14.46}$$

Then we identify the perturbatively corrected quantities with the measured ones. That is, we say

$$m^2 = m_0^2(\Lambda) + \delta m_0^2(\Lambda) \tag{14.47}$$

is the finite measured or *renormalized mass*, and that $m_0^2(\Lambda)$ is the *bare mass*, with an Λ-dependence chosen to ensure that m^2 equals the measured value. This means that we must choose

$$m_0^2(\Lambda) = m^2 - \frac{\lambda_0}{2}\int_0^\Lambda \frac{1}{k^2+m_0^2}\frac{d^4k}{(2\pi)^4} \tag{14.48}$$

$$= m^2 - \frac{\lambda}{2}\int_0^\Lambda \frac{1}{k^2+m^2}\frac{d^4k}{(2\pi)^4}, \tag{14.49}$$

where I have replaced the bare mass and coupling by the physical mass and coupling with errors of higher order.

Likewise, we must go back to Eq. (14.44) and choose

$$\lambda_0(\Lambda) = \lambda + 3\lambda_0^2\int_0^\Lambda \frac{1}{(k^2+m_0^2)^2}\frac{d^4k}{(2\pi)^4} \tag{14.50}$$

$$= \lambda + 3\lambda^2\int_0^\Lambda \frac{1}{(k^2+m^2)^2}\frac{d^4k}{(2\pi)^4}, \tag{14.51}$$

where I have replaced the bare mass squared by the physical mass squared and λ_0^2 by λ^2 with errors of higher order.

Equations (14.49) and 14.51 specify the requisite bare mass $m_0^2(\lambda, m, \Lambda)$ *and bare coupling* $\lambda_0(\lambda, m, \Lambda)$ *corresponding to the experimentally determined values of* λ *and* m

for any given Λ. If we choose the bare parameters as above, we will end up with physical mass and coupling that are finite and independent of Λ, *to this order*.

What about the scattering amplitude for non-zero external momenta? What about its divergences? We find that

$$\Gamma(\boldsymbol{k}_1,\ldots,\boldsymbol{k}_4) = \lambda_0 - \lambda_0^2\left[\int_0^\Lambda \frac{d^4k}{(2\pi)^4}\frac{1}{(k^2+m_0^2)(|\boldsymbol{k}+\boldsymbol{k}_1+\boldsymbol{k}_3|^2+m_0^2)}\right.$$
$$\left. + \text{ two more contributions}\right] \tag{14.52}$$

is logarithmically divergent as $\Lambda \to \infty$. Don't panic yet! We first replace m_0^2 by m^2 everywhere, due to the λ_0^2 in front of the integral. Next, we use Eq. (14.51) to replace the first λ_0 by

$$\lambda_0 = \lambda + 3\lambda^2 \int_0^\Lambda \frac{1}{(k^2+m^2)^2}\frac{d^4k}{(2\pi)^4}, \tag{14.53}$$

and the λ_0^2 in front of the integral by λ^2 (with errors of higher order), to arrive at

$$\Gamma(\boldsymbol{k}_1,\ldots,\boldsymbol{k}_4) = \lambda + \lambda^2\left[\int_0^\Lambda\left[\frac{1}{(k^2+m^2)(|\boldsymbol{k}+\boldsymbol{k}_1+\boldsymbol{k}_3|^2+m^2)} - \frac{1}{(k^2+m^2)^2}\right]\frac{d^4k}{(2\pi)^4}\right.$$
$$\left. + \text{ two more contributions}\right]. \tag{14.54}$$

I have divided the $3\lambda^2$ term in Eq. (14.53) into three equal parts and lumped them with the three integrals in large square brackets.

The integrals are now convergent because as $k \to \infty$, the integrand in the diagram shown goes as

$$\frac{(q^2+2\boldsymbol{k}\cdot\boldsymbol{q})k^3}{k^6}, \tag{14.55}$$

where $\boldsymbol{q} = \boldsymbol{k}_1 + \boldsymbol{k}_3$ is the external momentum flowing in. Because the $\boldsymbol{k}\cdot\boldsymbol{q}$ term does not contribute due to rotational invariance, the integrand has lost two powers of k due to renormalization. The other two diagrams are also finite for the same reason. In short, *once* $\Gamma(0,0,0,0)$ *is rendered finite, so is* $\Gamma(\boldsymbol{k}_1,\ldots,\boldsymbol{k}_4)$.

The moral of the story is that, *to one-loop order*, the quantities considered so far are free of divergences when written in terms of the renormalized mass and coupling.

14.6 Wavefunction Renormalization

However, at next order a new kind of trouble pops up that calls for more renormalization. I will describe this in terms of

$$\Gamma(k) = G^{-1}(k). \tag{14.56}$$

In free-field theory,

$$\Gamma(k) = k^2 + m_0^2 \tag{14.57}$$

and, to one-loop order (consult Figure 14.4(a) and (b)),

$$\Gamma(k) = k^2 + m_0^2 + \delta m_0^2, \tag{14.58}$$

where δm_0^2 is the one-loop contribution that we encountered in Eq. (14.20).

To next order in the loop expansion, we see two more diagrams shown in Figure 14.4(c) and (d). (Check that the two-loop diagram has one more power of $\hbar$ than the one-loop diagram.) Let us represent their (divergent) contributions as follows:

$$\Gamma(k) = k^2 + m_0^2 + \delta m_0^2 + \lambda_0^2 A(m_0, \Lambda) + \lambda_0^2 B(m_0, \Lambda, k). \tag{14.59}$$

The term $\lambda_0^2 A(m_0, \Lambda)$, being k-independent, makes a contribution to mass renormalization and we can deal with it as before.

By contrast, B depends on the external momentum k and is given, up to constants, by

$$B(m_0^2, \Lambda, k) = \int_0^\Lambda \frac{d^4k_1 d^4k_2}{(k_1^2 + m_0^2)(k_2^2 + m_0^2)(|\mathbf{k}_1 + \mathbf{k}_2 + \mathbf{k}|^2 + m_0^2)}. \tag{14.60}$$

Consider the expansion of B in a series in k^2. The zeroth-order term $\lambda_0^2 B(m_0^2, \Lambda, 0)$ also contributes to mass renormalization.

The next term, proportional to k^2, modifies the k^2 term from free-field theory:

$$k^2 \to k^2 + \lambda_0^2 \frac{dB(m_0, \Lambda, k^2)}{dk^2}\bigg|_0 k^2 \equiv k^2\left(1 + c\lambda_0^2 \ln \frac{\Lambda^2}{m_0^2}\right) \equiv k^2 Z^{-1}\left(\lambda_0, \frac{\Lambda^2}{m_0^2}\right), \tag{14.61}$$

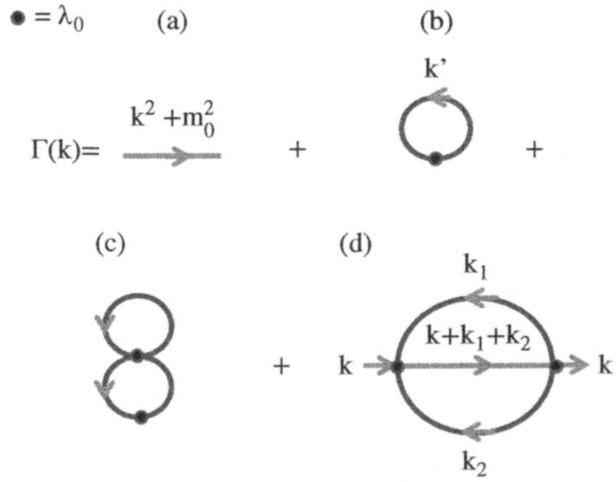

Figure 14.4 (a) $\Gamma(k)$ in free-field theory. (b) One-loop correction to m_0^2. (c) A k-independent two-loop correction $\lambda_0^2 A$ that renormalizes the mass. (d) A k-dependent two-loop correction $\lambda_0^2 B$ that renormalizes the k^2 part. The external legs have been amputated.

where c is some constant and I have introduced the *field renormalization factor*:

$$Z^{-\frac{1}{2}}\left(\lambda_0, \frac{\Lambda^2}{m_0^2}\right) = \left(1 + c\lambda_0^2 \ln \frac{\Lambda^2}{m_0^2}\right)^{\frac{1}{2}}. \tag{14.62}$$

Because Z^{-1} diverges, the k^2 term in $\Gamma(k)$ now has a divergent coefficient.

Let us first handle this divergence and then interpret our actions. We begin with

$$\Gamma(k) = Z^{-1}\left(\lambda_0, \frac{\Lambda^2}{m_0^2}\right) k^2 + k\text{-independent term } m_1^2 + \mathcal{O}(k^4). \tag{14.63}$$

Multiplying both sides by Z, we arrive at

$$Z\Gamma(k) = k^2 + Zm_1^2 + \mathcal{O}(k^4) \equiv k^2 + m^2 + \mathcal{O}(k^4), \tag{14.64}$$

where we have finally defined the quantity m^2 that is identified with the experimentally measured renormalized mass to this order.

The renormalized function

$$\Gamma_{\mathrm{R}} = Z\Gamma \tag{14.65}$$

now has a finite value and finite derivative at $k^2 = 0$:

$$\Gamma_{\mathrm{R}}(0) = m^2, \tag{14.66}$$

$$\left.\frac{d\Gamma_{\mathrm{R}}(k^2)}{dk^2}\right|_{k^2=0} = 1. \tag{14.67}$$

What does $\Gamma \to \Gamma_{\mathrm{R}}$ imply for G? Since $\Gamma = G^{-1}$, it follows that the *renormalized propagator*

$$G_{\mathrm{R}}(k) = Z^{-1}\left(\lambda_0, \frac{\Lambda^2}{m_0^2}\right) G(k) \tag{14.68}$$

is divergence free. As Z is independent of momentum we may also assert that the Fourier transform to real space given by

$$G_{\mathrm{R}}(\boldsymbol{r}) = Z^{-1}\left(\lambda_0, \frac{\Lambda^2}{m_0^2}\right) G(\boldsymbol{r}) \tag{14.69}$$

is also divergence free. But

$$G(\boldsymbol{r}) = \langle \phi(\boldsymbol{r})\phi(0)\rangle, \tag{14.70}$$

which means that

$$G_{\mathrm{R}}(\boldsymbol{r}) = \langle Z^{-\frac{1}{2}}\phi(\boldsymbol{r})Z^{-\frac{1}{2}}\phi(0)\rangle \equiv \langle \phi_{\mathrm{R}}(\boldsymbol{r})\phi_{\mathrm{R}}(0)\rangle \tag{14.71}$$

is divergence free. Above, we have defined a *renormalized field*

$$\phi_{\mathrm{R}} = Z^{-\frac{1}{2}}\phi \tag{14.72}$$

in coordinate or momentum space, which has divergence-free correlations when everything is expressed in terms of renormalized mass and coupling (except for the unavoidable momentum-conservation δ-function in front of $G(k)$). One refers to Eq. (14.72) as *field renormalization*.

Several questions arise at this point:

- Since our original task was to compute correlations of ϕ, what good is it to have correlations of ϕ_{R}, even if the latter are finite?
- Renormalization looks like a Ponzi scheme, wherein we keep shoving problems to higher and higher orders. How many more new infinities will arise as we go to higher orders in λ_0 and k^2 and consider correlation functions of more than two fields? Will all the infinities be removed by simply renormalizing the mass, coupling, and field?

As to the first point, it turns out that the overall scale of ϕ does not affect any physical quantity: one will infer the same particle masses and physical scattering matrix elements before and after rescaling. This is not obvious, and I will not try to show that here.

As for the second set of points, it is the central claim of renormalization theory that no more quantities need to be renormalized (though the amount of renormalization will depend on the order of perturbation theory), and that the renormalized correlation function of rescaled fields

$$\phi_{\mathrm{R}} = Z^{-\frac{1}{2}}\phi, \tag{14.73}$$

expressed in terms of the renormalized mass and coupling,

$$G_{\mathrm{R}}(\boldsymbol{k}_1,\ldots,\boldsymbol{k}_M,m,\lambda) = Z^{-M/2}G(\boldsymbol{k}_1,\ldots,\boldsymbol{k}_M,m_0,\lambda_0,\Lambda), \tag{14.74}$$

are finite and independent of Λ as $\Lambda \to \infty$.

(New divergences arise if the spatial arguments of any two or more ϕ's in G_{R} coincide to form the operators like ϕ^2. We will not discuss that here.)

The proof of renormalizability is very complicated. To anyone who has done the calculations, it is awesome to behold the cancellation of infinities in higher-loop diagrams as we rewrite everything in terms of quantities renormalized at lower orders. It seems miraculous and mysterious.

While all this is true for the theory we just discussed, ϕ^4 interaction in $d = 4$, referred to as ϕ^4_4, there are also *non-renormalizable theories*. For example, if we add a ϕ^6 interaction in $d = 4$, the infinities that arise cannot be fixed by renormalizing any finite number of parameters. Here it should be borne in mind that in quantum field theory one adds this term with a coefficient, $\lambda_6 = w_6/\mu^2$, where μ is some fixed mass (say 1 GeV) introduced to define a dimensionless w_6. In the post-Wilson era one adds the ϕ^6 term with coupling $\lambda_6 = w_6/\Lambda^2$, which is more natural. Its impact is benign and will be explained later.

What is the diagnostic for renormalizability? The answer is that *any interaction that requires a coupling constant with inverse dimensions of mass is non-renormalizable.* The couplings of ϕ^2 and ϕ^4 have dimensions m^2 and m^0, while a ϕ^6 coupling would have dimension m^{-2} in $d = 4$. These dimensions are established (in units of $\hbar = 1 = c$) by demanding that the kinetic term $\int (\nabla\phi)^2 d^d x$ be dimensionless and using that to fix the dimension of ϕ as

$$[\phi(x)] = \left(\frac{d}{2} - 1\right). \tag{14.75}$$

I invite you to show that

$$[\lambda] = 4 - d, \tag{14.76}$$

which means that λ is marginal in $d = 4$ and renormalizable in $d < 4$. Likewise, try showing that λ_6, the coupling for the ϕ^6 interaction, has dimension

$$[\lambda_6] = 6 - 2d, \tag{14.77}$$

which makes it non-renormalizable in $d = 4$ but renormalizable for $d \leq 3$.

You must have noticed the trend: *The renormalizable couplings are the ones which are relevant or marginal at the Gaussian fixed point.*

That the Gaussian fixed point plays a central role is to be expected in all old treatments of QFT because they were based on perturbation theory about the free-field theory. These topics are treated nicely in many places; a sample [1–6] is given at the end of this chapter. The relation between relevance and renormalizability can be readily understood in Wilson's approach to renormalization, which I will now describe. His approach gives a very transparent non-perturbative explanation of the "miracle" of canceling infinities in renormalizable theories.

14.7 Wilson's Approach to Renormalizing QFT

Compared to the diagrammatic and perturbative proof of renormalization in QFT, Wilson's approach [7, 8] is simplicity itself.

Recall our goal: to define a QFT in the continuum with the following properties:

- All quantities of physical significance – correlation functions, masses, scattering amplitudes, and so on – must be finite.
- There should be no reference in the final theory to a lattice spacing a or an ultraviolet momentum cut-off Λ.

Of course, at intermediate stages a cut-off will be needed and the continuum theory will be defined as the $\Lambda \to \infty$ limit of such cut-off theories.

Wilson's approach is structured around a fixed point of the RG. Every relevant direction will yield an independent parameter.

It is assumed that we know the eigenvectors and eigenvalues of the flow near this fixed point.

Even if we cannot find such fixed points explicitly, the RG provides a *framework* for understanding renormalizability, just as it provides a framework for understanding critical phenomena and demystifying universality in terms of flows, fixed points, scaling operators, and so on, even without explicit knowledge of these quantities.

Consider a scalar field theory. By assumption, we are given complete knowledge of a fixed point action S^* that lives in some infinite-dimensional space of *dimensionless couplings* such as r_0, u_0, and so forth. The values of these couplings are what we previously referred to as $\boldsymbol{K}^*$. Let the fixed point have one relevant direction, labeled by a coordinate t. As t increases from 0, the representative point moves from S^* to $S^* + tS_{\text{rel}}$, where S_{rel} is the relevant perturbation, a particular combination of ϕ^2,ϕ^4, and so on. Once we go a finite distance from S^* the flow may not be along the direction of the relevant eigenvector at S^*, but along its continuation, a curve called the *renormalized trajectory* (RT).

Let us say that our goal is to describe physics in the 1 GeV scale using a continuum theory. (In terms of length, 1 GeV corresponds to roughly 1 fermi, a natural unit for nuclear physics. More precisely, $1\,\text{GeV} \cdot 1\text{fermi} \simeq 5 \simeq 1$ in units $\hbar = c = 1$.) Although we limit our interest to momenta within the cut-off of 1 GeV, we want the correlations to be exactly those of an underlying theory with a cut-off that approaches infinity, a theory that knows all about the extreme-short-distance physics. The information from very short distances is not discarded, but encoded in the renormalized couplings that flow under the RG.

Notice the change in language: we are speaking of a very large cut-off Λ. We are therefore using laboratory units in contrast to the Wilsonian language in which the cut-off is always unity. (For example, when we performed decimation, the new lattice size a served as the unit of length in terms of which the dimensionless correlation length ξ was measured.)

To make contact with QFT, we too will carry out the following discussion *in fixed laboratory units*. In these units the allowed momenta will be reduced from a huge sphere of radius Λ GeV to smaller and smaller spheres of radius Λ/s GeV. The surviving momenta will range over smaller and smaller values, and they will be a small subset of the original set $k < \Lambda$.

We have had this discussion about laboratory versus running units before in discussing the continuum limit of a free-field theory. If we want the continuum correlation to fall by $1/e$ over a distance of 1 fermi, we fix the two points a fermi apart in the continuum and overlay lattices of smaller and smaller sizes a. As $a \to 0$, the number of lattice sites within this 1 fermi separation keeps growing and the dimensionless correlation length has to keep growing at the same rate to keep the decay to $1/e$.

So, we are not going to rescale momenta as modes are eliminated. How about the field? In the Wilson approach the field gets rescaled even in free-field theory because k gets rescaled to $k' = sk$. We will not do that anymore. However, we will rescale by the factor Z introduced in connection with the renormalized quantities Γ_R and G_R. This Z was needed in perturbation theory to avert a blow-up of the k^2 term in Γ due to the loop correction. [Recall the appearance of Z in the two-loop diagram, Eq. (14.61)]. In the Wilsonian RG there will also be a correction to the k^2 term from loop diagrams (now integrated over the eliminated modes), and these will modify the coefficient of the k^2 term. We will bring in a Z to keep the coefficient of k^2 fixed at 1. The reason is not to cancel divergences, for there are none, but because *the strength of the interaction is measured relative to the free-field term*. For example, in a ϕ^4 theory if we rescale $\phi(x)$ by 5 this will boost the coefficients ϕ^2 and $(\nabla\phi)^2$ by 25 and that of the quartic term by 625. But it is still the same theory. For this reason, to compare apples to apples, one always rescales the k^2 coefficient to unity, *even if there are no infinities*.

Let us now begin the quest for the continuum theory.

Say we want a physical mass of 1 GeV or a correlation length of 1 fermi. First we pick a point t_0 on the RT where the dimensionless correlation length $\xi_0 = 2^0 = 1$, as indicated in Figure 14.5. We refer to the action at t_0 as $S(0)$.

No cut-off or lattice size has been associated with the point t_0, since everything is dimensionless in Wilson's approach. All momenta are measured in units of the cut-off, and the cut-off is unity at every stage in the RG. We now bring in *laboratory units* and assign to t_0 a momentum cut-off of $\Lambda_0 = 2^0 = 1$ GeV.

What is the mass corresponding to this ξ_0 in GeV? For this, we need to recall the connection between ξ and m:

$$G(r) \simeq e^{-mr} = \exp\left[-\frac{r}{a\xi}\right] = \exp\left[-\frac{r\Lambda}{\xi}\right], \tag{14.78}$$

which means that the mass is related to the cut-off and ξ as follows:

$$m = \frac{\Lambda}{\xi}. \tag{14.79}$$

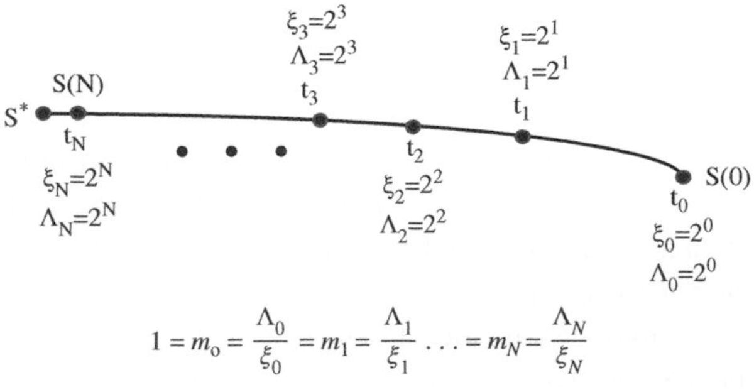

Figure 14.5 Points on the renormalized trajectory emanating from the fixed point S^*. To end up at the theory with cut-off $\Lambda_0 = 1$ GeV and action $S(0)$ after N RG steps of factor of 2 reduction of Λ, we must begin with the point labeled N, cut-off $\Lambda_N = 2^N$ GeV, $\xi_N = 2^N$ (dimensionless), and action $S(N)$. The sequence of points $S(N)$, $N \to \infty$ defines the continuum limit.

Thus, the mass corresponding to $S(0)$ is

$$m_0 = \frac{\Lambda_0}{\xi_0} = \frac{1\,\text{GeV}}{1} = 1\,\text{GeV}. \tag{14.80}$$

Imagine that we got to the point t_0 by performing N RG steps of size 2, starting with the point t_N where $\xi_N = 2^N$ and $\Lambda_N = 2^N$ GeV. *At every stage, the dimensionful mass is* 1 *GeV:*

$$m_N = \frac{\Lambda_N}{\xi_N} = 1. \tag{14.81}$$

Thus we have a sequence of actions, $S(n) : n = 0, 1, \ldots, N$, defined on smaller and smaller length scales or larger and larger momentum cut-offs, which produce the requisite physical mass. Not only is the mass fixed, the complete interaction is fixed to be $S(0)$. We have reverse-engineered it so that the theory at 1 GeV stays fixed at $S(0)$ while the underlying theory is defined on a sequence of actions $S(N)$ for which $\xi_N = 2^N$, and cut-off 2^N GeV, with $N \to \infty$. *We can make N as large as we like because ξ diverges as we approach S^*.*

We have managed to renormalize the theory by providing for each cut-off 2^N an action $S(t_N) \equiv S(N)$ that yields the theory $S(0)$ at low energies. This is the *continuum limit*.

This discussion also makes it obvious how to obtain a theory with a cut-off of 2 GeV: we just stop the RG one step earlier, at $S(1)$.

We can be more explicit about the continuum limit by invoking our presumed knowledge of ν. Near the fixed point we know that

$$\xi = t^{-\nu}. \tag{14.82}$$

This means that

$$2^N = t_N^{-\nu} \tag{14.83}$$

$$t_N = 2^{-N/\nu}, \tag{14.84}$$

which specifies the bare coupling or action $S(t_N) \equiv S(N)$ as a function of the cut-off $\Lambda_N = 2^N$ and the critical exponent ν. Just to be explicit: the bare action for cut-off $\Lambda = 2^N$ GeV is $S = S^* + 2^{-N/\nu} S_{\text{rel}}$, where S_{rel} is the relevant eigenoperator (some linear combination of ϕ^2, ϕ^4, etc.) that moves us along the RT starting at S^*.

We have managed to send the cut-off of the underlying theory to 2^N GeV with $N \to \infty$ holding fixed the action $S(0)$ for a theory with a cut-off of 1 GeV, but we need more. We need to ensure that not only does the low-energy action have a limit $S(0)$, as $\Lambda_N \to \infty$, but so do all the M-point correlation functions $G(\boldsymbol{k}_1, \boldsymbol{k}_2, \ldots, \boldsymbol{k}_M)$ defined by

$$\langle \phi(\boldsymbol{k}_1)\phi(\boldsymbol{k}_2)\cdots\phi(\boldsymbol{k}_M)\rangle = (2\pi)^d \delta\left(\sum_i \boldsymbol{k}_i\right) G(\boldsymbol{k}_1, \boldsymbol{k}_2, \ldots, \boldsymbol{k}_M). \tag{14.85}$$

Since we measure momentum in fixed laboratory units, the surviving momenta and fields $\phi(k)$ in the $\Lambda_0 = 1$ GeV theory are a subset of the momenta and fields in the underlying $\Lambda_N = 2^N$ GeV theory.

This may suggest that

$$G(\boldsymbol{k}_1,\ldots,\boldsymbol{k}_M,S(N)) = G(\boldsymbol{k}_1,\ldots,\boldsymbol{k}_M,S(0)). \tag{14.86}$$

However, Eq. (14.86) is incorrect. The reason is that the fields that appear in $S(0)$ are different from the ones we began with in $S(N)$, because we rescale the field to keep the coefficient of the k^2 term fixed in the presence of higher-loop corrections.

So, at every RG step we define a renormalized ϕ_R as follows:

$$\phi_\mathrm{R}(\boldsymbol{k}) = Z^{-\frac{1}{2}}\phi(\boldsymbol{k}), \tag{14.87}$$

and write S in terms of that field. If there are N steps in the RG the same equation would hold, with Z being the product of the Z's from each step. *So, the fields entering $S(0)$ are rescaled versions of the original fields entering $S(N)$.*

This means that, for the M-point correlation,

$$G(\boldsymbol{k}_1,\ldots,\boldsymbol{k}_M,S(N)) = Z(N)^{\frac{M}{2}}G(\boldsymbol{k}_1,\ldots,\boldsymbol{k}_M,S(0)), \tag{14.88}$$

where $Z(N)$ is the net renormalization factor after N RG steps starting with cut-off 2^N.

Look at the $G(\boldsymbol{k}_1,\ldots,\boldsymbol{k}_M,S(N))$ on the left-hand side. This is the correlation function of a theory with a growing cut-off. The coupling is chosen as a function of cut-off that grows like 2^N. If G is finite as $N \to \infty$, we have successfully renormalized. The equation above expresses G as the product of two factors. The second factor is a correlation function evaluated in a theory with action $S(0)$ which remains fixed as $N \to \infty$ by construction. It has a finite non-zero mass and a finite cut-off, and is thus free of ultraviolet and infrared divergences. So we are good there. But, this need not be true of the Z-factor in front, because it is the result of (product over) Z's from N steps, with $N \to \infty$. Let us take the Z factor to the left-hand side:

$$Z(N)^{-\frac{M}{2}}G(\boldsymbol{k}_1,\ldots,\boldsymbol{k}_M,S(N)) = G(\boldsymbol{k}_1,\ldots,\boldsymbol{k}_M,S(0)). \tag{14.89}$$

The left-hand side is now finite as $N \to \infty$, namely $G(\boldsymbol{k}_1,\ldots,\boldsymbol{k}_M,S(0))$. In other words, *the correlation functions of the renormalized fields are finite and cut-off independent as the cut-off approaches* ∞. This is the continuum limit.

In this approach it is obvious how, by choosing just one coupling (the initial value t_N of the distance from the fixed point along the RT) as a function of the cut-off ($\Lambda = 2^N$), we have an expression for finite correlation functions computed in terms of the finite renormalized interaction $S(0)$. Renormalizability is not a miracle if we start with an RG fixed point with a relevant coupling (or couplings) and proceed as above.

14.7.1 Possible Concerns

You may have some objections or concerns at this point.

What about $t < 0$? Is there not a flow to the left of S^*? There is, and it defines another continuum theory. In the magnetic case the two sides would correspond to the ordered and disordered phases. However, the rest of the discussion would be similar. (There are some cases, like Yang–Mills theory, where the fixed point is at the origin and the region of negative coupling is unphysical [9, 10].)

You may object that we have found a smooth limit for the correlation of the renormalized fields, whereas our goal was to find the correlations of the original fields. Have we not found a nice answer to the wrong question? No. As mentioned earlier (without proof), the physical results of a field theory – masses, scattering amplitudes, and so on – are unaffected by such a k- and x-independent rescaling of the fields. So what we have provided in the end are finite answers to all physical questions pertaining to the low-energy physics in the continuum.

Another very reasonable objection is that the preceding diagram and discussion hide one important complexity. Even though the flow along the RT is one-dimensional, it takes place in an infinite-dimensional space of all possible couplings. As we approach the fixed point S^* along the RT, we have to choose the couplings of an infinite number of terms like the ϕ^2, ϕ^4, ϕ^6, $\phi^2(\nabla\phi)^2$, and so on of the short-distance interaction. This seems impractical. It also seems to have nothing to do with standard renormalization, where we vary one or two couplings to banish cut-off dependence.

14.7.2 Renormalization with Only Relevant and Marginal Couplings

We resolve this by bringing in the irrelevant directions and seeing what they do to the preceding analysis. Look at Figure 14.6.

Besides the RT, I show one irrelevant trajectory that flows into the fixed point. This is a stand-in for the *entire multidimensional critical surface*, which includes every critical system of this class. Somewhere in the big $\boldsymbol{K}$ space is an axis describing a simple coupling, which I call r_0. It could be the nearest-neighbor coupling K of an Ising model or some combination of the elementary couplings $r_0\phi^2$ and $u_0\phi^4$ of a scalar field theory which can be varied to attain criticality. We will see how to define the continuum limit by taking a sequence of points on the r_0 axis.

Though the interaction is simple, we can hit criticality by varying its strength. The critical point, where the r_0 axis meets the critical surface, is indicated by r^*.

Now, r^* is a *critical* point while S^* is a *fixed point*. The two differ by irrelevant terms. This means that the correlation functions at r^* will not have the scaling forms of S^* in general. To see the ultimate scaling forms associated with the fixed point S^*, we do not have to renormalize: if we evaluate the correlation functions at r^* in the limit $k \to 0$ or $r \to \infty$, they will exhibit these laws. For example, at the Ising critical point, $G(k) \simeq 1/k^{2-\eta}$ will result as $k \to 0$, or $G(r) \simeq 1/r^{\frac{1}{4}}$ will follow as $r \to \infty$, despite being formulated on a lattice with just the symmetry of a square.

Of course, we can understand this in terms of the RG. If we limit ourselves to $k \to 0$, we are permitted to trade our initial theory with a large Λ for one with $\Lambda \simeq k$, which is related by RG flow to S^*.

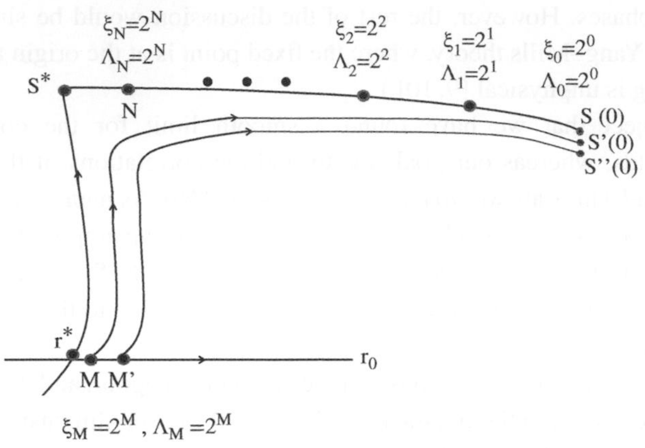

Figure 14.6 Flow with one relevant direction (the RT) and one irrelevant direction, which is a stand-in for the entire critical surface. The axis labeling the simple coupling r_0 (which could stand for $r_0\phi^2$) cuts the critical surface at r^*. Look at the points on the trajectory emanating from the point M on the r_0 axis. At point M, $\Lambda_M = 2^M$ and $\xi_M = 2^M$. We will end up at the theory with cut-off $\Lambda_0 = 1$ GeV and action $S'(0)$ after M RG steps of factor of 2 reduction of Λ. The sequence of points $S(M)$, $M \to \infty$ defines the continuum limit defined using just a single simple relevant coupling like r_0. If we start at M' we will reach $S''(0)$ (equivalent in the infrared to $S(0)$ and $S'(0)$) after $M-1$ steps. This is how one renormalizes in quantum field theory, by choosing simple couplings as a function of cut-off. The coupling M' corresponds to $\Lambda = 2^{M-1}$.

To define the continuum theory starting on this axis corresponding to a simple coupling, we pick a point M such that after M RG steps (of powers of 2) we arrive at the point $S'(0)$ that differs from $S(0)$, the theory generated from S^*, by a tiny amount *in the irrelevant direction*. The tiny irrelevant component will vanish asymptotically, and even when it is non-zero will make negligible corrections in the infrared. This result is inevitable given the irrelevance of the difference between r^* and S^*. We can go to the continuum limit by starting closer and closer to the critical surface (raising M) and reaching the target $S'(0)$ after more and more steps. As $M \to \infty$, our destination $S'(0)$ will coalesce with $S(0)$, which lies on the RT.

As a concrete example, consider Figure 13.4. Look at the dotted line parallel to the r_0 axis that comes straight down and crosses the critical line joining the Gaussian and WF fixed points. By starting closer and closer to the critical point where the dotted line crosses the critical line, we can renormalize the continuum theory based on the WF fixed point. The flow will initially flow toward the WF fixed point, and eventually will run alongside the RT. We can arrange to reach a fixed destination on the RT (the analog of $S(0)$) by starting at the appropriate distance from the critical line. You can also vary u_0 at fixed (negative) r_0 to approach the critical line with the same effect.

Now we can see the answer to a common question: how does a field theorist manage to compensate for a change in cut-off by renormalizing (i.e., varying with Λ) one or two couplings, whereas in Wilson's scheme, it takes a change in an infinite number of couplings? In other words, when we flow along the RT, i.e., vary one parameter t, we are actually varying an infinite number of elementary couplings in $\boldsymbol{K}$-space. How can a field theorist achieve the same result varying one or two couplings? The answer is that the field theorist does not really compensate for all the changes a changing cut-off produces. This is simply impossible. Whereas in Wilson's approach all correlation functions right up to the cut-off are preserved under the RG, in the field theory, *only correlations in the limit* $k/\Lambda \to 0$ *are preserved.*

Let us dig a little deeper into this. Suppose we begin at the point M, where $\Lambda = 2^M$, and reach the point $S'(0)$ in the figure after M RG steps of size 2. Say we ask what bare coupling with a cut-off 2^{M-1} will reproduce the answers of M with $\Lambda = 2^M$. It does not exist in general. Suppose, however, that we ask only about correlations in the infrared limit, $k/\Lambda \to 0$. Now we may trade the initial couplings for those on the RG trajectory. The point M flows to $S'(0)$ after M steps, i.e., when $\Lambda = 1$. The difference between $S(0)$ and the $S'(0)$ are technically and literally irrelevant in the infrared limit. If we start on the r_0 axis at M', at a suitably chosen point a little to the right of M, we can, after $M-1$ steps, reach the point $S''(0)$ that agrees with $S(0)$ and $S'(0)$ up to irrelevant corrections. It follows that if we reduce the cut-off by 2 we must change M to M', and if we increase the cut-off by 2 we must change M' to M. In other words, for each cut-off 2^M there is a point on the r_0 axis that has the same long-distance physics as the point M does with $\Lambda = 2^M$. This is how one renormalizes in QFT.

In QFT, one does not apologize for considering only the limit $k/\Lambda \to 0$ because there, Λ is an artifact that must be sent to ∞ at the end. So, $k/\Lambda \to 0 \,\forall\, k$.

Suppose I add a tiny irrelevant coupling, say $w_6\phi^6$, to the simple interaction of the starting point M. (Imagine the point is shifted slightly out of the page by w_6.) After M steps, the representative point again has to end up close to the RT. It may now end up slightly to the left or right of $S'(0)$ (ignore the component outside the page, which must have shrunk under the RG). Say it is to the right. This is what would have happened had we started with no w_6 but with a slightly bigger r_0 (a little to the right of M). A similar thing is true if the end point with w_6 in the mix is to the left of S'_0. In either case, *the effect of an irrelevant perturbation is equivalent to a different choice of the initial relevant coupling.*

It is understood above that w_6 is finite in units of the cut-off, and hence is very small in laboratory units, scaling as Λ^{-2} in $d = 4$. Had it been of order μ^{-2}, where μ is some fixed mass, it would not have been possible to absorb its effects by renormalization because it could correspond to an infinite perturbation in the natural units, namely Λ. But this is what field theorist tend to do in declaring it a non-renormalizable theory.

14.8 Theory with Two Parameters

Consider next the Gaussian fixed point in $d < 4$ when it has *two* relevant directions. Look at the flow in Figure 14.7. A generic point near the fixed point (the origin) will run away

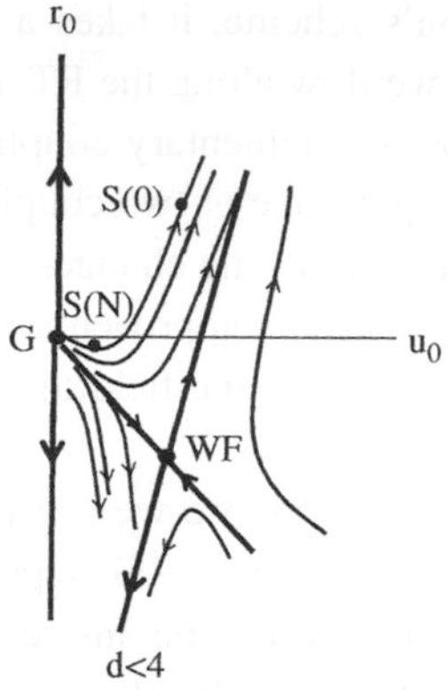

Figure 14.7 The situation in $d < 4$ when the Gaussian fixed point has two relevant directions. One can arrange to end up with a continuum theory with action $S(0)$, containing *two* free parameters, by starting closer and closer to the origin on the RT that passes through the point $S(0)$. Two bare couplings will have to be tuned, based on two relevant exponents that describe their growth under RG.

along the curves shown. We can make any point on any of those flow lines our destination $S(0)$ describing the continuum theory with 1 GeV cut-off, and arrange to get there after N RG steps by a suitable choice of initial coordinates $S(N)$ close to the origin. The flow away from the fixed point will now be controlled by two eigenvalues. We get a two-parameter family of continuum theories here. (The discussion near the fixed point is in terms of the simple interactions $r_0\phi^2$ and $u_0\phi^4$ because $\boldsymbol{K}^*$ is at the origin.)

In the relevant space of the Gaussian fixed point, there is a line connecting it to the Wilson–Fisher fixed point WF. If you pick a generic point on that line you will flow to WF and end up with its exponents. (In laboratory units, such a starting point will have a very large dimensionful coupling $\lambda = \Lambda^{4-d}u_0$.) But there is a way to fight that flow to WF: start closer and closer to G in such a way that after N steps you reach a fixed destination on the line. That would be a continuum theory that is massless but has one free parameter. (This is not a *natural* theory because the bare coupling is unnaturally small, being of order μ^{4-d}, where μ is some fixed mass, rather than of order Λ^{4-d}.)

14.8.1 Triviality of ϕ_4^4

Finally, consider the ϕ_4^4 theory based on the Gaussian fixed point that has one relevant coupling (mass or r_0) and interaction u_0 which is marginal at tree level but flows logarithmically slowly to the Gaussian fixed point at one loop. For this reason, this is not a suitable fixed point for constructing an interacting theory in the continuum. But suppose we try anyway. Since we can always make a theory massive, let us focus on getting an interacting field theory. So we begin with a point on the marginally attractive direction depicted in Figure 14.8.

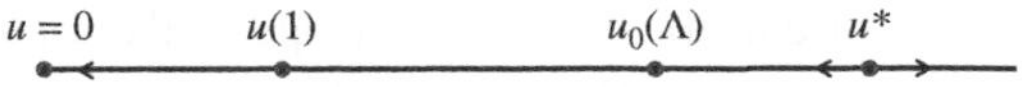

Figure 14.8 The flow of coupling in ϕ_4^4. The origin is marginally attractive. To end up at some $u(1)$ in a theory with, say, a 1 GeV cut-off, we need to begin at *larger* bare values $u_0(\Lambda)$, which in fact diverge as $\Lambda \to \infty$. It has been shown numerically by Wilson that any $u(\Lambda)$, including $u(\Lambda) = \infty$, flows to the origin, rendering the continuum theory trivial. The only way to define an interacting ϕ_4^4 is to construct one based on a strong coupling fixed point u^*. If u^* is the bare coupling, it will not move under the RG and define an interacting massless theory. We can also arrange to end up with u at a fixed distance to the left or right of u^* by starting out closer and closer to it in a way determined by the relevant eigenvalue at u^*.

We can parametrize this point by u_0 and assume r_0 is adjusted to put us on the critical line. Say we want a final coupling $u(1)$ in a 1 GeV cut-off theory. Let this target coupling be in the weak coupling regime where we have established marginal irrelevance. To get to $u(1)$ in the long-distance theory, we need to begin with a *larger* bare value $u(\Lambda)$ because the coupling is irrelevant. In perturbation theory, the desired bare value grows without limit as $\Lambda \to \infty$ (see Eq. (14.51)), rendering perturbation theory meaningless. The only legitimate way to construct an interacting ϕ_4^4 theory is to base it on a fixed point u^*, if we can find one. If we begin there at the bare level, we will stay there under the RG and define a massless interacting theory. We could also begin slightly to its left so as to end at our target value $u(1)$ after N steps, starting closer and closer to u^* as $N \to \infty$. This would be an interacting theory. (We can also get a theory with a fixed u to the right of u^*.) So now we are back to relevant flow coming out of the strong coupling fixed point u^*. However, Wilson has verified by thorough numerical analysis that such a fixed point does not exist anywhere on the u axis, including the point at infinity. The general consensus now is that ϕ_4^4 is trivial, i.e., non-interacting.

14.9 The Callan–Symanzik Equation

I will now provide a very brief introduction to this equation due to Callan [11] and Symanzik [12]. It is used extensively in quantum field theory as well as critical phenomena. It is mostly used to study the behavior of correlation functions in some extreme kinematical region: large momenta to describe asymptotic freedom in QCD [9, 10] or small momenta to describe critical phenomena. In the latter case it is the only practical way to deal with higher orders in the ε expansion.

14.9.1 Basis for the Callan–Symanzik Equation

Recall that in the Wilson approach, by construction, a theory with a cut-off Λ and couplings $\boldsymbol{K}(1)$ is equivalent to a theory with a cut-off Λ/s and couplings $\boldsymbol{K}(s)$ as long we ask questions below the new cut-off Λ/s. (We use laboratory units in which the cut-off shrinks by a factor s and momenta are not rescaled.) Correlation functions are, however, not invariant under this change of cut-off due to the change in the scale of the field to keep the k^2 term fixed after every iteration. The original ϕ we started with is related to the ϕ_R that appears in the theory with the new cut-off as

$$\phi(\boldsymbol{k}) = Z^{\frac{1}{2}}\phi_R(\boldsymbol{k}). \tag{14.90}$$

Consequently,

$$Z(N)^{-\frac{M}{2}}G(\boldsymbol{k}_1,\ldots,\boldsymbol{k}_M,S(N)) = G(\boldsymbol{k}_1,\ldots,\boldsymbol{k}_M,S(0)), \tag{14.91}$$

where the action $S(0)$ and the corresponding coupling $\boldsymbol{K}(0)$ are reached after N RG steps of cut-off reduction by 2.

The Callan–Symanzik equation is derived in quantum field theory from a similar relation which, however, holds only in the limit $\Lambda \to \infty$, or more precisely $k/\Lambda \to 0$, where k is any fixed momentum. The reason for the restriction is that a cut-off change can be compensated by changing a handful of (marginal and relevant) couplings only in this limit, in which irrelevant corrections vanish as positive powers of k/Λ. The Callan–Symanzik equation is not limited to the study of correlation functions as $\Lambda \to \infty$ in QFT. We can also use it in critical phenomena where Λ is some finite number $\Lambda \simeq 1/a$, provided we want to study the limit $k/\Lambda \to 0$, i.e., at distances far greater than the lattice size a. All that is required in both cases is that $k/\Lambda \to 0$.

We begin with the central claim of renormalization theory that the correlations of

$$\phi_R = Z^{-\frac{1}{2}}\phi, \tag{14.92}$$

expressed in terms of the renormalized mass and coupling,

$$G_R(\boldsymbol{k}_1,\ldots,\boldsymbol{k}_M,m,\lambda) = \lim_{\Lambda\to\infty} Z^{-M/2}(\lambda_0,\Lambda/m_0)G(\boldsymbol{k}_1,\ldots,\boldsymbol{k}_M,m_0(\Lambda),\lambda_0(\Lambda),\Lambda), \tag{14.93}$$

are finite and independent of Λ.

For a theory with a mass m we have seen that the renormalized *inverse* propagator Γ and four-point amplitude $\Gamma_R(\boldsymbol{k}_1,\ldots,\boldsymbol{k}_4)$ can be made to obey

$$\Gamma_R(0) = m^2, \tag{14.94}$$

$$\left.\frac{d\Gamma_R(k)}{dk^2}\right|_{k=0} = 1, \tag{14.95}$$

$$\Gamma_R(0,0,0,0) = \lambda. \tag{14.96}$$

We are going to study a critical (massless) theory in what follows. Although we can impose

$$\Gamma_R(0) = 0 \tag{14.97}$$

to reflect zero mass, we cannot impose Eqs. (14.95) and (14.96). This is because in a massless theory both these quantities have infrared divergences at $k = 0$. These are physical, just like the diverging Coulomb cross section. So we pick some point $k = \mu > 0$ where these quantities can be finite, and demand that

$$\left.\frac{d\Gamma_{\mathrm{R}}(k)}{dk^2}\right|_{k=\mu} = 1, \tag{14.98}$$

$$\Gamma_{\mathrm{R}}(\mu,\mu,\mu,\mu) = \lambda = \mu^{\varepsilon} u_{\mathrm{R}}. \tag{14.99}$$

This calls for some explanation.

First, μ is arbitrary, and any choice of μ can be used to specify a theory. If you change μ you will have to change λ accordingly if you want to describe the same theory.

Next, we are working in $d = 4 - \varepsilon$ dimensions, where λ has dimension ε. It is expressed as the product of a dimensionless parameter u_{R} and the factor μ^{ε}, which restores the right engineering dimension.

Finally, $\Gamma(\mu,\mu,\mu,\mu)$ is a schematic: it stands for a symmetric way to choose the momenta all of the scale μ:

$$\boldsymbol{k}_i \cdot \boldsymbol{k}_j = \frac{\mu^2}{3}(4\delta_{ij} - 1). \tag{14.100}$$

We will not need this expression from now on.

It is to be noted that the theory is not renormalizable in $d = 4 - \varepsilon$ due to the power-law infrared divergences that arise. However, if we expand everything in a double series in u and ε, the infinities (which will be logarithmic) can be tamed order by order, i.e., renormalized away. This double expansion will be understood from now on.

14.9.2 Massless $M = 2$ Correlations in $d = 4 - \varepsilon$

I will illustrate the Callan–Symanzik approach with the case $M = 2$, that is, two-point correlations, and study just the critical (massless) case in $d = 4 - \varepsilon$. Consider the system at point P in Figure 14.9 lying on the critical line joining the Gaussian and WF fixed points. It has a cut-off Λ and a coordinate $u(\Lambda) \equiv u$. We are interested in $\Gamma(k,u,\Lambda)$ in the limit $k/\Lambda \to 0$. We cannot use simple perturbation theory, even if u is small, because the expansion parameter will turn out to be $u \ln \frac{\Lambda}{k}$. The trick is to move the cut-off to a value of the order of k, thereby avoiding large logarithms, and work with the coupling $u(k)$ rather than $u = u(\Lambda)$. It is during this cut-off reduction that the coupling will flow from $u(\Lambda)$ to $u(k)$. We expect that $u(k) \to u^*$, the WF fixed point, as $k \to 0$.

It is convenient to work with the *inverse* propagator $\Gamma = G^{-1}$, which obeys

$$\Gamma_{\mathrm{R}}(k,u_{\mathrm{R}},\mu) = \lim_{\Lambda\to\infty}\left[Z^1(u(\Lambda),\Lambda/\mu)\Gamma(k,u(\Lambda),\Lambda)\right]. \tag{14.101}$$

The key to the Callan–Symanzik equation approach is the observation that since the left-hand side is independent of Λ (in the limit $\Lambda \to \infty$), so must be the right-hand side,

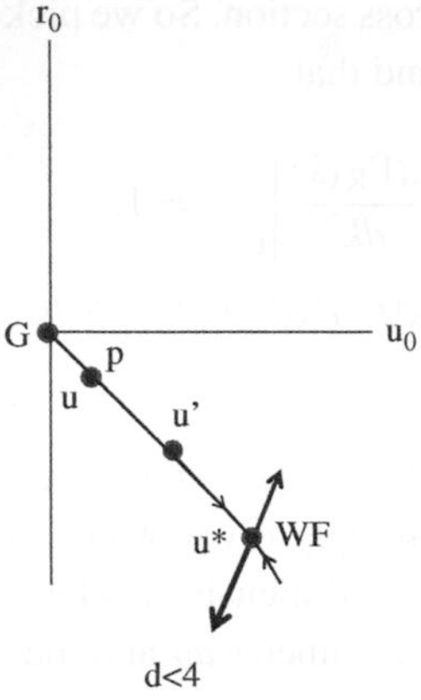

Figure 14.9 We want $\Gamma(k,u(\Lambda),\Lambda)$ as $k \to 0$ at a point P with coupling $u(\Lambda) = u$. The RG flow takes us to the WF fixed point u^* via the point u'.

which means

$$\lim_{\Lambda\to\infty} \Lambda \frac{d}{d\Lambda}\left[Z(u(\Lambda),\Lambda/\mu)\Gamma(k,u(\Lambda),\Lambda)\right] = 0. \tag{14.102}$$

Writing out the explicit and implicit Λ derivatives, we find that

$$\left[\Lambda\frac{\partial}{\partial\Lambda} + \beta(u,\Lambda/\mu)\frac{\partial}{\partial u} - \gamma(u,\Lambda/\mu)\right]\Gamma(k,u(\Lambda),\Lambda) = 0, \tag{14.103}$$

where

$$\beta(u,\Lambda/\mu) = \Lambda\frac{\partial u(\Lambda)}{\partial\Lambda}\bigg|_{\mu,u_R}, \tag{14.104}$$

$$\gamma(u,\Lambda/\mu) = -\Lambda\frac{\partial \ln Z(u(\Lambda),\Lambda/\mu)}{\partial\Lambda}\bigg|_{\mu,u_R}. \tag{14.105}$$

Next, we argue that since μ does not enter Γ, it cannot enter the dimensionless functions γ or β, which must therefore be functions only of $u(\Lambda)$. Thus we arrive at the *Callan–Symanzik equation*:

$$\left[\Lambda\frac{\partial}{\partial\Lambda} + \beta(u)\frac{\partial}{\partial u} - \gamma(u)\right]\Gamma(k,u(\Lambda),\Lambda) = 0. \tag{14.106}$$

The solution, derived by the method of characteristics, is

$$\Gamma(k,u(\Lambda_1),\Lambda_1) = \exp\left[\int_{\ln\Lambda_2}^{\ln\Lambda_1}\gamma(u(\ln\Lambda))d\ln\Lambda\right]\Gamma(k,u(\Lambda_2),\Lambda_2). \tag{14.107}$$

The solution is readily understood in Wilson's picture. The correlation function with cut-off Λ_2 is the same as that with Λ_1, provided we use the renormalized coupling in going from Λ_1 to Λ_2 and account for the field rescaling factor Z. Imagine doing the mode elimination in stages. Each stage will contribute a factor to Z, and the final Z will be a product of the Z's in each step depending on the coupling u at that stage. We reason as follows:

$$\Gamma(\Lambda_1)Z(\Lambda_1) = \Gamma(\Lambda_2)Z(\Lambda_2) = \Gamma_{\rm R} \tag{14.108}$$

$$\Gamma(\Lambda_1) = \frac{Z(\Lambda_2)}{Z(\Lambda_1)}\Gamma(\Lambda_2) \tag{14.109}$$

$$= e^{(\ln Z(\Lambda_2) - \ln Z(\Lambda_1))}\Gamma(\Lambda_2) \tag{14.110}$$

$$= \exp\left[\int_{\ln\Lambda_1}^{\ln\Lambda_2} \frac{d\ln Z}{d\ln\Lambda} d\ln\Lambda\right]\Gamma(\Lambda_2) \tag{14.111}$$

$$= \exp\left[\int_{\ln\Lambda_2}^{\ln\Lambda_1} \gamma(u(\ln\Lambda))d\ln\Lambda\right]\Gamma(\Lambda_2), \text{ with} \tag{14.112}$$

$$\gamma = -\frac{d\ln Z}{d\ln\Lambda}. \tag{14.113}$$

We verify that the solution Eq. (14.107) satisfies Eq. (14.106) by taking $\Lambda_1 \frac{\partial}{\partial\Lambda_1}$ of both sides:

$$\Lambda_1 \frac{\partial\Gamma(k, u(\Lambda_1), \Lambda_1)}{\partial\Lambda_1} + \beta(u(\Lambda_1))\frac{\partial\Gamma(k, u(\Lambda_1), \Lambda_1)}{\partial u(\Lambda_1)} = \gamma(u(\ln\Lambda_1))\Gamma(k, u(\Lambda_1), \Lambda_1). \tag{14.114}$$

Sometimes Eq. (14.107) is written in terms of an integral over the running coupling $u(\Lambda)$:

$$\Gamma(k, u(\Lambda_1), \Lambda_1) = \exp\left[\int_{u_2 \equiv u(\Lambda_2)}^{u_1 \equiv u(\Lambda_1)} \gamma(u)\frac{du}{\beta(u)}\right]\Gamma(k, u(\Lambda_2), \Lambda_2). \tag{14.115}$$

This version comes in handy if the integral over u is dominated by a zero of the β-function. We will have occasion to use it.

14.9.3 Computing the β-Function

The first step in using the Callan–Symanzik equation is the computation of β, which we will do to one loop. We begin with the renormalization condition,

$$u_{\rm R}\mu^{\varepsilon} = \Lambda^{\varepsilon}\left[u(\Lambda) - \frac{3u^2(\Lambda)}{16\pi^2}\ln\frac{\Lambda}{\mu}\right], \tag{14.116}$$

where the right-hand side was encountered earlier for the case $d = 4$ where $\varepsilon = 0$. Now we have to introduce the Λ^{ε} in front as part of the definition of the coupling. Setting to zero the $\ln\Lambda$-derivative of both sides (at fixed μ and u_R), we have (keeping only terms of order εu and u^2),

$$0 = \varepsilon u(\Lambda) + \underbrace{\frac{du(\Lambda)}{d\ln\Lambda}}_{\beta(u)} - \frac{3u^2(\Lambda)}{16\pi^2}. \tag{14.117}$$

(We anticipate that β will be of order εu or u^2, and do not take the $\ln\Lambda$-derivative of the $3u^2$ term, for that would lead to a term of order u^3 or $u^2\varepsilon$.) The result is

$$\beta(u) = -\varepsilon u + \frac{3u^2}{16\pi^2}. \tag{14.118}$$

The way β is defined, as Λ increases (more relevant to QFT), u flows toward the origin, while if Λ decreases (more relevant to us), it flows away and hits a zero at

$$u^* = \frac{16\varepsilon\pi^2}{3}. \tag{14.119}$$

That is,

$$\beta(u^*) = 0. \tag{14.120}$$

This is the WF fixed point. For future use, note that the slope of the β-function at the fixed point is

$$\omega = \left.\frac{d\beta(u)}{du}\right|_{u^*} = -\varepsilon + \frac{6u^*}{16\pi^2} = \varepsilon. \tag{14.121}$$

This irrelevant exponent $\omega = \varepsilon$ determines how quickly we approach the fixed point as we lower the cut-off. Here are the details.

14.9.4 Flow of $u - u^*$

Let us write a variable cut-off as

$$\Lambda(s) = \frac{\Lambda}{s}, \qquad s > 1. \tag{14.122}$$

It follows that

$$\frac{d}{d\ln\Lambda} = -\frac{d}{d\ln s}. \tag{14.123}$$

The coupling

$$u(s) \equiv u(\Lambda/s) \tag{14.124}$$

flows as follows:

$$\frac{du(s)}{d\ln s} = -\frac{du(\Lambda)}{d\ln\Lambda} = \varepsilon u(s) - \frac{3u_s^2}{16\pi^2} \equiv \bar{\beta}(u) = -\beta(u). \tag{14.125}$$

Integrating the flow of the coupling as a function of s, starting from $u(1) = u$, gives

$$\int_{u(1)=u}^{u(s)} \frac{du'}{\bar{\beta}(u')} = \ln s. \tag{14.126}$$

Now we expand $\bar{\beta}$ near the fixed point:

$$\bar{\beta}(u') = \bar{\beta}(u^*) - \omega(u' - u^*) = 0 - \omega(u' - u^*) = (-\omega)(u' - u^*). \tag{14.127}$$

(The minus in front of ω reflects the switch from β to $\bar{\beta} = -\beta$.) Substituting this into the previous equation, we get

$$\int_{u(1)=u}^{u(s)} \frac{du'}{(-\omega)(u' - u^*)} = \ln s, \tag{14.128}$$

with the solution

$$u(s) - u^* = (u - u^*)s^{-\omega} = (u - u^*)s^{-\varepsilon}. \tag{14.129}$$

That is, *the initial deviation from the fixed point* $(u - u^*)$ *shrinks by a factor* $s^{-\varepsilon} = s^{-\omega}$ *under the RG transformation* $\Lambda \to \Lambda/s$. Equation (14.129) will be recalled shortly.

14.9.5 Computing γ

The function γ begins at two loops. Armed with the two-loop result

$$Z = 1 + \frac{u^2}{6(4\pi)^4} \ln \frac{\mu}{\Lambda} + \cdots, \tag{14.130}$$

we find

$$\gamma = -\frac{d \ln Z}{d \ln \Lambda} = \frac{u^2}{6(4\pi)^4}. \tag{14.131}$$

At the fixed point

$$u^* = \frac{16\pi^2 \varepsilon}{3}, \tag{14.132}$$

we have

$$\gamma(u^*) \equiv \gamma^* = \frac{\varepsilon^2}{54}. \tag{14.133}$$

For later use, note that

$$\gamma' = \left.\frac{d\gamma}{du}\right|_{u^*} = \frac{\varepsilon}{144\pi^2}. \tag{14.134}$$

14.9.6 Computing $\Gamma(k,u,\Lambda)$

Now we are ready to confront the correlation function $\Gamma(k,u,\Lambda)$, which is the two-point function on the critical line shown in Figure 14.9. We want to know its behavior as a function of k as $k \to 0$. We expect it to be controlled by the WF fixed point.

The equation obeyed by $\Gamma(k,u,\Lambda)$ is

$$\left[-\frac{\partial}{\partial \ln s} - \bar{\beta}(u(s))\frac{\partial}{\partial u} - \gamma(u(s))\right]\Gamma(k,u(s),\Lambda/s) = 0. \tag{14.135}$$

Suppose we are at the fixed point, where $\bar{\beta} = 0$ and

$$\gamma = \gamma(u^*) \equiv \gamma^*. \tag{14.136}$$

The equation to solve is

$$\frac{\partial \Gamma(k,u^*,\Lambda/s)}{\partial \ln s} = -\gamma(u^*)\Gamma(k,u^*,\Lambda/s), \tag{14.137}$$

with an obvious solution

$$\Gamma(k,u^*,\Lambda) = s^{\gamma^*}\Gamma(k,u^*,\Lambda/s). \tag{14.138}$$

By dimensional analysis,

$$\Gamma(k,u^*,\Lambda/s) = k^2 f\left(\frac{k}{\Lambda/s}\right) = k^2 f\left(\frac{ks}{\Lambda}\right). \tag{14.139}$$

Substituting this into Eq. (14.138), we arrive at

$$\Gamma(k,u^*,\Lambda) = s^{\gamma^*}k^2 f\left(\frac{ks}{\Lambda}\right). \tag{14.140}$$

Now we choose

$$s = \frac{\Lambda}{k}, \tag{14.141}$$

which just means

$$\frac{\Lambda}{s} = k, \tag{14.142}$$

i.e., the new cut-off equals the momentum of interest. With this choice,

$$\Gamma(k,u^*,\Lambda) = \left(\frac{\Lambda}{k}\right)^{\gamma^*} k^2 f(1) \simeq k^{2-\gamma^*}. \tag{14.143}$$

Comparing to the standard form

$$\Gamma(k) \simeq k^{2-\eta}, \tag{14.144}$$

we find that

$$\eta = \gamma(u^*) \equiv \gamma^*. \tag{14.145}$$

In case you wondered how $\Gamma(k)$ can go as $k^{2-\eta}$ when it has engineering dimension 2, the answer is given above: $k^{-\eta}$ is really $\left(\frac{k}{\Lambda}\right)^{-\eta}$.

Finally, we ask how subleading corrections to the fixed point behavior arise if we start at some $u \neq u^*$ with a k that is approaching zero. For this, we return to the solution to the Callan–Symanzik equation

$$\Gamma(k, u(\Lambda_1), \Lambda_1) = \exp\left[\int_{\ln \Lambda_2}^{\ln \Lambda_1} \gamma(u(\ln \Lambda'))d\ln \Lambda'\right]\Gamma(k, u(\Lambda_2), \Lambda_2). \tag{14.146}$$

Let

$$\Lambda_1 = \Lambda, \tag{14.147}$$
$$\Lambda_2 = \Lambda/s, \tag{14.148}$$
$$u(\Lambda/s) \equiv u(s), \tag{14.149}$$
$$u(\Lambda) \equiv u(1). \tag{14.150}$$

Then

$$\Gamma(k, u(1), \Lambda) = \exp\left[\int_s^1 \gamma(u'(s'))\frac{-ds'}{s'}\right]\Gamma(k, u(s), \Lambda/s). \tag{14.151}$$

Corrections are going to arise from both the exponential factor and $\Gamma(k, u(s), \Lambda/s))$, due to the fact that at any non-zero $\frac{\Lambda}{s} = k$, the coupling $u(s)$ is close to, but not equal to, u^*, which is reached only asymptotically.

Consider first the exponential factor. Expanding γ near u^* as

$$\gamma(u') = \gamma^* + \gamma'(u' - u^*) + \cdots, \tag{14.152}$$
$$\gamma' = \frac{\varepsilon}{144\pi^2} \quad \text{[Eq. (14.134)]}, \tag{14.153}$$

we have, in the exponent,

$$\begin{aligned}\int_s^1 \gamma(u'(s'))\frac{-ds'}{s'} &= \int_1^s (\gamma^* + \gamma'(u(s') - u^*))\frac{ds'}{s'} \\ &= \gamma^* \ln s + \gamma'(u - u^*)\int_1^s (s')^{-\omega}\frac{ds'}{s'} \\ &= \gamma^* \ln s + \frac{\gamma'}{\omega}(u - u^*)(1 - s^{-\omega}).\end{aligned} \tag{14.154}$$

Thus the exponential factor becomes

$$\exp[\cdots] = s^{\gamma^*}\left(1 + \frac{\gamma'}{\omega}(u-u^*)(1-s^{-\omega})\cdots\right). \tag{14.155}$$

Next, consider

$$\begin{aligned}\Gamma(k,u(s),\Lambda/s)) &= \Gamma(k,u^*+u(s)-u^*,\Lambda/s)\\ &= k^2 f\left[\frac{ks}{\Lambda},u^*+(u(s)-u^*)\right]\\ &= k^2 f\left[\frac{ks}{\Lambda},u^*+(u-u^*)s^{-\omega}\right].\end{aligned} \tag{14.156}$$

If we now set

$$s = \frac{\Lambda}{k} \tag{14.157}$$

and recall that $\omega = \varepsilon$, we find, upon putting the two factors in Eqs. (14.155) and (14.156) together, an irrelevant correction of the form $\left(\frac{k}{\Lambda}\right)^{\varepsilon}$:

$$\Gamma(k,u(\Lambda),\Lambda) = k^2\left(\frac{\Lambda}{k}\right)^{\gamma^*}\left(a + c\left(\frac{k}{\Lambda}\right)^{\varepsilon}\right), \tag{14.158}$$

where a and c are some constants.

14.9.7 Variations of the Theme

The preceding introduction was aimed at giving you an idea of how the Callan–Symanzik machine works by focusing on Γ, corresponding to two-particle correlations, and only for the critical case. There are so many possible extensions and variations.

The first variation is to go to the non-critical theory, where, in addition to the marginal coupling u, we have a relevant coupling, denoted by t, which as usual measures deviation from criticality. It multiplies the operator ϕ^2, whose presence calls for additional renormalization. The final result will be quite similar: as $k/\Lambda \to 0$, the flow will first approach the fixed point and then follow the renormalized trajectory.

Next, we can go from correlations of two fields to M fields and work with $\Gamma(\boldsymbol{k}_1,\ldots,\boldsymbol{k}_M)$.

Finally, let us go back to the relation between bare and renormalized Γ's:

$$\Gamma_{\rm R}(k,u_{\rm R}(\mu),\mu) = \lim_{\Lambda\to\infty} Z(u(\Lambda),\Lambda/\mu)\Gamma(k,u(\Lambda),\Lambda). \tag{14.159}$$

We got the Callan–Symanzik equation by saying that since the $\Gamma_{\rm R}$ on the left-hand side had no knowledge of Λ, i.e., was cut-off independent, we could set the $\ln\Lambda$-derivative of the right-hand side to zero. This equation describes how the bare couplings and correlations have to change with the cut-off to keep fixed some renormalized quantities directly related to experiment.

Instead, we could argue that since Γ does not know about μ, the $\ln\mu$-derivative of the left-hand side must equal the same derivative acting on just the Z on the right-hand side (which has been expressed as $Z(u_R(\mu), \Lambda/\mu)$). The resulting equation,

$$\left[\frac{\partial}{\partial \ln\mu} + \beta\frac{\partial}{\partial u_R} - \gamma\right]\Gamma_R(k, u_R(\mu), \mu) = 0, \tag{14.160}$$

where

$$\beta(u_R) = \left.\frac{\partial u_R}{\partial \ln\mu}\right|_{u(\Lambda),\Lambda}, \tag{14.161}$$

$$\gamma(u_R) = \left.\frac{\partial \ln Z}{\partial \ln\mu}\right|_{u(\Lambda),\Lambda}, \tag{14.162}$$

dictates how the renormalized coupling and correlations must change with μ in order to represent the same underlying bare theory. (Again, the dimensionless functions β and γ cannot depend on μ/Λ because they are determined by Γ_R, which does not know about Λ.)

We can use either approach to get critical exponents, flows, and Green's functions (because Γ and Γ_R differ by Z, which is momentum and position independent), but there are cultural preferences. In statistical mechanics, the bare correlations are physically significant and describe underlying entities like spins. The cut-off is real and given by $\Lambda \simeq 1/a$. To particle physicists, the cut-off is an artifact, and the bare Green's functions and couplings are crutches to be banished as soon as possible so that they can work with experimentally measurable, finite, renormalized quantities defined on the scale μ. They prefer the second version based on Γ_R.

References and Further Reading

[1] C. Itzykson and J. B. Zuber, *Quantum Field Theory*, Dover (2005). Gives a more thorough treatment of QFT and renormalization.

[2] M. Le Bellac, *Quantum and Statistical Field Theory*, Oxford University Press (1992).

[3] J. Zinn-Justin, *Quantum Field Theory and Critical Phenomena*, Oxford University Press (1996).

[4] D. J. Amit, *Field Theory, Renormalization Group and Critical Phenomena*, World Scientific (1984).

[5] C. Itzykson and J. M. Drouffe, *Statistical Field Theory*, vol. I, Cambridge University Press (1989).

[6] N. Goldenfeld, *Lectures on Phase Transitions and the Renormalization Group*, Addison-Wesley (1992).

[7] K. G. Wilson, Reviews of Modern Physics, **47**, 773 (1975). The first few pages of this paper on RG and the Kondo problem are the best reference for renormalzing QFT.

[8] K. G. Wilson and J. R. Kogut, Physics Reports, **12**, 74 (1974). Provides a discussion of the triviality of ϕ_4^4.

[9] D. J. Gross and F. Wilczek, Physical Review Letters, **30**, 1343 (1973).

[10] H. D. Politzer, Physical Review Letters, **30**, 1346 (1973). In these two Nobel Prize winning works, these authors showed that quantum chromodynamics, the gauge theory of quarks and gluons, was *asymptotically free*, i.e., the coupling vanished

at very short distances or very large momenta and grew at long distances or small momenta. This allowed us to understand why quarks seemed free inside the nucleon in deep inelastic scattering and yet were confined at long distances. The β-function of this theory has a zero at the origin and the coupling grows as we move toward long distances. In all other theories like ϕ^4 or quantum electrodynamics, the behavior is exactly the opposite.
[11] C. Callan, Physical Review D, **2**, 1541 (1970).
[12] K. Symanzik, Communications in Mathematical Physics, **18**, 227 (1970).

15

Renormalization Group for Non-Relativistic Fermions: I

Consider the following problem that arose in two-dimensional ($d = 2$) high-T_c materials like the cuprates. It was emphasized early on by P. W. Anderson in 1987 [1,2] that when superconductivity is destroyed by the application of a strong magnetic field, the "normal" state one obtains is *not* normal. By this, one means that it does not correspond to *Landau's Fermi liquid* state, seen so often in three dimensions. For example, it has a resistivity linear in temperature, instead of quadratic. It does not have a *quasiparticle* pole in the real-time propagator $G(\omega, \boldsymbol{K})$ as $\boldsymbol{K}$ approaches the Fermi surface. It is dubbed a non-Fermi liquid for this and other reasons. Non-Fermi liquids are familiar in $d = 1$. In fact, in $d = 1$, once we turn on the smallest interaction, the free fermion becomes a non-Fermi liquid and the quasiparticle pole becomes a cut. So the question was this: Is $d = 2$ more like $d = 3$ or more like $d = 1$? Is it possible to have an interacting Fermi liquid in $d = 2$? I decided to let the RG be the arbiter and see what happened. In this chapter I illustrate my method in $d = 1$ and obtain the same result as the pioneering work of Solyom [3,4].

Let us go back to basics. Consider a system of non-interacting fermions of mass m in $d = 2$. Let them be spinless fermions, for spin can be easily incorporated. The Hamiltonian is

$$H = \int \psi^\dagger(\boldsymbol{K})\psi(\boldsymbol{K})\left[\frac{K^2}{2m} - \mu\right]\frac{d^2\boldsymbol{K}}{(2\pi)^2}. \tag{15.1}$$

The chemical potential μ will control the number of fermions in the ground state. All states within a circle of radius K_F, the *Fermi momentum*, will be filled, where

$$\frac{K_F^2}{2m} = \mu. \tag{15.2}$$

What are the low excitations of this system? *They are not near $K = 0$, as they are in the case of bosons.* If you hit this system with a small hammer, the fermions near $K = 0$ cannot respond since all nearby states are blocked by the Pauli principle. On the other hand, a fermion close to the Fermi surface can jump to an unoccupied state above K_F. This is a *particle–hole excitation*. The energy cost of this pair is

$$\varepsilon_{ph} = \varepsilon_e + \varepsilon_h, \tag{15.3}$$

where $\varepsilon_{\rm h} > 0$ is the *depth* below the Fermi surface of the negative-energy electron that was destroyed.

Since energies

$$\left[\frac{K^2}{2m} - \mu\right] = \frac{K^2}{2m} - \frac{K_{\rm F}^2}{2m}$$

are measured from the Fermi surface, states within the Fermi sea have negative energy. *Destroying* a *negative-energy* particle costs positive energy, just as in the case of the positron.

Since $\varepsilon_{\rm ph}$ can be made arbitrarily small, we say the system is gapless. It can respond to a DC voltage with a DC current.

The energy measured from the Fermi energy can be rewritten as follows:

$$\varepsilon = \frac{K^2}{2m} - \frac{K_{\rm F}^2}{2m} \tag{15.4}$$

$$= \frac{K+K_{\rm F}}{2m}(K-K_{\rm F}) \tag{15.5}$$

$$= \left[\frac{2K_{\rm F}+k}{2m}\right]k, \tag{15.6}$$

where I have introduced a very important symbol

$$k = K - K_{\rm F}, \tag{15.7}$$

the radial deviation from the Fermi circle of the vector $\boldsymbol{K}$. In the domain of low-energy excitations $k \ll K_{\rm F}$ we may write

$$\varepsilon = \left[\frac{K_{\rm F}}{m}\right]k \equiv v_{\rm F}k, \tag{15.8}$$

where $v_{\rm F} = K_{\rm F}/m$ is the *Fermi velocity*. Let us use units in which

$$v_{\rm F} = 1. \tag{15.9}$$

The Hamiltonian becomes (for states near the Fermi energy)

$$H = K_{\rm F}^2 \int \psi^\dagger(k,\theta)\psi(k,\theta)k\frac{dk d\theta}{(2\pi)^2}. \tag{15.10}$$

Consider next an excitation that changes the particle number: an electron at some $K > K_{\rm F}$ is injected. It will remain in that state forever.

What happens to all this when we turn on interactions? Landau [5–7] provided the answer with his unerring intuition. His arguments were buttressed by diagrammatic calculations by his students [8]. Landau argued for *adiabatic continuity*. In this picture, the ground state (the filled Fermi sea) would readjust itself to the ground state of the interacting fermions. A Fermi surface would still remain, since the non-analytic jump in occupation number as we entered the sea cannot suddenly disappear. It can only disappear adiabatically, though the jump in occupancy number could drop below 1. What about the particle injected at some $K > K_F$? It would retain its momentum (which is conserved), but would evolve from being a *bare particle* into a *dressed particle*, a superposition of non-interacting states consisting of one particle, plus one particle plus a particle–hole pair, and so on. This is Landau's *quasiparticle*. It will not live forever because it can decay, primarily into a particle and any number of particle–hole pairs. The most important channel is the one with two particles and one hole:

$$(\boldsymbol{K}_{\text{qp}}, \varepsilon_{\text{qp}}) \to (\boldsymbol{K}_1, \varepsilon_1) + (\boldsymbol{K}_2, \varepsilon_2) + (\boldsymbol{K}_{\text{h}}, \varepsilon_{\text{h}}). \tag{15.11}$$

The particle–hole pair is created by the interaction which scatters the initial particle at $\boldsymbol{K}_{\text{qp}}$ into one at $\boldsymbol{K}_1$ and scatters a particle below the Fermi sea at $\boldsymbol{K}_{\text{h}}$ into one at $\boldsymbol{K}_2$ above the sea (Figure 15.1).

The decay rate can be computed using Fermi's golden rule with the constraints

$$\varepsilon_{\text{qp}} = \varepsilon_1 + e_2 + \varepsilon_{\text{h}}, \tag{15.12}$$

$$\boldsymbol{K}_{\text{qp}} = \boldsymbol{K}_1 + \boldsymbol{K}_2 - \boldsymbol{K}_{\text{h}}, \tag{15.13}$$

where ε_{h} is the (positive) depth of the hole state below the sea. The decay width of the quasiparticle, proportional to the decay rate, goes as $\varepsilon_{\text{qp}}^2$. You can try to show this on your own, or wait for the next chapter (Exercise 16.1.1) for a derivation. Given this, the imaginary-time propagator changes from

$$G(\boldsymbol{K}, \varepsilon) = \frac{1}{i\omega - \varepsilon_{\text{qp}}} \tag{15.14}$$

to

$$G(\boldsymbol{K}, \varepsilon) = \frac{Z}{i\omega - (\varepsilon_{\text{qp}} - i\alpha\varepsilon_{\text{qp}}^2)}, \tag{15.15}$$

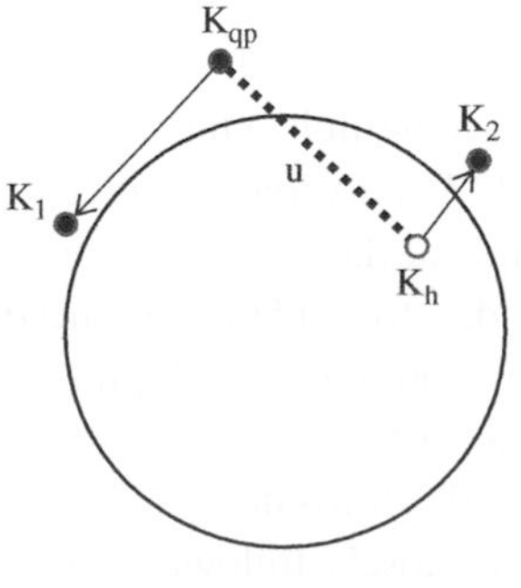

Figure 15.1 The instantaneous interaction u scatters the initial quasiparticle at $\boldsymbol{K}_{\text{qp}}$ to a quasiparticle at $\boldsymbol{K}_1$ and creates a hole at $\boldsymbol{K}_{\text{h}}$ by knocking the fermion out of the sea to a particle with momentum $\boldsymbol{K}_2$.

where α is some constant. The real-time propagator

$$G(\boldsymbol{K},\varepsilon) = \frac{1}{\omega - \varepsilon_{\text{qp}} + i\eta}, \tag{15.16}$$

where the imaginary part η is infinitesimal, changes to

$$G(\boldsymbol{K},\varepsilon) = \frac{Z}{\omega - (\varepsilon_{\text{qp}} - i\alpha\varepsilon_{\text{qp}}^2)}. \tag{15.17}$$

The main point is that the *imaginary part of the energy vanishes as the square of the real part*. The imaginary part is parametrically smaller than the real part. Thus, the quasiparticle becomes truly long-lived as we approach the Fermi surface. The key to its longevity is phase space, additionally restricted by the Pauli principle.

Landau went on to show that the entire low-energy (i.e., low-temperature) physics is governed by such quasiparticles, and all the response functions (like compressibility χ) are governed by a single function $F(\theta)$ (for spinless fermions) called the *Landau function* defined on the Fermi circle. Its harmonics $u_m \equiv F_m$,

$$F(\theta) = \sum_m u_m e^{im\theta} \equiv \sum_m F_m e^{im\theta}, \tag{15.18}$$

are called *Landau parameters*. Landau's view was that even if we could not calculate the F_m's *ab initio*, we could determine at least a few of them from some experiments and then use them to describe other experiments. For example, he showed that the compressibilty χ is related to χ_0 of the non-interacting system as follows:

$$\chi = \frac{\chi_0(1+F_1/3)}{1+F_0}. \tag{15.19}$$

He got his results by energy arguments, and members of his school established his results using Feynman diagrams [8]. This perturbative argument will fail only if perturbation theory (i.e., adiabatic continuity) fails.

Is this what happens in $d=2$? Do divergences creep in as we go from $d=3$ down to $d=2$, destroying the Fermi liquid for the smallest interaction, as in $d=1$?

Since we no longer have Landau to guide us, we need an approach that works for the rest of us, so we can better understand the Fermi liquid and its breakdown. I will show you how the RG may be adapted to do just that.

But first, in this chapter I will illustrate the RG and its power by applying it to non-relativistic fermions in $d=1$, closely following [9].

15.1 A Fermion Problem in $d = 1$

Let us consider the following specific Hamiltonian for a spinless fermion system on a $d = 1$ lattice labeled by an integer j:

$$H = H_0 + H_{\rm I} \tag{15.20}$$

$$= -\frac{1}{2}\sum_j \psi^\dagger(j+1)\psi(j) + \text{h.c.}$$

$$+U_0 \sum_j \left(\psi^\dagger(j)\psi(j) - \frac{1}{2}\right)\left(\psi^\dagger(j+1)\psi(j+1) - \frac{1}{2}\right), \tag{15.21}$$

where the fields obey

$$\{\psi^\dagger(j), \psi(j')\} = \delta_{jj'}, \tag{15.22}$$

with all other anticommutators vanishing.

The first term represents hopping. The second term represents nearest-neighbor repulsion of strength U_0. (The Pauli principle forbids the on-site repulsion term.) The role of the $\frac{1}{2}$'s subtracted from the charge densities n_j $(= \psi^\dagger{}_j \psi_j)$ and n_{j+1} is this: when we open up the brackets, it is readily seen that they represent a chemical potential

$$\mu = U_0. \tag{15.23}$$

This happens to be exactly the value needed to maintain half-filling in the presence of the repulsion U_0. To see this, make the change $\psi \leftrightarrow \psi^\dagger$ at all sites. This exchanges the site occupation number $n = \psi^\dagger\psi$ with $1 - n$, or changes the sign of $n - \frac{1}{2}$. Thus, both brackets in the interaction term change sign under this, and their product is unaffected. As for the hopping term, it changes sign under $\psi \leftrightarrow \psi^\dagger$. This can be compensated by changing the sign of ψ, $\psi^\dagger$ on just one sublattice (which preserves the anticommutation rules and does not affect the other term). Thus, H is invariant under exchanging particles with holes. If the ground state is unique and translation invariant, it will obey $\langle n\rangle = \langle 1 - n\rangle$, which in turn means $\langle n\rangle = \frac{1}{2}$. Another option, which is actually realized, is that there are two degenerate ground states that break translation invariance and are related by $\langle n\rangle \leftrightarrow \langle 1 - n\rangle$.

Let us understand this model in the limits $U_0 = 0$ and $U_0 = \infty$.

We begin by introducing momentum states via

$$\psi(j) = \int_{-\pi}^{\pi} \frac{dK}{2\pi}\psi(K)e^{iKj} \tag{15.24}$$

and the inverse transform

$$\psi(K) = \sum_j e^{-iKj}\psi(j). \tag{15.25}$$

Using

$$\sum_j = 2\pi\delta(0) \quad \text{(Exercise 15.1.1)}, \tag{15.26}$$

we can verify that

$$\{\psi(K), \psi^\dagger(K')\} = 2\pi\delta(K-K'). \tag{15.27}$$

Exercise 15.1.1 *Feed Eq. (15.25) into Eq. (15.24) (after changing dummy label j to j′) and demand that you get back $\psi(j)$. This should require*

$$\int_{-\pi}^{\pi} \frac{dK}{2\pi} e^{iK(j-j')} = \delta_{jj'}. \tag{15.28}$$

Feed Eq. (15.24) into Eq. (15.25) and demand that you get back $\psi(K)$. This should require

$$\sum_{j'} e^{i(K-K')j'} = 2\pi\delta(K-K'). \tag{15.29}$$

Setting $K = K'$ gives Eq. (15.26).

In terms of these operators,

$$H_0 = \int_{-\pi}^{\pi} \frac{dK}{2\pi}\psi^\dagger(K)\psi(K)E(K), \tag{15.30}$$

$$E(K) = -\cos K. \tag{15.31}$$

The Fermi sea is obtained by filling all negative energy states, i.e., those with $|K| \le K_{\rm F} = \pi/2$, which correspond to half-filling. The Fermi surface consists of just two points L and R at $|K| = \pm\pi/2$, as shown in Figure 15.2. It is clear that the ground state is a perfect conductor, since we can move a particle just below the Fermi surface to just above it at arbitrarily small energy cost.

Consider now the situation in the other extreme, $U_0 = \infty$. We now ignore the hopping term and focus on just the interaction. It is evident that the lowest-energy states are those in which no particle has a neighbor: thus, either the A-sublattice consisting of even sites is occupied, or the B-sublattice, made up of the odd sites, is occupied. This makes the product $(n_j - \frac{1}{2})(n_{j+1} - \frac{1}{2})$ negative on every bond. These two states, which break the translational symmetry of the lattice, are the *charge density wave* or CDW states. The order parameter, which measures the difference between the mean occupation of the odd and even sites, is maximal (unity). In the CDW state, the system is an insulator. Any excitation of the ground state requires us to move the charge, and this will cost an energy of order U_0. (This is clearest as $U_0 \to \infty$.) One expects that even for large but finite U_0, the symmetry would still be broken, but with a smaller order parameter.

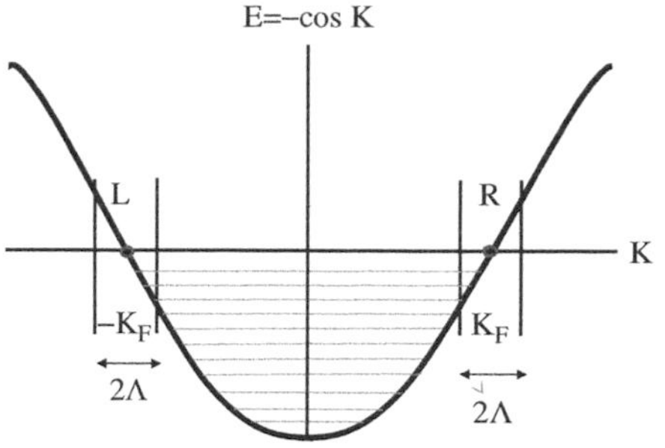

Figure 15.2 The shaded energy levels are filled in the ground state, the Fermi sea. For low-energy physics we can focus on states within $\pm\Lambda$ of the left (L) and right (R) Fermi points $\mp K_F$.

Here is the question we want to answer: Will the system develop the CDW order and gap for arbitrarily small repulsion, or will it remain a conductor up to some finite U_0?

15.2 Mean-Field Theory $d = 1$

We will use the RG to answer it. But first, let us see what a very standard tool, namely *mean-field theory*, can tell us [10–16]. In this approach, one assumes a CDW order parameter in the ground state and asks if the assumption is self-consistent and lowers the ground-state energy. Mean-field theory predicts that CDW will set in for any repulsion, however small. Here is a short description of the calculation.

Let us begin with Eq. (15.21) and make the *ansatz*

$$\langle n_j \rangle = \frac{1}{2} + \frac{1}{2}(-1)^j \Delta, \tag{15.32}$$

where Δ is the CDW order parameter and describes the modulation in charge density. We will now see if the ground-state energy of the system is lowered by a non-zero value of Δ. To this end, we will find the ground-state energy as a function of Δ and minimize it to see if the minimum occurs at a non-zero Δ. However, this last step will be done approximately since this is an interacting many-body system. The approximation is the following. We start with the Hamiltonian

$$H = -\frac{1}{2}\sum_j \psi^\dagger(j+1)\psi(j) + h.c + U_0 \sum_j \left(n_j - \frac{1}{2}\right)\left(n_{j+1} - \frac{1}{2}\right) \tag{15.33}$$

and make the substitution

$$n_j = \frac{1}{2} + \frac{1}{2}(-1)^j \Delta + :n_j:, \tag{15.34}$$

$$n_{j+1} = \frac{1}{2} + \frac{1}{2}(-1)^{j+1} \Delta + :n_{j+1}:, \tag{15.35}$$

where the normal-ordered operator $:n_j:$ has no expectation value in the true ground state and represents the fluctuations in number density. Upon making these substitutions and some rearrangements, we find that

$$H = -\frac{1}{2}\sum_j \psi^\dagger(j+1)\psi(j) + h.c$$
$$+U_0\left[\frac{1}{4}\sum_j \Delta^2 - \Delta\sum_j(-1)^j n_j\right] + U_0\sum_j :n_j::n_{j+1}: . \tag{15.36}$$

In the mean-field approximation we ignore the last term, quadratic in the fluctuations. The rest of the Hamiltonian is quadratic in the Fermi fields and solved by Fourier transformation. Due to the factor $(-1)^j$ that multiplies $n_j = \psi^\dagger(j)\psi(j)$, states with momentum K and

$$K' = K + \pi \tag{15.37}$$

will mix. The Hamiltonian becomes

$$\frac{H}{2\pi\delta(0)} = \int_0^\pi \frac{dK}{2\pi}\left[\psi^\dagger(K), \psi^\dagger(K')\right]\begin{bmatrix} E(K) & -U_0\Delta \\ -U_0\Delta & E(K') \end{bmatrix}\begin{bmatrix} \psi(K) \\ \psi(K') \end{bmatrix} + U_0\frac{\Delta^2}{4}. \tag{15.38}$$

Notice that we have halved the range of K integration, but doubled the number of variables at each K. The two-by-two matrix, which is traceless due to the relation

$$E(K') = -\cos(K+\pi) = -E(K), \tag{15.39}$$

is readily diagonalized. The one-particle energy levels come in equal and opposite pairs and we fill the negative energy states to obtain the following ground state energy per unit volume:

$$\frac{E_0}{2\pi\delta(0)} = \frac{\Delta^2 U_0}{4} - \int_0^\pi \frac{dK}{2\pi}\sqrt{E^2(K) + \Delta^2 U_0^2}, \tag{15.40}$$

where the integral comes from the filled sea. Minimizing with respect to Δ we obtain the relation

$$\Delta = \int_0^\pi \frac{dK}{\pi}\frac{U_0\Delta}{\sqrt{E^2(K) + \Delta^2 U_0^2}}. \tag{15.41}$$

Assuming $\Delta \neq 0$, we cancel it on both sides. For positive U_0, a non-trivial solution requires that

$$1 = U_0\int_0^\pi \frac{dK}{\pi}\frac{1}{\sqrt{E^2(K) + \Delta^2 U_0^2}}, \tag{15.42}$$

which is called the *gap equation*. On the left-hand side is the number 1, and on the right-hand side is something of order U_0. It appears that we will get a solution only above some minimum U_0. This is wrong. The integrand becomes very large near the Fermi point $K = K_F = \pi/2$, where $E(K)$ vanishes. Writing

$$E(K) = k, \tag{15.43}$$
$$k = |K| - K_F, \tag{15.44}$$

we approximate the gap equation as follows:

$$1 = U_0 \int_{-\Lambda}^{\Lambda} \frac{dk}{\pi} \frac{1}{\sqrt{k^2 + \Delta^2 U_0^2}} \simeq \frac{2U_0}{\pi} \ln \frac{2\Lambda}{\Delta \cdot U_0}, \tag{15.45}$$

where Λ, the upper cut-off on $|k|$, is not very important. What is important is that due to the logarithmic behavior of the integral near the origin in k, i.e., near the Fermi surface (or rather, Fermi point $K = K_F$), there will always be a solution to the gap equation given by

$$\Delta = \frac{2\Lambda}{U_0} e^{-\pi/2U_0}. \tag{15.46}$$

The logarithmic divergence is also reflected in the divergent susceptibility of the non-interacting system to a probe (or perturbation) at momentum π. (At second order in perturbation theory, the perturbation will link the ground state to states of arbitrarily low energy in which a particle just below the right (left) Fermi point is pushed to just above the left (right) Fermi point. The small energy denominators, summed over such states, will produce the logarithm.)

To see that the CDW state has lower energy than the Fermi sea we need to evaluate the expectation value of H in the two states. The CDW state will win because its energy goes as $E(\Delta) \simeq \Delta^2(a + b \ln \Delta)$, $a, b > 0$, and the $\ln \Delta$ ensures that $E(\Delta) < E(0)$ as $\Delta \to 0$.

Mean-field theory also predicts that the system will a have non-zero superconducting order parameter

$$\Delta = \langle \psi \psi \rangle \quad \text{and} \quad \Delta^* = \langle \psi^\dagger \psi^\dagger \rangle \tag{15.47}$$

for the smallest attractive coupling. (The two ψ's or $\psi^\dagger$'s are on adjacent sites in real space.) In the corresponding calculation, the instability will stem from the time-reversal symmetry of the problem: $E(K) = E(-K)$.

Unfortunately, both these predictions are wrong. The error comes from the neglected quartic operators like $:n_j::n_{j+1}:$. We know this because the Hamiltonian Eq. (15.21) was solved exactly by Yang and Yang [17]. The exact solution tells us that the system remains gapless for U_0 of either sign until it exceeds a minimum value of order unity, at which point a gap opens up. For $U_0 > 1$ positive, this is due to the CDW order alluded to. The state for $U_0 < 1$ will be described in Chapter 17.

We will now develop the RG approach to this problem and obtain results in harmony with this exact result.

15.3 The RG Approach for $d = 1$ Spinless Fermions

Our goal is to explore the stability of the non-interacting spinless fermions to weak interactions. We are not interested in the fate of just the model in Eq. (15.21) but in a whole family of models of which this will be a special case. Our strategy is to argue that at weak coupling, only modes within $\pm\Lambda$ of the Fermi points $\pm K_F$ (see Figure 15.2) will be relevant to the low-energy physics [3,4,9]. Thus, we will linearize the dispersion relation $E(K) = -\cos K$ near these points and work with a cut-off Λ:

$$H_0 = \sum_i \int_{-\Lambda}^{\Lambda} \frac{dk}{2\pi} \psi_i^\dagger(k)\psi_i(k)k, \tag{15.48}$$

where

$$k = |K| - K_F, \tag{15.49}$$

$$i = L,R \quad \text{(left or right)}. \tag{15.50}$$

Notice that H_0 is an integral over fermionic oscillators. The frequency Ω_0 of the oscillator at momentum k is simply k.

Pay attention to the way k is defined:

$$|K| = K_F + k. \tag{15.51}$$

Near the right Fermi point R, where K is positive, the relation is simply $K = K_F + k$. A positive (negative) k means a K larger than (smaller than) K_F. Near L, where $K < 0$, the relation is $K = -(K_F + k)$. In other words, $k > 0$ (or $k < 0$) means $|K| > K_F$ (or $|K| < K_F$). In terms of

$$\varepsilon_i = +1 \ (i = R), \quad -1 \ (i = L), \tag{15.52}$$

we may say that

$$\varepsilon_i(K_F + k) \textit{ is the actual momentum associated with } (k,i). \tag{15.53}$$

This way of defining k is aimed at higher dimensions, where its sign will tell us if we are above or below the Fermi surface. It is not ideal for $d = 1$, and we will abandon it later.

Next, we will write down a $T = 0$ partition function for the non-interacting fermions. This will be a Grassmann integral only over the degrees of freedom within the cut-off. We will then find an RG transformation that lowers the cut-off but leaves invariant S_0, the free-field action. With the RG so defined, we will look at the generic perturbations of this fixed point and classify them as usual. If no relevant operators show up, we will still have a scale-invariant gapless system. If, on the other hand, there are generic relevant perturbations, we will have to see to which new fixed point the system flows. (The new one could also be gapless.) The stability analysis can be done perturbatively. *In particular, if a relevant perturbation takes us away from the original fixed point, nothing at higher orders can ever bring us back to this fixed point.* The fate of the nearest-neighbor model will then be decided by asking if it had a relevant component in its interaction.

The partition function for our system of free fermions is

$$Z_0 = \int \prod_{i=L,R} \prod_{|k|<\Lambda} d\psi_i(\omega k) d\overline{\psi}_i(\omega k) e^{S_0}, \tag{15.54}$$

$$S_0 = \sum_{i=L,R} \int_{-\Lambda}^{\Lambda} \frac{dk}{2\pi} \int_{-\infty}^{\infty} \frac{d\omega}{2\pi} \overline{\psi}_i(\omega k)(i\omega - k)\psi_i(\omega k). \tag{15.55}$$

This is just a product of functional integrals for the Fermi oscillators at each momentum with energy $\Omega_0(k) = k$.

The first step in the RG transformation is to integrate out all $\psi(k\omega)$ and $\overline{\psi}(k\omega)$ with

$$\Lambda/s \le |k| \le \Lambda, \tag{15.56}$$

and *all* ω. Thus, our phase space has the shape of a rectangle, infinite in the ω direction, finite in the k direction. This shape will be preserved under the RG transformation. Since there is no real relativistic invariance here, we will make no attempt to treat ω and k on an equal footing. Allowing ω to take all values ensures locality in time and allows us to extract an effective Hamiltonian operator at any stage in the RG.

Since the integral is Gaussian, the result of integrating out fast modes is just a numerical prefactor, which we throw out. The surviving modes now have their momenta going from $-\Lambda/s$ to Λ/s. To make this action a fixed point, we define rescaled variables:

$$\begin{aligned} k' &= sk, \\ \omega' &= s\omega, \\ \psi_i'(k'\omega') &= s^{-3/2}\psi_i(k\omega). \end{aligned} \tag{15.57}$$

Ignoring a constant that comes from rewriting the measure in terms of the new fields, we see that S_0 is invariant under the mode elimination and rescaling operations.

We can now consider the effect of generic perturbations on this fixed point.

15.3.1 Quadratic Perturbations

First consider perturbations that are quadratic in the fields. These must necessarily be of the form

$$\delta S_2 = \sum_{i=L,R} \int_{-\Lambda}^{\Lambda} \frac{dk}{2\pi} \int_{-\infty}^{\infty} \frac{d\omega}{2\pi} \mu(k\omega)\overline{\psi}_i(\omega k)\psi_i(\omega k), \tag{15.58}$$

assuming symmetry between left and right Fermi points.

Since this action separates into slow and fast pieces, the effect of mode elimination is simply to reduce Λ to Λ/s in the integral above. Rescaling moments and fields, we find that

$$\mu'(\omega',k',i) = s \cdot \mu\left(\frac{\omega'}{s},\frac{k'}{s},i\right). \tag{15.59}$$

We get this factor s by combining the s^{-2} from rewriting the old momenta and frequencies in terms of the new and the s^3 from rewriting the old fields in terms of the new.

Let us expand μ in a Taylor series,

$$\mu(k,\omega) = \mu_{00} + \mu_{10}k + \mu_{01}i\omega + \cdots + \mu_{nm}k^n(i\omega)^m + \cdots, \tag{15.60}$$

and compare both sides. The constant piece is a relevant perturbation:

$$\mu_{00} \equiv \mu \longrightarrow s\mu \equiv s\mu_{00}. \tag{15.61}$$

This relevant flow reflects the readjustment of the Fermi sea to a change in chemical potential. The correct way to deal with this term is to include it in the free-field action by filling the Fermi sea to a point that takes μ_{00} into account. The next two terms are marginal and modify terms that are already present in the action. When we consider quartic interactions, it will be seen that mode elimination will produce terms of the above form even if they were not there to begin with, just as in ϕ^4 theory. The way to deal with them will be discussed in due course. Higher-order terms in Eq. (15.60) are irrelevant under the RG. This is, however, a statement that is correct at the free-field fixed point. We shall have occasion to discuss a term that is irrelevant at weak coupling but gets promoted to relevance as the interaction strength grows.

15.3.2 Quartic Perturbations: The RG at Tree Level

We now turn on the quartic interaction, whose most general form is

$$\delta S_4 = \frac{1}{2!2!}\int_{K\omega} \overline{\psi}(4)\overline{\psi}(3)\psi(2)\psi(1)u(4,3,2,1), \tag{15.62}$$

where

$$\overline{\psi}(i) = \overline{\psi}(K_i,\omega_i), \text{ etc.},$$

$$\int_{K\omega} = \left[\prod_{i=1}^{4}\int_{-\pi}^{\pi}\frac{dK_i}{2\pi}\int_{-\infty}^{\infty}\frac{d\omega_i}{2\pi}\right]\left[2\pi\overline{\delta}(K_1+K_2-K_3-K_4)2\pi\delta(\omega_1+\omega_2-\omega_3-\omega_4)\right], \tag{15.63}$$

and $\overline{\delta}$ enforces momentum conservation mod 2π, as is appropriate to any lattice problem. A process where lattice momentum is violated in multiples of 2π is called an *umklapp* process. The delta function containing frequencies enforces time translation invariance. The coupling function u is antisymmetric under the exchange of its first or last two arguments among themselves since that is true of the Grassmann fields that it multiplies.

Thus, the coupling u has all the symmetries of the full vertex function Γ with four external lines.

Let us now return to the general interaction, Eqs. (15.62) and (15.63), and restrict the momenta to lie within Λ of either Fermi point L or R. In the notation where L (left Fermi point) and R (right Fermi point) become a discrete label ($i = L$ or R) and 1–4 label the frequencies and momenta (measured from the appropriate Fermi points), Eqs. (15.62) and (15.63) become

$$\delta S_4 = \frac{1}{2!2!} \sum_{i_1 i_2 i_3 i_4 = L,R} \int_{K\omega}^{\Lambda} \overline{\psi}_{i_4}(4)\overline{\psi}_{i_3}(3)\psi_{i_2}(2)\psi_{i_1}(1) u_{i_4 i_3 i_2 i_1}(4,3,2,1), \tag{15.64}$$

where

$$\int_{K\omega}^{\Lambda} = \left[\int_{-\Lambda}^{\Lambda} \frac{dk_1 \cdots dk_4}{(2\pi)^4} \int_{-\infty}^{\infty} \frac{d\omega_1 \cdots d\omega_4}{(2\pi)^4}\right] [2\pi\delta(\omega_1 + \omega_2 - \omega_3 - \omega_4)]$$
$$\times \left[2\pi\overline{\delta}(\varepsilon_{i_1}(K_F + k_1) + \varepsilon_{i_2}(K_F + k_2) - \varepsilon_{i_3}(K_F + k_3) - \varepsilon_{i_4}(K_F + k_4))\right] \tag{15.65}$$

and

$$\varepsilon_i = \pm 1 \text{ for } R,\ L, \tag{15.66}$$

as explained in Eq. (15.53).

Let us now implement the RG transformation and do the order-u tree-level calculation for the renormalization of the quartic interaction. This gives us just Eq. (15.64) with $\Lambda \to \Lambda/s$. If we now rewrite this in terms of new momenta and fields, we get an interaction with the same kinematical limits as before, and we can meaningfully read off the coefficient of the quartic Fermi operators as the new coupling function. We find that

$$u'_{i_4 i_3 i_2 i_1}(k'_i, \omega'_i) = u_{i_4 i_3 i_2 i_1}(k'_i/s, \omega'_i/s). \tag{15.67}$$

Key to this analysis is the fact that K_F drops out of the momentum-conserving δ function in Eq. (15.65): either all the K_F's cancel in the non-umklapp cases, or get swallowed up in multiples of 2π (in inverse lattice units) in the umklapp cases due to the periodicity of the $\overline{\delta}$-function. As a result, the momentum δ functions are free of K_F and scale very nicely under the RG transformation:

$$\overline{\delta}(k) \to \overline{\delta}(k'/s) \tag{15.68}$$
$$= s\overline{\delta}(k'). \tag{15.69}$$

The expression for the renormalized δS_4 preserves its form when expressed in terms of the new field, k' and ω', which then allows us to show Eq. (15.67).

Another route to Eq. (15.67) is to first eliminate k_4 and ω_4 using the δ-functions and following through with the RG.

Turning now to Eq. (15.67), if we expand u in a Taylor series in its arguments and compare coefficients, we find that the constant term u_0 is marginal and the higher coefficients are irrelevant. Thus, u depends only on its discrete labels and we can limit the problem to just a few *coupling constants* instead of the coupling function we started with. Furthermore, all reduce to just one coupling constant:

$$u_0 = u_{LRLR} = u_{RLRL} = -u_{RLLR} = -u_{LRRL} \equiv u. \tag{15.70}$$

Other couplings corresponding to $LL \to RR$ are wiped out by the Pauli principle since they have no momentum dependence and cannot have the desired antisymmetry.

As a concrete example, consider the u that comes from the nearest-neighbor interaction in our original model Eq. (15.33),

$$\frac{u(4,3,2,1)}{2!2!} = U_0 \sin\left(\frac{K_1 - K_2}{2}\right) \sin\left(\frac{K_3 - K_4}{2}\right) \cos\left(\frac{K_1 + K_2 - K_3 - K_4}{2}\right),$$

and ask what sorts of couplings are contained in it.

If 1 and 2 are both from R, we find the following factor in the coupling:

$$\begin{aligned} \sin\left(\frac{K_1 - K_2}{2}\right) &= \sin\left(\frac{k_1 - k_2}{2}\right) \\ &\simeq \left(\frac{k_1 - k_2}{2}\right), \end{aligned} \tag{15.71}$$

which leads to the requisite antisymmetry but makes the coupling irrelevant (due to the k's). There will be one more power of k from 3 and 4, which must also come from near just one Fermi point so as to conserve momentum modulo 2π. For example, the umklapp process, in which $RR \leftrightarrow LL$, has a coupling

$$u(\text{umklapp}) \simeq (k_1 - k_2)(k_3 - k_4) \tag{15.72}$$

and is strongly irrelevant at the free-field fixed point.

On the other hand, if 1 and 2 come from opposite sides,

$$\sin\left(\frac{K_1 - K_2}{2}\right) \simeq \sin\left(\frac{\pi}{2} + O(k)\right), \tag{15.73}$$

and likewise for 3 and 4, we have a marginal interaction u_0 with no k's in the coupling. (In all cases, the cosine is of order 1.)

The tree-level analysis readily extends to couplings with six or more fields. All these are irrelevant, even if we limit ourselves to couplings independent of ω and k.

To determine the ultimate fate of the coupling $u_0 \equiv u$, marginal at tree level, we must turn to the RG at one loop.

15.3.3 RG at One Loop: The Luttinger Liquid

Let us begin with the action with the quartic interaction and do the mode elimination.

For the cut-off reduction

$$\Lambda \rightarrow \frac{\Lambda}{s} = \Lambda e^{-dt} \simeq \Lambda(1 - dt), \tag{15.74}$$

we must remember that $-\Lambda \leq k \leq \Lambda$, i.e., the allowed momenta lie above and below K_F, a possibility we have only with fermions because the cut-off is measured from the Fermi surface and not the origin. *We renormalize toward the Fermi surface and not the origin.* So there are four slivers we must chop off, as shown in Figure 15.3, and labeled $a, \ldots, d$.

To order u, this mode elimination leads to an induced quadratic term, represented by the tadpole graph in Figure 15.4. We have set $\omega = k = 0$ for the external legs and chosen them to lie at L, the left Fermi point. The loop momentum should be from the vicinity of the right Fermi point R, since u couples only opposite Fermi points. We have the following relation for the renormalization of the chemical potential under the infinitesimal scale transformation:

$$s = 1 + dt, \tag{15.75}$$

$$\mu' = (1 + dt)\left[\mu - u \int_{-\infty}^{\infty} \frac{d\omega}{2\pi} \int_{\Lambda(1-dt)<|k|<\Lambda} \frac{dk}{2\pi} e^{i\omega 0^+} \frac{1}{i\omega - k}\right]. \tag{15.76}$$

We do the ω integral by closing the contour in the upper half-plane, where $e^{i\omega 0^+}$ ensures convergence. We get a non-zero contribution from the pole at $\omega = -ik$ provided $k < 0$, which just means the incoming fermion interacts with fermions filling the sea. We take the contribution from the slice labeled c in Figure 15.3. The flow is

$$\frac{d\mu}{dt} = \mu - \frac{u}{2\pi}, \tag{15.77}$$

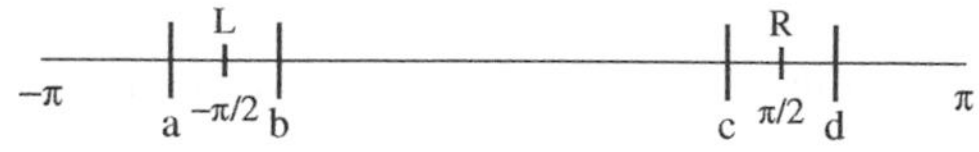

Figure 15.3 The slices of width $d\Lambda$ which are integrated above and below each Fermi point.

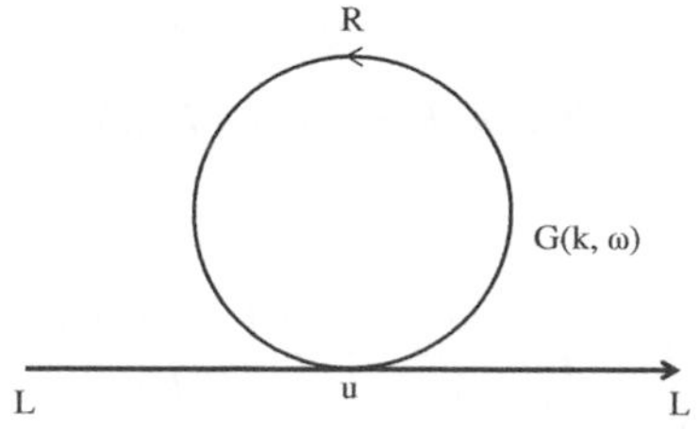

Figure 15.4 The tadpole graph which shifts the chemical potential.

assuming that we choose to measure μ in units of Λ. (Since we rescale momentum to keep Λ fixed, we can choose $\Lambda = 1$.) The fixed point of this equation is

$$\mu^* = \frac{u}{2\pi} = \frac{u^*}{2\pi}, \tag{15.78}$$

where I have set $u = u^*$, the fixed point value, since u is fixed (marginal) at tree level.

Bear in mind that we must fine-tune the chemical potential as a function of u, not to maintain criticality (as one does in ϕ^4 theory, where the bare mass is varied with the interaction to keep the system massless) but to retain the same particle density. To be precise, we are keeping fixed K_F, the momentum at which the one-particle Green's function (for real time and frequency) has its singularity. This amounts to keeping the density fixed [18, 19]. If we kept μ at the old value of zero, the system would flow away from the fixed point with $K_F = \frac{\pi}{2}$, not to a state with a gap, but to another gapless one with a smaller value of K_F. This simply corresponds to the fact that if the total energy of the lowest-energy particle that can be added to the system, namely μ, is to equal 0, the kinetic energy at the Fermi surface must be slightly negative so that the repulsive potential energy with the others in the sea brings the total to zero.

Let us now turn our attention to the order u^2 graphs that renormalize u. I label the diagrams ZS (zero-sound), ZS$'$, and BCS (for historical reasons) just so that I can refer to them individually. Unlike in ϕ^4 theory, the three diagrams make very different contributions even when all external momenta and frequencies vanish. Readers familiar with Feynman diagrams may obtain the final formula for du by drawing all the diagrams to this order in the usual Feynman graph expansion, but allowing the loop momenta to range only over the modes being eliminated.

In the present case, these modes are given by the four thick lines labeled a, b, c, and d in Figure 15.3. Each line stands for a region of width $d\Lambda$ located at the cut-off, i.e., a distance Λ from the Fermi points. The external momenta are chosen to be at the Fermi surface. All the external k's and ω's are set equal to zero since the marginal coupling u has no dependence on these. This has two consequences. First, the loop frequencies in the ZS and ZS$'$ graphs are equal, while those in the BCS graph are equal and opposite. Second, the momentum transfers at the left vertex are $Q = K_1 - K_3 = 0$ in the ZS graph, $Q' = K_1 - K_4 = \pi$ in the ZS$'$ graph, while the total momentum in the BCS graph is $P = K_1 + K_2 = 0$.

Using

$$E(-K) = E(K), \tag{15.79}$$

$$E(K' = K \pm \pi) = -E(K) \tag{15.80}$$

leads to

$$du(LRLR) = \int_{-\infty}^{\infty} \int_{d\Lambda} \frac{d\omega dK}{4\pi^2} \frac{u(KRKR)u(LKLK)}{(i\omega - E(K))(i\omega - E(K))}$$
$$- \int_{-\infty}^{\infty} \int_{d\Lambda} \frac{d\omega dK}{4\pi^2} \frac{u(K'LKR)u(RKLK')}{(i\omega - E(K))(i\omega + E(K))}$$

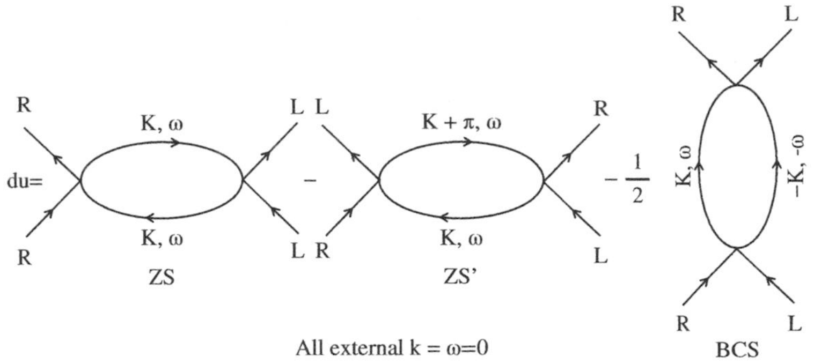

Figure 15.5 The three diagrams that renormalize the coupling u. All external lines are at the Fermi points, i.e. $k = 0$, and all external $\omega = 0$.

$$
\begin{aligned}
&-\frac{1}{2}\int_{-\infty}^{\infty}\int_{d\Lambda}\frac{d\omega dK}{4\pi^2}\frac{u(-KKLR)u(LR-KK)}{(i\omega - E(K))(-i\omega - E(K))}\\
&\equiv \mathrm{ZS} + \mathrm{ZS}' + \mathrm{BCS}, \qquad (15.81)
\end{aligned}
$$

where $\int_{d\Lambda}$ means the momentum must lie in one of the four slices in Figure 15.3.

In the ZS graph, the loop momentum K lies within a sliver of width $d\Lambda$ at the cut-off. Both propagators have poles at the same point, either $\omega = -iE(k = \Lambda)$ or $\omega = -iE(k = -\Lambda)$, because the external momentum and frequency transfer at the left and right vertices are zero. No matter which half-plane these poles lie in, we can close the contour the other way and the ω integral will vanish. So the ZS graph makes no contribution to the flow. The ZS also vanishes for another reason: since u only scatters particles from opposite Fermi points, whether K is from near L or R, there will be an RR or LL scattering at one or the other vertex. (In the first term on the right-hand side of Eq. (15.81), one factor in $u(KRKR)u(LKLK)$ will vanish whether we choose $K = R$ or $K = L$.)

Now for the ZS′ graph, Figure 15.5, Eq. (15.81). We see that K must lie near L since there is no RR scattering. As far as the coupling at the left vertex is concerned, we may set $K = L$ since the marginal coupling has no k dependence. Thus, $K + \pi = R$ and the vertex becomes $u(RLLR) = -u$. So does the coupling at the other vertex.

Look at the propagators. If K comes from the sliver a in Figure 15.3, which is a particle, $K + \pi$ is the sliver c, which is a hole. Likewise, if K is the sliver b, a hole, $K + \pi$ is d, a particle. Doing the ω integral (which is now non-zero since the poles are always on opposite half-planes), we obtain, upon using the fact that there are two shells (a and b in Figure 15.3) near L and that $|E(K)| = |k| = |\Lambda|$,

$$
\begin{aligned}
\mathrm{ZS}' &= u^2\int_{d\Lambda\in L}\frac{dK}{4\pi|E(K)|}\\
&= \frac{u^2}{2\pi}\frac{d|\Lambda|}{\Lambda}. \qquad (15.82)
\end{aligned}
$$

The BCS graph, Eq. (15.81), gives a non-zero contribution since the propagators have opposite frequencies, opposite momenta, but equal energies due to time-reversal invariance $E(K) = E(-K)$. Both slivers being integrated are either holes (b and c) or particles (a and d). Since the ω's are opposite in the two propagators, the poles lie on opposite half-planes again.

We notice that the factor of $\frac{1}{2}$ is offset by the fact that K can now lie in any of the four regions a, b, c, or d in Figure 15.3. We obtain a contribution of the same magnitude but opposite sign as ZS$'$, so that

$$du = \left(\frac{u^2}{2\pi} - \frac{u^2}{2\pi}\right)\underbrace{\frac{d|\Lambda|}{\Lambda}}_{dt}, \tag{15.83}$$

$$\frac{du}{dt} = \beta(u) = 0. \tag{15.84}$$

Thus, we find that u is *still* marginal. The flow to one loop for μ and u is

$$\frac{d\mu}{dt} = \mu - \frac{u}{2\pi}, \tag{15.85}$$

$$\frac{du}{dt} = 0. \tag{15.86}$$

There is a line of fixed points:

$$\mu = \frac{u^*}{2\pi}, \tag{15.87}$$

$$u^* \quad \text{arbitrary}. \tag{15.88}$$

Notice that β vanishes due to a cancellation between two diagrams, each of which by itself would have led to the CDW or BCS instability. When one does a mean-field calculation for CDW, one focuses on just the ZS$'$ diagram and ignores the BCS diagram. This amounts to taking

$$\frac{du}{dt} = \frac{u^2}{2\pi}, \tag{15.89}$$

which, if correct, would imply that any positive u grows under renormalization. If this growth continues, we expect a CDW. On the other hand, if just the BCS diagram is kept we will conclude a run-off for negative couplings leading to a state with $\langle\psi_R\psi_L\rangle \neq 0$.

What the β function does is to treat these competing instabilities simultaneously and predict a scale-invariant theory.

Is this the correct prediction for the spinless model? Yes, up to some value of u of order unity, where the CDW gap opens up according to the exact solution of Yang and Yang [17].

But if the RG analysis were extended to higher loops, we would keep getting $\beta = 0$ to all orders. This was shown by De Castro and Metzner [20] using the Ward identity in the cut-off continuum model, which reflects the fact that in this model, the number of fermions of type L and R are separately conserved.

How then do we ever reproduce the eventual CDW instability known to exist in the exact solution at strong coupling? The answer is as follows. As we move along the line of fixed points, labeled by u, the dimension of various operators will change from the free-field values. Ultimately, the umklapp coupling, $RR \leftrightarrow LL$, which was suppressed by a factor $(k_1 - k_2)(k_3 - k_4)$, will become marginal and then relevant, and produce the CDW gap. If we were not at half-filling, such a term would be ruled out for all u by momentum conservation and lead to the scale-invariant state, called the *Luttinger liquid* by Haldane [21], who developed the earlier work of Luttinger [18, 19].

As long as $\beta = 0$, we have the scale-invariant Luttinger liquid. The critical exponents vary with u. For example, G falls with a variable power in real space and momentum space. It has no quasiparticle pole, just a cut. We will be able to derive many of these results once we learn about bosonization in Chapters 17 and 18.

While this liquid provides us with an example where the RG does better than mean-field theory, it is rather special and seems to occur in $d = 1$ systems where the two Fermi points satisfy the conditions for *both* CDW and BCS instabilities. In higher dimensions we will find that any instability due to one divergent susceptibility is not precisely canceled by another.

References and Further Reading

[1] P. W. Anderson, Science, **235**, 1196 (1987).
[2] G. Baskaran, Z. Zou, and P. W. Anderson, Solid State Communications, **63**, 973 (1987).
[3] J. Solyom, Advances in Physics, **28**, 201 (1979).
[4] C. Bourbonnais and L. G. Caron, International Journal of Modern Physics B, **5**, 1033 (1991).
[5] L. D. Landau, Journal of Experimental and Theoretical Physics, **3**, 920 (1956).
[6] L. D. Landau, Journal of Experimental and Theoretical Physics, **8**, 70 (1957).
[7] L. D. Landau, Journal of Experimental and Theoretical Physics, **5**, 101, (1959).
[8] A. A. Abrikosov, L. P. Gorkov, and I. E. Dzyaloshinski, *Methods of Quantum Field Theory in Statistical Mechanics*, Dover (1963). Gives a diagrammatic proof of Landau theory.
[9] R. Shankar, Reviews of Modern Physics, **66**, 129 (1994).
[10] P. Coleman, *Introduction to Many-Body Physics*, Cambridge University Press (2015).
[11] S. Simon, *The Oxford Solid State Basics*, Oxford University Press (2013).
[12] M. L. Cohen and S. G. Louie, *Fundamentals of Condensed Matter Physics*, Cambridge University Press (2016).
[13] G. Giuliani and G. Vignale, *Quantum Theory of the Electron Liquid*, Cambridge University Press (2008).
[14] G. Baym and C. Pethik, *Landau Fermi Liquid Theory*, Wiley (1991).
[15] G. D. Mahan, *Many-Body Physics*, Plenum (1981).
[16] P. Nozieres, *Interacting Fermi Systems*, Benjamin (1964).

[17] C. N. Yang and C. P. Yang, Physical Review **150**, 321 (1976).
[18] J. M. Luttinger, Physical Review **119**, 1153 (1960).
[19] D. C. Mattis and E. H. Lieb, Journal of Mathematical Physics, **6**, 304 (1965). The model posed by Luttinger was correctly interpreted and solved in this paper.
[20] C. De Castro and W. Metzner, Physical Review Letters **67**, 3852 (1991).
[21] F. D. M. Haldane, Journal of Physics C, **14**, 2585 (1981).

16

Renormalization Group for Non-Relativistic Fermions: II

Now we turn to the central problem of fermions in $d = 2$. The treatment will follow my review [1]. A selected list of contemporaneous works is given in [2–7].

16.1 Fermions in $d = 2$

The low-energy manifold is shown in Figure 16.1. It is an annulus of thickness 2Λ symmetrically situated with respect to the Fermi circle $K = K_F$.

The Hamiltonian within this thin annulus has already been encountered in Eq. (15.10):

$$H = \int \psi^\dagger(k,\theta)\psi(k,\theta)k\frac{dkd\theta}{(2\pi)^2}. \tag{16.1}$$

(A factor of K_F has been absorbed in the field.) It differs from the answer in $d = 1$,

$$H_0 = \sum_{i=L,R}\int_{-\Lambda}^{\Lambda}\frac{dk}{2\pi}\psi_i^\dagger(k)\psi_i(k)k, \tag{16.2}$$

in the replacement of the discrete Fermi point labels L and R by an angle θ going around the annulus. The partition function follows:

$$Z_0 = \int\prod_\theta\prod_{|k|<\Lambda} d\psi(\omega k\theta)d\overline{\psi}(\omega k\theta)e^{S_0}, \tag{16.3}$$

$$S_0 = \int\frac{d\theta}{2\pi}\int_{-\Lambda}^{\Lambda}\frac{dk}{2\pi}\int_{-\infty}^{\infty}\frac{d\omega}{2\pi}\overline{\psi}(\omega k\theta)(i\omega - k)\psi(\omega k\theta). \tag{16.4}$$

Since θ is an "isospin" label like R or L, in that the energy is determined only by k, the scaling that preserves the action is the same as in $d = 1$ but for the switch $(R,L) \to \theta$:

$$k' = sk,$$

$$\omega' = s\omega,$$

$$\psi'(k'\omega'\theta) = s^{-3/2}\psi(k\omega\theta). \tag{16.5}$$

The quadratic term behaves exactly as in $d = 1$, and the relevant part in it is the chemical potential μ, which must be varied with the interaction to keep the Fermi surface fixed at K_F.

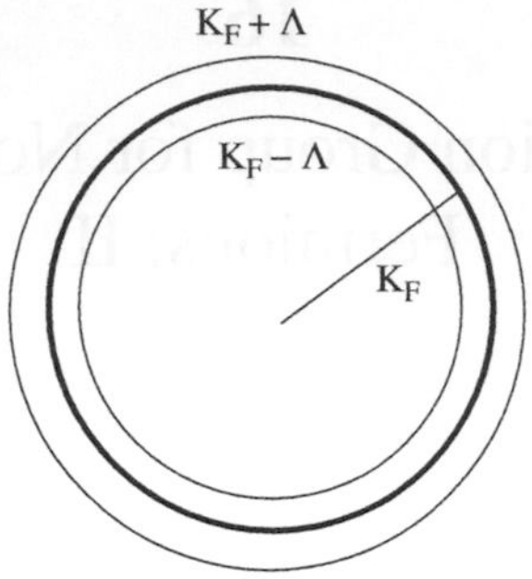

Figure 16.1 The region of low-energy physics lies within the annulus of thickness 2Λ straddling the Fermi circle $K = K_{\rm F}$ (dark line).

Let us then move on to the more interesting quartic interaction. This has the general form

$$\delta S_4 = \frac{1}{2!2!}\int_{K\omega\theta} \bar{\psi}(4)\bar{\psi}(3)\psi(2)\psi(1)u(4,3,2,1), \tag{16.6}$$

where

$$\bar{\psi}(i) = \bar{\psi}(K_i,\omega_i,\theta_i), \text{ etc.}, \tag{16.7}$$

$$\int_{K\omega\theta} = \left[\prod_{i=1}^{3}\int_0^{2\pi}\frac{d\theta_i}{2\pi}\int_{-\Lambda}^{\Lambda}\frac{dk_i}{2\pi}\int_{-\infty}^{\infty}\frac{d\omega_i}{2\pi}\right]\theta(\Lambda - |k_4|), \tag{16.8}$$

$$k_4 = |\boldsymbol{K}_4| - K_{\rm F}. \tag{16.9}$$

We have eliminated ω_4 and $\boldsymbol{K}_4$ using the conservation laws imposed by the δ-functions. The ω-integral is easy: since all ω's are allowed, the condition $\omega_4 = \omega_1 + \omega_2 - \omega_3$ is always satisfied for any choice of the first three frequencies. The same would be true for the momenta if all momenta in the plane were allowed. But they are not: they are required to lie within the annulus of thickness 2Λ around the Fermi circle. But if one freely chooses the first three momenta from the annulus, the fourth could have a length as large as $3K_{\rm F}$. The role of $\theta(\Lambda - |k_4|)$ in Eq. (16.8) is to prevent this.

We do not bother with $\theta(\Lambda - |k_4|)$ when we analyze the ϕ^4 theory because the fourth momentum can at most be as big as 3Λ if the other three come from a ball of radius Λ. A similar thing happened in the $d = 1$ example, even though there was an additional large momentum $K_{\rm F}$: it either canceled out or was dropped modulo 2π. In fact, if we wanted, we could could have easily kept the $\theta(\Lambda - |k_4|)$ factor in the analysis because it preserves its form under the RG:

$$\theta(\Lambda - |k_4|) \to \theta\left(\frac{\Lambda}{s} - |k_4|\right) = \theta(\Lambda - s|k_4|) = \theta\left(\Lambda - |k_4'|\right), \tag{16.10}$$

which in turn allows us to read off u' in terms of u.

In the present case, the role of K_F is not benign. The problem is that k_4 is not a function of just the other three little k's but also of K_F:

$$k_4 = |(K_F + k_1)\mathbf{\Omega}_1 + (K_F + k_2)\mathbf{\Omega}_2 - (K_F + k_3)\mathbf{\Omega}_3| - K_F, \tag{16.11}$$

where $\mathbf{\Omega}_i$ is a unit vector in the direction of $\boldsymbol{K}_i$:

$$\mathbf{\Omega}_i = \boldsymbol{i}\cos\theta_i + \boldsymbol{j}\sin\theta_i, \tag{16.12}$$

where θ_i is the orientation of $\boldsymbol{K}_i$. (In $d=2$, we will use $\mathbf{\Omega}_i$ and θ_i interchangeably.)

It is now easy to check that if we carry out the manipulation that led to Eq. (16.10) we will find:

$$\theta(\Lambda - |k_4(k_1, k_2, k_3, K_F)|) \longrightarrow \theta(\Lambda - |k_4'(k_1', k_2', k_3', sK_F)|). \tag{16.13}$$

Thus, the θ function after the RG transformation is not the same function of the new variables as the θ function before the RG transformation, because $K_F \to sK_F$. How are we to say what the new coupling is if the integration measure does not come back to its old form?

The best way to handle this [1] is to replace the hard cut-off $\theta(\Lambda - |k_4|)$ by a soft cut-off that exponentially suppresses the kinematical region that violates this condition:

$$\theta(\Lambda - |k_4|) \longrightarrow e^{-|k_4|/\Lambda}. \tag{16.14}$$

Now, couplings with $k_4 \gg \Lambda$ are not disallowed but suppressed as we renormalize. We may interpret the exponential decay as part of the renormalization of u. (Nothing is gained by using a soft cut-off for the other k's.)

Here are the details of the soft cut-off. Recall that

$$k_4 = |K_F \underbrace{(\mathbf{\Omega}_1 + \mathbf{\Omega}_2 - \mathbf{\Omega}_3)}_{\mathbf{\Delta}} + k_1\mathbf{\Omega}_1 + k_2\mathbf{\Omega}_2 - k_3\mathbf{\Omega}_3| - K_F, \tag{16.15}$$

where $\mathbf{\Omega}_i$ is a unit vector in the direction of $\boldsymbol{K}_i$. In what follows, we shall keep just the $\mathbf{\Delta}$ piece and ignore the $O(k)$ terms inside the absolute value. This is because the only time the $O(k)$ terms are comparable to the $\mathbf{\Delta}$ piece is when both are of order Λ, in which case $k_4 \simeq K_F$, and this region is exponentially suppressed by the smooth cut-off Eq. (16.14) anyway.

It is easily shown that [1]

$$u'(k'\omega'\theta) = e^{-((s-1)K_F/\Lambda)||\Delta|-1|} u\left(\frac{k'}{s}\ \frac{\omega'}{s}\ \theta\right). \tag{16.16}$$

We may conclude that the only couplings that survive the exponential suppression without any decay as $\Lambda \to 0$ correspond to the cases where

$$|\mathbf{\Delta}| = |\mathbf{\Omega}_1 + \mathbf{\Omega}_2 - \mathbf{\Omega}_3| = 1. \tag{16.17}$$

In $d = 2$, this equation has only three solutions:

$$\text{Case I}: \; \boldsymbol{\Omega}_3 = \boldsymbol{\Omega}_1 \;\; (\text{hence} \;\; \boldsymbol{\Omega}_2 = \boldsymbol{\Omega}_4); \tag{16.18}$$

$$\text{Case II}: \; \boldsymbol{\Omega}_3 = \boldsymbol{\Omega}_2 \;\; (\text{hence} \;\; \boldsymbol{\Omega}_1 = \boldsymbol{\Omega}_4); \tag{16.19}$$

$$\text{Case III}: \; \boldsymbol{\Omega}_1 = -\boldsymbol{\Omega}_2 \;\; (\text{hence} \;\; \boldsymbol{\Omega}_3 = -\boldsymbol{\Omega}_4). \tag{16.20}$$

We can understand these results as follows. Imagine Λ has been reduced to a negligible value, equal to the thickness of the circle in the middle third of Figure 16.2. It is clear that if two points 1 and 2 are chosen from the circle to represent the incoming lines in a four-point coupling, the outgoing ones, also chosen from this circle, are forced to be equal to them *not just in their sum, but individually* (up to a permutation, which is irrelevant for spinless fermions). This is the content of Eqs. (16.18) and (16.19).

There is just one exception, shown in the right third of Figure 16.2. If the incoming momenta are equal and opposite, so must be the outgoing ones, though the outgoing pair is free to move back-to-back in any direction, independent of the incoming pair. This is called the *Cooper channel* of BCS theory.

Of course, at any stage, Λ is small but not zero and the initial pair only fixes the final pair to within an angular uncertainty of order Λ/K_F. *There is room for non-forward scattering in this theory at every stage.* (This wiggle room will prove very important in computing the response functions like compressibility in the cut-off theory.) The situation is depicted in Figure 16.3. We see there two annuli whose centers are separated by $\boldsymbol{K} = \boldsymbol{K}_1 + \boldsymbol{K}_2$, the total incoming momentum. This must equal the sum of the outgoing momenta $\boldsymbol{K}_3 + \boldsymbol{K}_4$, each of which must also lie within the annuli bounded by $K_F \pm \Lambda$. This singles out the two intersection regions I and II for the terminal points of $\boldsymbol{K}_3$ and $\boldsymbol{K}_4$. Consequently, the angles of $\boldsymbol{K}_3$ and $\boldsymbol{K}_4$ are required to lie within $\simeq \Lambda/K_F$ of $\boldsymbol{K}_1$ and $\boldsymbol{K}_2$ (up to a permutation).

Exercise 16.1.1 (Quasiparticle lifetime) *Consider the phase space for the decay of the quasiparticle of momentum $\boldsymbol{K}_{qp}$ and energy ε_{qp} to a hole of momentum $\boldsymbol{K}_h$ and two particles of momentum $\boldsymbol{K}_1$ and $\boldsymbol{K}_2$ as indicated in Figure 15.1. We will now prove the assertion made*

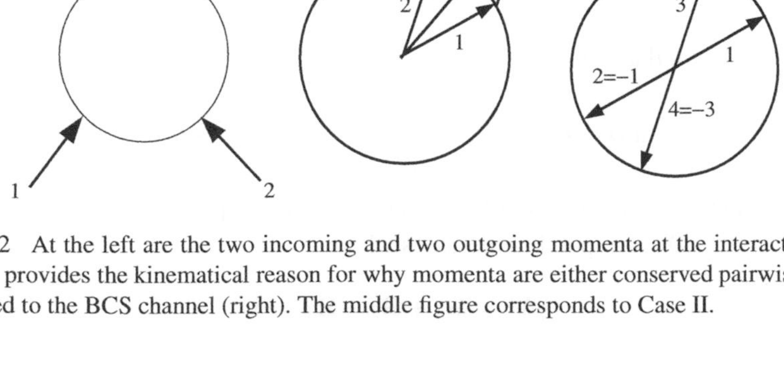

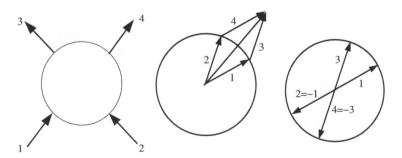

Figure 16.2 At the left are the two incoming and two outgoing momenta at the interaction vertex. The figure provides the kinematical reason for why momenta are either conserved pairwise (middle) or restricted to the BCS channel (right). The middle figure corresponds to Case II.

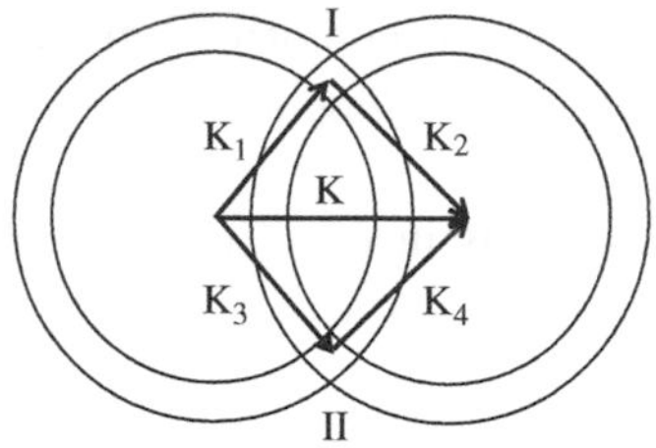

Figure 16.3 The requirement that the momenta come from the annulus and must be conserved forces the final directions of $\boldsymbol{K}_3$ and $\boldsymbol{K}_4$ to lie within Λ/K_F of $\boldsymbol{K}_1$ and $\boldsymbol{K}_2$ (up to a permutation). The two permutations correspond to the two allowed diamond-shaped intersection regions I and II. The situation shown in II corresponds to $\boldsymbol{\Omega}_3 = \boldsymbol{\Omega}_2$ and $\boldsymbol{\Omega}_1 = \boldsymbol{\Omega}_4$.

earlier that the phase space for decay goes as $\varepsilon_{\rm qp}^2$. The constraints are

$$\boldsymbol{K}_{\rm qp} + \boldsymbol{K}_{\rm h} = \boldsymbol{K}_1 + \boldsymbol{K}_2, \tag{16.21}$$

$$\varepsilon_{\rm qp} = \varepsilon_{\rm h} + \varepsilon_1 + \varepsilon_2, \tag{16.22}$$

where $\varepsilon_{\rm h}$ is measured as positive going into the Fermi sea (see Figure 15.1). In Figure 16.3, let the distance between the centers of the circles be $\boldsymbol{K}_{\rm qp} + \boldsymbol{K}_{\rm h}$. The energies ε_1 and ε_2 associated with $\boldsymbol{K}_1$ and $\boldsymbol{K}_2$ are bounded by $\varepsilon_{\rm qp}$ since $\varepsilon_{\rm h} > 0$. So, imagine two annuli of thickness $\varepsilon_{\rm qp}$ above the Fermi surface. Their intersection region (which is approximately a square of sides $\varepsilon_{\rm qp}$) limits the phase space for $\boldsymbol{K}_1$ and $\boldsymbol{K}_2$. It evidently has an area proportional to $\varepsilon_{\rm qp}^2$, which is the desired result.

To be more precise, describe the line in this square of size $\varepsilon_{\rm qp}$ (along whose sides we measure ε_1 and ε_2) on which the allowed values of ε_1 and ε_2 lie for a given value of $\varepsilon_{\rm h}$. (The quasiparticle energy is fixed at $\varepsilon_{\rm qp}$.) Show that the integral over all values of $\varepsilon_{\rm h}$ (which ranges from 0 to $\varepsilon_{\rm qp}$) is roughly half the area of the square.

16.2 Tree-Level Analysis of *u*

Now return to Eq. (16.16) for the quartic coupling. For angles obeying Eqs. (16.18) and (16.19), we can forget the exponential suppression factor due to $\theta(k_4 - \Lambda)$ and conclude that

$$u'(k,\omega,\theta) = u\left(\frac{k'}{s}\ \frac{\omega'}{s}\ \theta\right). \tag{16.23}$$

This is once again the evolution of the *coupling function*. As in $d = 1$, we expand the function in a Taylor series (schematic):

$$u = u_0 + ku_1 + k^2 u_2 + \cdots, \tag{16.24}$$

where k stands for all the k's and ω's. An expansion of this kind is possible since couplings in the Lagrangian are non-singular in a problem with short-range interactions.

If we now make such an expansion and compare coefficients, we find that u_0 *is marginal and the rest are irrelevant, as is any coupling of more than four fields. We can ignore the dependence of u on k and* ω.

Now, this is exactly what happens in ϕ_4^4. The difference here is that we still have dependence on the angles on the Fermi surface:

$$u_0 = u(\theta_1, \theta_2, \theta_3, \theta_4). \tag{16.25}$$

Unlike in ϕ_4^4, where the low-energy manifold reduces to a point ($k = 0$) and the quartic coupling to a number u_0, here the low-energy manifold is a circle and the coupling is a function of the angles parameterizing it. Therefore, in this theory you are going to get coupling functions and not coupling constants even as $\Lambda \to 0$.

But if k is irrelevant, we may evaluate u_0 with all four legs on the Fermi circle. In this case, we have the restriction

$$\theta_1 = \theta_3, \quad \theta_2 = \theta_4. \tag{16.26}$$

(We can forget the other permutation because of the symmetries of u under particle exchange.) Thus, we have

$$u_0 \equiv u(\theta_1, \theta_2, \theta_3, \theta_4) \tag{16.27}$$
$$= u(\theta_1, \theta_2, \theta_1, \theta_2) \tag{16.28}$$
$$\equiv F(\theta_1, \theta_2) \tag{16.29}$$
$$= F(\theta_1 - \theta_2) \equiv F(\theta) \text{ using rotational invariance.} \tag{16.30}$$

Thus, in the end we have just one function of the angular difference, $F(\theta)$. It is called *Landau's F function* because over 50 years ago Landau deduced that a Fermi system at low energies would be described by one function defined on the Fermi surface. He did this without the benefit of the RG, and for that reason, some of the leaps were hard to understand. Later, detailed diagrammatic calculations by Abrikosov *et al.* (referred to in the last chapter) justified this picture.

The RG provides yet another way to understand it. Indeed, I did not I know I had stumbled into the Landau function until my more experienced colleagues told me that was what F was. But the RG goes beyond F. It tells us that the final angles are not always slaved to the initial ones. If the former are exactly opposite, as in the right third of Figure 16.2, the final momenta can be pointing in any direction as long as they are mutually opposite. This leads to one more set of marginal couplings in the BCS channel:

$$u(\theta_1, -\theta_1, \theta_3, -\theta_3) = V(\theta_1 - \theta_3) \equiv V(\theta). \tag{16.31}$$

In terms of Figure 16.3, the two annuli coincide when $\boldsymbol{K}_1 + \boldsymbol{K}_2 = \boldsymbol{K} = 0$, in which case $\boldsymbol{K}_3$ and $\boldsymbol{K}_4$ can assume any angle, as long as they are mutually opposite.

16.3 One-Loop Analysis of u

Having established that F and V are marginal at tree level, we have to go to one loop to see if marginality survives. To this end we draw the usual diagrams, shown in Figure 16.4. We eliminate infinitesimal momentum slices of thickness $d\Lambda$ at $k = \pm\Lambda$.

These diagrams resemble the ones in any quartic field theory, but each one behaves differently from the others.

Consider the contribution of the first one (called ZS) to the renormalization of F. The external legs can be chosen to have $\omega = k = 0$, since dependence of u on these is irrelevant. Therefore, the momentum and frequency transfer at the left vertex are exactly zero and the

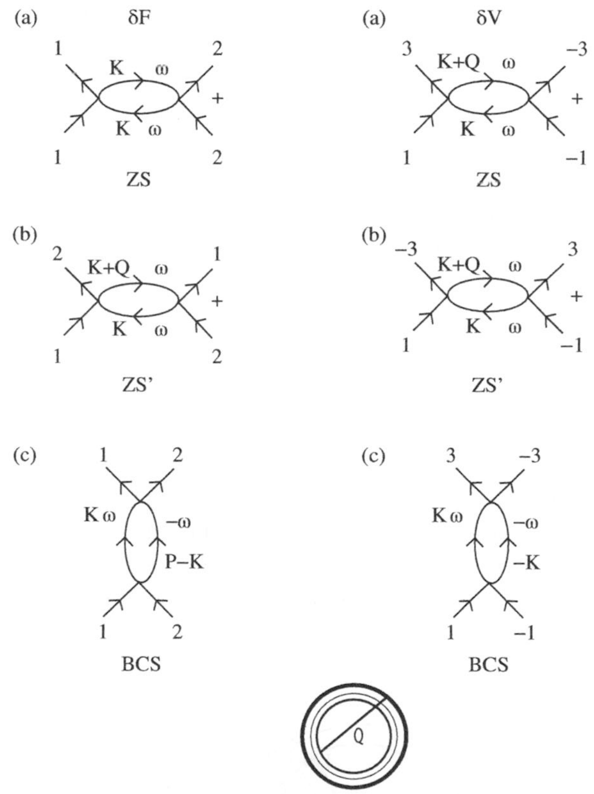

Figure 16.4 *Left:* The flow of F. ZS contributes nothing since poles are on the same half-plane due to vanishing frequency and momentum transfer at the interaction vertices. ZS$'$ does have a particle–hole contribution, but of order $d\Lambda^2$ because a large Q can be absorbed only at one of two special values of $\boldsymbol{K}$, one of which is shown in the bottom middle. The BCS diagram is equally negligible. *Right*: The flow of V. Only the BCS diagram contributes to δV because there is no restriction on the loop angle θ and the particle–particle and hole–hole contributions are non-zero. This is because the ω's in the two propagators are negatives of each other, while the energies are equal by time-reversal symmetry: $E(\boldsymbol{K}) = E(-\boldsymbol{K})$.

integrand has the following schematic form:

$$\delta F \simeq \int_0^{2\pi} d\theta \int_{d\Lambda} dk \int_{-\infty}^{\infty} d\omega \left(\frac{1}{(i\omega - \varepsilon(\boldsymbol{K}))} \frac{1}{(i\omega - \varepsilon(\boldsymbol{K}))} \right). \tag{16.32}$$

The loop momentum $\boldsymbol{K}$ lies in one of the two shells being eliminated. Since there is no energy difference between the two propagators, the poles in ω lie in the same half-plane and we get zero upon closing the contour in the other half-plane. In other words, this diagram can contribute only if it is a particle–hole diagram, but given zero momentum transfer we cannot convert a hole at $-\Lambda$ to a particle at $+\Lambda$ or vice versa. (Indeed, even a non-zero $q < 2\Lambda$ cannot connect particle and hole states which lie at the cut-off.)

In the ZS′ diagram, we do have a large momentum transfer, shown in the inset at the bottom. This **Q** *can* connect particle and hole. However, Q is typically of order K_{F} and much bigger than the radial cut-off, a phenomenon unheard of in, say, ϕ^4 theory, where all momenta and momentum transfers are of order Λ. This in turn means that the loop momentum is not only restricted in the radial direction to a width $d\Lambda$, but also in the angular direction to $d\Lambda/K_{\mathrm{F}}$. The bottom part of Figure 16.4 shows how a large **Q** can be absorbed only at one of two special angles of $\boldsymbol{K}$ if it is to connect particle and hole states being eliminated. This is further illustrated in Figure 16.5.

The thick circles represent shells of thickness $d\Lambda$. The allowed regions I, II, III, and IV have one loop momentum in the shell below K_{F} (hole) and one above (particle). The angular widths of the allowed regions go as $d\Lambda/K_{\mathrm{F}}$, and the radial widths go as $d\Lambda$. So, we have $du \simeq dt^2$, where $dt = d\Lambda/\Lambda$.

The same goes for the BCS diagram. Thus, both diagrams *individually vanish* due to phase space rather than produce non-zero canceling contributions as in $d = 1$.

Let us now turn to the renormalization of V. The first two diagrams are useless ($dV \simeq dt^2$) for the same reasons as before, but the last one is different. Since the total incoming momentum is zero, the loop momenta are equal and opposite and *no matter what direction* $\boldsymbol{K}$ *has,* $-\boldsymbol{K}$ *is guaranteed to lie in the same shell because* $E(\boldsymbol{K}) = E(-\boldsymbol{K})$ *or, equivalently,* $\varepsilon(\boldsymbol{K}) = \varepsilon(-\boldsymbol{K})$. There is no restriction on the loop angle θ. It is the loop frequencies that are now equal and opposite, and the poles now lie in opposite half-planes. We now get a

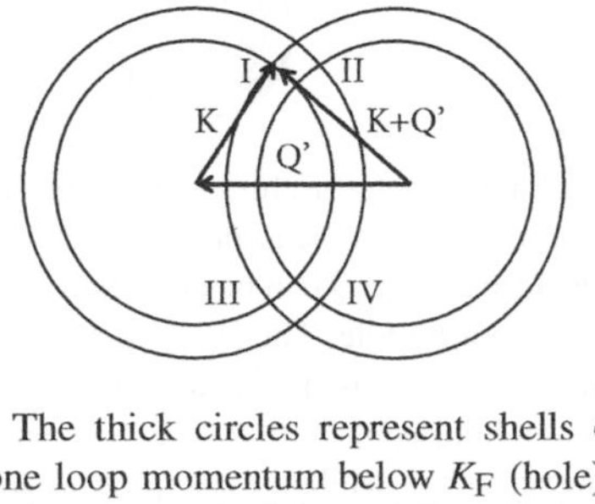

Figure 16.5 The ZS′ diagram. The thick circles represent shells of thickness $d\Lambda$. The allowed regions I, II, III, and IV have one loop momentum below K_{F} (hole) and one above (particle). The angular widths of the allowed regions go as $d\Lambda/K_{\mathrm{F}}$, and the radial widths go as $d\Lambda$.

flow (dropping constants and remembering that $\varepsilon(\boldsymbol{K}) = \varepsilon(-\boldsymbol{K})$):

$$\delta V(\theta_1 - \theta_3) \simeq \int_0^{2\pi} d\theta \int_{d\Lambda} dk \int_{-\infty}^{\infty} d\omega \left(\frac{V(\theta_1 - \theta) V(\theta - \theta_3)}{(i\omega - \varepsilon(\boldsymbol{K}))(-i\omega - \varepsilon(\boldsymbol{K}))} \right). \tag{16.33}$$

Closing the contour in either half-plane, we find (with constants restored) that

$$\frac{dV(\theta_1 - \theta_3)}{dt} = -\frac{1}{8\pi^2} \int d\theta \; V(\theta_1 - \theta) \; V(\theta - \theta_3). \tag{16.34}$$

This is an example of a *flow equation for a coupling function*. Upon expanding $V(\theta)$ in terms of angular momentum eigenfunctions

$$V(\theta) = \sum_m v_m e^{im\theta}, \tag{16.35}$$

we get an infinite number of flow equations for the coefficients v_m which denote the coupling of the Cooper pair with angular momentum m:

$$\frac{dv_m}{dt} = -\frac{v_m^2}{4\pi}, \tag{16.36}$$

with a solution

$$v_m(t) = \frac{v_m(0)}{1 + \frac{v_m(0)t}{4\pi}}. \tag{16.37}$$

What these equations tell us is that if the potential in angular momentum channel m is repulsive, it will get renormalized (logarithmically) down to zero, a result derived many years ago by Morel and Anderson [8], while if it is attractive, it will run off to large negative values signaling the BCS instability. This is the reason the V's are excluded in Landau theory, which assumes we have no phase transitions. (Remember that the sign of any given v_m is not necessarily equal to that of the microscopic interaction. Kohn and Luttinger [9] have shown that some of them will be always negative. Thus, the BCS instability is inevitable, though possibly at absurdly low temperatures or absurdly high angular momentum m.)

Not only did Landau say we could describe Fermi liquids with an F, he also managed to compute the soft response functions of the system in terms of the F function even when it was large, say 10 in dimensionless units. What is the small parameter that allows us to proceed given that it is not F?

The RG provides an answer. As we introduce a Λ and reduce it, a small parameter emerges, namely Λ/K_F.

Imagine that we have reduced Λ to a very tiny value and compute the *static compressibility* $\chi(0,\boldsymbol{q}) = \langle \rho(0,\boldsymbol{q})\rho(0,-\boldsymbol{q})\rangle = \langle \bar{\psi}\psi(0,\boldsymbol{q})\bar{\psi}\psi(0,-\boldsymbol{q})\rangle$, in the limit $\boldsymbol{q} \to 0$.

The diagrams are shown in Figure 16.6. Let us analyze them for the case $F_0^{\mathrm{s}} = F$, where the superscript s stands for the spin-symmetric channel.

The first diagram exists at free-field level. Remember, this is a Feynman diagram of the cut-off theory and not an RG diagram. The fermion momenta are not at the cut-off, but at any momentum up to the cut-off. For this ZS diagram to contribute, we need one line to be a hole and the other a particle, i.e., the poles in ω must be on different half-planes. An *infinitesimal* $\boldsymbol{q}$ can knock a hole into a particle and vice versa because the hole and particle states can be arbitrarily close to K_{F}. (In an RG diagram they would be separated by at least 2Λ.) The integral can be done by contour integration and gives a contribution proportional to the density of states.

The (geometric) sum of such bubbles gives the answer

$$\chi = \frac{\chi_0}{1+F_0}. \tag{16.38}$$

The response functions of Landau theory typically have this form of the geometric sum, which comes naturally from a self-consistent calculation in which χ appears on both sides of the equation.

We do not require F to be small. The small parameter is Λ/K_{F} and it tells us we can forget diagrams other than the iterated bubble. Diagrams with bubbles going any other way, like the crossed out one at the bottom of Figure 16.6, are suppressed by powers of Λ/K_{F}.

Likewise, one can compute Γ, the full four-point function or *physical scattering amplitude* in the cut-off theory, by adding to the tree-level contribution F the sum of bubbles connected by F. The amplitude Γ will have a very singular dependence on external ω and momentum transfer $\boldsymbol{q}$, unlike the *coupling* u. If $\omega \neq 0$ and $\boldsymbol{q} \equiv 0$, the bubbles will not contribute the δ-function at the Fermi energy, whereas in the opposite limit (which we encountered in the static compressibility) they will.

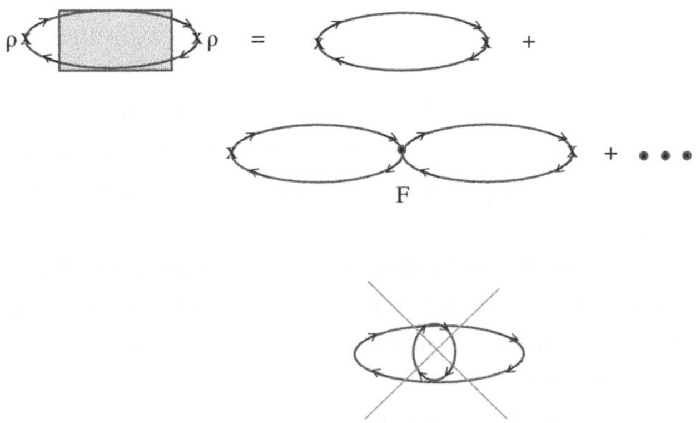

Figure 16.6 Computation of static compressibility $\chi(0,\boldsymbol{q}) = \langle \rho(0,\boldsymbol{q})\rho(0,-\boldsymbol{q})\rangle = \langle \bar{\psi}\psi(0,\boldsymbol{q})\bar{\psi}\psi(0,-\boldsymbol{q})\rangle$. When $\Lambda/K_{\mathrm{F}} \ll 1$, only the repeated bubble survives, and those like the crossed out one at the bottom are suppressed by powers of Λ/K_{F}.

Should this ZS diagram, which contributes to Γ, not contribute to the β-function of F if we differentiate it with respect to Λ and demand cut-off independence of Γ? It doesn't, because the entire contribution as $q \to 0$ comes from a δ-function at $k = 0$ and its k-integral is independent of the width of the integration region determined by Λ.

Thus, RG gives us a two-stage attack on the Fermi liquid:

- Eliminate modes down to a very thin shell of thickness $2\Lambda \ll K_F$. This theory is parametrized by $F(\theta)$.
- Use Λ/K_F as a small parameter to do the diagrammatic sum.

Readers familiar with the $1/N$ expansion (N being the number of species) will notice the familiar bubble sum which dominates the answer. It turns out that Λ/K_F plays the role of $1/N$. Recall that the action of the non-interacting fermions had the form of an integral over one-dimensional fermions, one for each direction. How many fermions are there? The answer is not infinity, since each direction has zero measure. It turns out that in any diagrammatic calculation, at any given Λ, the number of "independent" fermions is $N \simeq K_F/\Lambda$. If you imagine dividing the annulus into patches of size Λ in the angular direction also, each patch carries an index and contributes to one species. Landau theory is just the $N = \infty$ limit [1]. A similar result was obtained by Feldman *et al.* [10].

16.4 Variations and Extensions

Since [1] provides many details, this section will be brief.

First, consider $d = 2$. The theory applies to non-circular Fermi surfaces with just one modification: $F(\theta_1, \theta_2) \neq F(\theta_1 - \theta_2)$. Since E is not a function of K, we must first identify contours of constant energy for mode elimination. The BCS instability, which relies only on time-reversal symmetry $E(\boldsymbol{K}) = E(-\boldsymbol{K})$, survives, though $V(\theta_1, \theta_3) \neq V(\theta_1 - \theta_3)$ cannot be expanded in terms of angular harmonics.

Next, consider special *nested Fermi surfaces* in which a single momentum $\mathbf{Q}$ connects every point on the Fermi surface to another. An example, depicted in Figure 16.7, describes an anisotropic square lattice with spectrum

$$E = -\cos K_x - r\cos K_y, \quad \text{where } r \text{ is some constant.} \tag{16.39}$$

The energy obeys

$$E(\boldsymbol{K}) = -E(\boldsymbol{K} + \mathbf{Q}), \tag{16.40}$$

where $\mathbf{Q} = (\pi/2, \pi/2)$ connects not only states on the Fermi surface ($E = 0$) to other states on it, it connects a hole state of energy $-\varepsilon$ to a particle state of energy $+\varepsilon$. As a result, the process

$$\boldsymbol{K}_1 + \boldsymbol{K}_2 \to (\boldsymbol{K}_1 + \mathbf{Q}) + (\boldsymbol{K}_2 + \mathbf{Q}) \equiv \boldsymbol{K}_1' + \boldsymbol{K}_2' \tag{16.41}$$

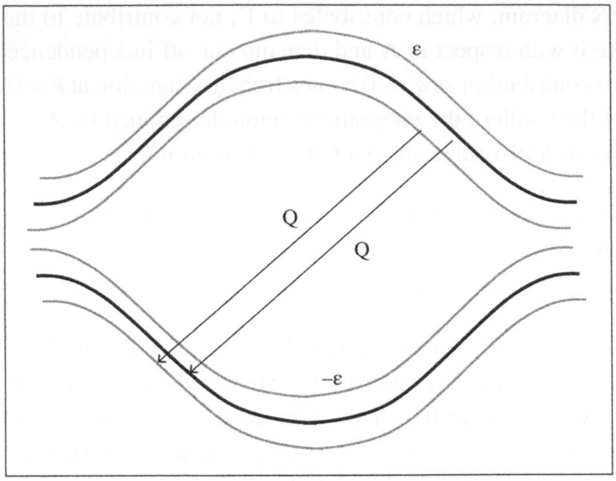

Figure 16.7 A nested Fermi surface (thick line) whose points are scattered back to the Fermi surface under the addition of a certain momentum, which is $\mathbf{Q} = (\pi/2, \pi/2)$ in this case. Further, $\mathbf{Q} = (\pi/2, \pi/2)$ connects a hole state of energy $-\varepsilon$ (thin line) to a particle state of energy $+\varepsilon$ (thin line) and vice versa. Thus, when we eliminate an energy shell at $\varepsilon = \pm\Lambda$, there is a non-zero contribution to the flow because the scattered momentum is in a shell $\varepsilon = \mp\Lambda$, which is also being eliminated.

never gets squeezed out as $\Lambda \to 0$. No matter how small Λ is, if the incoming energies $E(\boldsymbol{K}_1)$ and $E(\boldsymbol{K}_2)$ lie within the low-energy cut-off, so will $E(\boldsymbol{K}_1 + \mathbf{Q}) = -E(\boldsymbol{K}_1)$ and $E(\boldsymbol{K}_2 + \mathbf{Q}) = -E(\boldsymbol{K}_2)$. The corresponding amplitude W flows to large values under RG, correctly signaling the CDW instability at arbitrarily small W.

Thus, the RG is able to identify all the known instabilities (BCS and CDW) of the $d = 2$ system at weak coupling. Modulo these instabilities (due to V and W), the Fermi liquid will exist.

Next, consider $d = 3$, and a spherical Fermi surface. RG alone does not restrict the initial momenta to be equal to the final ones: the latter can rotate around a cone whose axis is the total incoming momentum. (Now we must replace intersecting annuli by intersecting spherical shells.) Thus, non-forward scattering amplitudes can be substantial. They are not irrelevant and they do contribute to the particle's lifetime. This contribution ($\propto \varepsilon^2$ due to phase space and the Pauli principle) is, however, RG-irrelevant under scaling. Landau's theory focuses on the forward scattering amplitude even in $d = 3$ because it controls the cost of deforming the Fermi surface and thereby determines the response functions in Landau's energy-based analysis. We, on the other hand, keep track of the non-forward amplitudes because we want the *Hamiltonian operator* for the low-energy theory, and not just the energies of the low-energy states.

Notice that in any number of spatial dimensions the low-energy theory is $(1 + 1)$ dimensional, with (ω, k) as the variables conjugate to time and space. Only the "isospin" labels change, depending on the dimensionality of the Fermi surface.

The previous results do not conflict with the fact that the actual systems (e.g., cuprates) are not Fermi liquids in $d = 2$. All that has been shown is that if by the time we renormalize down to a narrow region near the Fermi surface we still have a theory with interactions analytic in k and ω, a Fermi liquid will follow with a fixed-point interaction F. All I have shown is that a Fermi liquid can exist in $d = 2$ for small coupling, unlike in $d = 1$ where the smallest interaction can destroy it. However, a smooth road to the cut-off theory is not guaranteed for all coupling strengths. It is possible that beyond some interaction strength the sums of the diagrams renormalizing the interactions diverge. This is possible even though they individually are finite, being cut off in the ultraviolet and infrared. Another threat to the Fermi liquid is low-energy excitations (not of the particle–hole type), say magnons. The fermion can decay into these with a width that is not parametrically smaller than the energy [11–14]. Or, when on-site repulsion is very strong, the momentum states and Fermi surface may not be the right place to begin: one may have to begin in real space and find ways to limit or even forbid double occupancy. (These constraints are imposed by gauge fields, which can become dynamical and destabilize the fermion.) Yet another route to non-Fermi liquids is to couple one-dimensional Luttinger liquids transversely and hope to get a two-dimensional non-Fermi liquid [15–17]. Another way to obtain non-Fermi liquids is through impurities, as in the Kondo problem analyzed by Ludwig and Affleck [18, 19].

Sometimes one sees Lagrangians with singular interactions, typically coming from completely integrating out some gapless degrees of freedom, say spin waves. In such cases it is better to eliminate modes of all fields simultaneously. This will ensure that the β-function is always analytic, a feature greatly emphasized by Wilson and key to its utility. However, simultaneous mode elimination can be challenging. For example, if the fields are bosonic and fermionic, their low-energy regions will be near the origin and near the Fermi surface respectively. A large momentum will always knock a boson to a very high energy state but may scatter a fermion between two points near the Fermi surface.

References and Further Reading

[1] R. Shankar, Reviews of Modern Physics, **66**, 129 (1994).
[2] G. Benfatto and G. Gallavotti, Physical Review B, **42**, 9967 (1990). This is the earliest paper on RG for non-relativistic fermions.
[3] J. Feldman and E. Trubowitz, Helvetica Physica Acta, **63**, 157 (1990). This is more rigorous than my treatment.
[4] J. Feldman and E. Trubowitz, Helvetica Physica Acta, **64**, 213 (1991).
[5] R. Shankar, Physica A, **177**, 530 (1991). This was my first pass.
[6] J. Polchinski, *Proceedings of the 1992 TASI in Elementary Particle Physics*, eds. J. Polchinski and J. Harvey, World Scientific (1992). From the standpoint of effective field theories.
[7] S. Weinberg, Nuclear Physics B, **413**, 567 (1994).
[8] P. Morel and P. W. Anderson, Physical Review, **125**, 1263 (1962).
[9] W. Kohn and J. Luttinger, Physical Review Letters, **15**, 524 (1965).
[10] J. Feldman and E. Trubowitz, Europhysics Letters, **24**, 437 (1993).

[11] L. B. Ioffe and G. Kotliar, Physical Review **42**, 10348 (1990).
[12] P. Lee and N. Nagaosa, Physical Review Letters, **64**, 2450 (1990).
[13] J. Polchinski, Nuclear Physics B, **422**, 617 (1994).
[14] C. M. Varma, P. B. Littlewood, S. Schmidt-Rink, E. Abrahams, and A. E. Ruckenstein, Physical Review Letters, **63**, 1996 (1989).
[15] X. G. Wen, Physical Review B, **42**, 6623 (1990).
[16] H. J. Schulz, International Journal of Modern Physics, **43**, 10353 (1991).
[17] C. di Castro and W. Metzner, Physical Review Letters, **69**, 1703 (1992).
[18] I. Affleck and A. W. W. Ludwig, Physical Review Letters, **67**, 161 (1991).
[19] I. Affleck and A. W. W. Ludwig, Physical Review Letters, **68**, 1046 (1992).

17

Bosonization I: The Fermion–Boson Dictionary

"Bosonization" refers to the possibility of describing a theory of relativistic Dirac fermions obeying standard anticommutation rules by a boson field theory. While this may be possible in all dimensions, it has so far proved most useful in $d = 1$, where the bosonic version of the given fermionic theory is local and simple, and often simpler than the Fermi theory. This chapter should be viewed as a stepping stone toward a more thorough approach, for which references are given at the end.

In this chapter I will set up the bosonization machine, explaining its basic logic and the dictionary for transcribing a fermionic theory to a bosonic theory. The next chapter will be devoted to applications.

To my knowledge, bosonization, as described here, was first carried out by Lieb and Mattis [1] in their exact solution of the Luttinger model [2]. Later, Luther and Peschel [3] showed how to use it to find asymptotic (low momentum and energy) correlation functions for more generic interacting Fermi systems. It was independently discovered in particle physics by Coleman [4], and further developed by Mandelstam [5]. Much of what I know and use is inspired by the work of Luther and Peschel.

17.1 Preamble

Before getting into any details, I would first like to answer two questions. First, if bosonization applies only to relativistic Dirac fermions, why is it of any interest to condensed matter theory where relativity is not essential? Second, what is the magic by which bosonization helps us tame interacting field theories?

As for the first question, there are two ways in which Dirac fermions enter condensed matter physics. The first is in the study of two-dimensional Ising models, where we have already encountered them. Recall that if we use the transfer matrix approach and convert the classical problem on an $N \times N$ lattice to a quantum problem in one dimension we end up with a 2^N-dimensional Hilbert space, with a Pauli matrix at each of N sites. The two dimensions at each site represent the twofold choice of values open to the Ising spins. Consider now a spinless fermion degree of freedom at each site. Here too we have two choices: the fermion state is occupied or empty. There is some need for cleverness in going from the Pauli matrix problem to the fermion problem since Pauli matrices commute at different sites while fermions anticommute; this was provided by Jordan and Wigner. In the critical region the fermion is relativistic since one obtains all the symmetries of the continuum.

The second way in which Dirac fermions arise is familiar from our study of spinless fermions on a linear lattice, described by

$$H = -\sum_{n=-\infty}^{\infty} \psi^\dagger(n)\psi(n+1) + \text{h.c.} \tag{17.1}$$

In the above, the spinless fermion field obeys the standard anticommutation rules

$$\{\psi^\dagger(n), \psi(m)\} = \delta_{mn}, \tag{17.2}$$

with all other anticommutators vanishing.

Going to momentum states, the Hamiltonian becomes

$$H = -\int_{-\pi}^{\pi} \frac{dk}{2\pi} [\cos k]\, \psi^\dagger(k)\psi(k). \tag{17.3}$$

In the ground state we must fill all negative energy modes, that is, states between $\pm K_\text{F}$, where $K_\text{F} = \pi/2$. To study the low-energy properties of the system, we can focus on the modes near just the Fermi points, as shown in Figure 15.2. We find that they have $E = \pm k$, where k is measured from the respective Fermi points. These are the two components of the massless Dirac field. Any interaction between the primordial fermions can be described in terms of these two components at low energies.

Next, we ask how bosonization can make life easier. Say we have a problem where $H = H_0 + V$, where H_0 is the free Dirac Hamiltonian and V is a perturbation. Assume we can express all quantities of interest in terms of power series in V. In the interaction picture the series will involve the correlation functions of various operators evolving under H_0. Bosonization now tells us that the same series is reproduced by starting with $H = H_0^\text{B} + V^\text{B}$, where H_0^B is a massless free boson Hamiltonian and V^B is a bosonic operator that depends on V and is specified by the bosonization dictionary. Consider the special case $V = \rho^2$, where $\rho = \psi^\dagger(x)\psi(x)$, the Dirac charge density. This is a quartic interaction in the Fermi language and obviously non-trivial. But according to the dictionary, we must replace ρ by the bosonic operator $\frac{1}{\sqrt{\pi}}\partial_x\phi$, ϕ being the boson field. Thus, V is replaced by the *quadratic* interaction $\frac{1}{\pi}(\partial_x\phi)^2$. The bosonic version is trivial! I must add that this is not always the case; a simple mass term in the Fermi language becomes the formidable interaction $\cos\sqrt{4\pi}\phi$.

Let us now begin. I will first remind you of some basic facts about massless fermions and bosons in one dimension. This will be followed by the bosonization dictionary that relates interacting theories in one language to the other.

17.2 Massless Dirac Fermion

In one dimension, the Dirac equation

$$i\frac{\partial\psi}{\partial t} = H\psi \tag{17.4}$$

will have as the Hamiltonian

$$H = \alpha P + \beta m, \tag{17.5}$$

where P is the momentum operator, and

$$\alpha = \sigma_3 = \gamma_5, \tag{17.6}$$

$$\beta = \sigma_2 = \gamma_0. \tag{17.7}$$

Let us focus on the massless case. There is nothing to diagonalize now: $\psi_\pm$, the upper and lower components of ψ, called *right and left movers*, are decoupled. In terms of the field operators obeying

$$\{\psi_\pm^\dagger(x),\ \psi_\pm(y)\} = \delta(x-y), \tag{17.8}$$

the second quantized Hamiltonian is

$$H = \int \psi^\dagger(x)(\alpha P)\psi(x)dx \tag{17.9}$$

$$= \int \psi_+^\dagger(x)(-i\partial_x)\psi_+(x)dx + \int \psi_-^\dagger(x)(i\partial_x)\psi_-(x)dx. \tag{17.10}$$

In terms of the Fourier transforms

$$\psi_\pm(p) = \int_{-\infty}^{\infty} \psi_\pm(x)e^{ipx}dx \tag{17.11}$$

obeying

$$\{\psi_\pm^\dagger(p),\ \psi_\pm(q)\} = 2\pi\delta(p-q), \tag{17.12}$$

we find that

$$H = \int \psi_+^\dagger(p)\, p\, \psi_+(p)\frac{dp}{2\pi} + \int \psi_-^\dagger(p)\, (-p)\psi_-(p)\frac{dp}{2\pi}. \tag{17.13}$$

From the above, it is clear that the right/left movers have energies $E = \pm p$ respectively. The Dirac sea is thus filled with right movers of negative momentum and left movers with positive momentum, as shown in Figure 17.1.

The inverse of Eq. (17.11) is

$$\psi_\pm(x) = \int_{-\infty}^{\infty} \frac{dp}{2\pi}\psi_\pm(p)e^{ipx}e^{-\frac{1}{2}\alpha|p|}, \tag{17.14}$$

where α is a convergence factor that will be sent to 0 at the end.

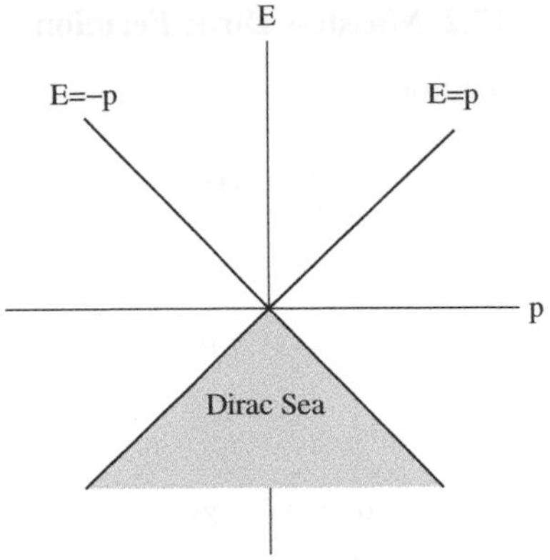

Figure 17.1 Relativistic fermion with right and left movers $E = \pm p$. The Dirac sea is filled with right movers of negative momentum and left movers with positive momentum. The two branches come from the linearized spectrum near the Fermi points $K = \pm K_F$ of the non-relativistic fermion. (Only states on the lines $E = \pm p$ are occupied in the Fermi sea.)

Since the fields have trivial time evolution in this free-field theory, we can write down the Heisenberg operators at all times:

$$\psi_{\pm}(x\, t) = \int_{-\infty}^{\infty} \frac{dp}{2\pi} \psi_{\pm}(p) e^{ip(x \mp t)} e^{-\frac{1}{2}\alpha|p|}. \tag{17.15}$$

Notice that $\psi_{\pm}$ is a function only of $x \mp t$.

Consider now the equal-time correlation function in the ground state:

$$\begin{aligned} \langle \psi_{+}(x) \psi_{+}^{\dagger}(0) \rangle &= \int_{-\infty}^{\infty} \frac{dp}{2\pi} e^{-\frac{1}{2}\alpha|p|} \int_{-\infty}^{\infty} \frac{dq}{2\pi} e^{-\frac{1}{2}\alpha|q|} e^{ipx} \underbrace{\langle \psi_{+}(p)\; \psi_{+}^{\dagger}(q) \rangle}_{2\pi\delta(p-q)\theta(q)} \\ &= \int_{0}^{\infty} \frac{dp}{2\pi} e^{ipx} e^{-\alpha|p|}. \end{aligned} \tag{17.16}$$

We have used the fact that a right mover can be created only for positive momenta since the Dirac sea is filled with negative momentum particles. So now we have

$$\langle \psi_{+}(x) \psi_{+}^{\dagger}(0) \rangle = \int_{0}^{\infty} \frac{dp}{2\pi} e^{-\alpha p} e^{ipx} \tag{17.17}$$

$$= \frac{1}{2\pi} \frac{1}{\alpha - ix}. \tag{17.18}$$

If we want the correlation function for unequal times, we just replace x by $x - t$ since we know that the right movers are functions of just this combination.

In the same way, we can show that

$$\langle \psi_\pm(x)\ \psi_\pm^\dagger(0)\rangle = \frac{\pm i/2\pi}{x \pm i\alpha}, \tag{17.19}$$

$$\langle \psi_\pm^\dagger(0)\ \psi_\pm(x)\rangle = \frac{\mp i/2\pi}{x \mp i\alpha}. \tag{17.20}$$

Note that

$$\langle \psi_\pm(x)\ \psi_\pm^\dagger(0) + \psi_\pm^\dagger(0)\ \psi_\pm(x)\rangle = \frac{\alpha/\pi}{x^2+\alpha^2} \tag{17.21}$$

$$\simeq \delta(x), \tag{17.22}$$

where in the last equation we are considering the limit of vanishing α.

Besides the Fermi field, there are bilinears in the field that occur often. Let us look at some key ones. The current density j_μ has components

$$j_0 = \psi^\dagger \psi \tag{17.23}$$

$$= \psi_+^\dagger(x)\psi_+(x) + \psi_-^\dagger(x)\psi_-(x), \tag{17.24}$$

$$j_1 = \psi^\dagger \alpha \psi \tag{17.25}$$

$$= \psi_+^\dagger(x)\psi_+(x) - \psi_-^\dagger(x)\psi_-(x). \tag{17.26}$$

The axial current is given by $j_\mu^5 = \varepsilon_{\mu\nu} j_\nu = (j_1, -j_0)$. The last bilinear is the "mass term"

$$\overline{\psi}\psi = \psi^\dagger(x)\beta\psi(x) \tag{17.27}$$

$$= -i\psi_+^\dagger(x)\psi_-(x) + i\psi_-^\dagger(x)\psi_+(x). \tag{17.28}$$

For later use, let us note that

$$\langle \overline{\psi}\psi(x)\ \overline{\psi}\psi(0)\rangle = \frac{1}{2\pi^2}\frac{1}{x^2+\alpha^2}. \tag{17.29}$$

The derivation of this result is left as an exercise. All you need are the anticommutation rules and the correlation functions from Eqs. (17.19) and (17.20).

17.2.1 Majorana Fermions

We close the section by recalling some facts about Majorana fermions. These may be viewed as Hermitian or real fermions. The Dirac field ψ_D can be expressed in terms of two *Hermitian* fields ψ and χ:

$$\psi_\mathrm{D} = \frac{\psi + i\chi}{\sqrt{2}}, \tag{17.30}$$

$$\psi_\mathrm{D}^\dagger = \frac{\psi - i\chi}{\sqrt{2}}. \tag{17.31}$$

(The components of the spinors ψ_D, ψ, and χ are implicit.) It is readily verified that

$$\{\psi_a(x), \psi_b(y)\} = \delta(x-y)\delta_{ab}, \tag{17.32}$$

where a and b label the two spinor components. There is a similar rule for χ. All other anticommutators vanish.

Exercise 17.2.1 *Show that*

$$\int \psi_D^\dagger(x)\psi_D(x)dx \equiv \int \psi_{Da}^\dagger(x)\psi_{Da}(x)dx = \tag{17.33}$$

$$= \int (i\psi_a\chi_a + \delta(0))dx \equiv \int (i\psi^T\chi + \delta(0))dx. \tag{17.34}$$

By computing the density of Dirac fermions in the vacuum, show that this means

$$\int : \psi_D^\dagger(x)\psi_D(x) : dx = \int (i\psi^T\chi)dx. \tag{17.35}$$

If we write the massive Dirac Hamiltonian in terms of the Majorana fields defined above, we will get, with $\alpha = \sigma_3$ and $\beta = \sigma_2$,

$$H_D = \int [\psi_D^\dagger(\alpha P + \beta m)\psi_D]dx \tag{17.36}$$

$$= \frac{1}{2}\int [\psi^T(\alpha P + \beta m)\psi + \chi^T(\alpha P + \beta m)\chi]dx \tag{17.37}$$

$$+ \frac{1}{2}\int \left[i\psi^T(\alpha P + \beta m)\chi - i\chi^T(\alpha P + \beta m)\psi\right]dx.$$

You may check that the cross terms add to zero. (To make contact with the Majorana fermions from Chapter 9, we should change the representation of the α matrix so that it equals Pauli's σ_1. This change of variables with real coefficients is consistent with the Hermitian nature of the Majorana fields.)

So remember: one free Dirac fermion equals two Majorana fermions, just as one charged scalar field equals two real fields (not just in degrees of freedom, but at the level of H).

Exercise 17.2.2 *Using $\alpha = \sigma_3$ and $\beta = \sigma_2$ and the components in explicit form, verify that the non-interacting Hamiltonian for one Dirac fermion is the sum of the Hamiltonians for two Majorana fermions.*

17.3 Free Massless Scalar Field

The Hamiltonian for a massless scalar field is

$$H_B = \frac{1}{2}\int (\Pi^2 + (\partial_x\phi)^2)dx, \tag{17.38}$$

where Π and ϕ obey

$$[\phi(x), \Pi(y)] = i\delta(x-y). \tag{17.39}$$

The Schrödinger operators are expanded as follows:

$$\phi(x) = \int_{-\infty}^{\infty} \frac{dp}{2\pi\sqrt{2|p|}} \left[\phi(p)e^{ipx} + \phi^\dagger(p)e^{-ipx}\right] e^{-\frac{1}{2}\alpha|p|}, \tag{17.40}$$

$$\Pi(x) = \int_{-\infty}^{\infty} \frac{dp|p|}{2\pi\sqrt{2|p|}} \left[-i\phi(p)e^{ipx} + i\phi^\dagger(p)e^{-ipx}\right] e^{-\frac{1}{2}\alpha|p|}, \tag{17.41}$$

where

$$[\phi(p), \phi^\dagger(p')] = 2\pi\delta(p-p'). \tag{17.42}$$

Due to the convergence factors, ϕ and Π will obey

$$[\phi(x), \Pi(y)] = \frac{i\alpha/\pi}{\alpha^2 + (x-y)^2} \tag{17.43}$$

$$\simeq i\delta(x-y). \tag{17.44}$$

The Hamiltonian now takes the form:

$$H = \int_{-\infty}^{\infty} \frac{dp}{2\pi} \phi^\dagger(p)\phi(p)|p|. \tag{17.45}$$

Exercise 17.3.1 *Verify Eq. (17.45).*

We now introduce right and left movers $\phi_\pm$:

$$\phi_\pm(x) = \frac{1}{2}\left[\phi(x) \mp \int_{-\infty}^{x} \Pi(x')dx'\right] \tag{17.46}$$

$$= \frac{1}{2}\int_{-\infty}^{\infty} \frac{dp}{2\pi\sqrt{2|p|}} e^{-\frac{1}{2}\alpha|p|} \left[\phi(p)(1 \pm |p|/p)e^{ipx} + \text{h.c.}\right]$$

$$= \pm\int_{0}^{\pm\infty} \frac{dp}{2\pi\sqrt{2|p|}} \left[e^{ipx}\phi(p) + \text{h.c.}\right] e^{-\frac{1}{2}\alpha|p|}. \tag{17.47}$$

I leave it to you to verify, using Eq. (17.46), that

$$\left[\phi_\pm(x), \phi_\pm(y)\right] = \pm\frac{i}{4}\varepsilon(x-y) \equiv \pm\frac{i}{4}\text{sgn}(x-y), \tag{17.48}$$

$$\left[\phi_+(x), \phi_-(y)\right] = \frac{i}{4}. \tag{17.49}$$

Exercise 17.3.2 *Verify Eqs. (17.48) and (17.49) starting with Eq. (17.46). If you started with Eq. (17.47), you would find that because of the convergence factors, a rounded-out step function will arise in place of $\varepsilon(x-y)$, and this will become a step function as $\alpha \to 0$.*

If we use the Heisenberg equations of motion for $\phi(p)$ and $\phi^\dagger(p)$, we will find that $\phi_\pm$ are functions only of $x \mp t$.

We must next work out some correlation functions in this theory. It is claimed that

$$G_\pm(x) = \langle \phi_\pm(x)\phi_\pm(0) - \phi_\pm^2(0)\rangle \tag{17.50}$$

$$= \frac{1}{4\pi}\ln\frac{\alpha}{\alpha \mp ix}, \tag{17.51}$$

$$G(x) = \langle \phi(x)\phi(0) - \phi^2(0)\rangle \tag{17.52}$$

$$= \frac{1}{4\pi}\ln\frac{\alpha^2}{\alpha^2 + x^2}. \tag{17.53}$$

I will now establish one of them, leaving the rest as exercises. Consider

$$G_+(x) = \int_0^\infty \frac{dp}{2\pi\sqrt{2|p|}}e^{-\frac{1}{2}\alpha|p|}\int_0^\infty \frac{dq}{2\pi\sqrt{2|q|}}e^{-\frac{1}{2}\alpha|q|}\langle(\phi(p)\phi^\dagger(q)\rangle(e^{ipx}-1)$$

$$= \int_0^\infty \frac{dp}{4\pi|p|}(e^{ipx}-1)e^{-\alpha p} \tag{17.54}$$

$$= \frac{1}{4\pi}\ln\frac{\alpha}{\alpha - ix}, \tag{17.55}$$

where the last line comes from looking up a table of integrals. If you cannot find this particular form of the result, I suggest you first differentiate both sides with respect to x, thereby eliminating the $1/|p|$ factor. Now the integral is easily shown to be $i/(4\pi(\alpha - ix))$. Next, integrate this result with respect to x, with the boundary condition $G_+(0) = 0$.

Finally, we consider a class of operators one sees a lot of in two-dimensional (spacetime) theories. These are exponentials of the scalar field. Consider first

$$G_\beta(x) \equiv \langle e^{i\beta\phi(x)}\, e^{-i\beta\phi(0)}\rangle. \tag{17.56}$$

For the correlator to be non-zero, the sum of the factors multiplying ϕ in the exponentials has to vanish. This is because the theory (the Hamiltonian of the massless scalar field) is invariant under a constant shift in ϕ. To evaluate this correlator, we need the following identity:

$$e^A \cdot e^B =: e^{A+B} : e^{\langle AB + \frac{A^2+B^2}{2}\rangle}, \tag{17.57}$$

where the normal-ordered operator $: A :$ has all its destruction operators to the right and creation operators to the left, as well as the fact that the vacuum expectation value of a normal-ordered exponential operator is just 1. All other terms in the series annihilate the vacuum state on the left or right or both. Thus,

$$\langle : e^\Omega : \rangle = 1. \tag{17.58}$$

Exercise 17.3.3 *If you want to amuse yourself by proving Eq. (17.57), here is a possible route. Start with the more familiar identity (which we will not prove):*

$$e^{A+B} = e^A e^B e^{-\frac{1}{2}[A,B]} \tag{17.59}$$

$$= e^B e^A e^{\frac{1}{2}[A,B]}, \tag{17.60}$$

provided $[A,B]$ *commutes with A and B. Using this, first write* $e^A = e^{A^+ + A^-}$, *where* $A^\pm$ *are the creation and destruction parts of A, in normal-ordered form. Now turn to* e^{A+B}, *and separate the exponentials using the identity above. Next, normal-order each part using this formula again, and finally normal-order the whole thing. (The last step is needed because :A::B: is not itself normal ordered.) Finally, remember that all commutators are c-numbers and therefore equal to their vacuum expectation values.*

We now use Eqs. (17.57) and (17.58) to evaluate G_β:

$$G_\beta(x) = \langle : e^{i\beta(\phi(x)-\phi(0))} : \rangle e^{\beta^2[\langle \phi(x)\phi(0) - \frac{\phi^2(0)+\phi^2(x)}{2}\rangle]} \tag{17.61}$$

$$= e^{\beta^2 \frac{1}{4\pi} \ln \frac{\alpha^2}{\alpha^2+x^2}} \tag{17.62}$$

$$= \left(\frac{\alpha^2}{\alpha^2 + x^2} \right)^{\beta^2/4\pi}. \tag{17.63}$$

Notice two things. First, by varying β we can get operators with a continuum of power-law decays of correlations. Next, as we send α to 0, the correlator vanishes. To avoid this we must begin with operators suitably boosted or renormalized. The thing to do in the above example is to consider the renormalized operator

$$[e^{i\beta\phi}]_{\mathrm{R}} = (\alpha\mu)^{\frac{-\beta^2}{4\pi}} e^{i\beta\phi}, \tag{17.64}$$

where μ is an arbitrary mass. This operator will have finite correlations in the limit of zero α: if we give it less of a boost, it dies; more, and it blows up.

One can similarly show, using Eqs. (17.50) and (17.53), that

$$\langle e^{i\beta\phi_\pm(x)} e^{-i\beta\phi_\pm(0)} \rangle = \left(\frac{\alpha}{\alpha \mp ix} \right)^{\beta^2/4\pi}. \tag{17.65}$$

17.3.1 The Dual Field θ

So far we have focused on the combination

$$\phi = \phi_+ + \phi_-. \tag{17.66}$$

In some calculations one needs correlations of the *dual field*,

$$\theta = \phi_- - \phi_+. \tag{17.67}$$

From Eq. (17.46),

$$\theta(x) = \int_{-\infty}^{x} \Pi(x')dx', \tag{17.68}$$

$$\Pi(x) = \frac{d\theta}{dx}. \tag{17.69}$$

The correlations of the dual field are just the same as those of ϕ:

$$\langle e^{i\beta\theta(x)} e^{-i\beta\theta(0)} \rangle = \left(\frac{\alpha^2}{\alpha^2 + x^2} \right)^{\beta^2/4\pi}. \tag{17.70}$$

Here is one way to derive Eq. (17.70):

$$\begin{aligned}
\langle e^{i\beta\theta(x)} e^{-i\beta\theta(0)} \rangle &= \langle e^{i\beta(\phi_-(x)-\phi_+(x))} e^{-i\beta(\phi_-(0)-\phi_+(0))} \rangle \\
&= \langle e^{i\beta\phi_-(x)} e^{-i\beta\phi_-(0)} \rangle \langle e^{-i\beta\phi_+(x)} e^{i\beta\phi_+(0)} \rangle \\
&= e^{\beta^2 G_-(x)} e^{\beta^2 G_+(x)} \\
&= \left(\frac{\alpha^2}{\alpha^2 + x^2} \right)^{\beta^2/4\pi},
\end{aligned} \tag{17.71}$$

where I have not shown the (canceling) phase factors coming from separating and recombining exponentials of $\phi_\pm$.

17.4 Bosonization Dictionary

So far we have dealt with massless Fermi and Bose theories and the behavior of various correlation functions in each. Now we are ready to discuss the rules for trading the Fermi theory for the Bose theory. The most important formula is this:

$$\psi_\pm(x) = \frac{1}{\sqrt{2\pi\alpha}} e^{\pm i\sqrt{4\pi}\phi_\pm(x)}. \tag{17.72}$$

This is not an operator identity: no combination of boson operators can change the fermion number the way ψ can. The equation above really means that any correlation function of the Fermi field, calculated in the Fermi vacuum with the given (α) cut-off, is reproduced by the correlator of the bosonic operator given on the right-hand side, if computed in the bosonic vacuum with the same momentum cut-off. Given this equivalence, we can replace any interaction term made out of the Fermi field by the corresponding bosonic counterpart. Sometimes this will require some care, but this is the general idea.

Substituting Eq. (17.46) in Eq. (17.72), we find

$$\psi_\pm(x) = \frac{1}{\sqrt{2\pi\alpha}} \exp\left[\pm i\sqrt{\pi} \left[\phi(x) \mp \int_{-\infty}^{x} \Pi(x')dx' \right] \right]. \tag{17.73}$$

The integral of Π plays the role of the Jordan–Wigner string that ensures the global anticommutation rules of fermions, as first shown by Mandelstam [5].

There are several ways to convince you of the correctness of the master formula Eq. (17.72). First, consider the correlation

$$\langle \psi_+(x)\psi_+^\dagger(0)\rangle = \frac{1}{2\pi}\frac{1}{\alpha - ix}. \tag{17.74}$$

Let us see this reproduced by the bosonic version:

$$\left\langle \frac{1}{\sqrt{2\pi\alpha}} e^{i\sqrt{4\pi}\phi_+(x)} \frac{1}{\sqrt{2\pi\alpha}} e^{-i\sqrt{4\pi}\phi_+(0)} \right\rangle \tag{17.75}$$

$$= \frac{1}{2\pi\alpha}\langle : e^{i\sqrt{4\pi}\phi_+(x)} e^{-i\sqrt{4\pi}\phi_+(0)} :\rangle e^{4\pi\langle \phi_+(x)\phi_+(0) - \phi_+^2\rangle} \tag{17.76}$$

$$= \frac{1}{2\pi\alpha} e^{4\pi G_+(x)} \tag{17.77}$$

$$= \frac{1}{2\pi\alpha}\frac{\alpha}{\alpha - ix}. \tag{17.78}$$

In the above we have used Eq. (17.58), the normal-ordering formula Eq. (17.57), the definition of G_+ from Eq. (17.50), and its actual value from Eq. (17.51).

It is possible to verify in the same spirit that the bosonized version of the Fermi field obeys all the anticommutation rules (with delta functions of width α). I leave this to the more adventurous ones among you. Instead, I will now consider some composite operators and show the care needed in dealing with their bosonization. The first of these is

$$\overline{\psi}\psi = -\frac{1}{\pi\alpha}\cos\sqrt{4\pi}\phi. \tag{17.79}$$

The proof involves just the use of Eq. (17.60), and goes as follows:

$$\overline{\psi}\psi(x) = -i\psi_+^\dagger(x)\psi_-(x) + \text{h.c.}$$

$$= \frac{1}{2\pi\alpha}\left[e^{-i\sqrt{4\pi}\phi_+(x)} e^{-i\sqrt{4\pi}\phi_-(x)}(-i) + \text{h.c.}\right]$$

$$= \frac{1}{2\pi\alpha}\left(e^{-i\sqrt{4\pi}\phi(x)} e^{\frac{1}{2}4\pi(-1)\frac{i}{4}}(-i) + \text{h.c.}\right) \tag{17.80}$$

$$= -\frac{1}{\pi\alpha}\cos\sqrt{4\pi}\phi. \tag{17.81}$$

The factor $\frac{i}{4}$ in the exponent arises from the commutator of the right and left movers, Eq. (17.49).

It can similarly be shown that

$$\overline{\psi}i\gamma^5\psi = -\left[\psi_+^\dagger(x)\psi_-(x) + \psi_-^\dagger(x)\psi_+(x)\right] \tag{17.82}$$

$$= \frac{1}{\pi\alpha}\sin\sqrt{4\pi}\phi. \tag{17.83}$$

In the above manipulations we brought together two operators at the same point. Each one has been judiciously scaled to give sensible matrix elements (neither zero nor infinite) acting on the vacuum. There is no guarantee that a product of two such well-behaved operators at the same point is itself well behaved. A simple test is to see if the product has a finite matrix element in the vacuum as the points approach each other. In the example above, this was the case; in fact, the mean value of the composite operator is zero since its factors create and destroy different (right- or left-moving) fermions. This is not the case for the next item: the operator $\psi_+^\dagger(x)\psi_+(x)$, say for $x = 0$. We define it by a limiting process called *point splitting* as follows:

$$\begin{aligned}
\psi_+^\dagger(0)\psi_+(0) &= \lim_{x\to 0} \frac{1}{2\pi\alpha} e^{-i\sqrt{4\pi}\phi_+(x)} e^{i\sqrt{4\pi}\phi_+(0)} \\
&= \lim_{x\to 0} \frac{1}{2\pi\alpha} : e^{-i\sqrt{4\pi}\phi_+(x)} e^{i\sqrt{4\pi}\phi_+(0)} : e^{4\pi G_+(x)} \\
&= \lim_{x\to 0} \frac{i}{2\pi(x+i\alpha)} : 1 - i\sqrt{4\pi}\frac{\partial\phi_+}{\partial x}x + \cdots : \qquad (17.84) \\
&= \lim_{x\to 0} \frac{i}{2\pi x} + \frac{1}{\sqrt{\pi}}\frac{\partial\phi_+}{\partial x} + \cdots \qquad (17.85)
\end{aligned}$$

These manipulations need some explanation. We perform a Taylor expansion only within the normal-ordering symbols because only the normal-ordered operators have nice (differentiable) matrix elements. Thus, terms of higher order in x and sitting within the symbol are indeed small and can be dropped as $x \to 0$. Consider next the $x + i\alpha$ in the denominator. Is it permissible to drop the α in comparison to x, even though x itself is being sent to 0? Yes. We must always treat any distance x in the continuum theory as being much larger than α, which is to be sent to 0 whenever possible. Finally, note that the density operator in question has an infinite c-number part which is displayed in front. This reflects the fact that the vacuum density of right movers is infinite due to the Dirac sea. If we define a normal-ordered density, i.e., take away the singular vacuum average from it, we obtain

$$: \psi_+^\dagger(x)\psi_+(x) : = \frac{1}{\sqrt{\pi}}\frac{\partial\phi_+}{\partial x}. \qquad (17.86)$$

A similar result obtains for the left-mover density. Combining the two, we get some very famous formulae in bosonization:

$$\begin{aligned}
j_0 &= \frac{1}{\sqrt{\pi}}\frac{\partial\phi}{\partial x}, \qquad (17.87) \\
j_1 &= \frac{1}{\sqrt{\pi}}\frac{\partial(\phi_+ - \phi_-)}{\partial x} \qquad (17.88) \\
&= -\frac{\partial_x\theta}{\sqrt{\pi}} = -\frac{\Pi}{\sqrt{\pi}}. \qquad (17.89)
\end{aligned}$$

For the Lagrangian formalism, we may assemble these into

$$j_\mu = \frac{\varepsilon_{\mu\nu}}{\sqrt{\pi}}\partial_\nu\phi. \qquad (17.90)$$

We close this section with two more results. First, a very useful but odd-looking relation:

$$\left[\frac{-1}{\pi\alpha}\cos\sqrt{4\pi}\phi\right]^2 = -\frac{1}{\pi}\left(\frac{\partial\phi}{\partial x}\right)^2 + \frac{1}{2\pi^2\alpha^2}\cos\sqrt{16\pi}\phi, \tag{17.91}$$

dropping c-numbers.

Here is a sketch of the derivation.

$$\left[\frac{-1}{\pi\alpha}\cos\sqrt{4\pi}\phi(0)\right]^2 = \frac{1}{4\pi^2\alpha^2}\lim_{x\to 0}\left[e^{i\sqrt{4\pi}\phi(x)} + cc\right]\cdot\left[e^{i\sqrt{4\pi}\phi(0)} + cc\right]. \tag{17.92}$$

Now we combine exponentials easily because everything commutes. We find that

$$\left[\frac{-1}{\pi\alpha}\cos\sqrt{4\pi}\phi(0)\right]^2 \tag{17.93}$$

$$= \frac{1}{2\pi^2\alpha^2}\lim_{x\to 0}\left[\cos(\sqrt{4\pi}(\phi(x)+\phi(0))) + \cos(\sqrt{4\pi}(\phi(x)-\phi(0)))\right]. \tag{17.94}$$

In the first cosine we can simply double the angle to $\sqrt{16\pi}\phi(0)$. In the second, we want to do a Taylor expansion, but can only do it within a normal-ordered operator. So we proceed as follows, using Eq. (17.57) along the way:

$$\begin{aligned}
&\frac{1}{2\pi^2\alpha^2}\lim_{x\to 0}\cos(\sqrt{4\pi}(\phi(x)-\phi(0))) \\
&= \lim_{x\to 0}\frac{1}{2\pi^2\alpha^2} : \cos(\sqrt{4\pi}(\phi(x)-\phi(0))) : \frac{\alpha^2}{x^2+\alpha^2} &(17.95)\\
&= \lim_{x\to 0}\frac{1}{2\pi^2\alpha^2} : 1 - \frac{x^2}{2}(4\pi)(\partial_x\phi)^2 + \cdots : \frac{\alpha^2}{x^2+\alpha^2} \\
&= -\frac{1}{\pi}\left(\frac{\partial\phi}{\partial x}\right)^2 + c\text{-number}, &(17.96)
\end{aligned}$$

where in the last line you must remember that $x \gg \alpha$ even at small x. Substituting this into Eq. (17.94), we arrive at Eq. (17.91).

Similar arguments lead to

$$\left[\frac{1}{\pi\alpha}\sin\sqrt{4\pi}\phi\right]^2 = -\frac{1}{\pi}\left(\frac{\partial\phi}{\partial x}\right)^2 - \frac{1}{2\pi^2\alpha^2}\cos\sqrt{16\pi}\phi. \tag{17.97}$$

In the field theory literature you will not see the second term mentioned. The reason is that at weak coupling this operator is highly irrelevant (or non-renormalizable). The reason for our keeping it is that in the presence of strong interactions it will become relevant.

Finally, having seen the dictionary reproduce various fermionic operators in terms of bosons, we may ask "What about the Hamiltonian?" Indeed, the dictionary may be used to show that

$$H_F = \int \left(\psi_+^\dagger(x)(-i\partial_x)\psi_+(x) + \psi_-^\dagger(x)(i\partial_x)\psi_-(x)\right) dx$$
$$= \frac{1}{2}\int (\Pi^2 + (\partial_x\phi)^2)dx = H_B. \tag{17.98}$$

Exercise 17.4.1 *Prove Eq. (17.98). I suggest you:*

- *use the symmetric derivatives; for example,*

$$\psi_+^\dagger(x)(\partial_x)\psi_+(x) = \lim_{\varepsilon\to 0}\psi_+^\dagger(x)\left(\frac{\psi_+(x+\varepsilon)-\psi_+(x-\varepsilon)}{2\varepsilon}\right); \tag{17.99}$$

- *expand bosonic exponentials to quadratic order in ε;*
- *remember that just as $x \gg \alpha$, so is $\varepsilon \gg \alpha$ in combinations like $\varepsilon \pm i\alpha$; drop total derivatives and c-numbers.*

17.5 Relativistic Bosonization for the Lagrangians

Often one uses bosonization in a relativistic theory. Here is the dictionary in *Euclidean* space with the notation defined for free fields:

$$Z_F = \int [d\bar\psi][d\psi]e^{-S_0(\psi)} = \int [d\bar\psi][d\psi]e^{-\int \bar\psi \partial\!\!\!/ \psi d^2x}, \tag{17.100}$$

$$Z_B = \int [d\phi]e^{-S_0(\phi)} = \int [d\phi]e^{-\int \frac{1}{2}(\nabla\phi)^2 d^2x}, \tag{17.101}$$

$$\bar\psi\partial\!\!\!/\psi \to \frac{1}{2}(\nabla\phi)^2 = \frac{1}{2}\left[(\partial_\tau\phi)^2 + \partial_x\phi)^2\right], \tag{17.102}$$

$$\bar\psi\gamma^\mu\psi \to \frac{\varepsilon^{\mu\nu}}{\sqrt{\pi}}\partial_\nu\phi \quad (=j^\mu), \tag{17.103}$$

$$\bar\psi\psi \to -\Lambda\cos\sqrt{4\pi}\phi, \tag{17.104}$$

$$\bar\psi i\gamma^5\psi \to \Lambda\sin\sqrt{4\pi}\phi, \tag{17.105}$$

$$(\bar\psi\psi)^2 = \left[-\Lambda\cos\sqrt{4\pi}\phi\right]^2 = -\frac{1}{2\pi}(\nabla\phi)^2. \tag{17.106}$$

Several points are worth noting:

- In the relativistic equations we make the replacement

$$\frac{1}{\pi\alpha} \to \Lambda, \tag{17.107}$$

where Λ is the cut-off in two-dimensional Euclidean momentum, in contrast to $1/\alpha$, which was a momentum cut-off on spatial momenta.
- In the last equation the highly irrelevant $\cos\sqrt{16\pi}\phi$ has been dropped and we have 2π and not π in the denominator because the point-splitting is done in space and time and there is a compensating sum over *two* squared derivatives.

- In Eq. (17.100), I integrate $e^{-S_0(\psi)}$ and not $e^{+S_0(\psi)}$ as in earlier chapters where I wanted to emphasize that the sign meant nothing for Grassmann actions. Here I use the e^{-S_0} for both to simplify the boson–fermion dictionary.

References and Further Reading

[1] D. C. Mattis and E. Lieb, Journal of Mathematical Physics, **6**, 304 (1965).
[2] J. M. Luttinger, Physical Review **119**, 1153 (1960).
[3] A. Luther and I. Peschel, Physical Review B, **12**, 3908 (1975).
[4] S. Coleman, Physical Review D, **11**, 2088 (1978).
[5] S. Mandelstam, Physical Review D, **11**, 3026 (1975).

18
Bosonization II: Selected Applications

We now pass from this rather sterile business of deriving the bosonization formulas to actually using them. Of the countless applications, I have chosen a few that I am most familiar with. While my treatment of the subject will not be exhaustive, it should prepare you to read more material dealing with the subject.

The applications are to the massless Schwinger and Thirring models, the uniform- and random-bond Ising models, and the Tomonaga–Luttinger and Hubbard models. There is an enormous body of literature devoted to these models. I will simply focus on those aspects that illustrate bosonization in the simplest possible terms.

18.1 Massless Schwinger and Thirring Models

The first two examples are the easiest since the Dirac fermion is present from the outset. In later examples it will arise after some manipulations and approximations.

18.1.1 The Massless Schwinger Model

This model was invented by Schwinger [1] to describe electrodynamics in two dimensions. The Euclidean Lagrangian density is

$$\mathcal{L} = \bar{\psi}\partial\!\!\!/\psi - e_0 j^{\mu} A_{\mu} + \frac{1}{2}(\varepsilon^{\mu\nu}\partial_{\mu}A_{\nu})^2. \tag{18.1}$$

Schwinger solved this by functional methods; we can now solve it by bosonization. Writing

$$j^{\mu}A_{\mu} = \frac{1}{\sqrt{\pi}}\varepsilon^{\mu\nu}\partial_{\nu}\phi A_{\mu} = -\phi\frac{1}{\sqrt{\pi}}\varepsilon^{\mu\nu}\partial_{\nu}A_{\mu}, \tag{18.2}$$

we can complete the square on the A integral and find the bosonic Lagrangian density

$$\mathcal{L} = \frac{1}{2}(\nabla\phi)^2 + \frac{e_0^2}{\pi}\phi^2. \tag{18.3}$$

This means that there is a scalar pole (in the Minkowski space propagator) at

$$m^2 = \frac{2e_0^2}{\pi}. \tag{18.4}$$

Here is what is going on. In one space dimension there is no photon. In the gauge $A_0 = 0$, we just have an instantaneous electrostatic potential A_1 between fermions. This Coulomb interaction yields a linear potential or constant force because the flux cannot spread out in $d = 1$. The density oscillations (sound) which would have been massless are now massive due to the long-range interaction. Schwinger's point was that gauge invariance did not guarantee a massless electromagnetic field.

Note for now that if we add a fermion mass term, the problem cannot be solved exactly because we are adding a $\cos\sqrt{4\pi}\phi$ term.

18.1.2 Massless Thirring Model

The Thirring model [2] describes a current–current interaction:

$$\mathcal{L} = \bar{\psi}\partial\!\!\!/\psi - \frac{g}{2}j^{\mu}j_{\mu}. \tag{18.5}$$

Upon bosonizing, this becomes

$$\mathcal{L} = \frac{1}{2}\left(1+\frac{g}{\pi}\right)(\nabla\phi)^2. \tag{18.6}$$

This model was a milestone because it exhibited correlation functions that decayed with a g-dependent power. Consider, for example, the $\bar{\psi}\psi - \bar{\psi}\psi$ correlation at equal time. In the non-interacting theory it has to fall as

$$\langle\bar{\psi}(r)\psi(r)\bar{\psi}(0)\psi(0)\rangle \simeq \frac{1}{r^2} \tag{18.7}$$

just from dimensional analysis: $[\psi] = \frac{1}{2}$ in momentum units. In the bosonized version this would be reproduced as follows (in the Hamiltonian version):

$$\begin{aligned}\langle\bar{\psi}(r)\psi(r)\bar{\psi}(0)\psi(0)\rangle &= \frac{1}{\pi^2\alpha^2}\langle\cos\sqrt{4\pi}\phi(r)\cos\sqrt{4\pi}\phi(0)\rangle \\ &= \frac{1}{2\pi^2\alpha^2}\left(\frac{\alpha^2}{r^2}\right)^{4\pi/4\pi} \simeq \frac{1}{r^2}.\end{aligned} \tag{18.8}$$

(In the path integral version Λ would replace $1/(\pi\alpha)$.) This formula is valid if the kinetic term has a coefficient of $\frac{1}{2}$, whereas now it is $\frac{1}{2}\left(1+\frac{g}{\pi}\right)$ due to interactions. So we define a new field

$$\phi' = \sqrt{\left(1+\frac{g}{\pi}\right)}\phi, \tag{18.9}$$

in terms of which

$$\mathcal{L} = \frac{1}{2}(\nabla\phi')^2, \tag{18.10}$$

$$\bar{\psi}\psi = -\frac{1}{\pi\alpha}\cos\sqrt{\frac{4\pi}{1+\frac{g}{\pi}}}\phi', \tag{18.11}$$

$$\langle\bar{\psi}(r)\psi(r)\bar{\psi}(0)\psi(0)\rangle = \frac{1}{2\pi^2\alpha^2}\left(\frac{\alpha^2}{r^2}\right)^{4\pi/4\pi(1+\frac{g}{\pi})} \simeq \frac{1}{r^\gamma}, \text{ where}$$

$$\gamma = \frac{2}{(1+\frac{g}{\pi})}. \tag{18.12}$$

The thing to notice is that the anomalous power or dimension of the correlation function varies continuously with the interaction strength. Once again, we see how in a massless theory the correlations can decay with a power not dictated by the engineering dimension of the operator: the cut-off, which has to be introduced to make sense of the theory (now in the guise of α), serves as the additional dimensional parameter.

The *massive Thirring* model is defined by adding $-m\bar{\psi}\psi$, which leads to the following bosonized Euclidean Lagrangian density:

$$\mathcal{L} = \frac{1}{2}\left(1+\frac{g}{\pi}\right)(\nabla\phi)^2 - \frac{m}{\pi\alpha}\cos\sqrt{4\pi}\phi \tag{18.13}$$

$$= \frac{1}{2}\left(1+\frac{g}{\pi}\right)(\nabla\phi)^2 - m\Lambda\cos\sqrt{4\pi}\phi. \tag{18.14}$$

We shall return to this model in the next chapter. We now move on to two applications of bosonization to condensed matter: the correlation functions of the Ising model at criticality and of the random-bond Ising model whose bonds fluctuate from site to site around their critical value. In the latter case we have to find the correlation function averaged over bond realizations. We will see how to do this using what is called the replica trick.

18.2 Ising Correlations at Criticality

Let us recall some key features of the $d = 2$ Ising model. The partition function is

$$Z = \sum_{s=\pm 1} e^{K\sum_{\langle ij\rangle} s_i s_j}, \tag{18.15}$$

where $\langle i,j\rangle$ tells us that the Ising spins $s_i = \pm 1$ and $s_j = \pm 1$ are nearest neighbors on the square lattice. The sum in the exponent is over bonds of the square lattice.

The correlation function

$$G(r) = \langle s_r s_0\rangle, \tag{18.16}$$

where 0 is the origin and r a point a distance r away, is known to fall at the critical point as

$$G(r) \simeq \frac{1}{r^\eta} = \frac{1}{r^{\frac{1}{4}}}. \tag{18.17}$$

This power is universal. This exponent of $\eta = \frac{1}{4}$ is rather difficult to derive and the reason will become clear as we go along. I will now describe a trick due to Itzykson and Zuber [3] that uses bosonization to circumvent this.

Let us recall the extreme anisotropic τ-continuum limit of Fradkin and Susskind [4],

$$K_x = \tau, \tag{18.18}$$

$$K_\tau^* = \lambda\tau \quad (\tau \to 0), \tag{18.19}$$

which leads to the transfer matrix

$$T = e^{-\tau H}, \quad \text{where} \tag{18.20}$$

$$H = -\lambda \sum \sigma_1(m) - \sum \sigma_3(m)\sigma_3(m+1). \tag{18.21}$$

The idea of Fradkin and Susskind is that anisotropy will change the metric but not the exponent for decay or any other universal quantity.

Next, we follow Schultz, Mattis, and Lieb [5] and trade the Pauli matrices for full-fledged Fermi operators defined by

$$\psi_1(n) = \frac{1}{\sqrt{2}} \left(\prod_{-\infty}^{n-1} \sigma_1 \right) \sigma_2(n), \tag{18.22}$$

$$\psi_2(n) = \frac{1}{\sqrt{2}} \left(\prod_{-\infty}^{n-1} \sigma_1 \right) \sigma_3(n). \tag{18.23}$$

We are now considering an infinite spatial lattice, and the "string" of σ_1's comes from the far left to the point $n-1$. The *Majorana* fermions obey

$$\{\psi_i(n), \psi_j(m)\} = \delta_{ij}\delta_{mn}. \tag{18.24}$$

Let us imagine that our lattice has a spacing a. Define continuum operators $\psi_c = \psi/\sqrt{a}$ that obey Dirac δ-function anticommutation rules as $a \to 0$. In terms of these, we get in the continuum limit the following continuum Hamiltonian $H_c = H/a$:

$$H_c = \frac{1}{2} \int \psi^T(\alpha P + \beta m)\psi \, dx, \qquad m = (1-\lambda)/a, \tag{18.25}$$

where α is now σ_1, as mentioned earlier.

We have seen in Chapter 8 how this quadratic Hamiltonian is diagonalized. By filling all the negative energy levels we get the ground-state energy E_0. This energy (per unit spatial volume) is essentially the free energy per site of the square lattice model.

Let us turn instead to the two-point correlation functions. Now, it may seem that in a free-field theory this should be trivial. But it is not, because we want the two-point function of the spins, which are non-local functions of the Fermi field.

Let us find the equal-time correlation of two spins a distance n apart in space. (The power law for decay should be the same in all directions even though length scales are not.) Thus, we need to look at

$$\begin{aligned}\langle 0|\sigma_3(0)\sigma_3(n)|0\rangle &= \langle 0|\sigma_3(0)\sigma_3(1)\sigma_3(1)\cdots\sigma_3(n-1)\sigma_3(n)|0\rangle \\ &= \langle 0|[2i\psi_1(0)\psi_2(1)\cdot 2i\psi_1(1)\psi_2(2)\cdots \\ &\quad \cdots\times 2i\psi_1(n-1)\psi_2(n)]|0\rangle. \end{aligned} \tag{18.26}$$

We find that *the two-point function of spins is a 2n-point function of fermions*. This becomes very hard to evaluate if we want the limit of large n: we must evaluate a Pfaffian of arbitrarily large size. We are, however, presently interested in obtaining just the power law of the asymptotic decay of the spin–spin correlation.

Bosonization cannot be invoked since it applies only to Dirac fermions, so we follow the trick of Itzykson and Zuber [3]. First, note that apart from the end factors, $\psi_1(0)$ at the left and $\psi_2(n)$ at the right, we have the product over sites of

$$2i\psi_2(i)\psi_1(i) = -ie^{\frac{i\pi}{2}[2i\psi_2(i)\psi_1(i)]} \tag{18.27}$$

$$= e^{\frac{i\pi}{2}[2i\psi_2(i)\psi_1(i)-1]}. \tag{18.28}$$

This equation follows from the fact that $2i\psi_2(i)\psi_1(i)$ is just like a Pauli matrix (with square unity) for which

$$\sigma_1 = (-i)e^{i\frac{\pi}{2}\sigma_1}. \tag{18.29}$$

The exponent in Eq. (18.28) is just $\frac{i\pi}{2}(\overline{\psi}\psi - 1)$. I will drop the 1 since it makes no difference to the decay of the correlation function. When we form the product over sites it becomes a sum, and in the continuum limit the integral of $\overline{\psi}\psi$ between 0 and R, where $R = na$ is the distance between the points in laboratory units. There is no simple way to evaluate

$$G(R) \simeq \langle 0|e^{\frac{i\pi}{2}\int_0^R \overline{\psi}(x)\psi(x)dx}|0\rangle. \tag{18.30}$$

Consider now an auxiliary problem, where we have made two non-interacting copies of the Ising system, with spins called s and t, and associated Pauli matrices σ and τ and Majorana fermions ψ and χ. It is clear that

$$\begin{aligned}\langle s_n t_n s_0 t_0\rangle &= \langle s_n s_0\rangle\langle t_n t_0\rangle \\ &= [G(n)]^2, \end{aligned} \tag{18.31}$$

since the thermal averages proceed independently and identically for the two sectors. The trick is to find G^2 and then take the square root. Let us see how this works. First, we will be dealing with products of the following terms:

$$2i\psi_2\psi_1 2i\chi_2\chi_1 = -[2i\chi_1\psi_1]\cdot[2i\chi_2\psi_2] \tag{18.32}$$

$$= e^{\frac{i\pi}{2}[2i\psi_1\chi_1 + 2i\psi_2\chi_2]} \tag{18.33}$$

$$= e^{i\pi :\psi_D^\dagger\psi_D:}. \tag{18.34}$$

The last step needs some explanation. Let us form a Dirac fermion

$$\psi_{\mathrm{D}} = \frac{\psi + i\chi}{\sqrt{2}} \tag{18.35}$$

and consider its charge density:

$$\psi_{\mathrm{D}}^{\dagger}\psi_{\mathrm{D}} = \frac{1}{2}(\psi_1 - i\chi_1)(\psi_1 + i\chi_1) + (1 \to 2) \tag{18.36}$$

$$= i\psi_1\chi_1 + i\psi_2\chi_2 + 1 \tag{18.37}$$

$$:\psi_{\mathrm{D}}^{\dagger}\psi_{\mathrm{D}}: = i\psi_1\chi_1 + i\psi_2\chi_2, \tag{18.38}$$

where I have used the fact that the vacuum density of the Dirac fermions is 1 per site: half for the right movers, half for the left movers. (Recall that in momentum space half the states are filled, which translates into half per site in real space.)

What about the fact that the Dirac fermion that comes out of the Ising model has a first quantized Hamiltonian $H = \alpha P + \beta m$, where $\alpha = \sigma_1$ and $\beta = \sigma_2$, whereas the one used in bosonization has $\alpha = \sigma_3$ and $\beta = \sigma_2$? It does not matter: the two are connected by a unitary transformation (a $\frac{\pi}{2}$ rotation generated by σ_2), and $\psi_{\mathrm{D}}^{\dagger}\psi_{\mathrm{D}}$ is invariant under this.

We now reveal our punch line: in view of the above,

$$G^2(R) = \langle 0|e^{i\pi\int_0^R :\psi_{\mathrm{D}}^{\dagger}(x)\psi_{\mathrm{D}}(x):dx}|0\rangle \tag{18.39}$$

$$= \langle 0|e^{\int_0^R i\sqrt{\pi}\partial_x\phi dx}|0\rangle \tag{18.40}$$

$$= \langle 0|e^{i\sqrt{\pi}\phi(R)}e^{-i\sqrt{\pi}\phi(0)}|0\rangle \tag{18.41}$$

$$\simeq \frac{1}{R^{\frac{1}{2}}}, \tag{18.42}$$

where I have recalled Eq. (17.63). *Thanks to bosonization, a non-local Green's function in the Fermi language has become a local two-point function in the bosonic language.* Several points of explanation are needed. First, we have used Eq. (17.87) in going from the first to the second equation in the above sequence. Next, we have used the fact that at the critical point the Fermi theory has no mass. Thus, the bosonic ground state in which the bosonic correlator is evaluated is the free-field vacuum. Lastly, we have used Eq. (17.63) to evaluate the desired two-point function. (I have ignored α compared to R in the denominator and dropped the power of α in the numerator since I just want the R dependence.) Taking the square root, we find the desired decay law $G(R) \simeq R^{-\frac{1}{4}}$.

I have been careless about the end points, where the product does not follow the pattern. If this is taken into account, one finds that we must use $\sin\sqrt{\pi}\phi$ in place of $e^{i\sqrt{\pi}\phi}$. This does not, however, change the critical exponent. If one tries the Itzykson–Zuber trick away from criticality one finds that one has to find the correlation function of the same operator but in the theory with an interaction $\cos\sqrt{4\pi}\phi$, which is the bosonized version of the harmless-looking mass term in the free Dirac theory.

To conclude, the following were the highlights of our derivation of $G(R)$:

- The critical theory of the Ising model in the extreme anisotropic τ-continuum limit is a massless Majorana theory.
- The two-point function of spins a distance R apart is given by the average of the exponential of the integral of a Majorana fermion bilinear from 0 to R.
- By considering the square of G, we made the integrand referred to above into the normal-ordered Dirac charge density.
- By bosonizing the latter into the derivative of ϕ, we got rid of the integral in the exponent and were left with just a two-point function of $e^{i\sqrt{\pi}\phi}$'s coming from the end points of the integration.
- By evaluating this in the free-field theory we found that G^2 falls off like $R^{-\frac{1}{2}}$. We then took the square root of this answer.

18.3 Random-Bond Ising Model

Consider an Ising model in which the coupling between neighboring spins is not uniformly K, but randomly chosen at each bond from an ensemble. This can happen in real systems due to vacancies, lattice imperfections, and so on. We should therefore imagine that each sample is different and translationally non-invariant. The study of the $d = 2$ Ising model with such a complication was pioneered by Dotsenko and Dotsenko [6] (referred to as DD hereafter) in a very influential paper. I will now describe their work, as well as further contributions by others. You will see bosonization at work once more.

First, let us understand what we want to calculate in a random system. The behavior of an individual system with bonds chosen in a sample-specific way from the ensemble of possibilities is not interesting, unless by luck we are dealing with a property that is sample independent. (The free energy per site in the infinite volume limit is one such object.) In general, what one wants are physical quantities, first calculated sample by sample *and then* averaged over samples. This is called a *quenched average*, and is a lot more difficult problem than the *annealed average* in which one treats the bond strength as another statistical variable in thermal equilibrium, just like the Ising spins themselves. Which one should one use? If the bonds are frozen into some given values over the period of the measurements, we must take them as a fixed external environment and do the quenched average. If they fluctuate ergodically over the period of measurement, we must do the annealed average. The DD problem deals with quenched averages. In this case one must work with the averaged free energy $\overline{f}$ obtained by averaging $\ln Z$ over all samples. The temperature-derivative of $\overline{f}$ gives the average internal energy, and so on. (As mentioned above, it is known that in the infinite-volume limit, each sample will give the same f. This is not true for all quantities.) Similarly, one can take two spins a distance R apart and find the correlator G sample by sample. This will depend on the absolute values of the coordinates, since there is no translational invariance. However, the ensemble average $\overline{G}$ will depend only on R. Besides these mean values, one can calculate the fluctuations around these mean values. Given the distance R and a temperature, there is a unique number $G(R)$ in a pure system describing the correlation. In our case there is probability $P(G(R))$ that $G(R)$ will have this or that value. We will return to this point at the end.

We have seen that the Ising model is described by a non-interacting Majorana field theory. We can take this Hamiltonian and write Z as a Euclidean path integral over Grassmann numbers as follows [Eq. (9.71)]:

$$Z_M(K) = \int [d\psi] \exp\left[-\int \frac{1}{2}\bar{\psi}(\partial\!\!\!/ + m)\psi \, d^2x \right]. \tag{18.43}$$

In the above, the mass m is determined by λ or equivalently the temperature. It vanishes at the critical temperature. We are assuming we are close enough to criticality for this continuum theory to be valid. Suppose now that the bonds, instead of being uniform, vary from point to point on the two-dimensional lattice, never straying too far from criticality. This means that $m = m(x)$ varies with the two-dimensional coordinate x, and $Z_M = Z_M(m(x))$ is therefore a functional of $m(x)$. Let us assume that the probability distribution for m is a Gaussian at each site:

$$P(m(x)) = \prod_x e^{-(m(x)-m_0)^2/2g^2}. \tag{18.44}$$

Hereafter we will focus on the case of zero mean: $m_0 = 0$. Thus, each bond fluctuates symmetrically around the critical value. To find $\bar{f}$ we must calculate

$$\bar{f} = \int P(m(x)) \ln Z_M(m(x)) dm(x). \tag{18.45}$$

Since we are averaging $\ln Z$ and not Z, we see that the problem is not as easy as that of adding an extra thermal variable $m(x)$. We circumvent this using what is called the *replica trick*. We use

$$\ln Z = \lim_{n\to 0} \frac{Z^n - 1}{n}. \tag{18.46}$$

In what follows, we will drop the minus one in the numerator since it adds a constant to the answer, and also drop the factor of inverse n since it multiplies the answer by a factor without changing any of the critical properties. In short, in Eq. (18.45) we can replace $\ln Z$ by Z^n (and of course send n to zero at the end). But Z^n is just the partition function of n replicas of the original model. Thus,

$$\begin{aligned}\bar{f} &= \int \left[\prod_1^n d\psi_i\right] \exp\left[-\int \sum_1^n \frac{1}{2}\bar{\psi}_i(\partial\!\!\!/ + m(x))\psi_i d^2x \right] e^{-m^2(x)/2g^2} dm(x) \\ &= \int \left[\prod_1^n d\psi_i\right] \exp\left[\int \left(-\frac{1}{2}\sum_1^n \bar{\psi}_i(\partial\!\!\!/)\psi_i + \frac{g^2}{8}\left(\sum_1^n \bar{\psi}_i\psi_i\right)^2 \right) d^2x \right].\end{aligned} \tag{18.47}$$

Thus, the random model has been traded for an interacting but translationally invariant theory, called the n-component Gross–Neveu model [7, 8]. The above is a shortened derivation of the DD result. It is understood that all calculations are performed for general

n, and that in any analytic expression where n occurs, the limit $n \to 0$ is taken. The value of the DD work is that it shows in detail that this crazy replica procedure is indeed doing the ensemble average we want to do.

Now DD proceed to deduce two results:

- The specific heat will have a $\ln \ln$ divergence instead of the $\ln$ divergence of the pure system. To derive this, one must also explore the case $m_0 \neq 0$.
- The average two-point function $\overline{G(R)}$ falls essentially like R^0 as compared to the $R^{-1/4}$ in the pure system.

While the first result seemed reasonable, the second did not for the following reason. It is known (and we will see) that when $n = 0$, the Gross–Neveu model is essentially a free-field theory at large distances, the interactions falling logarithmically. It is known in that in such asymptotically free theories correlations are usually that of a free field up to logarithms. Thus, we can accept the change in the specific heat from log to log-log, but not the change of the decay exponent from $\frac{1}{4}$ to 0. It was, however, difficult to see what had gone wrong in the rather formidable calculation of DD, which involved an average like Eq. (18.30), difficult enough in free-field theory, in an interacting theory.

I decided to approach the problem in a different way [9]. Recall how, in the pure case, by considering the square of the correlation, we could convert the problem, via bosonization, to the evaluation of a two-point function. Let us try the same trick here. Consider *any one sample* with some given set of bonds. On it, imagine making two copies of the Ising system. Then, following the reasoning from the last section.

$$G^2(0,R,m(x)) = \frac{\int [d\psi_{\mathrm{D}}][d\overline{\psi}_{\mathrm{D}}] \exp\left[-\int d^2x \overline{\psi}_{\mathrm{D}}(\not{\partial} + m(x))\psi_{\mathrm{D}}\right] \exp\left[i\pi \int_0^R : \psi_{\mathrm{D}}^{\dagger}\psi_{\mathrm{D}} : dx\right]}{Z_{\mathrm{D}}(m(x))}. \tag{18.48}$$

In the above, G remembers that one spin was at the origin and the other at R (in both copies). In principle one must move this pair over the lattice maintaining this separation R. However, this is obviated by the subsequent replica averaging which restores translation and rotational invariance.

The good news is that, due to the doubling, we have a Dirac fermion. The bad news is that the normalizing partition function downstairs is itself a functional of $m(x)$, which makes it hard to average G^2. So, we multiply top and bottom by Z_{D}^{n-1} and set $n = 0$. This gets rid of the denominator and adds $n-1$ copies upstairs. We then have

$$G^2(0,R,m(x)) \tag{18.49}$$

$$= \int \left[\prod_1^n d\psi_i d\overline{\psi}_i\right] \exp\left[-\int \sum_1^n \overline{\psi}_i(\not{\partial} + m(x))\psi_i d^2x\right] e^{i\pi \int_0^R \psi_1^{\dagger}\psi_1 dx}, \tag{18.50}$$

where the subscript 1 labels the species we started with and all fermions are understood to be Dirac. If we now do the Gaussian average over $m(x)$, we just complete the squares on the mass term and obtain

$$\overline{G^2(R)} \tag{18.51}$$

$$= \int \left[\prod_1^n d\psi_i \overline{\psi_i}\right] \exp\left[\int \left[\sum_1^n -\overline{\psi}_i(\partial\!\!\!/)\psi_i + \frac{g^2}{2}\left(\sum_1^n \overline{\psi}_i\psi_i\right)^2\right] d^2x\right] e^{i\pi \int_0^R \psi_1^\dagger \psi_1 dx}. \tag{18.52}$$

Let us now bosonize this theory using the results from Section 17.5 to obtain:

$$\overline{G^2(R)}$$
$$= \int \prod_{i=1}^n d\phi_i \exp\left[\int d^2x \sum_1^n -\frac{1}{2}(\nabla\phi_i)^2 + \frac{g^2\Lambda^2}{2}\left[\sum_1^n \cos(\sqrt{4\pi}\phi_i)\right]^2\right]$$
$$\times \exp\left[i\sqrt{\pi}(\phi_1(R) - \phi_1(0))\right]. \tag{18.53}$$

Consider the square of the sum over cosines. The diagonal terms can be lumped with the free-field term using Eq. (17.106):

$$\left[\Lambda \cos\sqrt{4\pi}\phi\right]^2 = -\frac{1}{2\pi}(\nabla\phi)^2. \tag{18.54}$$

(The relativistic formula ignores the $\cos\sqrt{16\pi}\phi$ term, which is fortunately highly irrelevant in the present weak coupling analysis.) In terms of the new field

$$\phi' = \left(1 + \frac{g^2}{2\pi}\right)^{\frac{1}{2}} \phi, \tag{18.55}$$

once again called ϕ in what follows,

$$\overline{G^2(R)} = \left\langle \exp\left[i\sqrt{\frac{\pi}{1+g^2/2\pi}}\phi_1(R)\right] \exp\left[-i\sqrt{\frac{\pi}{1+g^2/2\pi}}\phi_1(0)\right]\right\rangle_g, \tag{18.56}$$

where the subscript g tells us that the average is taken with respect to the vacuum of an interacting field theory with action

$$S = \int d^2x \left(\sum_1^n -\frac{1}{2}(\nabla\phi_i)^2 + \frac{g^2\Lambda^2}{2}\left[\sum_i \sum_{j\neq i} \cos\left(\sqrt{\frac{4\pi}{1+g^2/2\pi}}\phi_i\right) \cos\left(\sqrt{\frac{4\pi}{1+g^2/2\pi}}\phi_j\right)\right]\right). \tag{18.57}$$

Unlike in the homogeneous Ising model, where we had a two-point function to evaluate in a free-field theory, we have here an interacting theory. Since g^2 measures the width of the bond distribution, perhaps we can work first with small g in a perturbation expansion? For example, if $g^2 = 0.001$ we could read off the answer using perturbation theory. Unfortunately this is not possible. The problem is that the coupling in this theory cannot be a constant, it has to be function $g(\Lambda)$ because there are ultraviolet divergences. These divergences give Λ-dependent answers for quantities of interest, and to neutralize this unwanted dependence we must choose g as a function of Λ, i.e., we must renormalize.

The first step is to compute the β-function:

$$\beta(g) = \frac{dg}{d\ln\Lambda}. \tag{18.58}$$

This computation, best done in the fermionic version, involves finding, to any given order, the contributions the eliminated modes make to the interaction between the surviving modes. To second order in g^2 one draws the three possible one-loop graphs and integrates the loop momenta from the old Λ to the new. For the n-component Gross–Neveu model [7, 8], one knows that

$$\beta(g) = (1-n)\frac{g^3}{2\pi} + \text{higher order}. \tag{18.59}$$

Typically $n \geq 2$, and this leads to a theory where the coupling grows in the infrared, but here, with $n = 0$, it is the opposite. If the initial bare coupling is $g(a)$, where $a = 1/\Lambda$ is the lattice size, the coupling at scale R is obtained by integrating

$$\frac{dg}{d\ln\Lambda} = \frac{g^3}{2\pi} \tag{18.60}$$

from $\Lambda = 1/a$ to $\Lambda = 1/R$ to obtain

$$g^2(R) = \frac{g^2(a)}{1 + \frac{g^2(a)}{\pi}\ln(R/a)} \tag{18.61}$$

$$\simeq \frac{\pi}{\ln R/a} \quad \text{for } R/a \to \infty. \tag{18.62}$$

This means the following. If we naively perturb the theory defined at scale a to find a quantity like the correlation function at scale R, we will find that the effective parameter is not $g(a)$ but $g^2(a)\ln R/a$. This is because the result Eq. (18.61) will appear as a badly behaved power series,

$$g^2(R) = g^2(a)\left(1 - \frac{g^2(a)}{\pi}\ln(R/a) + \cdots\right). \tag{18.63}$$

What the RG does for us is to sum the series and allow us to use a coupling at scale R that is actually very small for large R/a. Not only will the effective coupling be small, there will be no large logs when we describe physics at scale R.

To exploit this, we have to follow the familiar route of integrating the Callan–Symanzik equations to relate $\overline{G^2(R, g(\Lambda), \Lambda = 1/a)}$ computed with the initial coupling and cut-off to $\overline{G^2(R, g(1/R), \Lambda = 1/R)}$.

So let us recall the solution given in Eqs. (14.108)–(14.113):

$$\overline{G^2(R, g(\Lambda), \Lambda)} = \exp\left[\int_{g(\Lambda)}^{g(1/R)} \frac{\gamma(g)}{\beta(g)}dg\right]\overline{G^2(R, g(1/R), 1/R)}. \tag{18.64}$$

Now, the dimensionless function $\overline{G^2(R, g(1/R), 1/R)} = \overline{\langle s_R s_0 \rangle^2}$ is a function only of $R \cdot \Lambda(R) = R \cdot R^{-1} = 1$ and $g(R) \simeq 1/\ln R$, which vanishes as $R \to \infty$. So the leading R dependence is in the exponential integral.

We already have $\beta(g)$, and we just need $\gamma(g)$ defined as

$$\gamma(g) = \frac{d \ln Z(g(\Lambda), \Lambda)}{d \ln \Lambda}, \tag{18.65}$$

where Z is the factor that multiplies the given correlation function and makes it independent of Λ.

The correlation function of interest is

$$\overline{G^2(R)} = \left\langle \exp\left[i\sqrt{\frac{\pi}{1+g^2/2\pi}}\phi_1(R)\right] \exp\left[-i\sqrt{\frac{\pi}{1+g^2/2\pi}}\phi_1(0)\right]\right\rangle_g, \tag{18.66}$$

where the subscript g means it is evaluated with coupling g. In particular, let $g(\Lambda)$ be the coupling in the lattice of size $a = 1/\Lambda$. Now,

$$\langle e^{i\beta(\phi(x)-\phi(0))} \rangle = \left(\frac{\alpha^2}{\alpha^2 + x^2}\right)^{\beta^2/4\pi} = \left(\frac{1}{\pi^2 x^2 \Lambda^2}\right)^{\beta^2/4\pi}, \tag{18.67}$$

which follows from Eq. (17.63) and $\alpha = 1/(\Lambda\pi)$.

If we ignore the interaction term [the double sum over cosines in Eq. (18.57)], we find that

$$\overline{G^2(R)} = \left[\frac{1}{\pi R \Lambda}\right]^{\frac{1}{2(1+g^2/2\pi)}} \cdot (1 + O(g^4)). \tag{18.68}$$

The term in square brackets comes from using Eq. (18.67), valid for the free-field theory, and its g dependence comes from explicit factors of g in the definition of the operators. The corrections due to the interactions begin at order g^4 because the diagonal terms in the double sum have been pulled out and the off-diagonal terms do not contribute to correlation in question due to the constraint that the sum of all the exponents must add up to zero for each boson.

We see that $\overline{G^2(R)}$ can be made independent of the cut-off if we pick some arbitrary mass μ and multiply it by

$$\left[\frac{\Lambda}{\mu}\right]^{\frac{1}{2(1+g^2/2\pi)}} \simeq \left[\frac{\Lambda}{\mu}\right]^{\frac{1}{2}-\frac{g^2}{4\pi}} \quad \text{to order } g^2 \tag{18.69}$$

$$\equiv \left[\frac{\Lambda}{\mu}\right]^{\gamma}, \quad \text{which means} \tag{18.70}$$

$$\gamma = \frac{1}{2} - \frac{g^2}{4\pi}. \tag{18.71}$$

Doing the integral in Eq. (18.64), it is easy to obtain (dropping corrections that fall as inverse powers of $\ln R$)

$$\overline{G^2(R)} = \overline{\langle s_R s_0 \rangle^2} \sim \frac{(\ln R)^{1/4}}{R^{1/2}}, \tag{18.72}$$

where the $\frac{1}{2}$ and the $-\frac{g^2}{4\pi}$ in γ [Eq. (18.71)] contribute to $R^{-\frac{1}{2}}$ and $(\ln R)^{1/4}$ respectively.

Exercise 18.3.1 *Do the g integral in Eq. (18.64) using the known expressions or $\beta(g)$ and $\gamma(g)$, and derive Eq. (18.72).*

We now use the fact that the mean of the square is an upper bound on the square of the mean to obtain

$$\overline{\langle s_R s_0 \rangle} \le \frac{(\ln R)^{1/8}}{R^{1/4}}. \tag{18.73}$$

Thus, we find that the DD formula $G(R) \simeq R^0$ cannot be right since it violates this bound. It is also nice to see the kind of logs you expect in an asymptotically free theory.

Several developments have taken place since this work was done. First, I learned that Shalayev [12] had independently done this, without using bosonization. Next, in my paper I had claimed that if my arguments were repeated for higher moments one would find that the average of the $2n$th power of G would be the nth power of $\overline{G^2}$. A. W. W. Ludwig pointed out [10, 11] that this was wrong: the error came from using the $e^{i\sqrt{\pi}\phi}$ in place of $\sin\sqrt{\pi}\phi$. Although this made no difference to the preceding derivation of $\overline{G^2}$, it does affect the higher moments. Ludwig in fact carried out the very impressive task of obtaining the full probability distribution $P(G(R))$.

Andreichenko and collaborators did a numerical study [13] to confirm the correctness of my bound and some additional predictions made by Shalayev. For more technical details of my derivation given above, see the excellent book by Itzykson and Drouffe [14].

18.4 Non-Relativistic Lattice Fermions in $d = 1$

We now turn to a family of problems where the fermion is present from the beginning instead of arising from a treatment of Ising spins. However, the fermion is non-relativistic to begin with and the Dirac fermion arises in the low-energy approximation. Some excellent sources are Emery [15], Sachdev [16], and Giamarchi [17].

Here is a road map for what follows so you don't fail to see the forest because of the trees.

We will explore many aspects of the following model of non-relativistic fermions hopping on a lattice in $d = 1$:

$$H = H_0 + H_I$$

$$= -\frac{1}{2}\sum_j \psi^\dagger(j+1)\psi(j) + \text{h.c.}$$
$$+\Delta \sum_j \left(\psi^\dagger(j)\psi(j) - \frac{1}{2}\right)\left(\psi^\dagger(j+1)\psi(j+1) - \frac{1}{2}\right). \tag{18.74}$$

I will refer to this as the Tomonaga–Luttinger (TL) model, although these authors [21, 22] only considered the low-energy continuum version of it. It is also related by the Jordan–Wigner transformation to what is called the XXZ spin chain.

Let me remind you of what we know from our previous encounter with this model in Section 15.3.

In real space it was clear that as $\Delta \to \infty$, the particles would occupy one or the other sublattice to avoid having nearest neighbors. Any movement of charge would produce nearest neighbors and cost an energy of order Δ. This is the gapped CDW state.

For weak coupling, we went to momentum space using

$$\psi(j) = \int_{-\pi}^{\pi} \psi(K)e^{iKj}\frac{dK}{2\pi}, \tag{18.75}$$

and found the kinetic energy

$$H_0 = \int_{-\pi}^{\pi} \psi^\dagger(K)\psi(K)(-\cos K)\frac{dK}{2\pi}. \tag{18.76}$$

The Fermi sea was made of filled negative energy states with $-\frac{\pi}{2} < K < \frac{\pi}{2}$. The Fermi "surface" was made of two points $R = K_F = \frac{\pi}{2}$ and $L = -K_F = -\frac{\pi}{2}$. I will limit myself to half-filling, where $K_F = \frac{\pi}{2}$, until I turn to the Hubbard model.

Keeping only modes within $\pm\Lambda$ of the Fermi points $\pm K_F$ (see Figure 15.2), we found that

$$H_0 = \sum_{i=L,R} \int_{-\Lambda}^{\Lambda} \frac{dk}{2\pi}\psi_i^\dagger(k)\psi_i(k)k, \tag{18.77}$$

where

$$k = |K| - K_F, \tag{18.78}$$
$$i = L, R \quad \text{(left or right)}. \tag{18.79}$$

(Notice that the k above is the magnitude $|K|$ minus K_F. This definition is most suited for going to higher dimensions, where the energy grows with the radial momentum. Soon we will trade this for a k measured from the nearest Fermi point.)

We then found an RG transformation that left the corresponding action S_0 invariant. We considered the most general interaction and found that there remained just one marginal interaction at tree level, namely u, which scattered particles from opposite Fermi points. (In our specific model, $u \propto \Delta$.) A one-loop calculation showed $\beta(u) = 0$ due to the cancellation between the ZS′ and BCS diagrams describing CDW and superconducting instabilities. It was then stated that $\beta(u)$ vanished to all orders, implying a line of fixed points. It was not clear from that analysis how the system would ever escape the fixed line and reach the gapped CDW state, as it had to at strong coupling.

We now resume that tale. We rederive some of these old results of the fermionic RG *using bosonization*, and then go beyond. In particular, we

- compute the varying exponents in several correlation functions along the fixed line;
- explain how we escape the fixed line as some operators that were irrelevant in the fermionic weak coupling RG become relevant;
- explain the nature of the gapped states to which these relevant perturbations take us; and
- map the model to that of a spin chain using the Jordan–Wigner transformation and interpret these results in spin language.

Since this chapter is long, I will also furnish a synopsis of the details to follow so that as you go through the material, you know where we are in our odyssey.

The first step to bosonization is to unearth a Dirac fermion. We will do this by focusing on the states near the Fermi surface, but this time in real space, by truncating the expansion Eq. (18.75) to states within $\pm\Lambda$ of $K = \pm K_F = \pm\frac{\pi}{2}$:

$$\psi(j) = \int_{-\pi}^{\pi} \psi(K) e^{iKj} \frac{dK}{2\pi} \tag{18.80}$$

$$\simeq \int_{-\Lambda}^{\Lambda} \psi(K_F + k) e^{iK_F j} e^{ikj} \frac{dk}{2\pi} + \int_{-\Lambda}^{\Lambda} \psi(-K_F + k) e^{-iK_F j} e^{ikj} \frac{dk}{2\pi} \tag{18.81}$$

$$\equiv a^{\frac{1}{2}} \left[e^{iK_F j} \psi_+(x = aj) + e^{-iK_F j} \psi_-(x = aj) \right] \tag{18.82}$$

$$= a^{\frac{1}{2}} \left[e^{i\frac{\pi}{2}j} \psi_+(x) + e^{-i\frac{\pi}{2}j} \psi_-(x) \right] \text{ since } K_F = \tfrac{\pi}{2} \text{ here.} \tag{18.83}$$

Observe that the k above is measured from the Fermi points $\pm K_F$. The subscript $\pm$ labels the Fermi point (R or L) on which the low-energy field is centered. The lattice spacing a converts position j on the lattice to position $x = ja$ in the continuum, and the factor $a^{\frac{1}{2}}$ relates continuum Fermi fields $\psi_\pm$ with Dirac-δ anticommutators to lattice fermions with Kronecker-δ anticommutators. The fields $\psi_\pm(x)$ have only small momenta ($|k| < \Lambda$) in their mode expansion. For the field $\psi_\pm$, the energy goes up (down) with k.

Substituting in Eq. (18.74), we will find that

$$H_c = \frac{H_0}{a} = \int dx \left[\psi_+^\dagger(x)(-i\partial_x)\psi_+(x) + \psi_-^\dagger(x)(i\partial_x)\psi_-(x) \right]$$
$$= \int \psi^\dagger(x) \alpha P \psi(x) dx, \tag{18.84}$$

where H_c is the continuum version of H. This paves the way for bosonization.

We will then express the interaction in terms of $\psi_\pm$ and bosonize it to get the *sine-Gordon model*:

$$H_c K = \int dx \left(\frac{1}{2}\left[K\Pi^2 + \frac{1}{K}(\partial_x \phi)^2 \right] + \frac{y}{2\pi^2\alpha^2}\cos\sqrt{16\pi}\,\phi \right), \tag{18.85}$$

$$K = \frac{1}{\sqrt{1+\frac{4\Delta}{\pi}}}, \tag{18.86}$$

$$y = K \cdot \Delta = \frac{\Delta}{\sqrt{1+\frac{4\Delta}{\pi}}}. \tag{18.87}$$

The interpolating steps will soon be provided in pitiless detail.

We will then analyze this bosonized Hamiltonian. *Although at this stage its two parameters y and K are functions of Δ, we will consider a two-parameter family of models in which y and*

$$x = 2 - 4K \tag{18.88}$$

are independent. The RG flows in the (x,y) plane will describe the fate of each starting point. The original model will be described by a one-parameter curve $(x(\Delta), y(\Delta))$ of starting points. The curve is reliably known only for small Δ. The nature of various fixed points, the fixed line, and phases will be examined.

We will then interpret the same flows and fixed points in terms of the spin-$\frac{1}{2}$ *Heisenberg chain*,

$$H = \sum_j S_x(j+1)S_x(j) + S_y(j+1)S_y(j) + \Delta \cdot S_z(j+1)S_z(j), \tag{18.89}$$

related to our spinless fermion Hamiltonian of Eq. (18.74) by a Jordan–Wigner transformation.

Finally, we will consider the *Hubbard model* with on-site interaction of spin-up and spin-down fermions. We will find that the inclusion of spin is far from being a cosmetic change. It will dramatize the gruesome fate of the fermion, which gets torn limb from limb when interactions are turned on.

18.4.1 Deriving the Sine-Gordon Hamiltonian

The first essential ingredient in bosonization is the massless Dirac fermion, which is lurking within our non-relativistic fermion. To extract it, we first write the non-interacting Hamiltonian in terms of the low-energy Dirac fields $\psi_\pm$ using Eq. (18.83):

$$H_0 = -\frac{1}{2}\sum_j \psi^\dagger(j+1)\psi(j) + \text{h.c.} \tag{18.90}$$

$$= -\frac{1}{2}a\sum_j\left[-ie^{-i\frac{\pi}{2}j}\psi_+^\dagger(x=ja+a)+ie^{i\frac{\pi}{2}j}\psi_-^\dagger(x=ja+a)\right]$$

$$\times\left[e^{i\frac{\pi}{2}j}\psi_+(x=ja)+e^{-i\frac{\pi}{2}j}\psi_-(x=ja)\right]+\text{h.c.} \tag{18.91}$$

$$= \frac{a}{2}\sum_j\left[i\psi_+^\dagger(x)\psi_+(x)-i\psi_-^\dagger(x)\psi_-(x)+ia\frac{\partial\psi_+^\dagger(x)}{\partial x}\psi_+(x)-ia\frac{\partial\psi_-^\dagger(x)}{\partial x}\psi_-(x)\right]$$

$$+\text{h.c.}+\text{ignorable terms and terms oscillating at }\pm 2K_\text{F}, \tag{18.92}$$

$$H_{0\text{c}} = \frac{H_0}{a} = \int dx\left[\psi_+^\dagger(x)(-i\partial_x)\psi_+(x)+\psi_-^\dagger(x)(i\partial_x)\psi_-(x)\right], \tag{18.93}$$

where $H_{0\text{c}}$ is the non-interacting continuum Hamiltonian and I have integrated by parts and used $a\sum_j \to \int dx$.

Now look at the interaction

$$H_\text{I} = \Delta\sum_j\left(\psi^\dagger(j)\psi(j)-\frac{1}{2}\right)\left(\psi^\dagger(j+1)\psi(j+1)-\frac{1}{2}\right) \tag{18.94}$$

$$\equiv \Delta\sum_j :\psi^\dagger(j)\psi(j)::\psi^\dagger(j+1)\psi(j+1):. \tag{18.95}$$

We may set

$$\psi^\dagger(j)\psi(j)-\frac{1}{2} = :\psi^\dagger(j)\psi(j):, \tag{18.96}$$

because we have half a fermion per site in the vacuum. Let us combine all this with the expansion of the lattice fields in terms of the smooth continuum fields for $K_\text{F}=\frac{\pi}{2}$ [Eq. (18.83)] to obtain

$$H_{\text{Ic}} = \frac{H_\text{I}}{a}$$

$$= a\Delta\sum_j\left[:\psi_+^\dagger(x)\psi_+(x)+\psi_-^\dagger(x)\psi_-:+(-1)^j(\psi_+^\dagger(x)\psi_-(x)+\psi_-^\dagger(x)\psi_+(x))\right]$$

$$\times\left[:\psi_+^\dagger(x)\psi_+(x)+\psi_-^\dagger(x)\psi_-(x):-(-1)^j(\psi_+^\dagger(x)\psi_-(x)+\psi_-^\dagger(x)\psi_+(x))\right] \tag{18.97}$$

$$= a\Delta\sum_j\left[\frac{1}{\sqrt{\pi}}\partial_x\phi\right]^2-\left[\psi_+^\dagger(x)\psi_-(x)+\psi_-^\dagger(x)\psi_+(x)\right]^2+\ (-1)^j\text{ oscillations}$$

$$= \Delta\int dx\left[\frac{(\partial_x\phi)^2}{\pi}-\left[\frac{1}{\pi\alpha}\sin\sqrt{4\pi}\phi)\right]^2\right] \tag{18.98}$$

$$= \Delta\int dx\left[\frac{2(\partial_x\phi)^2}{\pi}+\frac{1}{2\pi^2\alpha^2}\cos\sqrt{16\pi}\phi)\right]\quad\text{using Eq. (17.97).} \tag{18.99}$$

Notice that we ignore the change in $\psi(x)$ from site j to $j+1$ (down by a power of a), but not that of the factor $(-1)^j$, which oscillates on the lattice scale. We are also using the fact that at half-filling, the potentially oscillatory factor $e^{4K_{\mathrm{F}}j}$, which comes from the product of the second terms in each of the brackets in Eq. (18.97), becomes $(-1)^{2j} = 1$. This is the umklapp term which describes the process $RR \leftrightarrow LL$ with momentum change equal to a reciprocal lattice vector.

This brings us to the continuum Hamiltonian in bosonized form,

$$H_{\mathrm{c}} = \int dx \left(\frac{1}{2} \left[\Pi^2 + \left(1 + \frac{4\Delta}{\pi} \right) (\partial_x \phi)^2 \right] + \frac{\Delta}{2\pi^2\alpha^2} \cos\sqrt{16\pi}\phi \right). \tag{18.100}$$

At this stage we introduce the *Luttinger parameter*

$$K = \left[1 + \frac{4\Delta}{\pi} \right]^{-\frac{1}{2}}, \tag{18.101}$$

in terms of which

$$H_{\mathrm{c}} K = \int dx \left(\frac{1}{2} \left[K\Pi^2 + \frac{1}{K} (\partial_x \phi)^2 \right] + \frac{y}{2\pi^2\alpha^2} \cos\sqrt{16\pi}\phi \right), \tag{18.102}$$

$$y = K \cdot \Delta = \frac{\Delta}{\sqrt{1 + \frac{4\Delta}{\pi}}}. \tag{18.103}$$

The rescaling of H_{c} by K, which we ignore, can be easily incorporated as another parameter, a velocity.

We will take the view that y and K are two free parameters, rather than functions of a single underlying Δ. The TL model will be a one-parameter curve in this two-dimensional plane.

Let us now define a new field and momentum:

$$\phi' = \frac{1}{\sqrt{K}} \phi, \tag{18.104}$$

$$\Pi' = \sqrt{K}\Pi, \tag{18.105}$$

which still obey canonical commutation rules because they were scaled oppositely. By contrast, the ϕ is a c-number in the path integral and can be rescaled as Eq. (18.55). The Hamiltonian now becomes (upon dropping the primes)

$$H_{\mathrm{c}} = \int dx \left[\frac{1}{2} \left[\Pi^2 + (\partial_x \phi)^2 \right] + \frac{y}{2\pi^2\alpha^2} \cos\sqrt{16\pi K}\phi \right], \tag{18.106}$$

which is a special case of the *sine-Gordon model* whose canonical form is

$$H_{\mathrm{SG}} = \int dx \left[\frac{1}{2} \left[\Pi^2 + (\partial_x \phi)^2 \right] + \frac{y}{2\pi^2\alpha^2} \cos\beta\phi \right]. \tag{18.107}$$

In the Luttinger model analysis,

$$\beta^2 = 16\pi K. \tag{18.108}$$

We will also use a related parameter (unfortunately also called x),

$$x = 2 - 4K = 2\left(1 - \frac{\beta^2}{8\pi}\right), \tag{18.109}$$

because the physics changes dramatically with the sign of x. It is most natural to envisage the physics in the (x, y) plane.

18.4.2 Renormalization Group Analysis of the Sine-Gordon Model

We see that the model describes a massless scalar field plus the cosine interaction due to the umklapp process ($RR \leftrightarrow LL$). It is parametrized by K and y. We need to know what the umklapp term does to the massless boson.

The answer depends on K, which determines whether or not the umklapp term is relevant. For the RG analysis it is convenient to go from the Hamiltonian in Eq. (18.106) to the Euclidean action

$$S = \int \left(\frac{1}{2}(\nabla\phi)^2 + \frac{y\Lambda^2}{2}\cos\beta\phi\right) d^2x \tag{18.110}$$

and the path integral over $e^{-S(\phi)}$. Notice that we use the Lorentz-invariant bosonization formulas of Section 17.5. The replacement

$$\frac{1}{\pi\alpha} = \Lambda \tag{18.111}$$

trades the spatial momentum cut-off $1/\alpha$ for Λ, the cut-off on k, the magnitude of the two-dimensional Euclidean momentum $\boldsymbol{k}$. The evolution of y will be found by integrating out a thin shell of momenta near the cut-off $k = \Lambda$.

Let us write ϕ as a sum of slow and fast modes,

$$\phi = \phi_s + \phi_f \equiv \phi(0 \le k \le \Lambda(1 - dt)) + \phi(\Lambda(1 - dt) < k \le \Lambda). \tag{18.112}$$

The free-field action separates as well:

$$S_0 = \int \left[\frac{1}{2}(\nabla\phi_s)^2 + \frac{1}{2}(\nabla\phi_f)^2\right] d^2x. \tag{18.113}$$

The RG that leaves S_0 invariant involves integrating out ϕ_f, followed by the rescaling of spacetime coordinates:

$$d^2x = s^2 d^2x', \tag{18.114}$$

$$\frac{d}{dx} = \frac{1}{s}\frac{d}{dx'}, \tag{18.115}$$

$$\phi(x) = \phi'(x'). \tag{18.116}$$

Now we introduce the interaction, integrate out ϕ_{f} as usual, and see happens to the coupling y of the slow modes that remain. Here is the abridged analysis:

$$Z=\int d\phi_{\mathrm{s}}\int d\phi_{\mathrm{f}}\exp\left[-\int\left[\frac{1}{2}(\nabla\phi_{\mathrm{s}})^2+\frac{1}{2}(\nabla\phi_{\mathrm{f}})^2\right]d^2x-\frac{y\Lambda^2}{2}\int d^2x\cos\beta(\phi_{\mathrm{s}}+\phi_{\mathrm{f}})\right]$$

$$=\int d\phi_{\mathrm{s}}\exp\left[-\int\frac{1}{2}(\nabla\phi_{\mathrm{s}})^2d^2x\right]\left\langle\exp\left[-\frac{y\Lambda^2}{2}\int d^2x\cos\beta(\phi_{\mathrm{s}}+\phi_{\mathrm{f}})\right]\right\rangle_{\mathrm{f}} \tag{18.117}$$

$$\simeq\int d\phi_{\mathrm{s}}\exp\left[-\int\left(\frac{1}{2}(\nabla\phi_{\mathrm{s}})^2+\frac{y\Lambda^2}{2}\cos\beta\phi_{\mathrm{s}}\langle\cos\beta\phi_{\mathrm{f}}\rangle_{\mathrm{f}}\right)d^2x\right], \tag{18.118}$$

where $\langle\cdots\rangle_{\mathrm{f}}$ is the average over fast modes and we are using the leading term in the cumulant expansion ($\langle e^A\rangle\simeq e^{\langle A\rangle}$); the $\sin\beta\phi_{\mathrm{s}}\sin\beta\phi_{\mathrm{f}}$ term is ignored because it has zero average over fast modes. The average $\langle\cdots\rangle_{\mathrm{f}}$ above is *only over the sliver of width* Λdt.

To perform the average we first set $A=i\beta\phi,\ B=0$ in Eq. (17.57) to deduce that

$$\langle e^{i\beta\phi}\rangle=e^{-\frac{1}{2}\beta^2\langle\phi^2\rangle}. \tag{18.119}$$

Using this result, we find that

$$\langle\cos(\beta\phi_{\mathrm{f}})\rangle=e^{-\frac{1}{2}\beta^2\langle\phi_{\mathrm{f}}^2\rangle} \tag{18.120}$$

$$=\exp\left[-\frac{\beta^2}{2}\int_{\Lambda(1-dt)}^{\Lambda}\frac{kdkd\theta}{4\pi^2}\frac{1}{k^2}\right] \tag{18.121}$$

$$=1-\frac{\beta^2}{4\pi}dt. \tag{18.122}$$

Now we rescale the coordinates as per Eq. (18.114),

$$d^2x=s^2d^2x'=(1+2dt)d^2x', \tag{18.123}$$

to obtain (on dropping primes)

$$\frac{y\Lambda^2}{2}\int d^2x\cos\beta\phi\rightarrow\frac{y\Lambda^2}{2}\left(1+\left(2-\frac{\beta^2}{4\pi}\right)dt\right)\int d^2x\cos\beta\phi,$$

$$\frac{dy}{dt}=\left[2-\frac{\beta^2}{4\pi}\right]y \tag{18.124}$$

$$=(2-4K)y\ \text{ because} \tag{18.125}$$

$$\beta^2=16\pi K\quad\text{in the Luttinger model.} \tag{18.126}$$

Thus, we find that the umklapp term is

$$\text{irrelevant for } K>\tfrac{1}{2}\text{ or }\beta^2>8\pi, \tag{18.127}$$

$$\text{relevant for } K<\tfrac{1}{2}\text{ or }\beta^2<8\pi. \tag{18.128}$$

We rescaled x but not Λ, which just stood there. Are we not supposed to rescale all dimensionful quantities when we change units? The short answer is that in the Wilson approach the cut-off remains fixed because we use the cut-off as the unit of measurement. We could call it Λ or we could call it 1. If we begin with the ball of radius 10^{10} GeV and keep integrating away, in *laboratory units* then of course Λ_{lab} is being steadily reduced, but in rescaled units it will be fixed. It is this fixed value we are denoting by Λ above.

As a check, consider a Gaussian theory with action

$$S = \int d^2x \left[\frac{1}{2}(\nabla\phi_\Lambda)^2 + \frac{1}{2}m^2\phi_\Lambda^2 \right], \tag{18.129}$$

where m is the mass in lab units and Λ is the cut-off on the momentum content of ϕ_Λ. Suppose we integrate out modes between Λ/s and Λ. We are left with

$$S = \int d^2x \left[\frac{1}{2}(\nabla\phi_{\Lambda/s})^2 + \frac{1}{2}m^2\phi_{\Lambda/s}^2 \right], \tag{18.130}$$

which tells us that in lab units the theory with the reduced cut-off Λ/s continues to describe a particle of the same mass m, and asymptotic correlations will fall as e^{-mx}. There has been no change of units.

Let us now repeat this, but starting with the mass term expressed in terms of some initial cut-off Λ and a dimensionless parameter r_0:

$$S = \int d^2x \left[\frac{1}{2}(\nabla\phi_\Lambda)^2 + \frac{1}{2}r_0\Lambda^2\phi_\Lambda^2 \right]. \tag{18.131}$$

Upon mode elimination this becomes

$$S = \int d^2x \left[\frac{1}{2}(\nabla\phi_{\Lambda/s})^2 + \frac{1}{2}r_0\Lambda^2\phi_{\Lambda/s}^2 \right]. \tag{18.132}$$

We now change units:

$$k = \frac{k'}{s}, \tag{18.133}$$

$$x = sx', \tag{18.134}$$

$$\frac{d}{dx} = \frac{1}{s}\frac{d}{dx'}. \tag{18.135}$$

In these new units the momentum now goes all the way to Λ and we end up with

$$S = \int d^2x' \left[\frac{1}{2}(\nabla'\phi_\Lambda)^2 + \frac{1}{2}r_0 s^2\Lambda^2\phi_\Lambda^2 \right] \tag{18.136}$$

$$\stackrel{\text{def}}{=} \int d^2x' \left[\frac{1}{2}(\nabla'\phi_\Lambda)^2 + \frac{1}{2}r_{0s}\Lambda^2\phi_\Lambda^2 \right]. \tag{18.137}$$

We see that, under the RG,

$$r_0 \to r_{0s} = r_0 s^2. \tag{18.138}$$

(We could also lump the s^2 with Λ^2 in Eq. (18.136) and identify s^2 times Λ^2 in the new units with the Λ^2_{lab} original laboratory units, thereby showing that the m^2 in laboratory units is fixed at $r_0\Lambda^2_{\text{lab}}$.)

18.4.3 Tomonaga–Luttinger Liquid: ($K > \frac{1}{2}, y = 0$)

We consider the line of fixed points $y = 0$ and focus on the sector $K > \frac{1}{2}$ where the perturbation $y\cos\sqrt{16\pi K}\phi$ is irrelevant. In terms of a variable

$$x = 2 - 4K, \tag{18.139}$$

the region where the cosine is irrelevant is

$$x = 2 - 4K < 0. \tag{18.140}$$

Not only does this line $y = 0$ for $x < 0$ describe the models with $y = 0$, it also describes models which flow to $y = 0$ under the RG. Later we will see what range of y will flow into this line under RG. In studying this line we are studying all systems in the basin of attraction of this line. Remember, however, that if you begin at some (K, y) in this basin, you will end up at $(K^*, 0)$, where $K^* \neq K$ in general. (Equivalently, $(x, y) \to (x^*, 0)$ after the RG.) So the K in what follows is in general the final K^* of a system that started away from the fixed line and got sucked into it.

For $x > 0$, the line is unstable to perturbations and the system must be tuned to stay on it. Also bear in mind that we have assumed exactly half-filling; otherwise, the umklapp term is not allowed: $e^{4iK_{\text{F}}n}$ oscillates and averages to zero unless $K_{\text{F}} = \frac{\pi}{2}$. What if we are just a little off $K_{\text{F}} = \frac{\pi}{2}$? Then the oscillations will be very slow in space to begin with, but after a lot of RG iterations, the oscillations will become rapid in the new lattice units and the seemingly relevant growth will fizzle away.

The line of fixed points $(K > \frac{1}{2}, y = 0) \equiv (x < 0, y = 0)$ is ubiquitous and appears in many guises and with different interpretations. Here it describes a fermionic liquid state called the Tomonaga–Luttinger (TL) liquid. The name was coined by Haldane [19, 20], who explored its properties and exposed the generality of the notion. It is the $d = 1$ version of Landau theory. Recall that Landau's Fermi liquid is parametrized by the F function, or its harmonics $u_m \equiv F_m$. Even if we cannot calculate the u_m from some underlying theory, we can measure them in some experiments and use them to describe others in terms of these measured values. The main point is that many low-energy quantities can be described by a few Landau parameters. Likewise, K and a velocity parameter, which I have suppressed, fully define all aspects of the fermionic system – response functions, thermodynamics, correlation functions – in the infrared.

The line of fixed points has one striking property: exponents that vary continuously with K. (This is not so for the Landau Fermi liquid, which has canonical power laws as F varies.) I will show this now, and as a by-product, establish the claim made earlier that the fermion pole at $\omega = k$ (in Minkowski space) is immediately destroyed by the smallest interaction, i.e., the smallest departure from $K = 1$.

Consider $\langle \psi^\dagger(x)\psi(0)\rangle$. Without interactions, we had

$$H = \int dx \left[\frac{1}{2}\Pi^2 + \frac{1}{2}(\partial_x\phi)^2\right] dx, \tag{18.141}$$

$$\psi_\pm(x) = \frac{1}{\sqrt{2\pi\alpha}} e^{\pm i\sqrt{4\pi}\phi_\pm(x)}, \quad \text{where} \tag{18.142}$$

$$\phi_\pm(x) = \frac{1}{2}\left[\phi(x) \mp \int_{-\infty}^{x} \Pi(x')dx'\right] \equiv \frac{1}{2}(\phi \mp \theta), \tag{18.143}$$

and where the *dual field*

$$\theta(x) = \int_{-\infty}^{x} \Pi(x')dx'. \tag{18.144}$$

With interactions, we had

$$H = \int dx \left[\frac{K}{2}\Pi^2 + \frac{1}{2K}(\partial_x\phi)^2\right] dx. \tag{18.145}$$

Introducing the rescaled variables of the interacting theory,

$$\phi = K^{\frac{1}{2}}\phi', \quad \Pi = K^{-\frac{1}{2}}\Pi', \quad \theta = K^{-\frac{1}{2}}\theta', \tag{18.146}$$

in terms of which the kinetic energy has the standard coefficient of $\frac{1}{2}$, and recalling that

$$\phi = \phi_+ + \phi_-, \tag{18.147}$$

$$\theta = \phi_- - \phi_+, \tag{18.148}$$

one finds that

$$\psi_\pm(x) = \frac{1}{\sqrt{2\pi\alpha}} \exp \pm i\sqrt{\pi}\left[(K^{\frac{1}{2}} \pm K^{-\frac{1}{2}})\phi'_+ + (K^{\frac{1}{2}} \mp K^{-\frac{1}{2}})\phi'_-\right]. \tag{18.149}$$

Exercise 18.4.1 *Derive Eq. (18.149).*

It is now a routine exercise to show that

$$\langle \psi^\dagger{}_\pm(x)\psi_\pm(0)\rangle \simeq \left[\frac{1}{\alpha \mp ix}\right]^{\frac{(K\pm1)^2}{4K}} \cdot \left[\frac{1}{\alpha \pm ix}\right]^{\frac{(K\mp1)^2}{4K}} \tag{18.150}$$

$$= \frac{1}{\alpha \mp ix} \cdot \left[\frac{1}{\alpha^2 + x^2}\right]^\gamma, \tag{18.151}$$

$$\gamma = \frac{(K-1)^2}{4K}. \tag{18.152}$$

Exercise 18.4.2 *Derive Eq. (18.150).*

For unequal-time correlations, we just need to remember that $\psi_\pm$ are functions of $x \mp t$ to obtain

$$\langle \psi^\dagger{}_\pm(x,t)\psi_\pm(0)\rangle \simeq \frac{1}{\alpha \mp i(x \mp t)} \cdot \left[\frac{1}{\alpha^2 + x^2 - t^2}\right]^\gamma. \tag{18.153}$$

We see that the decay power varies with K. Upon Fourier transforming to (ω, k), we see that as soon as $K \neq 1$, the pole (in Minkowski space)

$$G(\omega,k) \simeq \frac{1}{\omega - k} \tag{18.154}$$

morphs into a cut using just dimensional analysis: $G(\omega,k)$ has fractional dimension in ω or k:

$$G \simeq (\omega,k)^{\frac{K^2-4K+1}{2K}}. \tag{18.155}$$

There is a huge body of literature on the response functions at non-zero T, ω, and q that you are now ready to explore. For example, one can show that in the TL liquid the occupation number $n(k)$ has not a jump at k_F, but a kink:

$$n(k) = n(k_\mathrm{F}) + c\ \mathrm{sgn}(k - k_\mathrm{F})|k - k_\mathrm{F}|^\delta, \tag{18.156}$$

$$\delta = \frac{K + K^{-1} - 2}{4}. \tag{18.157}$$

18.5 Kosterlitz–Thouless Flow

Let us now find the basin of attraction of the fixed TL line in the (x,y) plane and the manner in which a transition to a gapped phase occurs when we cross the boundary of this basin. We have seen from

$$\frac{dy}{dt} = (2 - 4K)y \tag{18.158}$$

that on the axis labeled by

$$x = (2 - 4K) \tag{18.159}$$

y is relevant or irrelevant for $x > 0$ or $x < 0$ respectively. So in the (x,y) plane we expect flow lines to terminate on or leave the x-axis in the y-direction as x goes from being negative to positive. The flow slows down as we approach $K = \frac{1}{2}$ ($x = 0$) and then reverses sign, as depicted in Figure 18.1. How do these lines change direction as we cross this point? What is the full story in the (x,y) plane?

For this we turn to the celebrated RG flow devised by Kosterlitz and Thouless [26] in their analysis of the phase transition in the XY model of planar spins. Recall from Chapter 10 that there too we have a line of fixed points with a T-dependent exponent. As $T \to \infty$ the decay had be exponential based on the high-T series. This decay cannot be brought about by spin waves, the small fluctuations about the constant field described by a

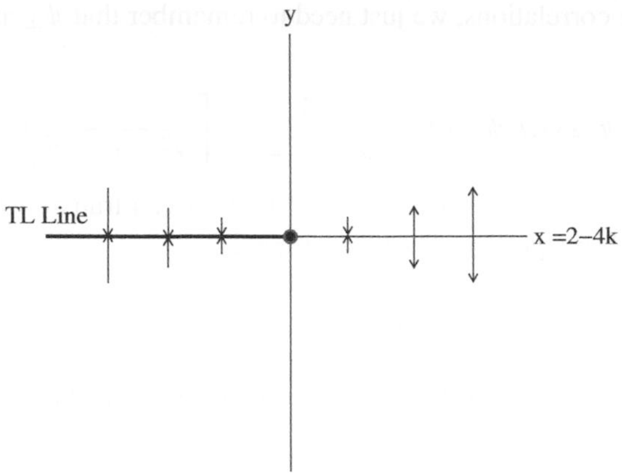

Figure 18.1 The line of fixed points of the sine-Gordon model as a function of β in $\cos\beta\phi$, or the Luttinger parameter K defined by $\beta^2 = 16\pi K$, or the parameter $x = 2-4K = 2\left(1-\frac{\beta^2}{8\pi}\right)$. The cos is increasingly irrelevant for $x < 0$ and increasingly relevant for $x > 0$. The size of the arrows indicates the rate of flow into or away from the fixed line. The dark line describes the Luttinger liquid.

Gaussian action that ignores the periodic nature of the angle θ. The transition to the phase with exponential decay is driven by vortices and antivortices, which are configurations in which the angle θ changes by $\pm 2\pi$ as we go around their cores. At low T these are tightly bound into vortex–antivortex pairs. The fugacity (likelihood of appearing in the sum over configurations) for free vortices and antivortices is described by the cosine interaction with coupling y, which changes (with T) from irrelevance to relevance and vortices and antivortices go from being bound to being free. The same sine-Gordon model describes this transition.

To zero in on the point $(x = 0, y = 0)$, let us rewrite Eq. (18.158) as

$$\frac{dy}{dt} = xy \tag{18.160}$$

and observe that the flow is quadratic in small quantities, so we need the flow of x, or essentially K, to the same order. The only way to renormalize K is by field renormalization, which begins at second order in y. Dropping all constants, we begin with the pair

$$\frac{dy}{dt} = xy, \tag{18.161}$$

$$\frac{dx}{dt} = y^2. \tag{18.162}$$

In this flow, one easily finds that

$$y^2(t_1) - x^2(t_1) = y^2(t_2) - x^2(t_2), \tag{18.163}$$

that is, the flow is along hyperbolas. Of special interest are its asymptotes,

$$x = \pm y. \tag{18.164}$$

Looking at Figure 18.2, for $y > 0$, the line $x = -y$ in the second quadrant separates flows into the massless fixed line from the ones that flow to massive or gapped theories. The reflected asymptote $x = y$ in the third quadrant defines the basin of attraction of the TL line for $y < 0$. We focus on the $y > 0$ case since the mathematics is identical in the two cases. The physics is different, as will be explained later.

Let us start at the far left at a point

$$y^2(0) - x^2(0) = \delta \tag{18.165}$$

just above the separatrix $x = -y$. This means that at any generic t,

$$y^2(t) - x^2(t) = \delta. \tag{18.166}$$

We want to know how $\xi(\delta)$ diverges as we approach the separatrix that flows into the fixed point at the origin.

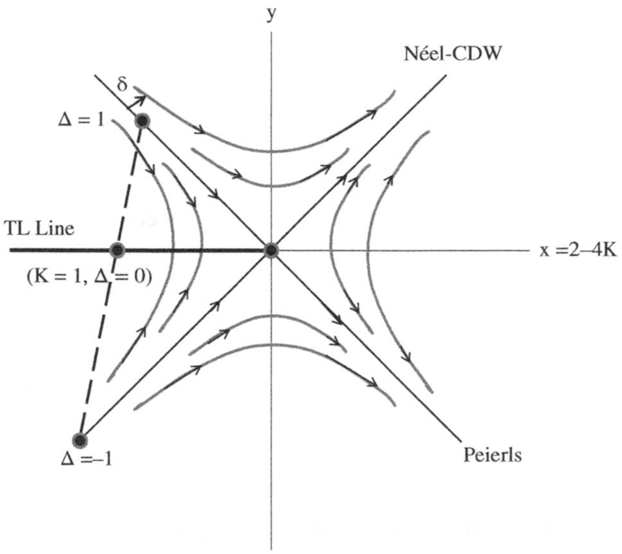

Figure 18.2 The Kosterlitz–Thouless (KT) flow. The origin is at ($x \stackrel{\text{def}}{=} 2 - 4K = 0, y = 0$). The TL liquid is found on the x-axis for $x < 0$ or $K > \frac{1}{2}$. The point $K = 1, y = 0$ describing a free fermion lies on this line. The dotted line passing through it is a *schematic* of the TL model as its sole parameter Δ is varied. That the point $\Delta = 1$ is the last of the massless phase and flows under RG to the origin we know from the exact solution. Larger values of Δ approach this end point and veer away to a gapped CDW phase. The point $\Delta = -1$ marks the other end of the gapless phase after which the flow is to the Peierls phase. The correlation length diverges as $e^{\pi/\sqrt{\delta}}$ when we approach either separatix.

This is determined by the flow

$$\frac{dx}{dt} = y^2(t) = (\delta + x^2(t)), \tag{18.167}$$

with a solution

$$t = \frac{1}{\sqrt{\delta}}\left[\arctan\frac{x(t)}{\delta} - \arctan\frac{x(0)}{\delta}\right], \tag{18.168}$$

$$t \simeq \frac{\pi}{\sqrt{\delta}}, \tag{18.169}$$

assuming that we start at the far left and end at the far right.

Since the rescaling factor $s = e^t$, it follows that $\xi(t) = e^{-t}\xi(0)$. Assuming that for large t the correlation length $\xi(t) \to \mathcal{O}(1)$ (because we are essentially on the line $x = y$, far from the critical point at the origin),

$$\xi(0) = e^t\xi(t) \simeq \exp\left[\frac{\pi}{\sqrt{\delta}}\right], \tag{18.170}$$

implying the *exponential* divergence of the correlation length as $\delta \to 0$.

What if we start *on* the line $x = y$? The solution to

$$\frac{dx}{dt} = x^2 \tag{18.171}$$

is

$$x(t) = \frac{x(0)}{1 - x(0)t} = -\frac{|x(0)|}{1 + |x(0)|t} \simeq -\frac{1}{t}, \tag{18.172}$$

which is the logarithmic, marginally irrelevant flow we saw earlier in ϕ_4^4. On the other side, if we begin on the separatrix $x = y$, the solution

$$x(t) = \frac{x(0)}{1 - x(0)t} \tag{18.173}$$

will grow to large values because now $x(0) > 0$. At some point the weak coupling analysis will fail.

Besides these flows, there are the relevant lines flowing away from the fixed line for $x > 0$. The behavior of various regions is shown in Figure 18.2. Although the analysis was for small x and y, it is assumed that the overall topology will survive, though the flow lines could deviate from what was shown above.

18.6 Analysis of the KT Flow Diagram

Figure 18.2 is worth more than the usual thousand words. As mentioned before, the separatrix $y = -x$ in the second quadrant, flowing into the origin, defines the domain of attraction of the fixed line for $y > 0$, $x < 0$. When we cross it, $\xi \simeq e^{1/\sqrt{\delta}}$, where the deviation δ is shown in the figure. If we start just above the separatrix, we initially flow along it toward the origin and then veer away along the separatrix $x = y$, to a state with a hefty gap. If we follow the original model along a curve parameterized by Δ, the point where it intersects the separatrix $y = -x$ is when $\Delta = 1$. This because we know from the exact solution that the gap develops for $\Delta > 1$.

What is behind this gap? The state we are headed for has a large positive y and that means we want, based on Eq. (18.102),

$$\cos\sqrt{16\pi}\phi = \frac{1}{2}(1 - 2\sin^2\sqrt{4\pi}\phi) \tag{18.174}$$

to be maximally negative, i.e.,

$$\sin^2\sqrt{4\pi}\phi = 1 \tag{18.175}$$

$$\sin\sqrt{4\pi}\phi = \pm 1. \tag{18.176}$$

Thus there are two ground states. In them,

$$\langle \sin\sqrt{4\pi}\phi \rangle \simeq \langle \psi_+^\dagger(x)\psi_-(x) + \psi_-^\dagger(x)\psi_+(x) \rangle = \langle i\bar{\psi}\gamma^5\psi \rangle = \pm \mathcal{D}_{\text{CDW}}, \tag{18.177}$$

where the CDW order parameter $\mathcal{D}_{\text{CDW}}$ describes a variable that connects the left and right Fermi points, and oscillates as $(-1)^j$. Indeed, from Eq. (18.99),

$$:\psi^\dagger(x)\psi(x): = :\psi_+^\dagger(x)\psi_+(x) + \psi_-^\dagger(x)\psi_-(x): + (-1)^j(\psi_+^\dagger(x)\psi_-(x) + \psi_-^\dagger(x)\psi_+(x)), \tag{18.178}$$

we see that the fermion charge density has one part that is smooth and one that oscillates as $(-1)^j$, and it is the latter which has developed a condensate or expectation value. This was our early conclusion based on looking at the nearest-neighbor interaction at very large coupling. As $\Delta \to \infty$, one sublattice is occupied and the other is empty to get rid of the nearest-neighbor repulsion. In such a state it costs energy to move the charge, forcing it to have a nearest neighbor. That is the gap.

Suppose we turn on a negative Δ on the fixed line. There is another separatrix $x = y$ in the third quadrant that defines the domain of attraction of the TL fixed line. The exact solution tells us that this end point corresponds to $\Delta = -1$. If we go below, we first flow toward the origin and then off to large negative values of y. This takes us to the *Peierls state*. What happens here?

Because $y < 0$, we want

$$\cos\sqrt{16\pi}\phi = \frac{1}{2}(-1 + 2\cos^2\sqrt{4\pi}\phi) \tag{18.179}$$

to be maximally positive, i.e.,

$$\cos^2\sqrt{4\pi}\phi = 1 \tag{18.180}$$

$$\cos\sqrt{4\pi}\phi = \pm 1. \tag{18.181}$$

Thus there are two "Peierls" ground states.

To interpret the physics of the Peierls state we recall that

$$\begin{aligned}\langle\psi^\dagger(j+1)\psi(j)+\text{h.c.}\rangle &= \langle(\psi_R^\dagger(x)(-i)^j+\psi_L^\dagger(x)(i)^j)(j\to j+1)\rangle+\text{h.c.}\\ &=(-1)^j\langle(-i\psi_R^\dagger(x)\psi_L(x)+\text{h.c.})\rangle+\text{NOP} \qquad (18.182)\\ &=\frac{(-1)^j}{2\pi\alpha}\langle\cos\sqrt{4\pi}\phi\rangle\equiv(-1)^j\mathcal{D}_{\text{P}}. \qquad (18.183)\end{aligned}$$

I have dropped the non-oscillatory part (NOP) $\psi_R^\dagger(-i\partial_x)\psi_R+\psi_L^\dagger(+i\partial_x)\psi_L$ and emphasized only that in the Peierls state the kinetic energy alternates as $(-1)^j\mathcal{D}_{\text{P}}$.

The dotted line in the figure shows our original model with just one parameter Δ. For small Δ we can start at a reliably known point in the (x,y) plane and follow the flow to the x-axis. As $y\to 0$, x will move to the right. In general, we cannot precisely relate Δ to the parameters K (or x) and y due to renormalization effects. However, we can say, based on the Yang and Yang solution [23], that the gapless liquid phase is bounded by $|\Delta|<1$.

Finally, on the $x>0$ side, we can go directly to the CDW and Peierls phases starting with arbitrarily small y, as shown in Figure 18.2.

The $y\to -y$ symmetry of the KT flow diagram is consistent with the fact that

$$\begin{aligned}H(\Delta)=&-\frac{1}{2}\sum_j\psi^\dagger(j+1)\psi(j)+\text{h.c.}\\ &+\Delta\sum_j\left(\psi^\dagger(j)\psi(j)-\frac{1}{2}\right)\left(\psi^\dagger(j+1)\psi(j+1)-\frac{1}{2}\right)\end{aligned} \qquad (18.184)$$

is unitarily equivalent to $-H(-\Delta)$:

$$U^\dagger H(\Delta)U=-H(-\Delta), \qquad (18.185)$$

where, under U,

$$\psi(j)\to(-1)^j\psi(j). \qquad (18.186)$$

This reverses the sign of the hopping term leading to Eq. (18.185). Despite this unitary equivalence under $\Delta\to-\Delta$, the physics can be very different: e.g., CDW versus Peierls as $|\Delta|\to\infty$.

The transition from a metal to insulator driven by interaction is generally very hard to analyze with any exactitude. The preceding model is one of the rare examples, albeit in $d=1$. Since Yang and Yang and Baxter have established many exact results (such as the expression for the CDW order parameter as a function of Δ), we can interpret them in the light of the metal insulator transition. One such study is [28]. Despite the use of continuum methods, many exact results are derived about conductivity as well as some surprising results on the effect of a random potential. Other illustrations of bosonization can be found in [29–32]; the list is not exhaustive or even representative – however, once you get your hands on these you can follow the leads given therein to find more. For the application of bosonization to a single impurity problem, see Kane and Fisher [33].

18.7 The XXZ Spin Chain

The model of spinless fermions we have solved is mathematically identical to the spin-$\frac{1}{2}$ chain with

$$H_{\mathrm{XXZ}} = \sum_j \left[S_x(j)S_x(j+1) + S_y(j)S_y(j+1) + \Delta S_z(j)S_z(j+1) \right]. \tag{18.187}$$

The following Jordan–Wigner transformation relates the two:

$$S_z(j) = \psi^\dagger(j)\psi(j) - \frac{1}{2}, \tag{18.188}$$

$$S_+(j) = (-1)^j \psi^\dagger(j) \exp\left[i\pi \sum_{k<j} \psi^\dagger(k)\psi(k) \right] = S_-^\dagger(j), \tag{18.189}$$

where the $(-1)^j$ is introduced to give the kinetic term the same sign as in the Luttinger model, with a minimum at zero momentum.

We can bodily lift our results from the fermion problem to the spin chain. In particular, both have a gapless region that gives way to broken symmetry states with an order parameter at momentum $2K_{\mathrm{F}} = \pi$. The gapless region is bounded by $\Delta = \pm 1$, as we know from the exact solutions of Yang and Yang [23] and Baxter [24, 25], who solved the XYZ model with different couplings for the three terms by relating H_{XYZ} to the transfer matrix of the eight-vertex model. (I remind you once again that we can relate K to Δ only at weak coupling. As we begin with larger values of Δ, the parameters $K(\Delta)$ or $x(\Delta)$ will get renormalized as the irrelevant coupling y renormalizes to 0. The flow in the (x, y) planes is not vertical, not known exactly, and the definition of y is sensitive to how we cut off the theory, i.e., α.)

In the spin language, the CDW state when $y \to +\infty$ corresponds to a state with $\langle S_z \rangle \simeq (-1)^j$ because $S_z(j) = n_j - \frac{1}{2}$. In the limit $y \to -\infty$, we have the *spin-Peierls* state in which the average bond energy $\langle S_+(j)S_-(j+1) + \mathrm{h.c.} \rangle$ oscillates as $(-1)^j$.

While we can borrow these results from the mapping to the TL model, correlation functions are a different matter. Whereas $S_z - S_z$ correlations are easy because S_z is just a fermion bilinear, correlation functions of $S_\pm$ are non-local in the fermion language and involve the dual field θ.

Consider, for example, the simplest case $\Delta = 0$ and the correlator

$$\langle S_+(0)S_-(j) \rangle$$
$$= \langle \psi^\dagger(0) \exp\left[i\pi \sum_{k=0}^{j-1} \psi^\dagger(k)\psi(k) \right] \psi(j) \rangle \tag{18.190}$$

$$\simeq (-1)^j \psi^\dagger(0) e^{i\sqrt{\pi}(\phi(x)-\phi(0))+ik_{\rm F}x}\psi(x) \tag{18.191}$$

$$= a(-1)^j \left[\psi_+^\dagger(0)+\psi_-^\dagger(0)\right] e^{i\sqrt{\pi}(\phi(x)-\phi(0))+ik_{\rm F}x}\left[\psi_+(x)e^{ik_{\rm F}x}+\psi_-(x)e^{-ik_{\rm F}x})\right], \tag{18.192}$$

where I have canceled the string to the left of $j=0$ and used

$$i\pi\sum_k^{j-1}\psi^\dagger(k)\psi(k) = i\int_0^x \sqrt{\pi}\,\partial_x\phi dx + \left[\frac{i\pi j}{2} = iK_{\rm F}j = ik_{\rm F}x\right], \tag{18.193}$$

where $k_{\rm F} = K_{\rm F}/a$ is the dimensional Fermi momentum.

Now we have, from Eq. (18.189) (upon ignoring the factor $\frac{a}{2\pi\alpha}$),

$$\begin{aligned}\langle S_+(0)S_-(j)\rangle &\simeq (-1)^j\left[e^{-i\sqrt{\pi}(\phi(0)-\theta(0))}+e^{i\sqrt{\pi}(\phi(0)+\theta(0))}\right]\\ &\quad\times e^{i\sqrt{\pi}(\phi(x)-\phi(0))}e^{ik_{\rm F}x}\left[e^{i\sqrt{\pi}(\phi(x)-\theta(x))}e^{ik_{\rm F}x}+e^{-i\sqrt{\pi}(\phi(x)+\theta(x))}e^{-ik_{\rm F}x}\right]\\ &= (-1)^j\langle e^{i\sqrt{\pi}(\theta(0)-\theta(x))}\rangle\left\langle\left(1+e^{-i\sqrt{4\pi}\phi(0)}\right)\left(1+e^{i\sqrt{4\pi}\phi(x)}e^{2ik_{\rm F}x}\right)\right\rangle\\ &= (-1)^j\left[\frac{\alpha^2}{\alpha^2+x^2}\right]^{1/4}\left[1+\frac{(-1)^j}{x^2}\right].\end{aligned} \tag{18.194}$$

At $K \neq 1$, the leading term will be

$$\langle S_+(0)S_-(j)\rangle \simeq (-1)^j\frac{1}{x^{(1/2K)}}, \tag{18.195}$$

whereas to leading order the $S_z - S_z$ correlation that goes as $(-1)^j$ is

$$\langle S_z(0)S_z(j)\rangle \simeq (-1)^j\frac{1}{x^{2K}}. \tag{18.196}$$

We see that at $K=\frac{1}{2}$, we have the isotropic Heisenberg chain, described by the origin in Figure 18.2. (This result does not follow from weak-coupling bosonization, which is reliable only near $K=1$. Rather, we take K as a phenomenological parameter.) The main message is that the origin describes the isotropic Heisenberg antiferromagnet as we approach it from the second quadrant on the separatrix $y=-x$. This problem was originally solved by Bethe, who introduced the famous Bethe ansatz.

18.8 Hubbard Model

Now we consider fermions with spin. Usually, the inclusion of spin causes some predictable changes. This is not so here.

The Hubbard model has a non-interacting part,

$$H_0 = -\frac{1}{2}\sum_{s,n}\left[\psi_s^\dagger(n)\psi_s(n+1)+\text{h.c.}\right]+\mu\sum_{s,n}\psi_s^\dagger(n)\psi_s(n), \tag{18.197}$$

where $s = \uparrow, \downarrow$ are two possible spin orientations. *We do not assume* $K_{\mathrm{F}} = \frac{\pi}{2}$ *at this point*, and use a general chemical potential μ.

Following the usual route, we get two copies of the spinless model:

$$H_0 = \sum_s \int_{-\pi}^{\pi} (\mu - \cos k) \psi_s^\dagger(k) \psi_s(k) \frac{dk}{2\pi}, \tag{18.198}$$

and the continuum version

$$H_{\mathrm{c}} = \sin K_{\mathrm{F}} \sum_s \int dx (\psi_{s-}^\dagger(x)(i\partial_x)\psi_{s-}(x) + \psi_{s+}^\dagger(x)(-i\partial_x)\psi_{s+}(x)). \tag{18.199}$$

Let us now turn on the Hubbard interaction,

$$H_{\mathrm{int}} = U \sum_n \psi_\uparrow^\dagger(n)\psi_\uparrow(n)\psi_\downarrow^\dagger(n)\psi_\downarrow(n), \tag{18.200}$$

where $\psi_\uparrow$, $\psi_\downarrow$ stand for the original non-relativistic fermion. The Hubbard interaction is just the extreme short-range version of the screened Coulomb potential between fermions. Due to the Pauli principle, only opposite-spin electrons can occupy the same site. One can extend the model to include nearest-neighbor interactions, but we won't do so here.

Let us now express this interaction in terms of the Dirac fields. We get, in obvious notation,

$$\begin{aligned} &\psi_\uparrow^\dagger(n)\psi_\uparrow(n)\psi_\downarrow^\dagger(n)\psi_\downarrow(n) \\ &= (\psi_{\uparrow+}^\dagger(n)\psi_{\uparrow+}(n) + \psi_{\uparrow-}^\dagger(n)\psi_{\uparrow-}(n) + (\psi_{\uparrow+}^\dagger(n)\psi_{\uparrow-}(n)e^{-2iK_{\mathrm{F}}n} + \mathrm{h.c.})) \\ &\quad \times (\uparrow \to \downarrow). \end{aligned} \tag{18.201}$$

If we expand out the products and keep only the parts with no rapidly oscillating factors (momentum conservation), we will, for generic K_{F}, get the following terms:

$$H_{\mathrm{int}} = U(j_{0\uparrow} j_{0\downarrow}) + U(\psi_{\uparrow+}^\dagger(n)\psi_{\uparrow-}(n)\psi_{\downarrow-}^\dagger(n)\psi_{\downarrow+}(n) + \mathrm{h.c.}). \tag{18.202}$$

If we now bosonize these terms as per the dictionary, we get, in the continuum (dropping the subscript c for continuum),

$$H = \int dx \frac{1}{2}\left[\Pi_\uparrow^2 + (\partial\phi_\uparrow)^2 + (\uparrow \to \downarrow)\right] + U\left[\frac{\partial\phi_\uparrow \partial\phi_\downarrow}{\pi} + \frac{1}{\pi^2\alpha^2}\cos\sqrt{4\pi}(\phi_\uparrow - \phi_\downarrow)\right]. \tag{18.203}$$

We can now separate the theory into two parts by introducing charge and spin fields ϕ_c and ϕ_s:

$$\phi_{c/s} = \frac{\phi_\uparrow \pm \phi_\downarrow}{\sqrt{2}}. \tag{18.204}$$

This will give us

$$H = H_c + H_s, \tag{18.205}$$

$$K_c \cdot H_c = \int \frac{1}{2}\left[K_c \Pi_c^2 + \frac{1}{K_c}(\partial\phi_c)^2\right] dx, \tag{18.206}$$

$$K_s \cdot H_s = \int \left(\frac{1}{2}\left[K_s \Pi_s^2 + \frac{1}{K_s}(\partial\phi_s)^2\right] + \frac{U}{\pi^2\alpha^2}\cos\sqrt{8\pi}\,\phi_s\right) dx, \tag{18.207}$$

$$K_{c/s}^2 = \frac{1}{1 \pm \frac{U}{\pi}}. \tag{18.208}$$

It is obvious that the charge sector is gapless and described by a quadratic Hamiltonian. This means that there will be no gap to creating charge excitations, the system will be metallic. The fate of the spin sector needs some work. Upon rescaling the kinetic term to standard form we find the cosine interaction

$$\cos\beta\phi_s = \cos\sqrt{\frac{8\pi}{\sqrt{1-U/\pi}}}\phi_s. \tag{18.209}$$

We can now see that for weak positive U, this interaction does not produce any gap because $\beta^2 > 8\pi$, while for weak negative U, it does because $\beta^2 < 8\pi$. The exact solution of Lieb and Wu [18] and the following physical argument explain the spin gap for $U < 0$. If there is an attraction between opposite spin electrons, they will tend to form on-site, singlet pairs. To make a spin excitation, we must break a pair, and this will cost us, i.e., there will be a gap in the spin sector.

The fact that $K_s \neq K_c$ means that charge and spin move at different velocities. This *spin–charge separation* cannot be understood in terms of interacting electrons whose charge and spin would be irrevocably bound. This is more evidence of the demise of the quasiparticle, adiabatically connected to the primordial fermion.

In the special case of half-filling, another term comes in. If we look at Eq. (18.201), we see that in the case of half-filling, since $K_F = \pi/2$, the factors $e^{\pm 4iK_F n}$ are not rapidly oscillating, but simply equal to unity. Thus, two previously neglected terms in which two right movers are destroyed and two left movers are created, and vice versa, come into play. (This is an umklapp process, in which lattice momentum is conserved modulo 2π [19,20]). I leave it to you to verify that the bosonized form of this interaction, after rescaling of the charge field in the manner described above for the spin field, is another $\cos\beta_c\phi_c$, with

$$\beta_c = \sqrt{\frac{8\pi}{\sqrt{1+U/\pi}}}. \tag{18.210}$$

Thus we find that the situation is exactly reversed in the charge sector: there is a gap in repulsive case, and no gap in the attractive case. To see what is happening, think of very large positive U. Now there will be one electron per site at half-filling, unable to move without stepping on someone else's toes, i.e., there is a charge gap of order U if you try to move the charge. But the spin can do whatever it wants with no cost. If U were very large and negative, there would be tightly bound pairs on half the sites. These doubly charged objects can be moved without cost. There will, however, be a cost for breaking the spin pair.

18.9 Conclusions

I have tried to show you how to use bosonization to solve a variety of problems. The formalism is straightforward, but has some potential pitfalls which I avoided because I know of them. So before I let you go, I need to inform you.

In this treatment we always work in infinite volume from the beginning and are cavalier about boundary conditions at spatial infinity. The Fermi fields expressed in terms of boson fields are meant to be used for computing correlation functions and not as operator identities. After all, no combinations of bosonic operators ϕ or Π can change the fermion number the way ψ or $\psi^\dagger$ can. But of course, this was never claimed.

There is a more comprehensive and careful development in which such an operator correspondence may be set up, starting with finite volume. In these treatments the mode expansions for $\phi(x)$ and $\Pi(x)$ have additional terms (of the form $\frac{x}{L}$) that vanish as the system size $L \to \infty$. Next, in our scheme we had $[\phi_+(x), \phi_-(y)] = \frac{i}{4}$, which was needed to ensure some anticommutators, while in the more careful treatments $[\phi_+(x), \phi_-(y)] = 0$, a feature that is central to conformal field theory, which treats right and left movers completely independently. In these treatments there are compensating *Klein factors*, which are operators tacked on to ensure that different species of fermions anticommute. (We did not need them in the problems I discussed since the factors come in canceling pairs.)

The excellent article by van Delft and Schoeller [34] devotes an appendix to the differences between what is presented here (called the field-theory approach) and what they call the constructive approach. Other online articles I have benefited from are due to Voit [35], Schulz [36], and Miranda [37]. A rigorous treatment may be found in Heidenreich *et al.* [38]. A more intuitive review is due to Fisher and Glazman [39].

In addition, I have found lucid introductions in the books by Itzykson and Drouffe (vol. 1) [14], Fradkin [40], Sachdev [16], Giamarchi [17], and Guiliani and Vignale [41].

There is a development called *non-Abelian bosonization*, due to Witten [42], in which the internal symmetries of the model are explicitly preserved. For example, if we are considering an N-component Gross–Neveu model, the $U(N)$ symmetry is not explicit if we bosonize each component with its own field ϕ_i. In non-Abelian bosonization, $U(N)$ group elements replace the ϕ_i and the symmetry is explicit. For a review, see [43].

Haldane expanded bosonization to $d = 2$ [44]. For an application, see [45].

References and Further Reading

[1] J. Schwinger, Physical Review B, **12**, 3908 (1975).
[2] W. Thirring, Annals of Physics, **82**, 664 (1951).
[3] C. Itzykson and J. B. Zuber, Physical Review D, **15**, 2875 (1977).
[4] E. Fradkin and L. Susskind, Physical Review D, **17**, 2637 (1978).
[5] T. D. Schultz, D. Mattis, and E. H. Lieb, Reviews of Modern Physics, **36**, 856 (1964).

[6] S. Dotsenko and V. S. Dotsenko, Advances in Physics, **32**, 129 (1983).
[7] D. J. Gross and A. Neveu, Physical Review D, **10**, 3235 (1974).
[8] W. Wetzel, Physics Letters B, **153**, 297 (1985). This describes the β-function to two loops.
[9] R. Shankar, Physical Review Letters, **58**, 2466 (1987).
[10] A. W. W. Ludwig, Physical Review Letters, **61**, 2388 (1988).
[11] R. Shankar, Physical Review Letters, **61**, 2390 (1988). Gives a response to the comment in [10].
[12] B. N. Shalayev, Soviet Physics (Solid State), **26**, 1811 (1984).
[13] V. B. Andreichenko, Vl. S. Dotsenko, W. Selke, and J. J. Wang, Nuclear Physics B, **344**, 531 (1990).
[14] C. Itzykson and J. M. Drouffe, *Statistical Field Theory*, vol. 2, Cambridge University Press (1990).
[15] V. J. Emery, in *Highly Conducting One-Dimensional Solids*, eds. J. T. Devreese, R. P. Evrard, and V. E. van Doren, Plenum (1979).
[16] S. Sachdev, *Quantum Phase Transitions*, Cambridge University Press (1999).
[17] T. Giamarchi, *Quantum Physics in One Dimension*, Clarendon (2004).
[18] E. H. Lieb and F. Y. Wu, Physical Review Letters, **20**, 1145 (1968).
[19] F. D. M. Haldane, Physical Review Letters, **45**, 1358 (1980).
[20] F. D. M. Haldane, Journal of Physics C: Solid State Physics, **14**, 2585 (1981).
[21] S. Tomonaga, Progress in Theoreticl Physics, **5**, 544 (1950).
[22] J. M. Luttinger, Journal of Mathematical Physics, **4**, 1154 (1963).
[23] C. N. Yang and C. P. Yang, Physical Review, **150**, 321 (1976).
[24] R. J. Baxter, *Exactly Solvable Models in Statistical Mechanics*, Academic Press (1982).
[25] M. P. M. den Nijs, Physical Review B, **23**, 6111 (1981).
[26] J. M. Kosterlitz and D. J. Thouless, Journal of Physics C: Solid State Physics, **6**, 1181 (1973).
[27] F. D. M Haldane, Journal of Physics C, **14**, 2585 (1981).
[28] R. Shankar, International Journal of Modern Physics B, **4**, 2371 (1990).
[29] T. Giamarchi and H. J. Schulz, Physical Review B, **37**, 325 (1988).
[30] S. Sachdev and R. Shankar, Physical Review, **38**, 826 (1988).
[31] R. Shankar, Physical Review Letters, **63**, 203 1989.
[32] Y. K. Ha, Physical Review D, **29**, 1744 (1984). Gives an early review of field-theory bosonization.
[33] C. L. Kane and M. P. A. Fisher, Physical Review B, **46**, 15233 (1992).
[34] J. van Delft, Annalen der Physik, **4**, 225 (1998).
[35] J. Voit, Reports on Progress in Physics, **58**, 977 (1994).
[36] H. J. Schulz, in *Mesoscopic Quantum Physics: Les Houches Summer School Proceedings*, vol. 61, eds. E. Akkermans, G. Montambaux, J.-L. Pichard, and J. Zinn-Justin, North Holland (1995).
[37] E. Miranda, Brazilian Journal of Physics, **33**, 3 (2003).
[38] R. Heidenreich, B. Schroer, R. Seiler, and D. Uhlenbrock, Physics Letters A, **54**, 119 (1975).
[39] M. P. A. Fisher and L. I. Glazman, in *Mesoscopic Electron Transport*, eds. L. L. Sohn, L. P. Kouwenhoven, and G. Schön, NATO ASI series, Kluwer (1997). Very physical, explains tunneling and fractionalization.
[40] E. Fradkin, *Field Theories of Condensed Matter Physics*, 2nd edition, Cambridge University Press (2013).

[41] G. F. Guiliani and G. Vignale, *Quantum Theory of the Electron Liquid*, Cambridge University Press (2005).
[42] E. Witten, Communications in Mathematical Physics, **92**, 455 (1984).
[43] I. Affleck, in *Fields, Strings, and Critical Phenomena: Les Houches Summer School Proceedings*, eds. E. Brezin and J. Zinn-Justin, North Holland (1991).
[44] F. D. M. Haldane, in *Proceedings of the International School of Physics "Enrico Fermi," Course CXXI "Perspectives in Many-Particle Physics"*, eds. R. Broglia and J. Schrieffer, North Holland (1994).
[45] A. Houghton, H. J. Kwon, and J. B. Marston, Physical Review B, **50**, 1351 (1994).

19

Duality and Triality

Duality is a very powerful and deep notion that plays a fundamental role in statistical mechanics and field theory. As for triality, I will describe the only significant and striking example I know of.

19.1 Duality in the $d = 2$ Ising Model

Let us begin with and extend the case we have already encountered: Kramers–Wannier duality of the Ising model. I urge you to consult Sections 7.3 and 8.10 as and when needed.

The $d = 2$ Ising model has the partition function

$$Z = \sum_{s_i} \exp\left[K \sum_{<ij>} s_i s_j\right], \tag{19.1}$$

where the parameter

$$K = J/kT \tag{19.2}$$

is small in the high-T region and large in the low-T region.

When K is small, it is possible to expand Z in the high-temperature series in $\tanh K$. For a lattice of $\mathcal{N}$ sites,

$$\frac{Z_{\mathrm{H}}}{2^{\mathcal{N}}(\cosh K)^{2\mathcal{N}}} = 1 + \mathcal{N}\tanh^4 K + 2\mathcal{N}\tanh^6 K + \cdots \equiv Z_{\mathrm{H}}(K). \tag{19.3}$$

In this series, which is the sum over closed non-intersecting loops, each bond in the loops contributes a factor $\tanh K$. For example, the $\tanh^4 K$ term comes from elementary squares and $\tanh^6 K$ from rectangles of perimeter 6 with two possible orientations. This series breaks down at the Ising transition $K_c = 0.4407\ldots$

At low temperatures, we organize the sum over configurations around the fully ordered state. (There are two of them, all up or all down. We pick one, say all up, and go with it. The two sectors do not mix to any order in perturbation theory.) Configurations are now labeled by the down spins in this sea of up spins. A single down spin breaks four bonds and costs a factor e^{-8K}. Two adjacent flipped spins cost a factor e^{-12K} and count twice because the separation can be along one or the other direction. This leads to the low-temperature series

$$\frac{Z_{\rm L}}{e^{2\mathcal{N}K}} = 1 + \mathcal{N}e^{-8K} + 2\mathcal{N}e^{-12K} + \cdots \equiv Z_{\rm L}(K). \tag{19.4}$$

We expect this series to work for small e^{-2K} or large K.

Kramers and Wannier showed that the series $Z_{\rm H}$ and $Z_{\rm L}$ for high and low temperatures had the same coefficients $\mathcal{N}$, $2\mathcal{N}$, ..., and differed only in the exchange

$$\tanh K \leftrightarrow e^{-2K}. \tag{19.5}$$

The functions $Z_{\rm H}$ and $Z_{\rm L}$ are different but related: if $Z_{\rm H}(K)$ has a certain value at $\tanh K = 0.1234$, then $Z_{\rm L}(K)$ has the same value when $e^{-2K} = 0.1234$.

Consider the function (which is its own inverse)

$$K^*(K) = -\frac{1}{2}\ln\tanh K, \tag{19.6}$$

or, equivalently,

$$e^{-2K^*} = \tanh K. \tag{19.7}$$

Notice that when K is small, K^* is large, and vice versa.

Given Eq. (19.7), it follows that

$$Z_{\rm H}(K) = Z_{\rm L}(K^*(K)). \tag{19.8}$$

The functions $Z_{\rm H}$ and $Z_{\rm L}$ differ from the standard Z by some totally analytic factors. If we insist on keeping track of them, we can derive the equivalent result [Eq. (7.17)],

$$\frac{Z(K)}{(\sinh 2K)^{\frac{\mathcal{N}}{2}}} = \frac{Z(K^*)}{(\sinh 2K^*)^{\frac{\mathcal{N}}{2}}}, \tag{19.9}$$

and refer to the two sides as $Z_{\rm H}$ and $Z_{\rm L}$ from now on.

Since this duality relates the Ising model at small K or small coupling to the *same* model at large K or large coupling, it is called *self-duality*. In general, duality relates a model at weak coupling to a possibly different model at strong coupling. For example, the gauge Ising model in $d = 3$ at small K is dual to the ordinary Ising model at large K (with the same relation between K and K^*). On the other hand, the gauge Ising model in $d = 4$ is self-dual.

The equality $Z_{\rm H}(K) = Z_{\rm L}(K^*)$ does not mean the physics is the same at K and $K^*(K)$. One describes a disordered phase, and the other the ordered phase. However, given duality we may infer that if there is a singularity at K, there is one at $K^*(K)$. If there is only one critical point, it must lie at $K = K^*$.

As K increases, the small-K expansion converges more and more slowly. We cannot access large K this way. Indeed, even if we had the entire series it would not converge beyond $K_{\rm c} = 0.4407\ldots$ Duality converts this insurmountable problem at large K to an easy one in terms of K^*. Kinematical regions that were inaccessible to perturbative analysis in the original version become accessible in the dual version.

There is another route to duality that involves a change of variables. Under this change, the model at some coupling goes to the model of the dual variables at the dual coupling. This change is, however, *non-local*, i.e., the dual variable at some point is a function of an arbitrary number of the original variables arbitrarily far away. I recall the details of the operator approach based on the transfer matrix in the τ-continuum limit, as discussed in Section 8.10:

$$T = e^{-H\tau}, \quad \text{where } \tau \to 0 \text{ and} \tag{19.10}$$

$$H = \sum_{1}^{N} [-\lambda\sigma_1(n) - \sigma_3(n)\sigma_3(n+1)]. \tag{19.11}$$

The parameter λ keeps track of temperature. When $\lambda = 0$, the second term forces ordering in σ_3 ($T = 0$), while if $\lambda \to \infty$, the first term chooses the eigenstate of σ_1, an equal mixture of spin up and down at each site ($T = \infty$). At small λ we can perturb in λ, but the series will diverge at $\lambda_c = 1$. (This will be shown shortly.)

Duality is established by the change to μ's:

$$\mu_1(n) = \sigma_3(n)\sigma_3(n+1), \tag{19.12}$$

$$\mu_3(n) = \prod_{-\infty}^{n} \sigma_1(l), \tag{19.13}$$

which have the same algebra as the σ's. The inverse transformation has the same form. In terms of μ,

$$H(\lambda) = \sum [-\lambda\mu_3(n-1)\mu_3(n) - \mu_1(n)] \tag{19.14}$$

$$= \lambda H\left(\frac{1}{\lambda}\right). \tag{19.15}$$

Remarkably, the Hamiltonian is a *local* function of μ, although μ is a non-local function of σ and also the same function as $H(\sigma)$ was of σ, except for the change $\lambda \leftrightarrow \frac{1}{\lambda}$.

From Eq. (19.15) we infer that the spectrum of H obeys

$$E(\lambda) = \lambda E\left(\frac{1}{\lambda}\right), \tag{19.16}$$

level by level. If the mass gap (the difference between the ground state and the first excited state) vanishes at some critical value λ_c, it must do so at $1/\lambda_c$ as well. If the transition is unique, it must occur at $\lambda_c = 1$.

Duality is seen to map the problem at λ to a problem at $\frac{1}{\lambda}$, i.e., weak coupling to strong coupling and vice versa.

Recall the nature of the dual variables. When $\mu_3(n)$ acts on the all-up state, it flips all the spins from $-\infty$ to n so that we now have the down vacuum from $-\infty$ to n and the up vacuum to the right of n:

$$\mu_3(n)|\cdots\uparrow\uparrow\uparrow\uparrow\uparrow\uparrow\cdots\rangle = |\cdots\downarrow\downarrow\downarrow\downarrow\downarrow\uparrow\uparrow\uparrow\uparrow\uparrow\cdots\rangle. \tag{19.17}$$

Thus, $\mu_3(n)$ creates a *kink* or *soliton* which interpolates between the two degenerate vacua. This is a general feature of kinks and solitons. (At any $0 < \lambda < 1$, the vacua are not simply all up or all down but only predominantly so, but the action of $\mu_3(n)$ is still to flip all spins up to n.) Even though an infinite number of spins is flipped by one kink, the energy cost is finite, namely 2 in the $\lambda = 0$ limit.

The expectation value of μ_3 vanishes in the ordered ground state:

$$\langle\lambda < 1|\mu_3|\lambda < 1\rangle = 0, \tag{19.18}$$

since the state with an infinite number of spins flipped is orthogonal to the state we began with.

The presence of kinks reduces the average magnetization. For example, with two kinks at n_1 and n_2, the spins in between oppose the majority. At small λ such pairs of kinks are rare, but as λ increases, the cost of producing kinks goes down and eventually vanishes at $\lambda = 1$. Beyond this point we have a state in which kinks have condensed, i.e., they are present with finite density. This state, which looks totally disordered in σ_3, actually has $\langle\mu_3\rangle \neq 0$.

Kadanoff dubbed the dual variables *disorder variables*. While μ_3 is the disorder variable with respect to σ_3, we can equally well argue the other way around, since $(\sigma, H(\sigma))$ and $(\mu, H(\mu))$ are algebraically identical.

Here are the features of duality that we will be looking for in its generalizations and extensions:

- A model at weak coupling is related to its dual (possibly a different model) at strong coupling.
- The relation between variables and their duals is non-local.
- Despite the previous point, the theory of the dual variables is also local. (In principle, we can find the dual of any $H(\sigma)$ and get a non-local $H(\mu)$. But $H(\mu)$ may not correspond to any problem of interest.)
- Acting on the original vacuum, a dual operator creates a kink interpolating between different vacua of the original theory. So kinks reduce order.
- A state that looks disordered in terms of one variable could be ordered in the dual, in that the dual variable will have an expectation value.

19.2 Thirring Model and Sine-Gordon Duality

The massless Thirring model is defined by the *Euclidean* Lagrangian density

$$\mathcal{L} = \bar{\psi}\partial\!\!\!/\psi - \frac{g}{2}j^{\mu}j_{\mu}. \tag{19.19}$$

Upon bosonizing, this becomes

$$\mathcal{L} = \frac{1}{2}\left(1 + \frac{g}{\pi}\right)(\nabla\phi)^2. \tag{19.20}$$

The *massive Thirring* model is defined by adding $m\bar{\psi}\psi$, which leads to the following bosonized theory:

$$\mathcal{L} = \frac{1}{2}\left(1 + \frac{g}{\pi}\right)(\nabla\phi)^2 - \frac{m}{\pi\alpha}\cos\sqrt{4\pi}\,\phi. \tag{19.21}$$

Rescaling the field as in Eq. (18.9),

$$\phi' = \sqrt{\left(1 + \frac{g}{\pi}\right)}\phi \tag{19.22}$$

(and dropping the prime on ϕ'), we end up with

$$\mathcal{L} = \frac{1}{2}(\nabla\phi)^2 - \frac{m}{\pi\alpha}\cos\sqrt{\frac{4\pi}{1+\frac{g}{\pi}}}\phi \tag{19.23}$$

$$= \frac{1}{2}(\nabla\phi)^2 - \frac{m}{\pi\alpha}\cos\beta\phi, \quad \text{where} \tag{19.24}$$

$$\beta = \sqrt{\frac{4\pi}{1+\frac{g}{\pi}}} \quad \text{or} \tag{19.25}$$

$$\beta^2 = \frac{4\pi^2}{g+\pi}. \tag{19.26}$$

Equation (19.24) is the *sine-Gordon model* in canonical form. The equivalence of the massive Thirring and sine-Gordon models was independently established by Luther and Coleman, whose work was referred to in the last chapter.

Countless books [1] and papers have been written on the sine-Gordon model, and with good reason: despite the formidable interaction term it is integrable, its S-matrix is known [2–4], and of the greatest importance to us, it has *solitons*. I will only discuss those aspects of this model that bear directly on the duality with the Thirring model and direct you to some papers at the end of this chapter.

Equation (19.26) is a classic example of duality relating strong and weak couplings. When $g \to \infty$, $\beta \to 0$, and we can expand the cosine in powers of β and do perturbation theory.

The duality of the Lagrangians is only the beginning. We need to know how the dual degrees of freedom are related. To this end, we consider the potential energy of the sine-Gordon model:

$$V(\phi) = -\frac{m}{\pi\alpha}\cos\beta\phi. \tag{19.27}$$

It admits multiple ground states:

$$\phi = \frac{2\pi n}{\beta} \quad n = 0, \pm 1, \pm 2, \ldots \tag{19.28}$$

A soliton interpolates between two such vacua, with different n. Let us assume that as $x \to -\infty$, $\phi \to 0$. Then the elementary soliton or antisoliton connects the $n = 0$ vacuum to the $n = \pm 1$ vacua. That these correspond to the fermion or antifermion of the Thirring model can be established in two equivalent ways.

The first is to begin with the bosonic expression for fermion charge density,

$$j^0 = \frac{1}{\sqrt{\pi}} \partial_x \phi, \tag{19.29}$$

which applies to Eq. (19.21). Upon the field rescaling of Eq. (19.22),

$$j^0 = \frac{1}{\sqrt{\pi(1 + \frac{g}{\pi})}} \partial_x \phi \tag{19.30}$$

$$= \frac{\beta}{2\pi} \partial_x \phi. \tag{19.31}$$

The fermion charge associated with the $n = 1$ soliton in which ϕ changes by $\frac{2\pi}{\beta}$ is

$$Q = \int_{-\infty}^{\infty} j^0 dx \tag{19.32}$$

$$= \frac{\beta}{2\pi} \int_{-\infty}^{\infty} \partial_x \phi dx \tag{19.33}$$

$$= \frac{\beta}{2\pi} \frac{2\pi}{\beta} = 1. \tag{19.34}$$

Thus, the $n = 1$ soliton is the fermion of the Thirring model! Likewise, the $n = -1$ soliton is the antifermion of the Thirring model. This was quite a bombshell when it first appeared. It may not shock you because you are familiar with bosonization, but that came later.

The second way to understand the soliton–fermion correspondence is through the bosonization formula (which is described in free-field theory, $\beta^2 = 4\pi$), by

$$\psi_\pm(x) = \frac{1}{\sqrt{2\pi\alpha}} \exp\left[\pm i\sqrt{\pi}\left[\phi(x) \mp \int_{-\infty}^{x} \Pi(x')dx'\right]\right]. \tag{19.35}$$

This tells us that when we create or destroy a fermion we shift the boson field by $\pm\sqrt{\pi}$ using the conjugate operator $\Pi(x')$ for $-\infty \le x' \le x$. Now, Eq. (19.28) tells us that when $\beta^2 = 4\pi$ this is exactly the change in ϕ in the one-soliton state. (This operator creates a point kink which gets dressed by interactions to the physical soliton, as first observed by Mandelstam.)

19.3 Self-Triality of the $SO(8)$ Gross–Neveu Model

The Thirring and sine-Gordon models exemplify duality, but not self-duality. I am not aware of any relativisitic self-dual field theories in two dimensions.

But I will now show you an example of a relativistic field theory in two dimensions that exhibits *self-triality* in the following manner. We will begin with a theory of eight Majorana fermions, ψ_i, $i = 1,\ldots,8$, which form an isovector of $SO(8)$. This theory has multiple vacua and there are 16 different kinks. These fall into two multiplets, R_i, $1 = 1,\ldots,8$ and L_i, $i = 1,\ldots,8$, where L and R are isospinors with half-integral weights. (All this will be explained shortly.) I will show that

- The original theory of ψ has *two* dual versions, written as local theories of the kinks L or R.
- The Lagrangian for the two dual versions is identical to the one we began with:

$$\mathcal{L}(\psi) = \mathcal{L}(L) = \mathcal{L}(R). \tag{19.36}$$

- Just as L and R are the kinks of $\mathcal{L}(\psi)$, so ψ and R are the kinks in the theory of $\mathcal{L}(L)$, and ψ and L the kinks in the theory of $\mathcal{L}(R)$.

19.3.1 The Group SO(8)

I follow here the analysis of Witten [5]. This group of rotations in 8 dimensions has 28 generators L_{ij} (which rotate in the (i,j) plane), and of which a maximum of 4 can be simultaneously diagonalized. They may be chosen to be L_{12}, L_{34}, L_{56}, and L_{78}. Their eigenvalues or weights could be integer (vectors and tensors) or half-integer (spinors), and correspond to the quantum numbers of the particles in various representations.

The group has an infinite number of irreducible representations, but we will need just two here.

The first is the vector representation whose components are real. Had we been talking about $SO(2)$, the vector would have two components V_1 and V_2. The eigenvectors of the sole generator L_{12} would be the combinations $\psi_1 \pm i\psi_2$. If V_1 and V_2 were Majorana fermions, i.e., real fermions, the complex eigenvectors $\psi_1 \pm i\psi_2$ would be Dirac fermions of charge ± 1, or, if you like, a complex fermion and its antifermion.

In the case of $SO(8)$, which I focused on in [7, 8], the states in the vector representation ψ have weights

$$V = [\pm 1,0,0,0],\ [0,\pm 1,0,0],\ [0,0,\pm 1,0],\ [0,0,0,\pm 1], \tag{19.37}$$

and correspond to the complex combinations $\psi_1 \pm i\psi_2, \ldots, \psi_7 \pm i\psi_8$.

We are also interested in the *spinor representation*, which is acted upon by the Dirac matrices γ_i, $i = 1,\ldots,8$, obeying

$$\{\gamma_i, \gamma_j\} = 2\delta_{ij}. \tag{19.38}$$

We can form these matrices by taking the tensor product of four copies of Pauli matrices. They are therefore 16-dimensional. (In the usual Euclidean Dirac theory in $d = 4$, there would be four of these γ matrices, which can be obtained as the tensor product of two copies of Pauli matrices.) The spinors on which the γ act are therefore 16-dimensional. However, they break up *under rotations* into two irreducible spinors R and L. These are eigenstates with eigenvalues ± 1 of $\gamma_9 = \gamma_1 \cdots \gamma_8$, which anticommutes with the eight γ's and commutes with the generators of rotation L_{ij},

$$L_{ij} = \frac{i}{4}[\gamma_i, \gamma_j]. \tag{19.39}$$

In the case of $SO(4)$, γ_5 plays the role of γ_9 and splits the Dirac spinor into two different irreducible representations R and L of two-component spinors. (In the standard model of weak interactions, the R and L fermions are treated asymmetrically.)

I will not discuss how the half-integral weights of these representations are derived, and simply state the result. The spinors R have an even number of positive weights, while the L have an odd number:

$$R = \pm\frac{1}{2}[+,+,+,+],\ \pm\frac{1}{2}[+,+,-,-],\ \pm\frac{1}{2}[+,-,-,+],\ \pm\frac{1}{2}[+,-,+,-]; \tag{19.40}$$

$$L = \pm\frac{1}{2}[+,+,+,-],\ \pm\frac{1}{2}[+,+,-,+],\ \pm\frac{1}{2}[+,-,+,+],\ \pm\frac{1}{2}[-,+,+,+]. \tag{19.41}$$

In the 4-dimensional weight space, the isovectors are unit vectors pointing along the positive and negative coordinate axes. The weights of R are also a unit distance from the origin (because $4 \cdot \left(\frac{1}{2}\right)^2 = 1$) and form a *rotated* orthonormal basis. The weights of L form another orthonormal basis. The three weight systems are perfectly symmetrical with respect to each other.

Just so you realize how special this is, look at Figure 19.1, which shows you the situation for $SO(4)$, where the weight space is two-dimensional. The figure shows one 4-dimensional

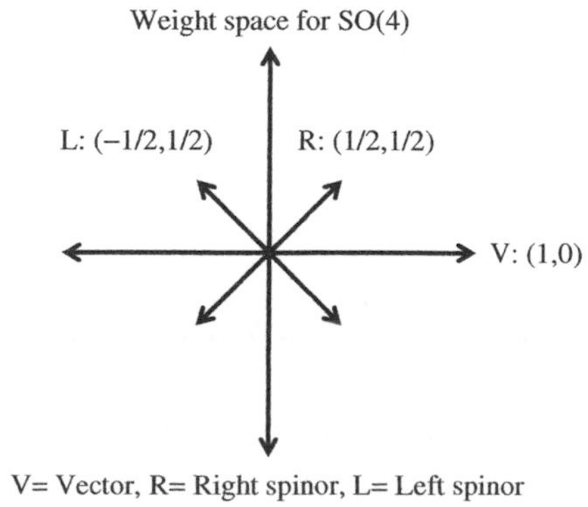

Figure 19.1 Weights of vector (V) and spinor representations R and L of $SO(4)$. Only the weights of one member of each multiplet are shown.

vector multiplet with unit integral weight and two isospinor representations (R and L) with half-integral weights and length $1/\sqrt{2}$.

19.3.2 Dynamics of the $SO(8)$ Model

The Lagrangian density of the model is

$$\mathcal{L}(\psi) = \frac{1}{2}\sum_{1}^{8}\overline{\psi}_i(\partial\!\!\!/)\psi_i - \frac{g^2}{8}\left(\sum_{1}^{8}\overline{\psi}_i\psi_i\right)^2, \tag{19.42}$$

where ψ is a Majorana isovector. The symmetry $\bar{\psi}\psi \to -\bar{\psi}\psi$ forbids a mass term. This symmetry is, however, dynamically broken because in the ground state

$$\langle\bar{\psi}\psi\rangle = \pm\Delta. \tag{19.43}$$

We refer to the two kinds of vacua as positive and negative vacua. We want to explore these vacua and the kinks that connect them.

That the kinks form isospinors for any even N was established by Witten [5] using bosonization, and further elucidated by Shankar and Witten in their derivation of the S-matrix of the kinks [4]. It can also be established by the methods of Jackiw and Rebbi [6].

My work focused on the remarkable case of $SO(8)$. What started out as a proof at the semiclassical level of the equality of kink and particle masses ended up with the exact property of self-triality. Here are the details.

Let us form 4 Dirac fermions from the 8 Majorana fermions:

$$\Psi_1 = \frac{\psi_1 + i\psi_2}{\sqrt{2}}, \tag{19.44}$$

$$\Psi_2 = \frac{\psi_3 + i\psi_4}{\sqrt{2}}, \tag{19.45}$$

$$\Psi_3 = \frac{\psi_5 + i\psi_6}{\sqrt{2}}, \tag{19.46}$$

$$\Psi_4 = \frac{\psi_7 + i\psi_8}{\sqrt{2}}, \tag{19.47}$$

in terms of which,

$$\bar{\Psi}_1\partial\!\!\!/\Psi_1 = \frac{1}{2}(\bar{\psi}_1\partial\!\!\!/\psi_1 + \bar{\psi}_2\partial\!\!\!/\psi_2), \tag{19.48}$$

$$\bar{\Psi}_1\Psi_1 = \frac{1}{2}(\bar{\psi}_1\psi_1 + \bar{\psi}_2\psi_2), \tag{19.49}$$

$$\vdots \tag{19.50}$$

$$\bar{\Psi}_4\Psi_4 = \frac{1}{2}(\bar{\psi}_7\psi_7 + \bar{\psi}_8\psi_8). \tag{19.51}$$

Bosonizing the 4 Dirac fields yields

$$\mathcal{L}(\psi) \leftrightarrow \mathcal{L}(\phi), \quad \text{where} \tag{19.52}$$

$$\mathcal{L}(\phi) = \sum_{1}^{4} \frac{1}{2}(\nabla\phi_i)^2 - \frac{g^2\Lambda^2}{2}\sum_{i=1}^{4}\sum_{j=1}^{4} \cos\sqrt{4\pi}\,\phi_i \cos\sqrt{4\pi}\,\phi_j. \tag{19.53}$$

The ground state is determined by minimizing the potential term. There are two kinds of vacua:

- Positive vacua where all cosines are positive:

$$\phi_i = 0,\ \pm\sqrt{\pi},\ \pm 2\sqrt{\pi},\ \pm 3\sqrt{\pi}, \ldots \tag{19.54}$$

- Negative vacua where all cosines are negative:

$$\phi_i = \pm\frac{1}{2}\sqrt{\pi},\ \pm\frac{3}{2}\sqrt{\pi},\ \pm\frac{5}{2}\sqrt{\pi}, \ldots \tag{19.55}$$

(Unfortunately, in the positive (negative) vacua $\bar{\Psi}\Psi < 0\ (> 0)$ because of our choice of $\bar{\Psi}\Psi = -\frac{1}{\pi\alpha}\cos\sqrt{4\pi}\phi$.)

Let us assume that as $x \to -\infty$, $\phi = 0$. The lightest solitons connect this vacuum to one where one of the fields, say $\phi_1 = \sqrt{\pi}$, and the rest vanish. I use the term soliton and not kink, which I reserve for configurations connecting opposite values of $\bar{\Psi}\Psi$.

What are the weights of this solitonic state? Given

$$\{\psi_i(x), \psi_j(x')\} = \delta_{ij}\delta(x - x'), \tag{19.56}$$

the Hermitian generator of rotations in the 12 plane is

$$L_{12} = \int_{-\infty}^{\infty} i\psi_1\psi_2 dx. \tag{19.57}$$

(The rotation operator is $e^{i\theta L_{12}}$. If we use the anti-Hermitian operator $\hat{L}_{12} = iL_{12}$, we must drop the i above and use as the rotation operator $e^{\theta\hat{L}_{12}}$.) Let us see if it does the job:

$$[L_{12}, \psi_1(x)] = \int i\left[\psi_1(x')\psi_2(x'), \psi_1(x)\right] dx = -i\psi_2(x),$$

$$[L_{12}, \psi_2(x)] = \int i\left[\psi_1(x')\psi_2(x'), \psi_2(x)\right] dx = i\psi_1(x),$$

which describes an infinitesimal rotation in the ψ_1–ψ_2 plane. With Dirac fermions in mind, we can equally well consider the action on complex combinations:

$$[L_{12}, \psi_1(x) + i\psi_2(x)] = -(\psi_1(x) + i\psi_2(x)), \quad \text{or} \tag{19.58}$$

$$[L_{12}, \Psi_1(x] = -\Psi_1(x), \tag{19.59}$$

which is correct because Ψ responds like a spin-1 object to the $U(1)$ rotation.

We now rewrite Eq. (19.57):

$$L_{12} = \int_{-\infty}^{\infty} i\psi_1\psi_2 dx \tag{19.60}$$

$$= \int_{-\infty}^{\infty} :\Psi_1^\dagger\Psi_1: dx \quad \text{[Eq. (17.34)]} \tag{19.61}$$

$$= \frac{1}{\sqrt{\pi}}\int_{-\infty}^{\infty} \frac{d\phi_1}{dx} dx, \tag{19.62}$$

$$\langle L_{12}\rangle = \left\langle \frac{\phi_1(\infty) - \phi_1(-\infty)}{\sqrt{\pi}} \right\rangle, \tag{19.63}$$

where $\langle\cdots\rangle$ is the average in any state. Similar equations hold for L_{34}, L_{56}, and L_{78}.

Now consider the average of these generators in the state with a soliton that connects the origin to the point $(\pm\sqrt{\pi},0,0,0)$. It is $(\pm 1,0,0,0)$. The rest of the vector multiplet is similarly described.

This is just like the sine-Gordon Thirring relation between the soliton of the former and the elementary fermion of the latter, done four times, once for each diagonal L_{ij}.

The same considerations tell us that the kinks connecting the origin to the points with *opposite* $\bar{\Psi}\Psi$, namely

$$(\phi_1,\phi_2,\phi_3,\phi_4) = \left(\pm\frac{\sqrt{\pi}}{2},\pm\frac{\sqrt{\pi}}{2},\pm\frac{\sqrt{\pi}}{2},\pm\frac{\sqrt{\pi}}{2}\right), \tag{19.64}$$

have the half-integral weights of the two spinor representations.

In summary, the 24 lightest particles may be visualized as follows in weight space or ϕ space. The 8 isovectors are represented by points at a unit distance from the origin along the positive and negative coordinate axes. The solitons that connect the origin to these points represent the original fermions and antifermions. The 8 spinors R are also a unit distance away but in the directions of another orthonormal tetrad. They have an even number of positive half-integral coordinates. The 8 spinors L are also a unit distance away and in the directions of another orthonormal tetrad. They have an odd number of positive half-integral coordinates. The isospinors connect the origin to states of opposite $\bar{\Psi}\Psi$. They are the real kink or disorder variables with respect to $\bar{\Psi}\Psi$.

What does the theory look like if written in terms of the isospinor kinks? Let us find out.

19.3.3 The Triality Transformation

Let us begin with the bosonic Lagrangian density (same as Eq. (18.57) with $n=4$):

$$\mathcal{L}(\phi) = \sum_1^4 \frac{1}{2}(\nabla\phi_i)^2 - \frac{g^2\Lambda^2}{2}\sum_{i=1}^4\sum_{j\neq i}^4 \cos\left(\sqrt{\frac{4\pi}{1+g^2/2\pi}}\phi_i\right)\cos\left(\sqrt{\frac{4\pi}{1+g^2/2\pi}}\phi_j\right). \tag{19.65}$$

We have used the identity

$$\left[\Lambda \cos \sqrt{4\pi}\phi_i\right]^2 = -\frac{1}{2}(\nabla\phi_i)^2 \tag{19.66}$$

to eliminate the $i = j$ terms in the double sum over cosines, and rescaled ϕ_i to bring the kinetic term to standard form. Since the $SO(8)$ Gross–Neveu model is free at short distances, i.e., the coupling $g \to 1/\ln\Lambda$ as $\Lambda \to \infty$,

$$\sqrt{\frac{4\pi}{1+g^2/2\pi}} \to \sqrt{4\pi} \tag{19.67}$$

in the continuum limit.

Let us perform a rigid rotation in weight space and trade the ϕ's for χ's:

$$\chi_1 = \frac{1}{2}(\phi_1+\phi_2+\phi_3+\phi_4), \tag{19.68}$$

$$\chi_2 = \frac{1}{2}(-\phi_1-\phi_2+\phi_3+\phi_4), \tag{19.69}$$

$$\chi_3 = \frac{1}{2}(\phi_1-\phi_2-\phi_3+\phi_4), \tag{19.70}$$

$$\chi_4 = \frac{1}{2}(-\phi_1+\phi_2-\phi_3+\phi_4). \tag{19.71}$$

Remarkably, *both the kinetic term and the interaction term* have the same form after this transformation:

$$\mathcal{L}(\chi) = \sum_1^4 \frac{1}{2}(\nabla\chi_i)^2 - \frac{g^2\Lambda^2}{2}\sum_{i=1}^4\sum_{j\neq i}^4 \cos\left(\sqrt{\frac{4\pi}{1+g^2/2\pi}}\chi_i\right)\cos\left(\sqrt{\frac{4\pi}{1+g^2/2\pi}}\chi_j\right). \tag{19.72}$$

If we now refermionize this theory, we will of course find that

$$\mathcal{L}(\chi) \leftrightarrow \mathcal{L}(R), \quad \text{where} \tag{19.73}$$

$$\mathcal{L}(R) = \frac{1}{2}\sum_1^8 \bar{R}_i(\not\partial)R_i - \frac{g^2}{8}\left(\sum_1^8 \bar{R}_iR_i\right)^2. \tag{19.74}$$

In this picture, R is an isovector with integer weights and ψ and L are isospinor kinks with half-integer weights that disorder the condensate $\langle\bar{R}R\rangle$ by interpolating between vacua of opposite $\langle\bar{R}R\rangle$.

Finally, we can rotate from ϕ_i to η_i, in terms of which L becomes an isovector and R and ψ the isospinor kinks that disorder the condensate $\langle\bar{L}L\rangle$.

In this manner we find a perfect triality between three descriptions, in which one field (say ψ) is the isovector that spontaneously breaks symmetry ($\bar{\psi}\psi \to -\bar{\psi}\psi$) and the other two are isospinor kinks that disorder the condensate.

Since the same g enters all three versions, we are at a point of self-triality. I have not found a way to move off this point.

There seems to be no simple way to pass between the three fermionic representations without first going to the bosonized versions.

This is the only theory I know of (modulo some rather artificial examples I found) in which there are three equivalent descriptions in each of which one field describes the particle and the other two the kinks, and all three have the same $\mathcal{L}$.

It seems unlikely that we will find more such examples of self-triality because of the unique properties of $SO(8)$ that were involved.

Here are three applications of triality. The first, pointed out by Witten [9–12], is in superstring theory. To demonstrate supersymmetry on the world sheet is very awkward if one begins with a fermion and boson that are both isovectors. But this is how they arise naturally in the superspace formalism. An isospinor fermion would be more natural. Witten used the transformation I introduced to convert the isovector to isospinor and established world sheet supersymmetry far more readily. However, Witten only used the triality of the free-field actions; the remarkable invariance of the quartic interaction does not seem to have a place in string theory. It will be wonderful to find instances where the interaction enters.

I know of just one, in the remarkable work of Fidkowski and Kitaev [13]. They asked if non-interacting topological insulators in different classes could be adiabatically connected in the presence of interactions. They answered in the affirmative: they found that a class made of 8 Majorana chains could morph into another one if the $(\bar{\psi}\psi)^2$ interaction were turned on at intermediate stages.

Finally, there is also an example where the $SO(8)$ theory enters condensed matter physics [14], and the role of the different represenations is very interesting.

19.4 A Bonus Result: The $SO(4)$ Theory

The following result, due to Witten [5], is now easily accessible. By a similar procedure, the $SO(4)$ model can be bosonized into

$$\mathcal{L}(\phi) = \frac{1}{2}(\nabla\phi_1)^2 + \frac{1}{2}(\nabla\phi_2)^2 - \frac{g^2}{2\pi^2\alpha^2}\cos\sqrt{4\pi^-}\phi_1 \cos\sqrt{4\pi^-}\phi_2. \tag{19.75}$$

In terms of

$$\phi_\pm = \frac{\phi_1 \pm \phi_2}{\sqrt{2}}, \tag{19.76}$$

$\mathcal{L}$ separates into two sine-Gordon models at $\beta^2 = 8\pi^-$:

$$\mathcal{L}(\phi) = \frac{1}{2}(\nabla\phi_+)^2 + \frac{1}{2}(\nabla\phi_-)^2 - \frac{g^2}{4\pi^2\alpha^2}\left(\cos\sqrt{8\pi^-}\phi_+ + \cos\sqrt{8\pi^-}\phi_-\right). \tag{19.77}$$

The $SO(4)$ symmetry, equivalent to $SO(3) \times SO(3)$, is realized by its action on the isospinor kinks R and L with weights $\pm(\frac{1}{2}, \frac{1}{2})$ and $\pm(-\frac{1}{2}, \frac{1}{2})$ shown in Figure 19.1. In studying the Kosterlitz–Thouless flow (Figure 18.2) it was stated that the point $\beta^2 = 8\pi$ describes the $SO(3)$ Heisenberg model, the XXZ chain at $\Delta = 1$. One may wonder how this non-Abelian symmetry can be realized by a scalar field. The answer is that the soliton and antisoliton form an $SO(3)$ or $SU(2)$ doublet. (For RG aficionados: To define a massive $SU(2)$-symmetric theory in the continuum, we must start at a point along the separatrix $x = y = (1/\ln\Lambda)$ infinitesimally close to the fixed point at the origin. Not surprisingly, both the sine-Gordon and Gross–Neveu models have bare couplings that vanish as $1/\ln\Lambda$, as they should if they are to describe the same physics.)

References and Further Reading

[1] R. Rajaraman, *Instantons and Solitons*, North Holland (1982). This is my favorite book.
[2] A. Zamolodchikov and Al. Zamolodchikov, Annals of Physics, **120**, 253 (1979). Gives a review of exact S-matrices.
[3] A. Zamolodchikov and Al. Zamolodchikov, Nuclear Physics B, **133**, 525 (1978). Gives the exact S-matrix of Gross–Neveu elementary fermions.
[4] R. Shankar and E. Witten, Physical Review D, **17**, 2134 (1978). Gives the S-matrix of the supersymmetric non-linear σ model. This paper provides ways to understand the determination of S-matrices, as does [1].
[5] E. Witten, Nuclear Physics B, **141**, 349 (1978).
[6] R. Jackiw and C. Rebbi, Physical Review D, **13**, 2298 (1976). Alternative derivation of the kink spectrum.
[7] R. Shankar, Physics Letters B, **92**, 333 (1980).
[8] R. Shankar, Physical Review Letters, **46**, 379 (1981).
[9] E. Witten, A. Weldon, P. Langacker, and P. Steinhardt (eds.), *Proceedings of the Fourth Workshop on Grand Unification*, Birkhäuser (1983). The following three papers alluded to my work before it passed into the public domain.
[10] R. Nepomichie, Physics Letters B, **178**, 207 (1986).
[11] D. Friedan, E. Martinec, and S. Shenker, Nuclear Physics B, **271**, 93 (1986);.
[12] P. Goddard and D. Olive, in *Kac–Moody and Virasoro Algebras: A Reprint Volume for Physicists*, World Scientific (1988).
[13] L. Fidkowski and A. Kitaev, Physical Review B, **81**, 134509 (2010).
[14] H. H. Lin, L. Balents, and M. P. A. Fisher, Physical Review B, **58**, 1794 (1998).

20

Techniques for the Quantum Hall Effect

The quantum Hall effect (QHE) has captivated the attention of theorists and experimentalists following its discovery. First came the astounding integer quantum Hall effect (IQHE) discovered by von Klitzing, Dorda, and Pepper in 1980 [1]. Then came the even more mysterious discovery of the fractional quantum Hall effect (FQHE) by Tsui, Störmer, and Gossard in 1982 [2]. Obviously I cannot provide even an overview of this vast subject. Instead, I will select two techniques that come into play in the theoretical description of the FQHE. Along the way I will cover some aspects of IQHE. However, of necessity, I will be forced to leave out many related developments, too numerous to mention. The books in [3–7] and online notes in [8] may help you with further reading.

The first technique is due to Bohm and Pines (BP) [9], and was used to describe an excitation of the electron gas called the *plasmon*. Since the introduction by BP of this technique in first quantization, it has been refined and reformulated in the diagrammatic framework. I will stick to the wavefunction-based approach because it is very beautiful, and because two of the great problems in recent times – the theory of superconductivity and the theory of the FQHE – were first cracked open by ingenious trial wavefunctions that captured all the essentials. I will introduce the BP approach in terms of the electron gas.

The second technique is Chern–Simons field theory. Originally a product of the imaginations of the mathematicians S. S. Chern and J. Simons, it first entered particle physics in the work of Deser, Jackiw, and Templeton [10], and then condensed matter [11–14]. I will describe its role in the FQHE after introducing the problem to you.

20.1 The Bohm–Pines Theory of Plasmons: The Goal

Consider a system of N spinless fermions experiencing the Coulomb interaction

$$V = \frac{1}{2}\sum_{i=1}^{N}\sum_{j\neq i,1}^{N} \frac{e^2}{4\pi\varepsilon_0}\frac{1}{|\boldsymbol{r}_i - \boldsymbol{r}_j|}. \tag{20.1}$$

Invoking the Fourier transformation (in unit spatial volume)

$$\frac{1}{|\boldsymbol{r}_i - \boldsymbol{r}_j|} = \sum_{\boldsymbol{q}} \frac{4\pi}{q^2} e^{-i\boldsymbol{q}\cdot(\boldsymbol{r}_i - \boldsymbol{r}_j)}, \tag{20.2}$$

we find that

$$V = \frac{e^2}{2\varepsilon_0} \sum_{i=1}^{N} \sum_{j \neq i,1}^{N} \sum_{\boldsymbol{q} \neq 0} e^{-i\boldsymbol{q}\cdot(\boldsymbol{r}_i - \boldsymbol{r}_j)} \frac{1}{q^2} \tag{20.3}$$

$$= \frac{e^2}{2\varepsilon_0} \sum_{\boldsymbol{q}} \frac{\rho(\boldsymbol{q})\rho(-\boldsymbol{q})}{q^2}, \quad \text{where} \tag{20.4}$$

$$\rho(\boldsymbol{q}) = \sum_j e^{-i\boldsymbol{q}\cdot\boldsymbol{r}_j} \tag{20.5}$$

is the density operator (in first quantization), and the $q = 0$ component is presumed to have been neutralized by some background charge.

If you like second quantization, you must recall that the one-body operator

$$O = \sum_{\alpha\beta} |\alpha\rangle O_{\alpha\beta} \langle\beta|, \tag{20.6}$$

where $|\alpha\rangle$ and $\langle\beta|$ are single-particle states, has the second-quantized version (for which I use the same symbol)

$$O = \psi_a^\dagger O_{\alpha\beta} \psi_b, \tag{20.7}$$

where $\psi_\alpha^\dagger$ (ψ_α) destroys (creates) a particle in state $|\alpha\rangle$.

Using this, you may show that, in second quantization,

$$\rho(\boldsymbol{q}) = \sum_{\boldsymbol{k}} c^\dagger(\boldsymbol{k} - \boldsymbol{q}) c(\boldsymbol{k}), \tag{20.8}$$

where $\boldsymbol{k}$ denotes momentum and I follow tradition and use $c(\boldsymbol{k})$ instead of $\psi(\boldsymbol{k})$. The density operator $\rho(\boldsymbol{q})$ is the sum of operators that scatter a particle of momentum $\boldsymbol{k}$ to one with $\boldsymbol{k} - \boldsymbol{q}$. Acting on the Fermi sea, they create particle–hole pairs.

However, their coherent sum $\rho(\boldsymbol{q})$ produces a sharply defined, long-lived *collective excitation*, called the *plasmon*. It is the quantized version of an oscillation that can be understood in classical terms as follows.

Imagine two very large rectangular solids of opposite charge density $\pm ne$ superposed on each other. If you displace them relative to each other by δx, you produce opposite surface charges on faces perpendicular to the displacement. The field that is established inside this "capacitor" is

$$E = \frac{\sigma}{\varepsilon_0} = \frac{n \cdot e \cdot \delta x}{\varepsilon_0}, \tag{20.9}$$

and the restoring force per unit displacement on the charges e,

$$k = \frac{eE}{\delta x}, \tag{20.10}$$

causes oscillations at the *plasma frequency* ω_0 given by

$$\omega_0 = \sqrt{\frac{ne^2}{m\varepsilon_0}}, \tag{20.11}$$

where m is the particle mass. (In CGS units, this becomes $\omega_p = \sqrt{\frac{4\pi ne^2}{m}}$.)

What I have described is the $q = 0$ plasmon. One could also have a space-dependent displacement as an excitation at frequency $\omega_0(q) = \omega_0 + \mathcal{O}(q^2)$. If this oscillator is quantized, the quanta are the plasmons with energy $E = \hbar\omega_0(q)$. (As we move away from $q = 0$, $\rho(\boldsymbol{q})$ also creates particle–hole pairs of much lower energy.)

The energy of the plasmon is much higher than the other particle–hole excitations near the Fermi surface or the thermal energy of order kT. Suppose we want to describe physics way below the plasmon scale. One option is to simply drop the plasmon, but a more responsible one would be to eliminate it using the RG and modify the low-energy physics to take into account its effect. However, this cannot be done using simple momentum-shell elimination because the plasmon is built out of momentum states *close* to the Fermi surface and owes its high energy to the Coulomb interaction that is singular as $q \to 0$. The BP method provides a way to freeze the plasmon in its ground state and retain its influence on the low-energy sector in the form of a *correlated wavefunction*. What does "correlated" mean? The filled Fermi sea is an uncorrelated wavefunction in which particles neither approach nor avoid each other in response to the interactions. Their behavior is uncorrelated except for the zeros in the wavefunction mandated by the Pauli principle. These zeros would exist even in the non-interacting case. A correlated wavefunction would go beyond this in trying to lower the interaction energy. Consider, for example, spinless fermions on a square lattice with nearest-neighbor repulsion. In any wavefunction, they would, of course, avoid double occupancy because of the Pauli principle, but a correlated wavefunction will additionally suppress configurations with nearest neighbors. The method of BP provides such a correlation factor, which multiplies the Fermi sea wavefunction.

20.2 Bohm–Pines Treatment of the Coulomb Interaction

The high energy of the plasmon comes from the long-range Coulomb interaction. So let us write

$$V = \frac{e^2}{2\varepsilon_0}\sum_{\boldsymbol{q}} \frac{\rho(\boldsymbol{q})\rho(-\boldsymbol{q})}{q^2} \tag{20.12}$$

$$= \frac{e^2}{2\varepsilon_0}\sum_{q<Q} \frac{\rho(\boldsymbol{q})\rho(-\boldsymbol{q})}{q^2} + \frac{e^2}{2\varepsilon_0}\sum_{q>Q} \frac{\rho(\boldsymbol{q})\rho(-\boldsymbol{q})}{q^2} \tag{20.13}$$

$$= V_{\mathrm{L}} + V_{\mathrm{S}}, \tag{20.14}$$

where L and S stand for long and short range, respectively. One can handle V_{S} within Fermi liquid theory because it is non-singular. Let us drop the short-range interaction V_{S} and the subscript on $V_{\rm L}$. It will be understood that

$$\sum_{q} = \sum_{q;q<Q} . \tag{20.15}$$

The Hamiltonian is

$$H = \sum_{j=1}^{N} \frac{|\boldsymbol{p}_j|^2}{2m} + \frac{e^2}{2\varepsilon_0} \sum_{q} \frac{\rho(\boldsymbol{q})\rho(-\boldsymbol{q})}{q^2}. \tag{20.16}$$

Now we introduce at each $q \leq Q$ an operator $a(\boldsymbol{q})$ and its conjugate momentum $P(\boldsymbol{q})$ obeying

$$a^{\dagger}(\boldsymbol{q}) = a(-\boldsymbol{q}), \tag{20.17}$$

$$P^{\dagger}(\boldsymbol{q}) = P(-\boldsymbol{q}), \tag{20.18}$$

$$\left[a(\boldsymbol{q}), P(-\boldsymbol{q}')\right] = i\delta_{\boldsymbol{q}\boldsymbol{q}'}. \tag{20.19}$$

The first two equations ensure the Hermiticity of the real-space fields $\boldsymbol{a}(\boldsymbol{r})$ and $\boldsymbol{P}(\boldsymbol{r})$,

$$\boldsymbol{a}(\boldsymbol{r}) = i\sum_{\boldsymbol{q}} \hat{\boldsymbol{q}} a(\boldsymbol{q}) e^{i\boldsymbol{q}\cdot\boldsymbol{r}}, \tag{20.20}$$

$$\boldsymbol{P}(\boldsymbol{r}) = -i\sum_{\boldsymbol{q}} \hat{\boldsymbol{q}} P(\boldsymbol{q}) e^{i\boldsymbol{q}\cdot\boldsymbol{r}}, \quad \text{where} \tag{20.21}$$

$$\hat{\boldsymbol{q}} = \frac{\boldsymbol{q}}{q}. \tag{20.22}$$

The fields $\boldsymbol{a}$ and $\boldsymbol{P}$ are called *longitudinal* because their Fourier coefficients at each $\boldsymbol{q}$ have no component transverse to $\boldsymbol{q}$.

We now introduce $a(\boldsymbol{q})$ into H, along with a corresponding constraint, so as to leave the physics unchanged:

$$H = \sum_{j=1}^{N} \frac{|\boldsymbol{p}_j|^2}{2m} + \frac{e^2}{2\varepsilon_0} \sum_{q} \left(a(\boldsymbol{q}) + \frac{\rho(\boldsymbol{q})}{q}\right)\left(a(-\boldsymbol{q}) + \frac{\rho(-\boldsymbol{q})}{q}\right), \tag{20.23}$$

$$a(\boldsymbol{q})|\ \rangle = 0, \tag{20.24}$$

where $|\ \rangle$ is any physical state.

We have gone to an enlarged Hilbert space acted upon by the particle coordinates $\boldsymbol{r}_j$ and $\boldsymbol{p}_j$ as well as $a(\boldsymbol{q})$ and $P(\boldsymbol{q})$. Since only $a(\boldsymbol{q})$ enters H, we can label states by the eigenvalue of $a(\boldsymbol{q})$. We pick the states obeying the constraint Eq. (20.24) to describe the original problem. The other states do not bear on the original problem and should be discarded as unphysical.

Why do we introduce $a(\boldsymbol{q})$, then? Consider the unitary operator

$$U = \exp\left[-i\sum_{q} P(-\boldsymbol{q})\frac{\rho(\boldsymbol{q})}{q}\right]. \tag{20.25}$$

Since $P(-\boldsymbol{q})$ is conjugate to $a(\boldsymbol{q})$, U translates $a(\boldsymbol{q})$. Under its action,

$$Ua(\boldsymbol{q})U^{\dagger} = a(\boldsymbol{q}) - \frac{\rho(\boldsymbol{q})}{q} \tag{20.26}$$

and the constraint Eq. (20.24) becomes

$$\left(a(\boldsymbol{q}) - \frac{\rho(\boldsymbol{q})}{q}\right)|\,\rangle = 0. \tag{20.27}$$

Let us see what happens to the Hamiltonian. The interaction term becomes

$$V \to UVU^{\dagger} = \frac{e^2}{2\varepsilon_0}\sum_q a(\boldsymbol{q})a(-\boldsymbol{q}). \tag{20.28}$$

In the kinetic term, we find that

$$\begin{aligned} U\boldsymbol{p}_j U^{\dagger} &= \boldsymbol{p}_j - \boldsymbol{P}(\boldsymbol{r}_j), \quad \text{where} \\ \boldsymbol{P}(\boldsymbol{r}) &= -i\sum_q \hat{q}P(\boldsymbol{q})e^{i\boldsymbol{q}\cdot\boldsymbol{r}}, \end{aligned} \tag{20.29}$$

so that

$$\begin{aligned} H \to UHU^{\dagger} &= \sum_{j=1}^{N} \frac{(\boldsymbol{p}_j - \boldsymbol{P}(\boldsymbol{r}_j))\cdot(\boldsymbol{p}_j - \boldsymbol{P}(\boldsymbol{r}_j))}{2m} + \frac{e^2}{2\varepsilon_0}\sum_{\boldsymbol{q}} a(\boldsymbol{q})a(-\boldsymbol{q}) \\ &= \sum_{j=1}^{N} \frac{|\boldsymbol{p}_j|^2 + \boldsymbol{P}(\boldsymbol{r}_j)\cdot\boldsymbol{P}(\boldsymbol{r}_j) + \boldsymbol{p}_j\cdot\boldsymbol{P}(\boldsymbol{r}_j) + \boldsymbol{P}(\boldsymbol{r}_j)\cdot\boldsymbol{p}_j}{2m} \\ &\quad + \frac{e^2}{2\varepsilon_0}\sum_q a(\boldsymbol{q})a(-\boldsymbol{q}). \end{aligned} \tag{20.30}$$

Exercise 20.2.1 *Prove Eq. (20.29).*

Bohm and Pines analyze this theory in great depth, but I will pluck from their findings just one result that comes in handy in the FQHE. Let us ignore the interaction terms between $\boldsymbol{p}$ and $\boldsymbol{P}$ to focus on just the non-interacting part:

$$H_0 = \sum_{j=1}^{N}\left(\frac{|\boldsymbol{p}_j|^2}{2m} + \frac{\boldsymbol{P}(\boldsymbol{r}_j)\cdot\boldsymbol{P}(\boldsymbol{r}_j)}{2m}\right) + \frac{e^2}{2\varepsilon_0}\sum_q a(\boldsymbol{q})a(-\boldsymbol{q}).$$

Consider the ground state. It will be a product wavefunction where the particles fill a Fermi sea. What about $\boldsymbol{a}$ and $\boldsymbol{P}$? Their Hamiltonian is

$$H(\boldsymbol{a},\boldsymbol{P}) = \sum_{j=1}^{N}\frac{\boldsymbol{P}(\boldsymbol{r}_j)\cdot\boldsymbol{P}(\boldsymbol{r}_j)}{2m} + \frac{e^2}{2\varepsilon_0}\sum_q a(\boldsymbol{q})a(-\boldsymbol{q}). \tag{20.31}$$

Look at the first piece:

$$H(\boldsymbol{P}) = \sum_{j=1}^{N} \frac{\boldsymbol{P}(\boldsymbol{r}_j)\cdot\boldsymbol{P}(\boldsymbol{r}_j)}{2m} \tag{20.32}$$

$$= -\frac{1}{2m}\sum_{j=1}^{N}\sum_{\boldsymbol{q}}\sum_{\boldsymbol{q}'}(P(\boldsymbol{q})\hat{\boldsymbol{q}}e^{i\boldsymbol{q}\cdot\boldsymbol{r}_j})\cdot(P(\boldsymbol{q}')\hat{\boldsymbol{q}}'e^{i\boldsymbol{q}'\cdot\boldsymbol{r}_j})$$

$$= -\frac{1}{2m}\sum_{\boldsymbol{q}}\sum_{\boldsymbol{q}'}P(\boldsymbol{q})P(\boldsymbol{q}')\hat{\boldsymbol{q}}\cdot\hat{\boldsymbol{q}}'F(\boldsymbol{q}+\boldsymbol{q}'),\ \text{where}$$

$$F(\boldsymbol{q}+\boldsymbol{q}') = \sum_{j=1}^{N} e^{i(\boldsymbol{q}+\boldsymbol{q}')\cdot\boldsymbol{r}_j}. \tag{20.33}$$

For generic $(\boldsymbol{q}+\boldsymbol{q}')$, F receives contributions that oscillate as we sum over $\boldsymbol{r}_j$. These are expected to cancel when we sum over a macroscopic number of particles. The exception is when $\boldsymbol{q} = -\boldsymbol{q}'$, in which case $F = N$. Keeping only this term is called the *random phase approximation* or RPA. Setting

$$F(\boldsymbol{q}+\boldsymbol{q}') = N\delta_{\boldsymbol{q},-\boldsymbol{q}'} \tag{20.34}$$

in Eq. (20.33), we arrive at

$$H(\boldsymbol{a},\boldsymbol{P}) = \sum_{q}\left[\frac{N}{2m}P(\boldsymbol{q})P(-\boldsymbol{q}) + \frac{e^2}{2\varepsilon_0}a(\boldsymbol{q})a(-\boldsymbol{q})\right]. \tag{20.35}$$

Since we are working in unit volume, we may set

$$N = n, \tag{20.36}$$

the number density.

Next, we trade $a(\boldsymbol{q})$ and $P(\boldsymbol{q})$ for the scalar fields $a(\boldsymbol{r})$ and $P(\boldsymbol{r})$ to obtain

$$H = \int d^2\boldsymbol{r}\left[\frac{n}{2m}P^2(\boldsymbol{r}) + \frac{e^2}{2\varepsilon_0}a^2(\boldsymbol{r})\right]. \tag{20.37}$$

Comparing to a single oscillator for which the Hamiltonian, frequency, and ground-state wavefunction are

$$H = \frac{p^2}{2M} + \frac{1}{2}kx^2, \tag{20.38}$$

$$\omega == \sqrt{\frac{k}{M}}, \tag{20.39}$$

$$\Psi_0(x) = e^{-\frac{1}{2}\sqrt{kM}x^2}, \tag{20.40}$$

we see that here,

$$“k” = \frac{e^2}{\varepsilon_0}, \tag{20.41}$$

$$“M” = \frac{m}{n}, \tag{20.42}$$

$$\omega = \sqrt{\frac{ne^2}{m\varepsilon_0}}, \tag{20.43}$$

$$\Psi_0(x) = \exp\left[-\frac{1}{2}\sqrt{\frac{me^2}{n\varepsilon_0}} \int d^2 r a^2(\mathbf{r}) \right] \tag{20.44}$$

$$= \exp\left[-\frac{1}{2}\sqrt{\frac{me^2}{n\varepsilon_0}} \sum_q a(\boldsymbol{q})a(-\boldsymbol{q}) \right]. \tag{20.45}$$

In the ground state of this system the oscillators at every $\boldsymbol{q}$ will be in their ground states. The combined ground state of the fermions and oscillators is, in the $a(\boldsymbol{q})$ basis,

$$\Psi(\boldsymbol{r}, a) = |\text{Fermi sea}\rangle \otimes \exp\left[-\sum_{q<Q} \frac{1}{2}\sqrt{\frac{me^2}{n\varepsilon_0}} a(\boldsymbol{q})a(-\boldsymbol{q}) \right]. \tag{20.46}$$

To obtain the physical wavefunction from the preceding one in the enlarged space we must impose the constraint

$$a(\boldsymbol{q}) = \frac{\rho(\boldsymbol{q})}{q}, \tag{20.47}$$

which gives us

$$\Psi(\boldsymbol{r}) = |\text{Fermi sea}\rangle \otimes \exp\left[-\sum^{q} \frac{1}{2}\sqrt{\frac{me^2}{n\varepsilon_0}} \frac{\rho(\boldsymbol{q})\rho(-\boldsymbol{q})}{q^2} \right]. \tag{20.48}$$

Had the sum over $\boldsymbol{q}$ been unrestricted we would have, upon going back to coordinate space,

$$\Psi(\boldsymbol{r}) = |\text{Fermi sea}\rangle \otimes e^{-\frac{V(r)}{\omega_0}}, \tag{20.49}$$

where

$$V(\boldsymbol{r}) = \frac{1}{2} \sum_{1=1}^{N} \sum_{j \neq i}^{N} \frac{e^2}{4\pi\varepsilon_0} \frac{1}{|\boldsymbol{r}_i - \boldsymbol{r}_j|} \tag{20.50}$$

is the electrostatic interaction energy. However, since actually $q < Q$, this expression holds only for separations $|\boldsymbol{r}_i - \boldsymbol{r}_j| > 1/Q$. For smaller separations we must replace V by $\tilde{V}$, where

$$\tilde{V}(\boldsymbol{r}) = \frac{e^2}{2\varepsilon_0} \sum_{i=1}^{N} \sum_{j \neq i,1}^{N} \sum_{q<Q} e^{-i\boldsymbol{q}\cdot(\boldsymbol{r}_i - \boldsymbol{r}_j)} \frac{1}{q^2}. \tag{20.51}$$

Equation (20.49) is a wonderful example of a correlated wavefunction: in addition to keeping particles away from each other by the Pauli zeros in |Fermi sea⟩, it discourages them from coming too close and increasing the Coulomb energy. The penalty for coming too close is a Boltzmann-like factor with energy V and "kT" $= \omega_0$. Why not suppress configurations of high potential energy even further? The answer is that in the present approximation, this is the state of lowest *total* energy.

This is the kind of result we will derive for the FQHE. There, too, we will encounter a magneto-plasmon at high energies, and by freezing it in its ground state we will obtain a correlated wavefunction for electrons.

The wavefunction is just one of the results from BP, who go on to consider the effect of the short-range potential V_S, the coupling between fermions and plasmons (the cross terms between $\boldsymbol{p}$ and $\boldsymbol{P}$), and the constraint that remains in the theory after fermions and plasmons are (approximately) decoupled. Nowadays the whole theory is couched in diagrams.

20.3 Fractional Quantum Hall Effect

The discussion that follows has a very limited goal: to focus on wavefunctions and Hamiltonians, and some of the ideas from Bohm–Pines that come into play. A comprehensive review of FQHE will not fit into a single chapter, or a single book. I will aim this section at a non-expert and start from the basics.

I will first introduce you to the phenomenon one is trying to explain, and show you why it defies many standard approaches like the RG, Hartree–Fock, or perturbation theory.

Then I will describe Laughlin's breakthrough: a trial wavefunction (for a class of FQHE states) that one could write on a T-shirt and which we knew right away had to be a cornerstone of the final theory. This was followed by Jain, who wrote down wavefunctions that described an even greater number of FQHE states. It relied on the notion of flux attachment, which I will describe in terms of wavefunctions.

In parallel with these, there were developments on another front. These began with the microscopic Hamiltonians and tried to arrive at a description of the key features following a sequence of (possibly uncontrolled) approximations. These approaches have great value because the human mind can never comprehend an exact solution unless it is a free-field theory in disguise. For us to "understand" the phenomenon, we necessarily need to work with caricatures and approximations. For example, our faith in the Newtonian description of planetary motion stems from approximations in which we consider just the Sun and one planet at a time. An exact solution involving all possible interactions between the planets would not even tell us we were on the right path because we would be unable to digest it. The same goes for the simplified description of atoms that gave us faith in quantum mechanics in its infancy. In general, it is better to start with an approximation that we can internalize and assimilate, and then seek embellishments via tools like perturbation theory.

The optimal tool for describing FQHE is Chern–Simons (CS) theory. Rather than tell you in advance what it is, I will analyze the problem in terms of Hamiltonians, wavefunctions, constraints, and so forth, and then reveal that what we have arrived at is really a CS theory.

I will conclude with a brief section on my work with Murthy on what we called the Hamiltonian formalism. I cannot judge its long-term impact. Unlike the RPA or Hartree–Fock, it is designed just for the FQHE. It walks like a duck and quacks like a duck, but maybe it is still not a duck. You can take it or leave it.

20.4 FQHE: Statement of the Problem

To study the Hall effect, one takes carriers of charge q (assumed positive) in the (x,y) plane, subject to a magnetic field $\boldsymbol{B}$ in the z-direction as shown in Figure 20.1.

A current density $\boldsymbol{j}$ ($\mathrm{A\,m^{-1}}$) is driven in the x direction and the voltage V_y is measured in the y direction. In classical theory, at equilibrium, an electric field $\boldsymbol{E}$ is set up in the $+y$ direction, which balances the Lorentz force per unit charge $v \times \boldsymbol{B}$ in the $-y$ direction. By definition, the Hall conductance is

$$\sigma_{xy} = \frac{I_x}{V_y} = \frac{j_x L_y}{E_y L_y} = \frac{j_x}{E_y} = \frac{nqv_x}{v_x B} = \frac{nq}{B}. \tag{20.52}$$

For electrons $q = -e$, but we can continue to use

$$\sigma_{xy} = \frac{ne}{B} \tag{20.53}$$

if we also choose $\boldsymbol{B} = -\hat{z}B$, i.e., pointing down the z-axis.

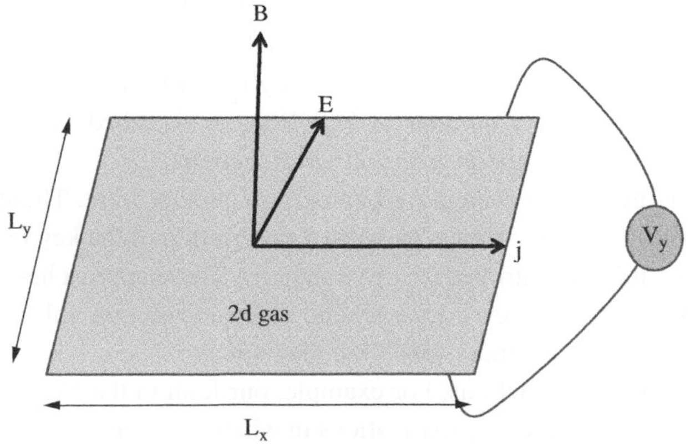

Figure 20.1 The two-dimensional carriers of charge q (assumed positive) lie in the (x,y) plane, occupying an area $L_x \times L_y$. There is a magnetic field $\boldsymbol{B}$ pointing along the z-axis. The current density $\boldsymbol{j}$ is along x, and the induced electric field $\boldsymbol{E}$ is along y.

Equation (20.52), independent of $\hbar$, is valid even in quantum theory and follows from just Galilean invariance. It holds in the presence of interparticle interaction, but not an external disorder potential $V(\boldsymbol{r})$. To derive this result, start with a charge density nq at rest in the (x,y) plane and a magnetic field $\boldsymbol{B}$ up the z-axis. Now boost to a frame at a velocity v in the *negative* x direction. In this frame there will be a current $J_x = -nqv_x = nq|v_x| > 0$ in the positive x direction and, by classical electromagnetic theory, also an electric field $\boldsymbol{E} = v \times \boldsymbol{B}$ in the $+y$ direction. Therefore, $\sigma_{xy} = j_x/E_y = nq/B$. As mentioned earlier, $\sigma_{xy} = ne/B$ holds for electrons if $\boldsymbol{B}$ points down the z-axis.

But experiments have a surprise in store. Instead of this Galilean invariant result of a straight line in the σ_{xy}–$\frac{ne}{B}$ plane, we find steps at some special values. The steps are flat to parts in a billion. Some of them are shown in Figure 20.2, which is schematic and not to scale. Not shown in the figure is $\sigma_{xx} = j_x/E_x$, which implies dissipation and is non-zero only at the jumps. In real life, the regions of non-zero σ_{xx} and the jumps between plateaus have non-zero widths, as depicted in [1] and [2].

The Galilean invariant result $\sigma_{xy} = ne/B$ agrees with the data only at special *intersection points*, indicated by dots in Figure 20.2, where the straight line crosses the steps. Away from these intersection points we need necessarily to bring in disorder, i.e., a random potential $V(\boldsymbol{r})$. While $V(\boldsymbol{r})$ can explain some of the steps even in the absence of the electron–electron interaction, others will require interactions for their explanation.

We shall focus on steps in σ_{xy} that occur at the following values:

$$\sigma_{xy} = \frac{e^2}{2\pi\hbar}\frac{p}{2ps+1} \equiv \sigma_0 \frac{p}{2ps+1}, \quad p = 0,1,2,\ldots, \quad s = 0,1,2,\ldots, \tag{20.54}$$

where

$$\sigma_0 = \frac{e^2}{2\pi\hbar} \tag{20.55}$$

is the standard *unit of conductance* in quantum transport.

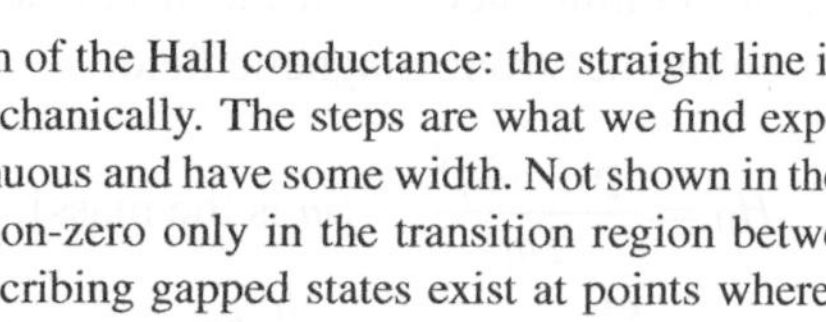

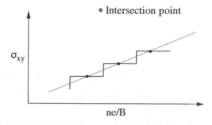

Figure 20.2 Schematic form of the Hall conductance: the straight line is expected in a clean system, classically and quantum mechanically. The steps are what we find experimentally. The actual steps (e.g., [1,2]) are not discontinuous and have some width. Not shown in the figure is $\sigma_{xx} = j_x/E_x$, which implies dissipation and is non-zero only in the transition region between plateaus. Translationally invariant wavefunctions describing gapped states exist at points where the straight line crosses the steps at *intersection points*, shown by dots.

The case $s = 0$, or $\frac{\sigma_{xy}}{\sigma_0} = p$, corresponds to the IQHE discovered by von Klitzing *et al.* [1]. The first example of the more general case of FQHE, where $\frac{\sigma_{xy}}{\sigma_0} = \frac{p}{2ps+1}$, $s,p \neq 0$, was first discovered by Tsui *et al.* [2] for $s = p = 1$, $\frac{\sigma_{xy}}{\sigma_0} = \frac{1}{3}$. Many other fractions followed.

The theoretical breakthroughs came at the intersection points where the steps crossed the straight line $\sigma_{xy} = ne/B$. These are the points where there is some chance of describing the state with a translationally invariant wavefunction. (Away from the intersection points, translation symmetry is manifestly violated and the disorder potential $V(\boldsymbol{r})$ *has* to be invoked.)

The translationally invariant wavefunctions for

$$p = 1, \ \frac{\sigma_{xy}}{\sigma_0} = \frac{1}{2s+1} = \frac{1}{3}, \frac{1}{5}, \ldots \tag{20.56}$$

were provided by Laughlin [15], while Jain [16, 17] did the same for general p when

$$\frac{\sigma_{xy}}{\sigma_0} = \frac{p}{2ps+1} = \frac{1}{3}, \frac{2}{5}, \frac{3}{7}, \ldots; \ \frac{1}{5}, \frac{2}{9}, \frac{3}{13}, \ldots \tag{20.57}$$

The Laughlin and Jain wavefunction described gapped ground states *only at the intersection points*. As we move off these intersections points by changing the density, particle-like excitations are nucleated in response. In other words, the change in density is not uniformly spread over the system but lumped into these particle-like excitations. The excitations get trapped or localized by disorder, thereby contributing to density but not σ_{xy}. At some point along the plateau it becomes energetically favorable to jump to the next plateau described by another ground-state wavefunction and *its* excitations.

Of course, once we bring in disorder to explain the plateaus, we also have to explain why it does not change σ_{xy} at the intersection points.

20.5 What Makes FQHE So Hard?

The answer will emerge naturally if we start from scratch, try the standard recipe described below, and fail.

- Find the single-particle energy states.
- Fill the states in the order of increasing energy to find the ground state or Fermi sea.
- Consider particle–hole excitations of the Fermi sea to get excited states.
- Add interactions and disorder (random potential $V(\boldsymbol{r})$) to taste.

So, let us begin with H_0, the Hamiltonian for one electron in a magnetic field. We ignore spin, assuming that it is frozen by the strong magnetic field:

$$H_0 = \frac{(\mathbf{p} + e\mathbf{A})^2}{2m} \quad (m \text{ is the mass}), \tag{20.58}$$

$$e\mathbf{A} = \frac{eB}{2}(y, -x) = -\frac{\hbar}{2l^2}\hat{z} \times \mathbf{r}, \tag{20.59}$$

$$l = \sqrt{\frac{\hbar}{eB}} \quad \text{is the magnetic length,} \tag{20.60}$$

$$\nabla \times e\mathbf{A} = -e\hat{z}B. \tag{20.61}$$

We choose a **B** pointing down the z-axis because the electron charge is $-e$.

In a two-dimensional world, the curl is a scalar and we don't really need the $\hat{z}$ in the last equation. It will be dropped except in cases like Eq. (20.59).

We have chosen the vector potential A in the *symmetric gauge*, which allows us to find states of definite angular momentum. Continuing,

$$H_0 = \frac{\hbar^2 \boldsymbol{\eta}^2}{2ml^4}, \quad \text{where} \tag{20.62}$$

$$\boldsymbol{\eta} = \frac{1}{2}\mathbf{r} + \frac{l^2}{\hbar}\hat{z} \times \mathbf{p}, \tag{20.63}$$

$$\left[\eta_x, \eta_y\right] = il^2. \tag{20.64}$$

One calls $\boldsymbol{\eta}$ the *cyclotron coordinate*, even though its two components form a *conjugate pair* like x and p.

Let us define canonical operators

$$\eta = \frac{\eta_x + i\eta_y}{\sqrt{2}l} \tag{20.65}$$

obeying

$$\left[\eta, \eta^\dagger\right] = 1. \tag{20.66}$$

Then, in terms of these operators,

$$H_0 = \left(\eta^\dagger \eta + \frac{1}{2}\right)\frac{eB\hbar}{m} = \left(\eta^\dagger \eta + \frac{1}{2}\right)\hbar\omega_0. \tag{20.67}$$

Exercise 20.5.1 *Starting with Eq. (20.58), derive Eq. (20.67).*

The energy levels of the oscillator are called Landau levels (LL's), and

$$\omega_0 = \frac{eB}{m} \tag{20.68}$$

is called the *cyclotron frequency*. It is the (radius-independent) frequency of classical orbits in the plane perpendicular to the field. Of special interest is the lowest Landau level (LLL), which has energy $\frac{1}{2}\hbar\omega$. In this state, $\langle \eta^2 \rangle_{n=0} = l^2$. This means that the ground-state wavefunction in the symmetric gauge will have size $\simeq l$.

One often considers the limit $\frac{B}{m} \to \infty$, when the scales of interactions and disorder are dwarfed by $\hbar\omega_0$. It is generally believed that the central issues of the FQHE can then be addressed by considering just the LLL. (There are some exceptions where higher LL's enter, but they are not our focus.)

Now, a one-dimensional oscillator spectrum, Eq. (20.67), for a two-dimensional problem suggests that the LL's must be degenerate. This is correct. The degeneracy is due to another canonical pair that does not enter the Hamiltonian and commutes with $\boldsymbol{\eta}$. This is the *guiding center coordinate* $\mathbf{R}$ defined by

$$\mathbf{R} = \frac{1}{2}\mathbf{r} - \frac{l^2}{\hbar}\hat{z} \times \mathbf{p}, \tag{20.69}$$

$$\left[R_x, R_y\right] = -il^2, \tag{20.70}$$

$$[\boldsymbol{\eta}, \boldsymbol{R}] = 0, \tag{20.71}$$

$$[H_0, \boldsymbol{R}] = 0. \tag{20.72}$$

The familiar coordinate $\boldsymbol{r}$ is given by

$$\mathbf{r} = \mathbf{R} + \boldsymbol{\eta}. \tag{20.73}$$

If $l^2 = 1/eB \to 0$, we can ignore the non-commutativity of the components of $\boldsymbol{R}$ and η and interpret Eq. (20.73) as follows: The location of the particle $\boldsymbol{r}$ is the vector sum of a macroscopic guiding center $\boldsymbol{R}$ around which the particle orbits in a microscopic circle of size $\eta \simeq l$ in the LLL. (The size grows with the LL index.)

Since $\eta \simeq l$ is a microscopic length, the range of $\boldsymbol{R}$ is the range of $\boldsymbol{r}$, which in turn is the sample area L_xL_y. However, because R_x and R_y are conjugate variables, the area L_xL_y is interpreted as the *phase space* and not the coordinate space for $\boldsymbol{R}$. Equation (20.70) tells us that the role of $\hbar$ is played by l^2. Consequently, the degeneracy of the Landau levels is, by Bohr–Sommerfeld quantization,

$$D = \frac{\text{Area of phase space}}{2\pi\text{``}\hbar\text{''}} = \frac{L_xL_y}{2\pi l^2} = \frac{eBL_xL_y}{2\pi\hbar} = \frac{\Phi}{\Phi_0}, \tag{20.74}$$

where Φ is the total flux and

$$\Phi_0 = \frac{2\pi\hbar}{e} \tag{20.75}$$

is the *flux quantum*.

The number of states in any Landau level equals the flux penetrating the sample measured in units of Φ_0, *i.e., the number of flux quanta.* This result, derived semiclassically above, is confirmed by explicit calculations.

Remember that for the electron these flux quanta point down the z-axis, and that the flux quanta are used only to measure the strength of the external field. The latter is perfectly uniform and not lumped into quantized flux tubes.

This enormous degeneracy of the LL's derails the game plan we started with, of finding the ground state by filling the single-particle states in the order of increasing energy. For example, if there are 200 states in the LLL and we have only 100 electrons, there are $\frac{200!}{(100!)^2} \simeq 10^{59}$ degenerate options. Even the finest book on quantum mechanics [18] does not prepare you to do perturbation theory in this case. *The only exceptions occur when we have just the right number of electrons to fill an integer number of Landau levels.* In this case there is one state per electron in each LL, and there is only one way to occupy them. The wavefunction is a Slater determinant of the occupied LL wavefunctions.

This motivates us to introduce the *filling factor* ν:

$$\nu = \frac{\text{number of particles}}{\text{number of states in an LL}} = \text{filled LL's} = \frac{N}{(\Phi/\Phi_0)} = \frac{2\pi n\hbar}{eB}, \tag{20.76}$$

where $n = N/L_xL_y$ is the particle density. Remember, ν need not be an integer.

Let us rewrite

$$\nu = \frac{2\pi n\hbar}{eB} \tag{20.77}$$

as

$$\frac{ne}{B} = \nu\frac{e^2}{2\pi\hbar} = \nu\sigma_0. \tag{20.78}$$

But ne/B is the value of σ_{xy} for a translation-invariant system or one at the intersection points of the steps with the translation-invariant answer. At the intersection points, where σ_{xy} is compatible with translation invariance,

$$\sigma_{xy} = \nu\sigma_0. \tag{20.79}$$

The filling fraction is the dimensionless Hall conductance given translation invariance.

If ν is an integer p, we simply fill p LL's to obtain a non-degenerate gapped ground state for the electrons with Hall conductance $p\sigma_0$.

What if we are at an intersection point where ν is a fraction? We have seen that the intersection points occur at fractions

$$\nu = \frac{p}{2ps+1}. \tag{20.80}$$

To see what is special about these points, consider

$$\nu^{-1} = \frac{(\Phi/\Phi_0)}{N} = \text{flux quanta per particle} = 2s + \frac{1}{p}. \tag{20.81}$$

If $s = 0$ we are back to the simple case of $\frac{1}{p}$ flux quanta per electron, and fill exactly p LL's to get a unique ground state. What happens if $2s$ is some non-zero even integer? It turns out that by transforming the electron to a *composite fermion* (CF) – in a manner to be described later – we arrive at a particle that sees only the $\frac{1}{p}$ part of ν^{-1}. The $2s$ quanta get canceled by the additional field the CF's themselves produce. The CF can then fill p LL's, which is a unique and gapped state. It is then possible to translate the results back to a unique gapped ground state for electrons. Of course, all this will be elaborated in due course.

Let us begin with the simpler case of integer $\nu = p$ first.

20.6 Integer Filling: IQHE

Let us focus on the case $s = 0$, or $\nu = p$, when there are enough electrons to fill exactly p Landau levels. We will only study $p = 1$, when we can exactly fill the LLL. Only occasionally is there new physics when we fill more than one LL.

20.6.1 Getting to Know the LLL

So far, we have the spectrum of the particle in a magnetic field and the degeneracy of each LL. Now it is time to look at the wavefunctions.

Let us begin with

$$H_0 = \left(\eta^\dagger \eta + \frac{1}{2}\right)\hbar\omega_0, \quad \text{where} \tag{20.82}$$

$$\eta = \frac{\eta_x + i\eta_y}{\sqrt{2}l}. \tag{20.83}$$

The LLL condition $\eta|\ \rangle = 0$ becomes, using Eq. (20.63),

$$\left[\frac{eBz}{4\hbar} + \frac{\partial}{\partial\bar{z}}\right]|\mathrm{LLL}\rangle = 0, \tag{20.84}$$

implying the following wavefunctions for the LLL states:

$$\psi = e^{-|z|^2/4l^2} f(z), \tag{20.85}$$

where $z = x + iy$ and $f(z)$ is *any* analytic function, $\frac{\partial f}{\partial \bar{z}} = 0$.

A basis for ψ is

$$\psi_m(z) = z^m\ e^{-|z|^2/4l^2}, \qquad m = 0, 1, \ldots \tag{20.86}$$

The state ψ_m has angular momentum $L_z = m\hbar$.

Only the LLL wavefunctions are analytic (up to the ubiquitous Gaussian), while the ones from higher LL's depend on z and $\bar{z}$.

If $\nu = 1$, there is a unique non-interacting determinantal ground state (which may then be perturbed by standard means):

$$\chi_1 = \det\left[z_j^i\right] e^{-\sum_j |z_j|^2/4l^2}, \tag{20.87}$$

where the *Vandermonde determinant*

$$\det\left[z_j^i\right] \equiv \begin{vmatrix} z_1^0 & z_1^1 & z_1^2 & \cdots \\ z_2^0 & z_2^1 & z_2^2 & \cdots \\ \vdots & \vdots & \vdots & \ddots \end{vmatrix} \tag{20.88}$$

can be written as

$$\det\left[z_j^i\right] = \prod_{i<j}(z_i - z_j). \tag{20.89}$$

Imagine taking one of the z's, say z_1, for a loop around the rest of the particles in the positive sense. The wavefunction changes its phase by $2\pi N$. This is what we expect, given that the sample is penetrated by N flux quanta (one per particle, in this case). Again, the flux points down the z-axis if the electron with charge $-e$ is to pick up this phase $+2\pi N$.

Back to the single-particle wavefunction. The probability density $|\psi_m(z)|^2$ is peaked at radius

$$r_m = \sqrt{2m}\, l. \tag{20.90}$$

If the sample is a disk of radius R, then the maximum value m^* obeys

$$r_{m^*} = \sqrt{2m^*}\, l = R, \tag{20.91}$$

which means that the number of states that fit into the sample is

$$m^* = \frac{R^2}{2l^2} = \frac{\pi R^2 Be}{2\pi\hbar} = \frac{\Phi}{\Phi_0}, \tag{20.92}$$

as was deduced earlier based on semiclassical quantization.

20.6.2 LL Redux: The Landau Gauge

We will now solve the free-particle problem in the Landau gauge

$$e\mathbf{A} = -\mathbf{j}eBx. \tag{20.93}$$

In this gauge,

$$H_0 = \frac{p_x^2}{2m} + \frac{(p_y - eBx)^2}{2m}. \tag{20.94}$$

Since y is cyclic, we may choose as the normalized wavefunctions

$$\psi(x,y) = \frac{1}{\sqrt{L_y}} e^{iky}\psi_k(x) \quad \text{with} \tag{20.95}$$

$$\int dx |\psi_k(x)|^2 = 1. \tag{20.96}$$

Since $p_y = \hbar k$ acting on this state, $\psi_k(x)$ obeys

$$\left[-\frac{\hbar^2}{2m}\frac{d^2}{d\bar{x}^2} + \frac{e^2B^2}{2m}\bar{x}^2\right]\psi_k(\bar{x}) = E\psi_k(\bar{x}), \quad \text{where} \tag{20.97}$$

$$\bar{x} = x - kl^2. \tag{20.98}$$

Obviously, $\psi_k(\bar{x})$ describes an oscillator centered at $x = kl^2$. For LLL physics, we put the oscillator in the ground state. In every LL the energy will be independent of k, which merely decides where the oscillators are centered.

In the Landau gauge, the wavefunctions labeled by k are extended momentum states in y and strips of width $\delta x \simeq l$ in the x direction, centered at $x_0(k) = kl^2$.

How many strips can we fit in an $L_x \times L_y$ sample?

Assuming that $k = 0$ is the left-most strip, the right-most strip must obey

$$k_{\max} l^2 = L_x. \tag{20.99}$$

Since

$$k = \frac{2\pi}{L_y} j, \quad j = 0, 1, 2, \ldots, \tag{20.100}$$

we have the following restriction on $j_{\max}$, the number of states per LL:

$$\frac{2\pi}{L_y} j_{\max} = \frac{L_x}{l^2} \tag{20.101}$$

$$j_{\max} = \frac{L_x L_y}{2\pi l^2} = \frac{\Phi}{\Phi_0}. \tag{20.102}$$

The Landau gauge is suited to problems with rectangular geometry, which is typically what we encounter in Hall measurements.

20.7 Hall Conductivity of a Filled Landau Level: Clean Case

For a clean system with p filled LL's, we may appeal to the result following from Galilean invariance, Eq. (20.79):

$$\sigma_{xy} = \nu \sigma_0 = p \sigma_0. \tag{20.103}$$

In particular, when $\nu = 1$,

$$\sigma_{xy} = \sigma_0 = \frac{e^2}{2\pi \hbar}. \tag{20.104}$$

This is quite remarkable if you consider the fact that a single channel carrying current in a wire has the same conductance as a planar sample! How can a macroscopic sample have the same conductance? To understand this, let us look at the current density operator in the Landau gauge:

$$j_y = -\left(\frac{\partial H_0}{\partial A_y}\right) \propto \frac{p_y + eA_y}{m} \propto \left(\frac{eBx - \hbar k}{m}\right) \propto (x - kl^2) = x - x_0(k), \tag{20.105}$$

where $x_0(k) = kl^2$ is the center of the oscillator wavefunction with y-momentum k. Consequently, as you sweep across a strip associated with the Landau gauge wavefunction, the current density changes sign at the center. The current density is equal and opposite on the two sides of the center. If you now glue together strips side by side to form a rectangular sample, the adjoining edges cancel each other out (as in Stokes's theorem), and what is left is just the current at the edges, near $x = 0$ and $x = L_x$. It follows that no matter how wide the sample is, you are going to get just the edge currents. This is a profound aspect of the QHE: all the action is at the edges.

Let us now get quantitative and compute the Hall conductance more directly, following Halperin [19, 20], who demystified the Hall effect by pointing out the role of the edges in transport. Suppose we confine the electronic Hall system by a barrier potential energy $-eV(x)$ that is flat in the bulk but sharply curved up at the edges, as shown in Figure 20.3. The LL's that are flat in the middle get curved as well. The figure shows the chemical potential μ that intercepts the LLL at x_L and x_R. The following analysis is for the LLL; the other levels are essentially the same.

As a prelude, let us understand what the barrier will do in semiclassical terms. Let $\mathcal{P}$ denote projection to the LLL. Then

$$\mathcal{P}: \mathbf{r} = \mathbf{R} + \eta \Rightarrow \mathbf{R}. \tag{20.106}$$

In other words, we are ignoring the tiny cyclotron orbit around $\boldsymbol{R}$. We saw that $\boldsymbol{R}$ commutes with H_0. This changes when we apply an external potential $V(\boldsymbol{r}) \simeq V(\boldsymbol{R})$. Since R_x and R_y are conjugate variables, $-eV(R_x, R_y)$ is like a Hamiltonian $H(x, p)$.

The Heisenberg equations of motion are:

$$\dot{R}_x = \frac{1}{i\hbar}\left[R_x, -eV(R_x, R_y)\right] \tag{20.107}$$

$$= \frac{il^2}{i\hbar}\frac{\partial eV}{\partial R_y} = \frac{1}{B}\frac{\partial V}{\partial R_y}, \tag{20.108}$$

$$\dot{R}_y = -\frac{1}{B}\frac{\partial V}{\partial R_x}, \tag{20.109}$$

$$\frac{d\boldsymbol{R}}{dt} = -\frac{1}{B}\hat{z} \times \nabla V = \frac{\boldsymbol{E} \times \boldsymbol{B}}{B^2}. \tag{20.110}$$

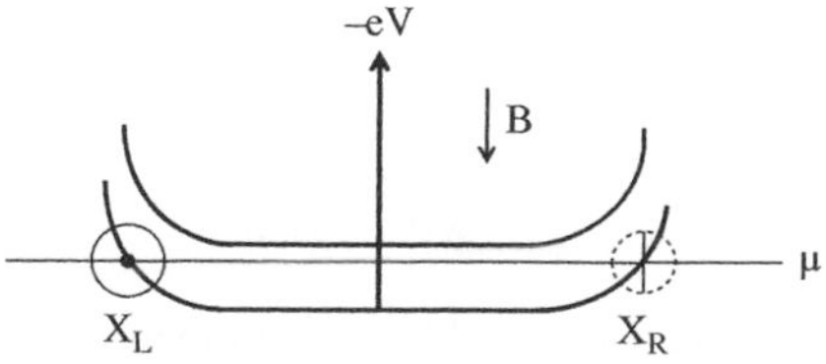

Figure 20.3 The electronic Hall system in a confining potential $-eV(x)$ that is flat in the bulk and curves up at the edges. The LL's also curve up. The chemical potential μ intersects the curved LLL at x_L and x_R. As explained in the text, the confining potential causes edge currents of equal magnitude going into and out of the page as shown. If the two edges are at slightly different potentials, a net Hall current flows into or out of the page, perpendicular to the applied voltage.

Exercise 20.7.1 *Derive Eq. (20.110). It might help to set $R_x = -il^2\frac{\partial}{R_y}$, which follows from the commutation rules.*

Because its velocity is perpendicular to the gradient, Eq. (20.110), *the particle moves along equipotentials*. In Figure 20.3, the gradient is non-zero only near the edge of the sample, and this leads to edge currents of equal magnitude going into and out of the page as indicated. The net current in the y direction, going into the page, is zero. Let us now apply an infinitesimal voltage difference δV between the two edges. The current density j_y is the product of the particle density and the group velocity $v = \frac{dE}{dk}$. In this problem, k and x are correlated because the wavefunction is centered at $x = x_0(k) = kl^2$. In the bulk $E(k)$ is independent of k, but due to the applied voltage $V(x)$ the energy varies as $-eV(x) = -eV(x(k)) = -eV(x = kl^2)$.

So the group velocity is

$$v = \frac{dE(k)}{\hbar dk} = \frac{dE(k(x))}{\hbar dx}\frac{dx}{dk} = \frac{dE(k(x))}{\hbar dx}l^2 = -\frac{edV(x)}{\hbar dx}l^2 = -\frac{1}{B}\frac{dV}{dx}, \tag{20.111}$$

which agrees with the $\boldsymbol{E} \times \boldsymbol{B}/B^2$ velocity we found in Eq. (20.110).

The current density due to one state centered at some $x = kl^2$ is

$$j_y(x,k) = -e \cdot v \cdot |\psi_k(x,y)|^2 = -e\frac{dE}{\hbar dk}\frac{1}{L_y}|\psi_k(x)|^2, \tag{20.112}$$

$$I_y(k) = \int j_y(x,k)dx = -e\frac{dE}{\hbar dk}\frac{1}{L_y}, \tag{20.113}$$

where I use the fact that $\int dx|\psi_k(x)|^2$ is normalized and assume that dE/dk is constant over the width of the wavefunction.

The total current along y is the sum over all k values, i.e., an integral over $\frac{L_y}{2\pi}dk$:

$$I_y = \frac{L_y}{2\pi}\int dk \cdot (-e) \cdot \frac{dE}{\hbar dk}\frac{1}{L_y} = -\frac{e}{2\pi\hbar}(E(\text{right edge}) - E(\text{left edge})) = \frac{e^2}{2\pi\hbar}\delta V, \tag{20.114}$$

which implies

$$\sigma_{xy} = \frac{I_y}{\delta V} = \frac{e^2}{2\pi\hbar}. \tag{20.115}$$

Here is another way to derive Eq. (20.115) for a clean, filled LL. It is worth learning because of its generality and its constant appearance in the thriving field of *topological insulators* [21]. The argument was originally espoused by Laughlin [22], and developed further by Halperin [19,20].

Consider our QHE sample rolled into a cylinder in the y direction by gluing $y = 0$ and $y = L_y$, as shown in Figure 20.4. Imagine adding a vector potential along the rolled-up y

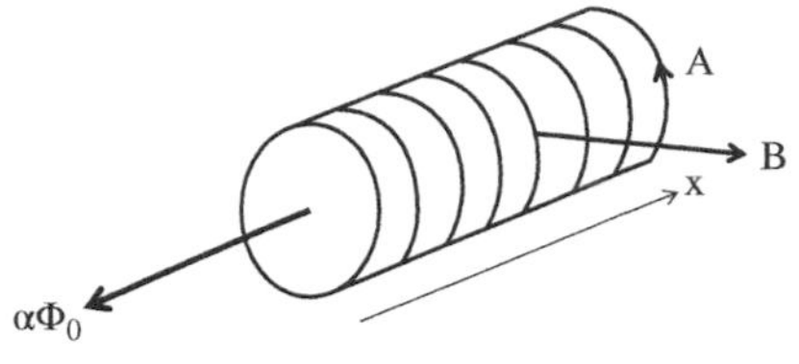

Figure 20.4 The Hall system of the electron (charge $-e$) in the Landau gauge and strip geometry (penetrated by a perpendicular magnetic field $\boldsymbol{B}$) folded in the y direction into a cylinder (on gluing $y=0$ and $y=L_y$). A flux $\alpha(t)\Phi_0$ is threaded as shown, by means of a vector potential $A_y=\frac{\alpha\Phi_0}{L_y}$.

direction:

$$A_y=\frac{\alpha(t)\Phi_0}{L_y}, \tag{20.116}$$

where $\alpha(t)$ grows slowly from 0 to 1 over a very long time. This means inserting a flux $\Phi=\alpha\oint A_y dy=\alpha\Phi_0$ along the axis of the rolled-up cylinder. At the end, we have introduced a flux quantum that can be gauged away by a non-singular gauge transformation. (For $\alpha<1$, the gauge function that is needed to undo A will be non-single-valued around the y direction and disallowed.) We may safely assume that the particles will follow the states they are in as the states evolve, because there is a gap $\hbar\omega_0$ to excitations.

In the Landau gauge, adding the flux is described by

$$H_0=\frac{p_x^2}{2m}+\frac{(p_y-eBx)^2}{2m}\rightarrow\frac{p_x^2}{2m}+\frac{(p_y-eBx+\alpha e\frac{\Phi_0}{L_y})^2}{2m}, \tag{20.117}$$

which means that, as α goes from 0 to 1,

$$p_j=\frac{2\pi j\hbar}{L_y}\rightarrow\frac{2\pi(j+1)\hbar}{L_y}. \tag{20.118}$$

Exercise 20.7.2 *Verify Eq. (20.118).*

Therefore, as α grows from 0 to 1, every strip state slides over to the right by one, carrying the particle with it. At the end we have one particle of charge $-e$ transferred from the left edge to the right. This is called *pumping*.

What does this imply for Hall conductance? Let it be some σ_{xy}. The electric field associated with A_y is

$$E_y=-\frac{d\alpha}{dt}\frac{\Phi_0}{L_y}, \tag{20.119}$$

and the current in the x direction is

$$I_x = L_y j_x = L_y \sigma_{xy} E_y = -\sigma_{xy} \Phi_0 \frac{d\alpha}{dt}, \tag{20.120}$$

$$Q_x = -\sigma_{xy} \Phi_0 \int d\alpha = -\sigma_{xy} \Phi_0 = -\sigma_{xy} \frac{2\pi\hbar}{e}. \tag{20.121}$$

But the charge transferred in the x direction is $Q_x = -e$, which gives $\sigma_{xy} = \frac{e^2}{2\pi\hbar}$.

Exercise 20.7.3 *Apply the flux insertion argument to a disc-shaped sample with a tiny hole in the center, with the filled LLL. Argue that if Φ_0 is slowly inserted down the hole, the wavefunctions in the symmetric gauge, peaked in concentric rings, will move outward by one, carrying the particles with them. Deduce that $\sigma_{xy} = \sigma_0$.*

20.7.1 Effect of Disorder

So far we have managed to establish in multiple ways that the filled LL of a clean system has one quantum of Hall conductance. If we want to sit precisely at this filling, we can ignore disorder. But the minute we move off the intersection point, the flat plateau on either side tells us that the answer violates translation invariance. Even if we do not understand how violating translation invariance can produce plateaus, we have to concede that it is violated.

This leaves us with two questions. First, why does disorder not modify the clean-limit answer at the intersection point? Second, how does it produce the plateau?

The short answer to the first question is that the edge states are *helical*, i.e., the particles travel in only one direction (in each edge). If the particle runs into an impurity, it would normally get backscattered and the conductance would drop. However, backscattering is not an option in the quantum Hall edges. The only way to backscatter is to tunnel across the entire sample (which is gapped) and reach the other edge, a possibility that is exponentially small in system width. (In an ordinary wire, the forward and backward channels coexist in space.)

Laughlin and Halperin also provide arguments to extend the flux insertion result to a disordered QHE sample, and show that each filled LL contributes a unit of conductance, unless the disorder is so strong that the sample has become insulating. The crux of the argument is that states localized in y cannot sense the threaded flux and will see it as due to an $\boldsymbol{A}$ that can be gauged away by a non-singular transformation. The previous pumping arguments then apply only to extended states that go around the cylinder and feel the flux.

20.7.2 Toplogical Insulators

The QHE systems are an example of *topological insulators*, a fertile area of research (see, for example, [21]). These systems are insulating in the bulk and conducting only at the boundaries. In the $d = 2$ case the boundaries are the edges. The $(d-1)$-dimensional boundary of a d-dimensional topological insulator can do things an isolated system of

$(d-1)$ dimensions cannot. For example, in the IQHE case each $d=1$ edge carries current in only one direction, whereas you cannot find an isolated one-dimensional system, e.g., a wire, that can carry current in only one direction. *The edge is able to do this by virtue of being the boundary of the QHE system.* Likewise, there are $d=3$ topological insulators whose $d=2$ surfaces do things that no isolated $d=2$ system can.

Here is an analogy. Suppose I only have dipoles but want to show you a positive charge, an electric monopole. I string a large number of dipoles up, tip to tail, and show you the one end with an uncanceled positive charge. Its opposite negatively charged partner is far away at the other end, which may be at infinity. But it has to be there. If I chop the string at some point, the uncanceled negative charge will appear there. The charges appear in opposite pairs, just like the edges in the Hall system.

Why the name "topological insulators?" Because there is a *topological number* $\mathcal{Q}$ associated with the QHE system. It has the value $\mathcal{Q}=1$ in the filled LL. Because it is quantized to be an integer, $\mathcal{Q}$ cannot change smoothly: the next available number is 0 or 2, not 1.001. The only way to change from 1 to, say, 0, is for the ground state to become non-analytic by encountering another degenerate state. This can happen only if the gap closes. In the presence of a gap, all changes are analytic. Chopping off the system into a Hall strip with edges puts it next to the vacuum, which has $\mathcal{Q}=0$. At the transition, namely the edges, the gap must close. These are the edge states. They are mandated by the topological nature of the bulk. Such states will also appear in the boundary between any two states of different $\mathcal{Q}$.

Now, the origin of $\mathcal{Q}$ for the QHE is somewhat complicated to explain. I will therefore illustrate it with a simpler case discussed by Xi *et al.* [23]. Consider a a translationally invariant, free-particle problem on a square lattice with two degrees of freedom per site. In momentum space, at each $\boldsymbol{k}$ on a Brillouin zone (BZ), which has the topology of a torus, we have a 2×2 Hamiltonian. Assume also that the energies are equal and opposite. Then we may write

$$H_0(\boldsymbol{k}) = -\boldsymbol{\sigma}\cdot\boldsymbol{B}(\boldsymbol{k}), \tag{20.122}$$

where $\boldsymbol{B}(\boldsymbol{k})$ is some smooth periodic function that I assume never vanishes in the BZ. The energy levels are $E_\pm(\boldsymbol{k}) = \pm|\boldsymbol{B}(\boldsymbol{k})|$. In the ground state of the system the lower energy state is occupied at each $\boldsymbol{k}$ in the entire BZ. With the lower-energy two-component spinor $\chi(\boldsymbol{k})$, we may associate a (pseudo)spin vector

$$\boldsymbol{n}(\boldsymbol{k}) = \chi^\dagger(\boldsymbol{k})\boldsymbol{\sigma}\chi(\boldsymbol{k}) \tag{20.123}$$

whose tip lies on a unit two-sphere S^2 and points in the direction of $\boldsymbol{B}(\boldsymbol{k})$.

Thus, each $\boldsymbol{k}$ in the BZ maps onto a point on this S^2:

$$\boldsymbol{k} \to \chi(\boldsymbol{k}) \to \boldsymbol{n}(\boldsymbol{k}) \in S^2. \tag{20.124}$$

The entire BZ gets mapped onto the surface of this sphere for a given choice of $\chi(\boldsymbol{k})$. For example, if $\boldsymbol{B}(\boldsymbol{k}) \equiv \boldsymbol{i}B_0$, then $\sqrt{2}\chi(\boldsymbol{k}) = [1,1]^{\mathrm{T}}\ \forall\boldsymbol{k}$, and the entire BZ maps onto the point

$(1,0,0)$ on S^2. If, however, $\boldsymbol{B}$ varies, the BZ will cover some part of the sphere. If it does not wrap around it, we can slowly deform the map (the spinor) so that all points in the BZ map onto one point on S^2. If, however, the map wraps around once, i.e., the torus covers the sphere once, there is no way to deform the map continuously to the constant map. (You cannot bunch the skin of an orange to one point unless you tear it.) This is a map with topological index $\mathcal{Q} = 1$. If this system, which is clearly gapped (since $|\boldsymbol{B}(\boldsymbol{k})| \neq 0$, or $E_- \neq E_+$), were to be truncated to a sample with edges, the topological number would have to drop to 0 in the vacuum, and this can happen only if the gap vanishes at the edges.

In short, a system with a topological charge, quantized (say to integers), cannot change its charge without encountering a degeneracy (gapless states) at the boundary.

20.7.3 Plateaus

I have argued that once we get off the intersection points we have to invoke disorder.

It was known at the time the QHE effect was discovered that in $d = 2$ any random potential $V(\boldsymbol{r})$ leads to a localized wavefunction that cannot conduct [24]. This result had to be amended in the presence of a strong magnetic field to explain the non-zero σ_{xy} of a filled LL. To see how this comes about, we consider the LLL because we know it so well. We will follow the compelling and easily visualized argument due to Trugman [25].

Imagine that $V(\boldsymbol{r})$ is smooth on the scale of the magnetic length l. In its absence, the LLL is flat. In its presence, $V(\boldsymbol{r})$ projected to the LLL becomes the Hamiltonian $V(\boldsymbol{R})$. We have seen that $\boldsymbol{R}$ moves along equipotentials. Now look at Figure 20.5. If you start with the empty LLL, you see mountains big and small ($V > 0$) and oceans of various depths ($V < 0$).

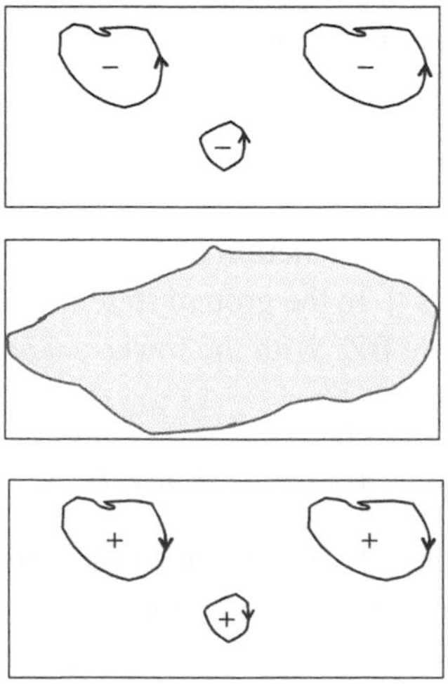

Figure 20.5 Top: Semiclassical orbits (lake shores) around deep minima of V at low-filling of the LL. Middle: The extended state at half-filling (assuming disorder is symmetric). Bottom: Orbits around a few island beaches associated with large maxima.

Now start adding electrons. The first will go around the lowest minimum in the ocean floor in a tight orbit chosen by Bohr–Sommerfeld quantization in the sense shown. Its energy will be roughly the depth of V in this orbit. The next electron will fill the orbit around the next-lowest minimum, which will generally be far away. Call these filled regions the lakes. The filled and localized orbits at the lake shores will not percolate or conduct if we connect the sample to leads. They don't know about the leads or the other orbits centered about other lake shores. As we add electrons there will be more orbits of bigger and bigger sizes. At the other extreme, most of the land will be submerged and a few islands will remain near the maxima, with their coastal orbits circulating the other way. These orbits will also not touch or conduct. The flat LLL has been broadened by disorder to a band of energies going roughly from the deepest sea crater to the tallest mountain.

The main point of Trugman is that somewhere along the way, when the orbits around lake shores morph into ones around island coasts, *there must come a stage when orbits go all the way across the sample*. These are the extended states, shown in the middle of Figure 20.5. If the disorder is symmetric between positive and negative values, this will occur at the middle of the LLL. Let us assume this for simplicity.

What about Hall conductance? It will start and remain at 0 as we populate the localized orbits of the broadened LLL from the bottom up. It will jump to $\sigma_{xy} = \sigma_0$ when we encounter the extended state(s) at half-filling. It will remain there until we reach the middle of the second LL, when it will jump to $2\sigma_0$, and so on, as indicated in Figure 20.6. So the reason we have plateaus as we move off the special points where the steps meet the straight line is that the particles that are added or removed are localized and do not contribute to σ_{xy}.

Initially one wonders how exact quantization of σ_{xy} (to one part in a billion) is possible in a dirty sample. Now we are driven to the dramatic, counterintuitive conclusion that *the*

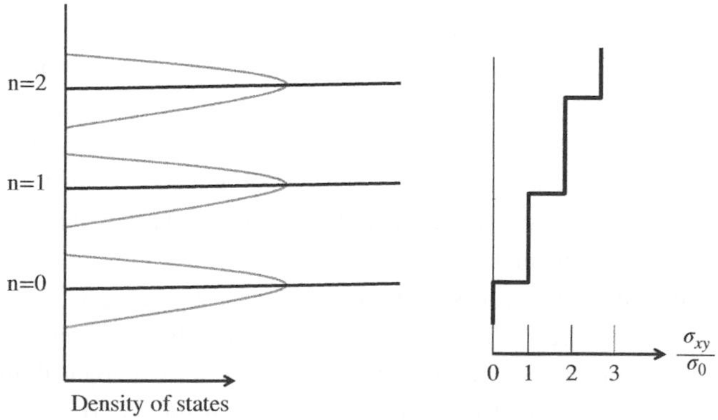

Figure 20.6 Due to disorder, the LL's get broadened into bands as shown. (The lobes indicate the density of states that used to be concentrated on sharp Landau levels in the absence of disorder.) Only state(s) at the middle of the LL are extended. As we add particles, σ_{xy} jumps by σ_0 every time we cross the extended state(s), which are assumed to be in the middle of the broadened band.

quantum Hall plateaus owe their flatness to the dirt that traps the added particles. Thus, disorder (within reason) is essential to the QHE!

20.8 FQHE: Laughlin's Breakthrough

We were able to explain the IQHE because we had a nice starting point: an integer number of filled LL's that could be filled in only one way. We then introduced disorder to explain the plateaus.

Now we come to a situation where we cannot even start – when there are enough electrons to fill only a fraction of an LL. A representative case is the $\frac{1}{3}$-filled LLL. We now have three states per electron in the LLL.

The breakthrough on this problem came from Laughlin, who wrote down the following trial wavefunction for $\nu = 1/(2s+1)$:

$$\Psi_{\frac{1}{(2s+1)}} = \prod_{i=1}^{N}\prod_{j<i}(z_i - z_j)^{2s+1} \exp\left(-\sum_i |z_i|^2/4l^2\right). \tag{20.125}$$

Let us now contemplate its many virtues:

- It lies in the LLL (analytic function times Gaussian).
- It has definite angular momentum (homogeneous in z's).
- It obeys the Pauli principle ($2s+1$ is odd, spin is assumed fully polarized).
- Particles avoid each other strongly because of the $2s+1$-fold zeros.
- If you take any variable, say z_1, on a loop encircling the whole sample, the phase of z_1^{2s+1} changes by $2\pi(2s+1)N$, showing that the sample encloses $(2s+1)N$ flux quanta. The filling is then $\nu = \frac{1}{2s+1}$.

Not obvious but true is that this wavefunction describes an *incompressible fluid*, which means a fluid that abhors density changes. Whereas a Fermi gas would increase (decrease) its density globally when compressed (decompressed), an incompressible fluid is wedded to a certain density and would first show no response and then suddenly nucleate a localized region of different density. These are the quasiparticles and quasiholes that get trapped by disorder and contribute to density but not conductance. (Just the way a Type II superconductor, in which a magnetic field is not welcome, will allow it to enter in quantized units in a region that turns normal.)

Laughlin proved the incompressibility of the state by a very clever mapping of expectation values computed in his state to statistical mechanics averages in a classical two-dimensional incompressible plasma whose charges felt a logarithmic interparticle potential. This was not the usual mapping of a quantum theory to a classical theory in one-higher dimension. It came from viewing $|\Psi|^2$ as a Boltzmann weight for a plasma problem in the (x, y) plane.

Laughlin also provided the wavefunction for an excited state, a state with a *quasihole*, which signifies a charge deficit. He motivates it as follows. Suppose we insert a tiny flux tube at a point z_0 and slowly increase the flux from 0 to $\frac{2\pi\hbar}{-e}$. In other words, the extra flux points in the same direction as the overall field, also down the z-axis. The electron's phase will change by $+2\pi$ as it encircles the negative quantum of flux.

By gauge invariance, the state must evolve into an eigenstate of H. This and analyticity imply the ansatz

$$\Psi_{\text{qh}} = \prod_i (z_i - z_0)\Psi_{2s+1}. \tag{20.126}$$

This is a quasihole. (There is a more complicated state with a quasiparticle.) The prefactor is a *vortex* at z_0, which is an analytic zero at z_0 for every coordinate. It denotes a hole near z_0, which we will now show corresponds to a charge deficit $1/(2s+1)$ (in electronic units).

As the flux $\Phi(t)$ is inserted, the induced field obeys

$$\boldsymbol{E} = \boldsymbol{e}_\phi E_\phi = -\boldsymbol{e}_\phi \frac{1}{2\pi r}\frac{d\Phi}{dt}. \tag{20.127}$$

The Hall current (for $\boldsymbol{B}$ pointing down the z-axis) is given by

$$\boldsymbol{j} = \sigma_{xy}\hat{z}\times\boldsymbol{E} = \sigma_{xy}\frac{\boldsymbol{e}_r}{2\pi r}\frac{d\Phi}{dt}. \tag{20.128}$$

The charge running off to infinity is

$$Q_\infty = \int (2\pi r j_r)dt = \int \sigma_{xy}\frac{d\Phi}{dt}dt = \sigma_{xy}\left[\frac{2\pi\hbar}{-e}\right] = -\nu e. \tag{20.129}$$

Thus, a fraction ν of the electron flows off to infinity. In the case $\nu = \frac{1}{3}$, a third of the electron charge flows to infinity leaving behind a quasihole of charge $\frac{e}{3}$. If an electron is removed, three quasiholes will be nucleated.

Notice that the fractional charge of the quasiparticles is due to fractional Hall conductance σ_{xy} and not vice versa.

Halperin [26] was able to show that the fractionally charged objects obeyed *fractional statistics*, by which one means that exchanging two of them leads to a factor $e^{i\theta}$, where θ is not-zero (bosons) or π (fermions). Arovas *et al.* [27,28] used Berry phase arguments to show the fractional statistics of Laughlin quasiparticles. A comprehensive quantum theory of the edge modes was given by Wen [29]. He showed the proper way to describe the chiral edge modes for all FQHE states, as well as the nature (charge and statistics) of the particles moving along the edge.

Finally, the plateaus are produced when you go off the magic fractions because the change in density appears in the form of quasiparticles (or quasiholes) that get localized (at least at small concentrations.)

20.8.1 Beyond Laughlin

Laughlin's work firmly established the nature of the FQHE state and its excitations. However, it marked not the end but the beginning of some intense research.

For the observed fractions not of the form $\frac{1}{2s+1}$, such as $\frac{2}{5}$, the Laughlin wavefunction, if blindly written down, would have non-analytic factors like $(z_i - z_j)^{5/2}$ and not obey the Pauli principle.

Jain found a way to construct trial wavefunctions for $\nu = \frac{p}{2ps+1}$ by mapping the electronic problem to that of *composite fermions*. This will be explained soon.

In parallel with the quest for trial wavefunctions were approaches that started with the underlying Hamiltonian and tried to solve the problem through a sequence of approximations.

Both approaches relied on mapping the electronic problem to one of the following problems with unique gapped ground states:

- Boson in zero magnetic field with repulsive interaction (become superfluids).
- Fermions that fill an integer number of Landau levels.

The mapping was thanks to the notion of *flux attachment*.

20.9 Flux Attachment

It was pointed out by Leinaas and Myrheim [30] that in $d = 2$ one could have particles (dubbed "anyons" by Wilczek [31]) that suffered a phase change $e^{i\theta}$ upon exchange, where θ was not restricted to 0 or π, corresponding to bosons and fermions respectively. To do this, one takes a fermion and drives through its center a point flux tube containing some arbitrary amount of flux. (We will see how exactly this impalement is carried out. For now, assume we can do this.) If the flux tube contains an even/odd number of flux quanta, the composite particle one gets is a fermion/boson.

How can adding a flux quantum make any difference, given that by definition a flux quantum is supposed to be invisible to a charge going around it? The answer is that an exchange involves only *half a circumnavigation*. Consider two particles, A at the origin and B at $x = R$. To exchange them we rotate B by π (not 2π) around A and shift both to the right by R.

Zhang, Hansson, and Kivelson [11–13] took the fermions of the Laughlin fraction, which saw $2s+1$ flux quanta each, and attached to each of them $2s+1$ flux quanta pointing opposite to the external flux. This turned the fermions into *composite bosons* (CB). In addition, when each boson went on a loop in the (x, y) plane, it thought that *on average* there was no flux because the external flux was canceled by the flux tubes attached to the particles. So, the bosons formed a gapped superfluid. Its vortices were the Laughlin quasiparticles.

Jain expanded this idea to the more general case $\nu^{-1} = 2s + \frac{1}{p}$, where each electron sees $2s + \frac{1}{p}$ flux quanta (down the z-axis). He turned the electrons into CF's by attaching to each electron $2s$ flux quanta pointing opposite to the external B, so that on average they saw $1/p$ flux quanta each and filled up exactly p LL's. Thus, the fermionic version offers more opportunities for finding gapped states. In addition, the fact that the composite particle is also a fermion is of some psychological value.

It is time to ask how exactly the flux is to be attached. One can do it in the language of wavefunctions (which we will discuss first), or use a Chern–Simons theory following Zhang *et al.* [11–13] for bosons and Lopez and Fradkin [32–34] for fermions.

Let us consider CF's. The CB case is very similar.

20.9.1 Wavefunction Approach to Composite Fermions

Let us start with just the non-interacting part of the electronic Hamiltonian that acts on the many-electron wavefunction Ψ_e. Focus on

$$(-i\hbar\nabla_i + e\boldsymbol{A}(\boldsymbol{r}_i))\psi_e,$$

the covariant derivative for particle i. We now introduce the CF wave function Ψ_c related by a gauge transformation:

$$\Psi_e = \prod_{i<j}\left(\frac{(z_i - z_j)}{|z_i - z_j|}\right)^{2s} \cdot \Psi_c(z,\bar{z}) \tag{20.130}$$

$$= e^{2is\sum_{i<j}\phi(\boldsymbol{r}_i-\boldsymbol{r}_j)}\Psi_c(z,\bar{z}), \tag{20.131}$$

where $\phi(\boldsymbol{r}_i - \boldsymbol{r}_j)$ is the azimuthal angle of the separation vector $(\boldsymbol{r}_i - \boldsymbol{r}_j)$ or the phase of $z_i - z_j$. (I switch freely between real coordinates $\boldsymbol{r}_i = (x_i, y_i)$ and complex coordinates $(z_i, \bar{z}_i)$. The distance between two particles is $|\boldsymbol{r}_i - \boldsymbol{r}_j| = |z_i - z_j|$.)

Look at Eq. (20.131). Pick some particle coordinate $\boldsymbol{r}_i$ in the exponential and have it encircle $\boldsymbol{r}_j$. You see the phase change $2\pi \cdot 2s$, which is what a $2s$-fold flux tube would produce. But if there really is a flux tube attached to the particle at $\boldsymbol{r}_j$, *every* particle going around it should pick up the same phase change. Indeed, this is what happens because the index i could have referred to any other particle.

Let us look at the Schrödinger equation obeyed by Ψ_c. We see that

$$(-i\hbar\nabla_i + e\boldsymbol{A}(\boldsymbol{r}_i))\psi_e = (-i\hbar\nabla_i + \boldsymbol{a}_c(\boldsymbol{r}_i) + e\boldsymbol{A}(\boldsymbol{r}_i))\Psi_c, \tag{20.132}$$

$$\boldsymbol{a}_c(\boldsymbol{r}_i) = (2\hbar s)\sum_j(\nabla_i\phi(\boldsymbol{r}_i - \boldsymbol{r}_j)), \tag{20.133}$$

where $\boldsymbol{a}_c$ is the gauge potential introduced by this gauge transformation. It is not an external potential, but one due to the particles themselves, due to the $2s$ flux tubes each one carries.

If we integrate $\boldsymbol{a}_c(\boldsymbol{r}_i)$ around a closed contour $\mathcal{C}$ that bounds an area $\mathcal{A}$, i.e., $\mathcal{C} = \partial\mathcal{A}$, the answer is $4\pi s\hbar$ times the number of particles enclosed:

$$\oint_{\mathcal{C}=\partial\mathcal{A}} \boldsymbol{a}_c(\boldsymbol{r}_i)\cdot d\boldsymbol{r}_i = 4\pi s\hbar \times \text{particles in } \mathcal{C} \tag{20.134}$$

$$\int_{\mathcal{A}} \nabla\times\boldsymbol{a}_c(\boldsymbol{r}_i)\cdot d\boldsymbol{S} = 4\pi s\hbar\int_{\mathcal{A}}\sum_j\delta(\boldsymbol{r}-\boldsymbol{r}_j)d^2\boldsymbol{r} \tag{20.135}$$

$$\nabla\times\boldsymbol{a}_c(\boldsymbol{r}) = 4\pi s\hbar\,\rho(\boldsymbol{r}), \quad \text{where} \tag{20.136}$$

$$\rho(\boldsymbol{r}) = \sum_j \delta(\boldsymbol{r} - \boldsymbol{r}_j). \tag{20.137}$$

Let us separate the density into its average n and fluctuations,

$$\rho = n + :\rho:, \tag{20.138}$$

so that

$$\nabla \times \boldsymbol{a}_{\rm c}(\boldsymbol{r}) = 4\pi s\hbar n + 4\pi s\hbar :\rho: . \tag{20.139}$$

Now, given that $\nabla \times e\boldsymbol{A} = -eB$ is a constant and so is $4\pi s\hbar n$ in Eq. (20.139), we conclude that the part of $\boldsymbol{a}_{\rm c}$ whose curl gives the constant flux density $4\pi ns\hbar$ is proportional to $e\boldsymbol{A}$. In other words, we see that

$$\nabla \times \boldsymbol{a}_{\rm c}(\boldsymbol{r}) = -\nabla \times \left[e\boldsymbol{A}\frac{4\pi s\hbar n}{eB}\right] + 4\pi \hbar s :\rho: \tag{20.140}$$

$$= -\nabla \times \left[e\boldsymbol{A}\frac{2ps}{2ps+1}\right] + 4\pi s\hbar :\rho: \quad \text{(using } 4\pi\hbar n = 2eB\nu\text{)}$$

$$\boldsymbol{a}_{\rm c} = -e\boldsymbol{A}\frac{2ps}{2ps+1} + (\nabla\times)^{-1} 4\pi s\hbar :\rho: \tag{20.141}$$

$$e\boldsymbol{A} + \boldsymbol{a}_{\rm c} = \frac{e\boldsymbol{A}}{2ps+1} + (\nabla\times)^{-1} 4\pi s\hbar :\rho: \tag{20.142}$$

$$\equiv e\boldsymbol{A}^* + (\nabla\times)^{-1} 4\pi s\hbar :\rho:, \tag{20.143}$$

where we have defined the reduced field $\boldsymbol{A}^*$ seen by the CF:

$$e\boldsymbol{A}^* = \frac{e\boldsymbol{A}}{2ps+1}. \tag{20.144}$$

The inverse curl $(\nabla\times)^{-1}$ is uniquely defined by its Fourier transform if we demand that the answer be transverse. In general, if

$$(\nabla\times)^{-1} f(\boldsymbol{r}) = \boldsymbol{a}, \tag{20.145}$$

then

$$\boldsymbol{a}(\boldsymbol{r}) = -\frac{1}{2\pi}\int d^2r' \frac{\hat{z}\times(\boldsymbol{r}-\boldsymbol{r}')}{|\boldsymbol{r}-\boldsymbol{r}'|^2} f(\boldsymbol{r}). \tag{20.146}$$

Let us return to the covariant derivative,

$$(-i\hbar\nabla_j + e\boldsymbol{A}(\boldsymbol{r}_j))\Psi_e = (-i\hbar\nabla_j + e\boldsymbol{A}^*(\boldsymbol{r}_j) + (\nabla\times)^{-1} 4\pi s\hbar :\rho:)\Psi_{\rm c}. \tag{20.147}$$

Let us now ignore the fluctuating part, $(\nabla\times)^{-1}4\pi\hbar s :\rho:$, and consider the magnetic field at mean-field level,

$$\nabla \times e\boldsymbol{A}^* = -\frac{eB}{2ps+1}. \tag{20.148}$$

Since the electrons saw $2s + \frac{1}{p}$ flux quanta each (down the z-axis) in the field due to $\boldsymbol{A}$, the CF's, which see the reduced potential $\boldsymbol{A}^* = \boldsymbol{A}/(2ps+1)$, will see

$$\frac{1}{2sp+1}\left(2s+\frac{1}{p}\right) = \frac{1}{p} \tag{20.149}$$

quanta each (down the z-axis) and fill p LL's. Thus, flux attachment has produced at the mean-field level (ignoring density fluctuations) a magnetic field just right for the CF to fill exactly p Landau levels.

This leads to Jain's initial trial wavefunction for the electron:

$$\Psi_{\frac{p}{2ps+1}} = \prod_{i<j}\left(\frac{(z_i - z_j)}{|z_i - z_j|}\right)^{2s} \cdot \chi_p(z,\bar{z}), \tag{20.150}$$

where χ_p is the CF wavefunction with p-filled CF LL's and the prefactor is the gauge transformation back to electrons. Remember that in general χ_P is a function of z and $\bar{z}$, and χ_1 from the LLL is an exception.

Flux attachment only gives the *phase* and not the zeros of the *Jastrow factor*

$$J(2s) = \prod_{i<j}(z_i - z_j)^{2s}, \tag{20.151}$$

which describes $2s$-fold vortices on particles. This is unavoidable given that the gauge parameter is real. It is possible to obtain the zeros in a field theory at mean-field level upon introducing a complex vector potential and a non-Hermitian Hamiltonian, as pointed out by Rajaraman and Sondhi [36].

In CS theory, the zeros emerge from the fluctuations about the mean field [13, 32–35]. Shortly I will show you another route based on the Bohm–Pines approach [37, 38].

Let us return to the quest for trial wavefunctions, wherein we are free to improve the guess in any way we want. Jain improved Eq. (20.150) in two ways and proposed:

$$\Psi_{\frac{p}{2ps+1}} = \prod_{i<j}(z_i - z_j)^{2s} \cdot \mathcal{P}\chi_p(z,\bar{z}). \tag{20.152}$$

- He replaced flux attachment by vortex attachment: $\prod_{i<j}(z_i - z_j)^{2s}$. Remember: a flux tube implies a phase as we go around it; a vortex, a zero as well, with some charge deficit. It is also good to get rid of the $|z_i - z_j|$ factors, which do not lie in the LLL.
- He did a further projection to the LLL of $\chi_p(z,\bar{z})$ using $\mathcal{P}$:

$$\mathcal{P} : \bar{z} \to 2l^2\frac{\partial}{\partial z} \quad \text{(recall that } [z,\bar{z}] = -2l^2 \text{ in the LLL).} \tag{20.153}$$

At $p = 1$, $\chi(z,\bar{z}) = \prod(z_i - z_j)$, and we do not need $\mathcal{P}$ to get Laughlin's answer. At $p > 1$, we have a concrete expression for Ψ in terms of electron coordinates.

The Jastrow factor

$$\prod_{i<j}(z_i - z_j)^{2s} \tag{20.154}$$

has two effects besides keeping particles apart and reducing repulsive energy.

First, when one CF goes around a loop, it effectively sees $\nu^{-1*} = 2s + \frac{1}{p} - 2s = \frac{1}{p}$ flux quanta per enclosed particle (down the z-axis), since the phase change due to encircling vortices attached to CF's neutralizes $2s$ of the flux quanta per particle (down the z-axis) of the Aharanov–Bohm phase of the external field.

Next, the $2s$ vortices carrying charge νe each, Eq. (20.129), reduce $-e$ down to the CF charge

$$e^* = -e\left[1 - \frac{2ps}{2ps+1}\right] = -\frac{e}{2ps+1}. \tag{20.155}$$

In summary, while degeneracy of the non-interacting problem is present for any $\nu < 1$, at the Jain fractions one can beat it by thinking in terms of composite fermions, which see a field that is just right to fill p CF LL's. As we move off the Jain fractions, the incremental CF's (particles or holes) get localized, giving rise to the plateaus.

20.9.2 The Unusual Case of $\nu = \frac{1}{2}$

This is a special fraction that is gapless, but one for which flux attachment works extremely well. By adding two flux quanta to the electron, we get CF's that see no net field on average and form a Fermi sea. This case was explored by Kalmeyer and Zhang [39], and subsequently Halperin, Lee, and Read squeezed out an amazing amount of information and testable predictions in what is a classic paper, [40]. Alas, we have to move on in this treatment focused on techniques.

20.10 Second Quantization

Let us write down our findings in second quantization so we can go beyond the trial wave functions. We will also set $\hbar = 1$ from now on.

The kinetic part of the electronic Hamiltonian density is

$$\psi_{\rm e}^\dagger(\mathbf{r})\frac{(-i\nabla + e\boldsymbol{A})^2}{2m}\psi_{\rm e}(\boldsymbol{r}), \tag{20.156}$$

where $\psi_{\rm e}$ is the electron destruction operator. The CS transformation is represented by the operator relation

$$\psi_{\rm e}(\mathbf{r}) = \exp\left[2is\int \phi\,(\mathbf{r}-\mathbf{r}')\rho(\mathbf{r}')d^2r'\right]\psi_{\rm c}(\mathbf{r}'), \tag{20.157}$$

where $\psi_{\rm c}$ is the field operator that annihilates the CS composite particle. (They are not to be confused with the wave functions Ψ).

The density is unaffected,

$$\rho(\mathbf{r}) = \psi_{\rm e}^{\dagger}(\mathbf{r})\psi_{\rm e}(\mathbf{r}) = \psi_{\rm c}^{\dagger}(\mathbf{r})\psi_{\rm c}(\mathbf{r}), \tag{20.158}$$

and the interaction may be introduced readily at a later stage. The non-interacting Hamiltonian density is

$$H_{\rm c} = \psi_{\rm c}^{\dagger}\frac{|-i\nabla + e\mathbf{A} + \boldsymbol{a}_{\rm c}|^2}{2m}\psi_{\rm c} \tag{20.159}$$

$$= \psi_{\rm c}^{\dagger}\frac{|-i\nabla + e\mathbf{A}^* + 4\pi s(\nabla\times)^{-1} :\rho: |^2}{2m}\psi_{\rm c}, \tag{20.160}$$

where $\mathbf{A}^*$ is the mean-field vector potential. The normal-ordered density is

$$:\rho: = \psi_{\rm c}^{\dagger}\psi_{\rm c} - n. \tag{20.161}$$

To tackle the nasty inverse curl, Murthy and I [37, 38] proceed as follows. We first enlarge the Hilbert space. Consider a disk (in momentum space) of radius Q. For each $\mathbf{q}$ in this disc, we associate a canonical pair of fields $(a(\mathbf{q}), P(\mathbf{q}))$ obeying

$$\left[a(\mathbf{q}), P(\mathbf{q}')\right] = i\delta_{q,-q'}. \tag{20.162}$$

The pair $(a(\mathbf{q}), P(\mathbf{q}))$ defines a pair of transverse and longitudinal vector fields

$$\mathbf{a}(\mathbf{q}) = -i\hat{\mathbf{z}}\times\hat{\mathbf{q}}\; a(\mathbf{q}), \qquad \mathbf{P}(\mathbf{q}) = i\hat{\mathbf{q}}\; P(\mathbf{q}). \tag{20.163}$$

The real-space counterparts are

$$\boldsymbol{a}(\boldsymbol{r}) = -i\sum_{q}\hat{\mathbf{z}}\times\hat{\mathbf{q}}\; a(\mathbf{q})e^{i\boldsymbol{q}\cdot\boldsymbol{r}}, \tag{20.164}$$

$$\boldsymbol{P}(\boldsymbol{r}) = i\sum_{q}\hat{\mathbf{q}}P(\mathbf{q})e^{i\boldsymbol{q}\cdot\boldsymbol{r}}. \tag{20.165}$$

The Hamiltonian density of Eq. (20.160) is completely equivalent to

$$H = \frac{1}{2m}\psi^{\dagger}{}_{\rm c}(-i\nabla + e\mathbf{A}^* + \mathbf{a}(\boldsymbol{r}) + (\nabla\times)^{-1}4\pi s :\rho:)^2\psi_{\rm c}, \tag{20.166}$$

provided we restrict ourselves to states in the larger space obeying

$$a(\mathbf{q})|\text{physical}\rangle = 0 \qquad q < Q. \tag{20.167}$$

In other words, $[H, a] = 0$ allows us to find simultaneous eigenstates of H and a, and of these, the eigenstates with $a = 0$ describe the original problem. The others are of no physical importance.

Now for why we introduce $\boldsymbol{a}$. Look at Eq. (20.166), which contains the very nasty inverse curl. We deal with it through the following unitary transformation:

$$U = \exp\left[\sum_q^Q iP(-q)\frac{4\pi s}{q} : \rho(q) :\right]. \tag{20.168}$$

Under the action of U, which is a translation operator on $a(\boldsymbol{q})$,

$$U^\dagger a(q) U = a(q) - \frac{4\pi s : \rho(q) :}{q} \tag{20.169}$$

$$U^\dagger (\boldsymbol{a}(\boldsymbol{r}) + (\nabla\times)^{-1} 4\pi s : \rho :) U = \boldsymbol{a}(\boldsymbol{r}) \tag{20.170}$$

$$U^\dagger \psi_c^\dagger(\boldsymbol{r})(-i\nabla)^2 \psi_c(\boldsymbol{r}) U = \psi_{\rm CP}^\dagger(x)(-i\nabla + 4\pi s\mathbf{P}(\boldsymbol{r}))^2\, \psi_{\rm CP}(x). \tag{20.171}$$

Now go back to the Hamiltonian for ψ_c in Eq. (20.160). After the unitary transformation U,

$$U^\dagger H_c U = \frac{1}{2m}\psi_{\rm CP}^\dagger(-i\nabla + e\mathbf{A}^* + \mathbf{a}(\boldsymbol{r}) + 4\pi s\mathbf{P}(\boldsymbol{r}) + \delta\mathbf{a})^2 \psi_{\rm CP} \tag{20.172}$$

$$0 = \left[a(\boldsymbol{q}) - \frac{4\pi s : \rho(\boldsymbol{q}) :}{q}\right] |\text{physical}\rangle, \quad 0 < q \le Q, \tag{20.173}$$

where $\delta\mathbf{a}$ is the *dependent* short-range vector potential $(\nabla\times)^{-1} 4\pi s : \rho :$ for $q > Q$ that did not get canceled by the unitary transformation, and $\psi_{\rm CP}$ refers to the unitarily transformed composite fermion field, which better describes the composite fermions alluded to in the literature than does ψ_c. Whereas the latter is associated with particles carrying just flux tubes, the former is associated with particles that carry flux tubes *and the correlation holes*, i.e., they describe electrons bound to vortices, as we shall see.

Exercise 20.10.1 *Prove Eqs. (20.170) and (20.171). Here is one possible route for the first part. To find the action of U on ψ_c, use the fact that, for a single fermionic oscillator with $H = \omega_0 \psi^\dagger \psi$,*

$$e^{iHt}\psi e^{-iHt} = \psi(t) = \psi e^{-i\omega_0 t}.$$

For the second part, write U in real space and work with a function $\tilde{P}(\boldsymbol{r})$ which is the Fourier transform of $P(-\boldsymbol{q})/q$.

We now explore our Hamiltonian Eq. (20.172), expanding it as follows (and dropping the subscript CP):

$$\begin{aligned} H &= \int d^2\boldsymbol{r}\left[\frac{1}{2m}|(-i\nabla + e\mathbf{A}^* + \mathbf{a} + 4\pi s\mathbf{P} + \delta\mathbf{a})\psi|^2\right] \\ &= \int d^2\boldsymbol{r}\left[\frac{1}{2m}|(-i\nabla + e\mathbf{A}^*)\psi|^2 + \frac{n}{2m}(a^2(\boldsymbol{r}) + 16\pi^2 s^2 P^2(\boldsymbol{r}))\right] \\ &\quad + \int d^2\boldsymbol{r}\left[(\mathbf{a} + 4\pi s\mathbf{P})\cdot\frac{1}{2m}\psi^\dagger(-i\overleftrightarrow{\nabla} + e\mathbf{A}^*)\psi\right] \end{aligned} \tag{20.174}$$

$$+\int d^2r \frac{:\psi^\dagger\psi:}{2m}(\mathbf{a}+4\pi s\mathbf{P})^2$$
$$+\int d^2r\left[\frac{\delta\mathbf{a}}{2m}\cdot\psi^\dagger(-i\overleftrightarrow{\nabla}+e\mathbf{A}^*)\psi+\frac{\psi^\dagger\psi}{2m}(2(\mathbf{a}+4\pi s\mathbf{P})+\delta\mathbf{a})\cdot\delta\mathbf{a}\right]$$
$$\equiv H_0+H_{\text{I}}+H_{\text{II}}+H_{\text{sr}} \tag{20.175}$$

in obvious notation, with H_{sr} being the terms associated with the non-dynamical short-range gauge field $\delta\mathbf{a}$ that was not eliminated. To all this we must add interactions and the constraint Eq. (20.173).

20.11 Wavefunctions for Ground States: Jain States

In this section we will see how the Hamiltonian above leads to the well-known correlated wave functions upon making the simplest approximation. Let us consider the Jain fractions, keeping only H_0:

$$H_0=\int d^2r\left[\frac{1}{2m}|(-i\nabla+eA^*)\psi|^2+\frac{n}{2m}\left[a^2+16\pi^2s^2P^2\right]\right], \tag{20.176}$$

where $a(\boldsymbol{r})$ and $P(\boldsymbol{r})$ are the scalar fields obtained from $a(\boldsymbol{q})$ and $P(\boldsymbol{q})$ by the inverse Fourier transformation. There are no cross terms between $\mathbf{a}$ and $\mathbf{P}$ since they are transverse and longitudinal respectively. You may check the result I have used, namely that

$$\int d^2r a^2(\boldsymbol{r})=\int d^2r|\boldsymbol{a}(\boldsymbol{r})|^2,$$

and similarly for $P^2(\boldsymbol{r})$ and $|\boldsymbol{P}(\boldsymbol{r})|^2$. Given that $(a\ ,P)$ are canonically conjugate, we see that the second term describes harmonic oscillators.

The wave function will factorize over oscillator and particle coordinates.

Following the same route as in Bohm–Pines, Eqs. (20.37)–(20.45), and invoking Eq. (20.79),

$$\nu=\frac{2\pi n\hbar}{eB},$$

we infer that the oscillators are at frequency

$$\frac{4\pi ns}{m}=\frac{2ps}{2ps+1}\omega_0=2\nu s\omega_0, \tag{20.177}$$

which is off the mark except at $2\nu s=1$. Fortunately, the correlated wavefunction they produce is independent of the oscillator energy and depends only on the magnetic length.

Their ground-state wavefunction in the $a(\boldsymbol{q})$ basis, following the BP route, is

$$\Psi_{\text{osc}}=\exp\left[-\sum_q\frac{a(\boldsymbol{q})a(-\boldsymbol{q})}{8\pi s}\right]. \tag{20.178}$$

Exercise 20.11.1 *Establish Eq. (20.178), starting with Eq. (20.176).*

The CF's are in the state χ_p, corresponding to p-filled CF LL's. The total wavefunction in the enlarged space is

$$\Psi_c = \exp\left[-\sum_{\boldsymbol{q}} \frac{a(\boldsymbol{q})a(-\boldsymbol{q})}{8\pi s}\right] \chi_p(z,\bar{z}). \tag{20.179}$$

We should project this to the physical sector by setting

$$a(\boldsymbol{q}) = \frac{4\pi s : \rho(\boldsymbol{q}) :}{q}. \tag{20.180}$$

The result is

$$\Psi_c = \exp\left[2\pi s \sum_{\boldsymbol{q}} \frac{:\rho(\boldsymbol{q})::\rho(-\boldsymbol{q}):}{q^2}\right] \chi_p(z,\bar{z}). \tag{20.181}$$

We now evaluate the oscillator part using, along the way,

$$\sum_{\boldsymbol{q}} \frac{e^{i\boldsymbol{q}\cdot\boldsymbol{r}}}{q^2} \to -\frac{1}{2\pi} \ln(\boldsymbol{r}) : \tag{20.182}$$

$$\Psi_{\rm osc} = \exp\left[-2\pi s \sum_{\boldsymbol{q}} \frac{:\rho(\boldsymbol{q})::\rho(-\boldsymbol{q}):}{q^2}\right] \tag{20.183}$$

$$= \exp\left[s \int d^2\boldsymbol{r} \int d^2\boldsymbol{r}' \left(\sum_i \delta(\boldsymbol{r}-\boldsymbol{r}_i) - n\right) \ln|\boldsymbol{r}-\boldsymbol{r}'| \left(\sum_j \delta(\boldsymbol{r}'-\boldsymbol{r}_j) - n\right)\right]$$

$$= \exp\left[s \sum_i \sum_{j\neq i} \ln|\boldsymbol{r}_i - \boldsymbol{r}_j| - 2ns\left(\int d^2\boldsymbol{r} \sum_j \ln|\boldsymbol{r}-\boldsymbol{r}_j|\right) + \text{constant}\right]$$

$$= \exp\left[2s \sum_i \sum_{j<i} \ln|\boldsymbol{r}_i - \boldsymbol{r}_j| - \sum_j \frac{r_j^2}{4l_v^2}\right] \tag{20.184}$$

$$= \prod_{i,j<i} |\boldsymbol{r}_i - \boldsymbol{r}_j|^{2s} \exp\left[-\sum_j \frac{r_j^2}{4l_v^2}\right], \quad \text{where} \tag{20.185}$$

$$\frac{1}{l_v^2} = \frac{1}{l^2} \frac{2ps}{2ps+1}. \tag{20.186}$$

(Since we cut off the oscillators at $q = Q$, the expression holds only for $|z_i - z_j| \gg 1/Q$.)

You might have two questions. How did we get to Eq. (20.184) from the line before it? What about the wrong magnetic length in the Gaussian factor?

The steps leading to the Gaussian factor are from Kane *et al.* [35]. The integral over $\boldsymbol{r}$ in the previous line may be interpreted as the potential energy of a point charge at $\boldsymbol{r}_j$, due to a uniform charge density $4\pi ns$ and the two-dimensional Coulomb potential $V(|\boldsymbol{r}-\boldsymbol{r}'|) = -\frac{1}{2\pi}\ln|\boldsymbol{r}-\boldsymbol{r}'|$. (To find the potential at $\boldsymbol{r}$, it helps to first find the radial electric field E_r by invoking the applicable form of Gauss's law, $2\pi r E_r(r) = \pi r^2 \cdot 4\pi ns$, and then integrating.)

Now for the Gaussian. The total Ψ_e is

$$\Psi_e = \exp\left[2s\sum_i\sum_{j<i}\ln|\boldsymbol{r}_i-\boldsymbol{r}_j| - \sum_j \frac{r_j^2}{4l_v^2}\right]\chi_p(z,\bar{z}), \tag{20.187}$$

where χ_p comes with its own Gaussian,

$$\exp\left[-\frac{r_j^2}{4l^{*2}}\right], \tag{20.188}$$

where l^* is the CF magnetic length. Given that

$$\frac{A^*}{A} = \frac{1}{2ps+1}, \tag{20.189}$$

it follows that

$$(l^*)^2 = (2ps+1)l^2. \tag{20.190}$$

We now use

$$\frac{1}{l_v^2} + \frac{1}{l^{*2}} = \frac{1}{l^2}\left(\frac{2ps}{2ps+1} + \frac{1}{2ps+1}\right) = \frac{1}{l^2}, \tag{20.191}$$

and obtain the right Gaussian.

The wavefunction at this point is

$$\Psi_{\text{CS}} = \prod_{i<j}|z_i - z_j|^{2s}\exp\left[-\sum_j \frac{|z_j|^2}{4l^2}\right]\chi_p(z,\bar{z}). \tag{20.192}$$

If we undo the gauge transformation, we get, for the electron,

$$\Psi_e = \prod_{i<j}(z_i - z_j)^{2s}\exp\left[-\sum_j \frac{|z_j|^2}{4l^2}\right]\chi_p(z,\bar{z}). \tag{20.193}$$

Let us pause to appreciate what the Bohm–Pines approach has done for us:

- The BP approach, which in the three-dimensional electrostatic interaction gave us a correlation factor $e^{-V(r)/\omega_0}$, gives in the QH problem the modulus of the Jastrow factor.
- The modulus converts the non-analytic phase factor in Ψ_e to the analytic Jastrow factor that lies in the LLL. Projection to the LLL has always remained a challenge, but one

worth conquering because of the universal belief that the physics of the FQHE can be captured entirely within the LLL by taking $\omega_0 \to \infty$. In the BP approach, it amounts to simply freezing the oscillator in its ground state.

- Generally, in a theory with constraints, it is very hard or even impossible to implement the constraints explicitly. Yet in this problem the constraint $a(q) \simeq \rho(q)/q$ was very easily implemented in projecting the wavefunction from the big space to the physical space.

The projection to the LLL of $\chi_p(z,\bar{z})$ (elimination of $\bar{z}$) does not arise naturally in this approach. This is a flaw except when $p = 1$ (Laughlin fractions) and χ_1 is already analytic.

Had we done exactly this analysis for the Laughlin fraction by attaching $2s+1$ flux tubes, we would convert the electrons to bosons in zero average field and find the result of Zhang *et al.*:

$$\Psi_e = \prod_{i<j}(z_i - z_j)^{2s+1} \exp\left[-\sum_j \frac{|z_j|^2}{4l^2}\right] \times 1, \tag{20.194}$$

where the 1 is the Bose condensate.

As Murthy and I show in [41], we can do a lot more in the enlarged space containing $(\boldsymbol{a}, \boldsymbol{P})$. For example, we can write down operators that create composite fermions (or bosons) by creating the electron and the correlation holes or vortices by using operators of the form $U(P)\psi^\dagger$, where U shifts the collective variables (charge density) to create the hole, while $\psi^\dagger$ creates the electron. Of course, once the state is created, it has to be projected. The point is this: We know the correlation hole around each electron is, of course, produced by electrons themselves, but by introducing an independent collective variable for moving charge around, we can accomplish this very easily in the big space and then project. This is our version of the *Read operator* [42] which condenses in the FQHE state. His was written in a hybrid notation of first-quantized factors and second-quantized operators. Ours is all second quantized, but in an enlarged space. In neither case does it seem possible to describe this operator easily in terms of just electrons.

20.12 Hamiltonian Theory: A Quick Summary

Following the derivation of the correlated wave function for FQHE using a variant of the Bohm–Pines technique, Murthy and I pursued the operator-based "Hamiltonian approach." Our findings were reported in [41] and a sequence of papers summarized in a long review [43].

Here is a telegraphic summary, with an invitation to consult these references if you are so inclined.

Let us begin with Eqs. (20.172) and (20.173) for the Hamiltonian and constraint:

$$H_0 = \frac{1}{2m}\psi_{\text{CP}}^\dagger(-i\nabla + e\mathbf{A}^* + \mathbf{a} + 4\pi s\mathbf{P} + \delta\mathbf{a})^2\psi_{\text{CP}}, \tag{20.195}$$

$$0=\left(a-\frac{4\pi s:\rho:}{q}\right)|\text{physical}\rangle, \quad 0<q\le Q. \tag{20.196}$$

This is the theory before interactions have been added.

A sign that more work was needed was provided by the wrong plasma frequency. This could be traced back to the coupling between the fermions and the oscillators, which could renormalize energies. We decided to decouple them by a canonical transformation. The following approximations were made:

- Drop the short-range interaction $\delta\boldsymbol{a}$.
- Drop the term proportional to $:\psi^\dagger\psi:$, the fluctuations in density, in Eq. (20.175).
- Keep the smallest possible powers of q.
- Use the RPA,

$$\sum_j e^{i(\boldsymbol{q}-\boldsymbol{q}')\cdot\boldsymbol{r}_j}\simeq n\delta\boldsymbol{q}\boldsymbol{q}'. \tag{20.197}$$

When the dust settles, we have the following expressions for the transformed current density, charge density, and constraint for the fractions with $s=1$ and

$$\nu=\frac{p}{2p+1}:$$

$$J_+(\boldsymbol{q})=\frac{\hat{\boldsymbol{q}}_+\omega_0 c}{\sqrt{2\pi}}A(\boldsymbol{q}) \quad \left(J_\pm=J_x\pm iJ_y,\ A(\boldsymbol{q})=\frac{1}{\sqrt{8\pi}}(a(\boldsymbol{q})+4\pi iP(\boldsymbol{q}))\right), \tag{20.198}$$

$$\rho(\boldsymbol{q})=\sum_j e^{-i\boldsymbol{q}\cdot\boldsymbol{r}}\left(1-\frac{il^2}{1+c}\boldsymbol{q}\times\boldsymbol{\Pi}_j\right)+\frac{qc}{\sqrt{8\pi}}(A(\boldsymbol{q})+A^\dagger(-\boldsymbol{q})), \tag{20.199}$$

$$\chi(\boldsymbol{q})=\sum_j e^{-i\boldsymbol{q}\cdot\boldsymbol{r}}\left(1+\frac{il^2}{c(1+c)}\boldsymbol{q}\times\boldsymbol{\Pi}_j\right), \tag{20.200}$$

$$\boldsymbol{\Pi}=\boldsymbol{p}+e\boldsymbol{A}^*=\boldsymbol{p}+\frac{e\boldsymbol{A}}{2p+1}, \tag{20.201}$$

$$c^2=\frac{2p}{2p+1} \quad (c \text{ is \textbf{not} the velocity of light } (=1)). \tag{20.202}$$

Note that:

- J depends *only* on the oscillator coordinate A and not on particle coordinates.
- The constraint χ depends *only* on the particle coordinates. This is good: it is no use decoupling the particle and oscillator Hamiltonians if the constraint mixes them up.
- The density ρ depends on both. For LLL physics we must drop the A part.

What about the Hamiltonian? Though we are only interested in LLL physics we have to keep the oscillators for a while to find the Hall response. This is because the current J depends on A and not the particle coordinates. The oscillator Hamiltonian coupled to an external potential $\Phi(\boldsymbol{q})$ is

$$H_{\text{oss}} = \sum_{\boldsymbol{q}}^{\mathrm{Q}} A^\dagger(\boldsymbol{q})A(\boldsymbol{q})\omega_0 - e\sum_{\boldsymbol{q}} \Phi(\boldsymbol{q})\rho(-\boldsymbol{q}). \tag{20.203}$$

Going back to Eq. (20.199) to pull out the part involving just A, we find that

$$H_{\text{oss}} = \sum_{\boldsymbol{q}} \omega_0 \left[A^\dagger(\boldsymbol{q}) - \frac{qec}{\sqrt{8\pi\,\omega_0}}\Phi(-\boldsymbol{q})\right]\left[A(\boldsymbol{q}) - \frac{qec}{\sqrt{8\pi\,\omega_0}}\Phi(\boldsymbol{q})\right] + \text{constant}. \tag{20.204}$$

Shifting the oscillator to its new minimum, we find

$$\langle -eJ_+\rangle = -e\frac{\hat{q}_+\omega_0 c}{\sqrt{2\pi}}\langle A(\boldsymbol{q})\rangle = -\frac{e^2}{2\pi(\hbar)}\nu q_+\Phi(\boldsymbol{q}), \tag{20.205}$$

where I have restored $\hbar$, which was set to unity to display the correct Hall conductance $\sigma_{xy} = \nu\sigma_0$.

The main point is that *the Hall current is carried by the oscillator and not the CF*. This is particularly relevant at $\nu = \frac{1}{2}$ when $e^* = 0$ and one is often asked who carries the current.

Now we are free to freeze the oscillator and focus on LLL physics. The particle Hamiltonian is entirely due to interactions,

$$H = \frac{1}{2}\sum_{\boldsymbol{q}} \bar{\rho}(\boldsymbol{q})v(\boldsymbol{q})\bar{\rho}(-\boldsymbol{q}), \tag{20.206}$$

where $\bar{\rho}$ is the ρ in Eq. (20.199) without the oscillator contribution which is frozen out:

$$\bar{\rho} = \sum_j e^{-i\boldsymbol{q}\cdot\boldsymbol{r}}\left(1 - \frac{il^2}{1+c}\boldsymbol{q}\times\boldsymbol{\Pi}_j\right). \tag{20.207}$$

The constraint is

$$\left[\chi(\boldsymbol{q}) = \sum_j e^{-i\boldsymbol{q}\cdot\boldsymbol{r}}\left(1 + \frac{il^2}{c(1+c)}\boldsymbol{q}\times\boldsymbol{\Pi}_j\right)\right]|\text{physical state}\rangle = 0. \tag{20.208}$$

20.13 The All-q Hamiltonian Theory

At this point the small-q theory has breathed its last. But before collapsing, it gave a clue that I decided to act on. *I assumed [44] that Eqs. (20.207) and (20.208) represent the beginnings of two exponential series and adopted the following expressions for charge and constraint*:

$$\bar{\rho} = \sum_j \exp\left(-i\mathbf{q}\cdot\left(\mathbf{r}_j - \frac{l^2}{1+c}\hat{\mathbf{z}}\times\Pi_j\right)\right) \equiv \sum_j e^{-i\mathbf{q}\cdot\mathbf{R}_{ej}}, \tag{20.209}$$

$$\bar{\chi} = \sum_j \exp\left(-i\mathbf{q}\cdot\left(\mathbf{r}_j + \frac{l^2}{c(1+c)}\hat{\mathbf{z}}\times\Pi_j\right)\right) \equiv \sum_j e^{-i\mathbf{q}\cdot\mathbf{R}_{vj}}. \tag{20.210}$$

Note that $\mathbf{R}_e$ *and* $\mathbf{R}_v$ *were fully determined by the two terms we did derive*. In what follows, the electronic variables will carry a subscript e, while the CF variables will carry none.

Here is an equivalent expression for $\boldsymbol{R}_e$:

$$\boldsymbol{R}_e = \boldsymbol{R} + \boldsymbol{\eta}c, \tag{20.211}$$

where $\boldsymbol{R}$ and $\boldsymbol{\eta}$ are CF variables guiding center and cyclotron variables describing a particle that sees a weaker magnetic field $B^* = \frac{B}{2p+1}$ and obeying

$$\left[R_x, R_y\right] = -il^{*2} = -i(2p+1)l^2, \tag{20.212}$$

$$\left[\eta_x, \eta_y\right] = +il^{*2} = i(2p+1)l^2, \tag{20.213}$$

$$[\boldsymbol{\eta}, \boldsymbol{R}] = 0. \tag{20.214}$$

I do not imply that an exact implementation of the last canonical transformation will lead to the above results. I have minimally extended the (derivable) small-ql theory to all ql, which we will see has a lot of mathematical consistency. The resulting short-distance physics could well be at odds with CF maxims, but isn't.

Now to reap the benefits. The assignment

$$\mathbf{R}_e = \mathbf{r} - \frac{l^2}{(1+c)}\hat{\mathbf{z}}\times\Pi = \boldsymbol{R} + \boldsymbol{\eta}c \ \text{ implies} \tag{20.215}$$

$$\left[R_{ex}, R_{ey}\right] = -il^2. \tag{20.216}$$

These describe the guiding center coordinates of a unit-charge object, the electron we began with, *but in terms of CF variables*. We will see that this is very important.

We can readily combine exponentials and show that $\bar{\rho}(\mathbf{q})$ obeys

$$\left[\bar{\rho}(\mathbf{q}), \bar{\rho}(\mathbf{q}')\right] = 2i\sin\left[\frac{l^2(\mathbf{q}\times\mathbf{q}')}{2}\right]\bar{\rho}(\mathbf{q}+\mathbf{q}'), \tag{20.217}$$

which is none other than the Girvin–MacDonald–Platzman (GMP) algebra [45]. These authors pointed out that if we took the density

$$\rho_e(\boldsymbol{q}) = \sum_j e^{-i\boldsymbol{q}\cdot\boldsymbol{r}_e}, \tag{20.218}$$

which commutes at different $\boldsymbol{q}$, and projected it to the LLL using

$$\mathcal{P} : \boldsymbol{r}_e = \boldsymbol{R}_e + \boldsymbol{\eta}_e \to \boldsymbol{R}_e, \tag{20.219}$$

the projected electronic charge density should obey the GMP algebra. It was very skillfully used by these authors to shed light on the FQHE in the early days.

(I refer to $\bar{\rho}$ as the projected density, although it is actually the *magnetic translation operator* and differs from the density by a factor $e^{-q^2l^2/4}$.)

Although the algebra follows from the commutation relation of $\boldsymbol{R}_e$, there is a key difference between GMP and what we have done: they worked with the electronic variable $\boldsymbol{r}_e$ and projected out $\boldsymbol{R}_e$, but our transformations (which brought in oscillators and then froze them out to get LLL physics) have managed to arrive at an expression for $\boldsymbol{R}_e$ *in terms of CF variables.*

What does this get us? In the electronic language the FQHE problem in the LLL is defined by the Hamiltonian, which is all potential energy:

$$\bar{H} = \frac{1}{2}\sum_{\boldsymbol{q}} e^{-q^2l^2/2}\bar{\rho}_e(\boldsymbol{q})v(q)\bar{\rho}_e(-\boldsymbol{q}), \tag{20.220}$$

$$\left[\bar{\rho}_e(\mathbf{q}), \bar{\rho}_e(\mathbf{q}')\right] = 2i\sin\left[\frac{l^2(\mathbf{q}\times\mathbf{q}')}{2}\right]\bar{\rho}_e(\mathbf{q}+\mathbf{q}'), \tag{20.221}$$

where

$$\bar{\rho}_e(\mathbf{q}) = \sum_j e^{-i\boldsymbol{q}\cdot\boldsymbol{R}_{ej}}. \tag{20.222}$$

(The Gaussian factor in the potential energy is due to the aforementioned difference between $\bar{\rho}$ and the actual density.)

This is a non-trivial problem because $\bar{\rho}$'s at different $\boldsymbol{q}$'s do not commute. We cannot even do Hartree–Fock because of the huge degeneracy at fractional filling. Suppose we now transcribe the problem to the CF variables and write

$$\bar{H} = \frac{1}{2}\sum_{\boldsymbol{q}} e^{-q^2l^2/2}\bar{\rho}(\boldsymbol{q})v(q)\bar{\rho}(-\boldsymbol{q}), \tag{20.223}$$

where $\bar{\rho}$ also obeys the GMP algebra but is now a function of CF variables,

$$\bar{\rho}(\boldsymbol{q}) = \sum_j e^{-i\boldsymbol{q}\cdot(\boldsymbol{R}_j + c\boldsymbol{\eta}_j)}. \tag{20.224}$$

Although the problem is algebraically the same as in Eq. (20.221), *now there is an obvious HF state* staring at us, the one with p-filled CF LL's. We can regain Jain's description but in terms of operators, and go beyond wavefunctions to correlation functions at non-zero $\boldsymbol{q}$, ω, and T.

But two matters of principle need to be resolved first:

- Two problems that are algebraically the same need not be physically the same. For example, the spin-$\frac{1}{2}$ and spin-1 Heisenberg chains are written in terms of generators obeying the same $SU(2)$ algebra, but behave very differently: the former is gapless and the latter is gapped.

- The CF version of the problem is in a Hilbert space that has twice as many variables per particle as the LLL: the latter has just $\boldsymbol{R}_e$, while the former is a full-blown Hilbert space with $\boldsymbol{R}$ and $\boldsymbol{\eta}$. This is why Jain has to project his answers to the LLL using $\bar{z} \to 2l^2 \frac{\partial}{\partial z}$.

While it is possible that two realizations of the same algebra could be physically inequivalent, that is not going to happen here. The reason is that there is a continuous path, involving unitary transformations, starting with the electronic problem and ending with the $\bar{H}$ above. True, some of the steps were approximate (small q, RPA, etc.), but a representation cannot change along the way.

The second issue of the larger Hilbert space is more complicated. Here is where the constraint

$$\bar{\chi}\,|\text{physical}\rangle = 0 \tag{20.225}$$

comes in. It is the transformed version of $\nabla \times \boldsymbol{a} = 4\pi n s\hbar : \rho :$. As pointed out earlier, it depends on only the particle coordinates.

Recall that $\bar{\chi}$ was defined as

$$\bar{\chi} = \sum_j e^{-i\boldsymbol{q}\cdot\boldsymbol{R}_v}, \tag{20.226}$$

where

$$\mathbf{R}_v = \mathbf{r} + \frac{l^2}{c(1+c)}\hat{\mathbf{z}} \times \Pi, \quad \text{or equivalently} \tag{20.227}$$

$$\boldsymbol{R}_v = \boldsymbol{R} + \boldsymbol{\eta}/c. \tag{20.228}$$

$$\left[R_{vx}, R_{vy}\right] = il^2/c^2. \tag{20.229}$$

These describe the guiding center coordinates of a particle whose charge is $-c^2$, namely the double vortex. It follows that the constraints $\bar{\chi}(q)$ also close to form a GMP algebra with $l^2 \to -l^2/c^2$ in the structure constants.

Additionally, we have

$$[\mathbf{R}_e, \mathbf{R}_v] = 0, \tag{20.230}$$

which means that

$$[\bar{\chi}, \bar{\rho}] = 0. \tag{20.231}$$

So the mathematical problem of the FQHE in the LLL takes the following form in CF variables:

$$\bar{H} = \frac{1}{2}\sum_{\boldsymbol{q}} e^{-q^2 l^2/2} \bar{\rho}(\boldsymbol{q}) v(q) \bar{\rho}(-\boldsymbol{q}), \tag{20.232}$$

$$\bar{\chi}(\boldsymbol{q}) \simeq 0. \tag{20.233}$$

Since $\boldsymbol{R}_v$ and $\bar{\chi}(\mathbf{q})$ do not appear in $\bar{H}$, $\bar{\chi}(\mathbf{q})$ does not have any dynamics, just like the longitudinal part of the vector potential in a gauge theory where the Hamiltonian is gauge invariant. As in the Yang–Mills case, the constraints form a non-Abelian algebra and commute with H.

The condition $\bar{\chi}(\mathbf{q}) \simeq 0$ means $\bar{\chi}(\mathbf{q})$ must vanish within correlation functions. (Since $\bar{\chi}(\mathbf{q})$ commutes with $\bar{H}$, this is a first-class constraint preserved by the equations of motion.)

20.14 Putting the Hamiltonian Theory to Work

The rationale for working with the CF was to get a unique ground state for the HF approximation.

There are at least two good reasons to expect that the naive HF result will require fairly strong corrections. First, if we compute the matrix element of the projected electron density between any two HF states, the answer will be linear in q, whereas in the exact theory, we know that within the LLL it must go as q^2 as per Kohn's theorem [46]. Secondly, as $ql \to 0$, the projected electronic density has unit contribution from each CF, while we would like it to be $e^*/e = \frac{1}{2p+1} = 1 - c^2$. Evidently the HF result will receive strong corrections that will renormalize these quantities until they are in line with these expectations. These renormalizations will occur once we pay attention to the constraint $\bar{\chi}(\mathbf{q}) \simeq 0$.

Now, Baym and Kadanoff [47, 48] invented the *conserving approximation* for improving the HF state with additional diagrammatic corrections (ladder sums) to enforce conservation laws. (The non-conservation comes from using Hartree–Fock self-energies for propagators while using bare vertices in the one-loop response functions, in violation of Ward identities.)

For $\nu = \frac{1}{2}$, Read [49] showed that this procedure applied to Eqs. (20.232) and (20.233) restores Kohn's theorem, exhibits the overdamped mode, reveals a dipolar structure for density–density correlations, and yields a compressible state. Murthy [50] used it to calculate density–density correlations in gapped fractions.

20.14.1 Preferred Charge: A Short Cut to the Constraint

We found that in many problems where there is a large enough gap, temperature, or disorder, there is a short cut to implementing the constraint in the infrared limit. Unlike the Baym–Kadanoff route, this one is peculiar to the FQHE and we do not fully understand why it works.

Suppose, in the Hamiltonian and elsewhere, that we replace $\bar{\rho}(\mathbf{q})$ by the *preferred combination*

$$\bar{\rho}^p(\mathbf{q}) = \bar{\rho}(\mathbf{q}) - c^2 \bar{\chi}. \tag{20.234}$$

In an exact calculation it makes no difference to the computation of anything physical whether the coefficient in front of $\bar{\chi}(\mathbf{q})$ is zero, or $-c^2$, or anything else, since $\bar{\chi}$ is essentially zero.

On the other hand, in the HF approximation (which does not respect $\bar{\chi}(\mathbf{q}) \simeq 0$) it certainly matters what coefficient we place in front of $\bar{\chi}(\mathbf{q})$. The preferred combination $\bar{\rho}^p(\mathbf{q})$ stands out as the sum of the electronic and vortex charge densities. But the reason we are forced to use it is that it helps us avoid violating Kohn's theorem within simple HF.

Consider its expansion in powers of ql:

$$\bar{\rho}^p = \sum_j e^{-i\mathbf{q}\cdot\mathbf{r}_j}\left(\frac{1}{2p+1} - il^2\boldsymbol{q}\times\Pi_j + 0\cdot(\boldsymbol{q}\times\Pi_j)^2 + \cdots\right). \tag{20.235}$$

The transition matrix elements are now of order q^2 between HF states because the coefficient of $\boldsymbol{q}$ is proportional to the CF guiding center coordinate $\boldsymbol{r} - l^{*2}\hat{z}\times\boldsymbol{\Pi}$, with no admixture of the CF cyclotron coordinate. This is more transparent if we use $\boldsymbol{R}$ and $\boldsymbol{\eta}$ to write

$$\begin{aligned}\bar{\rho}^p(\mathbf{q}) &= (1 - i\boldsymbol{q}\cdot(\boldsymbol{R}+c\boldsymbol{\eta}) + \cdots) - c^2(1 - i\boldsymbol{q}\cdot(\boldsymbol{R}+\boldsymbol{\eta}/c) + \cdots) \\ &= (1-c^2)(1 - i\boldsymbol{q}\cdot\boldsymbol{R} + \mathcal{O}(q^2)).\end{aligned} \tag{20.236}$$

The choice of $(-c^2)$ as the coefficient of $\bar{\chi}(\mathbf{q})$ in $\bar{\rho}^p(\mathbf{q})$, uniquely determined by compliance with Kohn's theorem (reflected in the vanishing coefficient of $\boldsymbol{\eta}$), is also the one that leads to two important collateral benefits:

- The electronic charge density associated with $\bar{\rho}^p(\mathbf{q})$ is $1 - c^2 = e^*/e$.
- We see from Eq. (20.235) that when $\nu = \frac{1}{2}$, the preferred density couples to an external electric field like a dipole of size $\boldsymbol{d}^* = l^2\hat{z}\times\boldsymbol{p}$, giving a precise operator expression of Read's picture [51,52].

The Hamiltonian $\bar{H}(\bar{\rho}^p)$ is *weakly gauge invariant*, that is,

$$\left[\bar{H}(\bar{\rho}^p), \bar{\chi}(\mathbf{q})\right] \simeq 0, \tag{20.237}$$

where the $\simeq 0$ symbol means that it vanishes in the subspace obeying $\bar{\chi}(\mathbf{q}) = 0$. Thus, neither $\bar{H}(\bar{\rho}^p)$ nor $\bar{\rho}^p$ will mix physical and unphysical states.

The significance of $\bar{H}(\bar{\rho}^p)$ is the following. If the constraint $\bar{\chi} = 0$ is imposed *exactly*, there are many equivalent Hamiltonians, depending on how $\bar{\chi}$ is insinuated into it. However, in the HF *approximation*, these are not equivalent and $\bar{H}(\bar{\rho}^p)$ best approximates, between HF states and at long wavelengths, the true Hamiltonian between true eigenstates. In contrast to a variational calculation where one searches among trial states for an optimal one, here the HF states are the same for a class of Hamiltonians (where $\bar{\chi}$ is introduced into $\bar{H}$ in any rotationally invariant form), and we seek the best Hamiltonian. This happens to be $\bar{H}(\bar{\rho}^p)$ because it encodes the fact that every electron is accompanied by a correlation hole of some sort, which leads to the correct e^* and $\boldsymbol{d}^*$, and obeys the all-important Kohn's theorem (q^2 matrix elements for the density projected to the LLL).

The preferred charge $\bar{\rho}^p(\mathbf{q})$ and preferred Hamiltonian $\bar{H}(\bar{\rho}^p)$ have been used to compute gaps, finite-temperature response functions (polarization, NMR rates), and even the effect of disorder. The results are in reasonable agreement (10%–20%) with computer simulations and real data [43].

When we use the preferred charge and Hamiltonian we make no further reference to constraints, and simply carry out the Hartree–Fock approximation. This is based on the expectation that even if we found some way to include the effect of constraints, it would make no difference in the small-ql region because the leading renormalization of e to e^* and suppression of $\mathcal{O}(q)$ matrix elements down to $\mathcal{O}(q^2)$ that are achieved by the conserving approximation are built-in here. Of course, errors at larger q will corrupt the actual numbers, say for gaps.

The short cut, however, fails in one important regard. For the gapless $\nu = \frac{1}{2}$ state at $T = 0$, since $\bar{\rho}^p(\mathbf{q})$ starts out linearly in $\boldsymbol{q}$, the CF couples like a dipole to the external potential, leading to a compressibility that vanishes as $q \to 0$. The only way to restore compressibility is to have some very low energy collective excitations that overcome the factors of q in the matrix elements. This was first pointed out to us by Halperin and Stern [53, 54], who used a toy model to make their point that respecting gauge invariance (or the constraint) is crucial. They went on to give a detailed analysis of the realistic model with additional coworkers [55]. Independently, Read [49] did the ladder sum on top of HF and obtained the overdamped mode, finite compressibility, and dipolar coupling.

The reader will recall that any *simple* picture of quasiparticles, whether it be in Landau's Fermi liquid theory, or in BCS theory, is best captured by approximate and not exact descriptions. The quasiparticles are all caricatures of some exact reality, and therein lies their utility. Similarly, the CF in our extended formalism appears only in the HF approximation to $\bar{H}(\bar{\rho}^p)$. Recall that we brought in the coordinate $\boldsymbol{R}_v$ to become the electron's partner in forming the CF. However, $\boldsymbol{R}_v$ was cyclic in the exact Hamiltonian $\bar{H}$. *Thus, the exact dynamics never demanded that $\boldsymbol{R}_v$ be bound to $\boldsymbol{R}_e$, or even be anywhere near $\boldsymbol{R}_e$.* However, in the HF approximation, since we wanted the right charge and transition matrix elements of the density operator (Kohn's theorem) to be manifest, we needed to replace $\bar{\rho}$ by $\bar{\rho}^p$, and trade $\bar{H}(\bar{\rho})$ for $\bar{H}(\bar{\rho}^p)$, the preferred Hamiltonian. In $\bar{H}(\bar{\rho}^p)$, $\boldsymbol{R}_v$ is coupled to $\boldsymbol{R}_e$. The HF approximation and this coupling go hand in hand. The exact eigenfunctions of the original $\bar{H}$ are factorized in the analytic coordinates z_e and z_v, and presumably reproduce the electronic correlations of the FQHE states. On the other hand, in the HF approximation to $\bar{H}(\bar{\rho}^p)$, the wavefunctions (e.g., p-filled LL's) mix up z_e and z_v, and $\bar{H}(\bar{\rho}^p)$, the preferred Hamiltonian, dynamically couples $\boldsymbol{R}_e$ and $\boldsymbol{R}_v$. The net result is that, at least at long wavelengths, these two wrongs make it right and mimic what happens in the exact solution in the small-q region.

Another advantage of $\bar{H}(\bar{\rho}^p)$ is that it gives an approximate formula for m^* originating entirely from interactions. This is best seen at $\nu = \frac{1}{2}$. When we square $\bar{\rho}^p$, Eq. (20.235), we get a double sum over particles whose diagonal part is the one-particle (free-field) term:

$$H^0_{\nu=\frac{1}{2}} = 2\sum_j \int \frac{d^2q}{4\pi^2} \sin^2\left[\frac{\mathbf{q}\times\mathbf{k}_j l^2}{2}\right] v(q) e^{-q^2l^2/2}. \tag{20.238}$$

This is not a Hamiltonian of the form $k^2/2m^*$. However, if the potential is peaked at very small q, we can expand the sine and read off an approximate $1/m^*$:

$$\frac{1}{m^*} = \int \frac{qdqd\theta}{4\pi^2} \left[(\sin^2\theta)\,(ql)^2\right] v(q)\, e^{-q^2l^2/2}, \tag{20.239}$$

which has its origin in electron–electron interactions. However, we can do more: we have the full Hamiltonian for CF dynamics, which is just the electron–electron interaction written in terms of CF variables.

20.14.2 *Another Argument for $\bar{\rho}^p(q)$*

Consider the combination $\bar{\rho}^p = \bar{\rho} - c^2\bar{\chi}$ at $c = 1$ or $\nu = \frac{1}{2}$:

$$\bar{\rho}^p = \bar{\rho} - \bar{\chi}, \tag{20.240}$$

and its commutator

$$\left[\bar{\rho}(\boldsymbol{q}) - \bar{\chi}(\boldsymbol{q}), \bar{\rho}(\boldsymbol{q}') - \bar{\chi}(\boldsymbol{q}')\right] = \left[\bar{\rho}(\boldsymbol{q}), \bar{\rho}(\boldsymbol{q}')\right] + \left[\bar{\chi}(\boldsymbol{q}), \bar{\chi}(\boldsymbol{q}')\right] \tag{20.241}$$

$$= 2i\sin\left[\frac{l^2(\mathbf{q}\times\mathbf{q}')}{2}\right](\bar{\rho}(\boldsymbol{q}+\boldsymbol{q}') - \bar{\chi}(\boldsymbol{q}+\boldsymbol{q}')). \tag{20.242}$$

The combination $\bar{\rho} - \bar{\chi}$ also obeys the GMP algebra like $\bar{\rho}$! In other words, if, instead of first finding $\boldsymbol{R}_e$ in terms of CF coordinates and then exponentiating it to find $\bar{\rho}$, we directly sought a representation of the projected charge, we have this second alternative. This singles out the preferred density over all other linear combinations of $\bar{\rho}$ and $\bar{\chi}$. (Away from $c = 1$, we find that $\bar{\rho}^p(\mathbf{q})$ obeys the GMP algebra to some very high order in ql, but not exactly.)

But recall the cautionary note: the same algebra does not mean the same problem. Unlike the derivation of $\bar{\rho}(\boldsymbol{q})$ from the original electronic problem by the (approximate) unitary transformation, this $\bar{\rho}^p$ has been pulled out of a hat. So it could also describe a different problem where this algebra appears. There is an argument due to Senthil [56, 57] that it might describe the $\nu = 1$ bosons studied by Read [49].

20.15 Chern–Simons Theory

Let us go back to our key equations, (20.172) and (20.173), repeated below:

$$H = \frac{1}{2m}\psi^\dagger_{\mathrm{CP}}(-i\nabla + e\mathbf{A}^* + \mathbf{a} + 4\pi s\mathbf{P} + \delta\mathbf{a})^2\psi_{\mathrm{CP}}, \tag{20.243}$$

$$0 = \left(a - \frac{4\pi s : \rho :}{q}\right)|\text{physical}\rangle, \quad 0 < q \leq Q. \tag{20.244}$$

I am going to rewrite this as follows in view of what is to come:

$$H = \frac{1}{2m}\psi^{\dagger}_{\rm CP}(-i\nabla + e\mathbf{A} + \tilde{\mathbf{a}} + 4\pi s\tilde{\mathbf{P}})^2\psi_{\rm CP}, \tag{20.245}$$

$$0 = \left(\frac{\nabla\times\tilde{\boldsymbol{a}}}{4\pi s} - \rho\right)|\text{physical}\rangle. \tag{20.246}$$

I have done the following:

- I have dropped $\delta\boldsymbol{a}$ because I imagine no cut-off on $\boldsymbol{q}$. We have in mind a continuum theory with continuum fields $\tilde{\boldsymbol{a}}$ and $\tilde{\boldsymbol{P}}$.
- I have gone backwards from $\boldsymbol{A}^*$ to $\boldsymbol{A}$ and undone the separation of ρ into n and $:\rho:$:

$$e\boldsymbol{A}^* = \frac{e\boldsymbol{A}}{2ps+1} \tag{20.247}$$

$$= e\boldsymbol{A} - e\boldsymbol{A}\left(\frac{2ps}{2ps+1}\right), \quad \text{so that} \tag{20.248}$$

$$e\boldsymbol{A}^* + \boldsymbol{a} = e\boldsymbol{A} + \boldsymbol{a} - e\boldsymbol{A}\left(\frac{2ps}{2ps+1}\right) \tag{20.249}$$

$$\equiv e\boldsymbol{A} + \tilde{\boldsymbol{a}}. \tag{20.250}$$

I leave it to you to verify that

$$\nabla\times\tilde{\boldsymbol{a}} = 4\pi ns\hbar + 4\pi\hbar s :\rho: = 4\pi ns\rho. \tag{20.251}$$

So, $\boldsymbol{A}$ is the full external $\boldsymbol{A}$ and not the mean-field $\boldsymbol{A}^*$, and $\nabla\times\tilde{\boldsymbol{a}}$ is proportional to ρ and not $:\rho:$. The tilde on $\tilde{\boldsymbol{P}}$ is for uniformity of notation.

Exercise 20.15.1 *Verify that Eqs. (20.245) and (20.246) follow from Eqs. (20.172) and (20.173) upon making the changes mentioned above.*

Now it turns out that there is a continuum theory, the Chern–Simons theory, in which Eqs. (20.245) and (20.246) arise naturally. It is based on the fact that in three dimensions, (x, y, x_0), there is, in addition to $F_{\mu\nu}F^{\mu\nu}$, another Lorentz-invariant action:

$$S_{\rm CS} = \int dxdydx_0\left[\frac{1}{8\pi s}\varepsilon^{\mu\nu\lambda}a_\mu\partial_\nu a_\lambda - a_\mu j^\mu\right], \tag{20.252}$$

where j^μ is a conserved current, say due to matter fields.

Let us check gauge invariance. Under

$$a_\mu \to a_\mu + \partial_\mu\Lambda, \tag{20.253}$$

$$\varepsilon^{\mu\nu\lambda}a_\mu\partial_\nu a_\lambda - a_\mu j^\mu \to \varepsilon^{\mu\nu\lambda}a_\mu\partial_\nu a_\lambda - a_\mu j^\mu + \varepsilon^{\mu\lambda\nu}\partial_\mu(a_\lambda\partial_\nu\Lambda) + \Lambda\partial_\mu j^\mu$$
$$= \varepsilon^{\mu\nu\lambda}a_\mu\partial_\nu a_\lambda - a_\mu j^\mu, \tag{20.254}$$

where I have used current conservation, dropped some terms using the antisymmetry of $\varepsilon_{\mu\nu\lambda}$, and done some integration by parts to drop the total derivative. Thus, the CS action is invariant under gauge transformations that vanish at infinity.

Deser *et al.* noticed that if one adds the F^2 term one gets a *massive* field, although the action is gauge invariant.

The trouble with the CS term is that it violates time reversal and parity. But that is just fine for the Hall effect, which does not have these symmetries. It also has one fewer derivatives than F^2 and is more relevant in the infrared.

The equation of motion comes from taking the a derivative of the action:

$$\varepsilon^{\mu\nu\lambda}\frac{\partial_\nu a_\lambda}{4\pi s} = j^\mu. \tag{20.255}$$

By taking ∂_μ of both sides, we find $\partial_\mu j^\mu = 0$. Next, in terms of components we have

$$\nabla \times \boldsymbol{a} = 4\pi s\rho, \quad \text{where } \boldsymbol{a} = (a_x, a_y), \tag{20.256}$$

$$\partial_y a_0 - \partial_0 a_y = 4\pi s j_x, \tag{20.257}$$

$$\partial_0 a_x - \partial_x a_0 = 4\pi s j_y. \tag{20.258}$$

The first is the constraint we want. It is not an equation of motion for $\boldsymbol{a}$, but a condition on its components fixed by ρ. Before looking at the other two, let us deal with this constraint. In quantum electrodynamics the analogous term would be $a_0(\nabla\cdot\boldsymbol{E} - \frac{\rho}{\varepsilon_0})$, which imposes Gauss's law in the functional integral over a_0.

We can keep going with the constraint in the path integral, but if we want a *Hamiltonian* description, we have to dump the freedom to do time-dependent gauge transformation. This is because a Hamiltonian is supposed to generate a unique future from the initial value data, and this cannot happen in a formalism where you can keep the initial value data fixed but alter the fields at future times using a time-dependent gauge transformation. So we can set $a_0 = 0$ in the action. But what about Gauss's law? It is imposed as a constraint on physical states. Being a constraint that commutes with the Hamiltonian, it will be obeyed at all times, if imposed initially.

We are going to do the same thing in the CS theory. We will impose Eq. (20.256) as a condition on physical states after making sure that the constraint commutes with the equations of motion, Eqs. (20.257) and (20.258):

$$\begin{aligned}\partial_0(\partial_x a_y - \partial_y a_x) - 4\pi s\partial_0\rho &= \partial_x\partial_0 a_y - \partial_y\partial_0 a_x - 4\pi s\partial_0\rho \\ &= \partial_x(\partial_y a_0 - 4\pi s j_x) - \partial_y(4\pi s j_y + \partial_x a_0) - 4\pi s\partial_0\rho \\ &= -4\pi s\partial_\mu j^\mu = 0.\end{aligned} \tag{20.259}$$

Now we may set $a_0 = 0$ in the action to obtain

$$S_{CS} = \int d^3x\left[\frac{1}{8\pi s}(a_y\partial_0 a_x - a_x\partial_0 a_y) + a_x j_x + a_y j_y\right]. \tag{20.260}$$

Comparing this to the phase-space path integral for quantum mechanics with the term $\frac{1}{2\hbar}(p\dot{q} - q\dot{p})$, which implies $[q,p] = i\hbar$, we deduce that a_x and a_y are canonically conjugate and obey

$$[a_x(\boldsymbol{r}), a_y(\boldsymbol{r}')] = 4\pi i s \delta^{(2)}(\boldsymbol{r} - \boldsymbol{r}'). \tag{20.261}$$

Consider the case where the current is due to a non-relativistic fermion. The corresponding Hamiltonian problem would be defined by

$$H = \int d^2x \psi^\dagger \frac{(-i\nabla + eA + \boldsymbol{a})^2}{2m} \psi, \tag{20.262}$$

$$0 = \left(\frac{\nabla \times \boldsymbol{a}}{4\pi s} - \rho \right) |\text{physical}\rangle, \tag{20.263}$$

which coincides with Eqs. (20.245) and (20.246) if we can make the identification

$$\boldsymbol{a} = \tilde{\boldsymbol{a}} + 4\pi s \tilde{\boldsymbol{P}}. \tag{20.264}$$

Remember: because $\tilde{\boldsymbol{P}}$ is longitudinal,

$$\nabla \times \tilde{\boldsymbol{a}} = \nabla \times (\tilde{\boldsymbol{a}} + 4\pi s \tilde{\boldsymbol{P}}) = \nabla \times \boldsymbol{a}. \tag{20.265}$$

Given that $\tilde{\boldsymbol{a}}$ and $\tilde{\boldsymbol{P}}$ are transverse and longitudinal respectively, it is plausible that together they could represent a full-fledged vector field $\boldsymbol{a}$. I invite you to show that

$$[a_x(\boldsymbol{r}), a_y(\boldsymbol{r}')] = 4\pi i s \delta^{(2)}(\boldsymbol{r} - \boldsymbol{r}') \tag{20.266}$$

follows from Eqs. (20.162)–(20.165).

Exercise 20.15.2 *Verify Eq. (20.266). (Assume Eq. (20.264), write the latter in* $\boldsymbol{q}$ *space, and project* $\hat{\boldsymbol{q}}$ *and* $-\hat{z} \times \hat{\boldsymbol{q}}$ *along* $\mathbf{i}$ *and* $\mathbf{j}$ *to find the expansions for* a_x *and* a_y*, then work out the commutators.)*

This completes the discussion of how the CS theory can be used to describe FQHE physics. Like so many other topics covered in this book, the discussion of CS theory had to be truncated. I have as usual focused on what is directly applicable to the theme. I strongly urge you to read more about this unique and powerful theory, not only applied to FQHE but to so many other areas of physics, like Witten's theory of knots. Once again, the notes of Nayak and Tong [8] are very readable, accessible online, and detailed.

20.16 The End

In a book of this type, there is no natural place to end. Since this was a book on techniques I was conversant with, the focus was on them and not on the detailed description of the illustrative and underlying physical problem. This is why many discussions were truncated at points where it would have been perfectly natural to continue the discussion. An example that comes to mind is the $\nu = \frac{1}{2}$ Hall effect, which was merely mentioned and not described in the detail it deserves as a physical problem. To me it was an example of using the CS theory and that was a topic I had already covered via other examples. It was a question of sticking to the game plan of writing a book of reasonable size that would not undergo gravitational collapse.

The best way to use this book is to see if a topic or technique you are interested in is mentioned in the table of contents, and see if you are able get an introduction to it from reading it. That should mark the beginning and not the end, for you should follow through and attack the rest of the topic using one of the numerous excellent books and papers referred to.

References and Further Reading

[1] K. von Klitzing, G. Dorda, and M. Pepper, Physical Review Letters, **45**, 494 (1980).
[2] D. Tsui, H. Stromer, and A. Gossard, Physical Review Letters, **48**, 1599 (1982).
[3] R. E. Prange and S. M. Girvin (eds.), *The Quantum Hall Effect*, Springer-Verlag (1990).
[4] T. Chakraborty and P. Pietiäïnen, *The Fractional Quantum Hall Effect: Properties of an Incompressible Quantum Fluid*, Springer Series in Solid State Sciences, vol. 85, Springer-Verlag (1988).
[5] A. H. MacDonald (ed.), *Quantum Hall Effect: A Perspective*, Kluwer (1989).
[6] A. Karlhede, S. A. Kivelson, and S. L. Sondhi, in *Correlated Electron Systems*, ed. V. J. Emery, World Scientific (1993).
[7] S. D. Sarma and A. Pinczuk, *Perspectives in Quantum Hall Effects*, Wiley (1997).
[8] There are lectures for entire courses on many websites, for example C. Nayak, "Quantum Condensed Matter Physics," UCLA notes; D. Tong, "Lectures on the Quantum Hall Effect," DAMPT, Cambridge University. I am sure there are more such gems to be found.
[9] D. Bohm and D. Pines, Physical Review, **92**, 609 (1953).
[10] S. Deser, R. Jackiw, and S. Templeton, Physical Review Letters, **48**, 975 (1982).
[11] S. C. Zhang, H. Hansson, and S. Kivelson, Physical Review Letters, **62**, 82 (1989).
[12] D.-H. Lee and S.-C. Zhang, Physical Review Letters, **66**, 1220 (1991).
[13] S. C. Zhang, International Journal of Modern Physics B, **6**, 25 (1992).
[14] E. Fradkin, *Field Theories in Condensed Matter Physics*, Addison-Wesley (1991).
[15] R. B. Laughlin, Physical Review Letters, **50**, 1395 (1983). Needs no explanation.
[16] J. Jain, Physical Review Letters, **63**, 199 (1989).
[17] J. K. Jain and R. Kamilla, in *Composite Fermions*, ed. O. Heinonen, World Scientific (1998).
[18] R. Shankar *Principles of Quantum Mechanics*, Springer (1994).
[19] B. I. Halperin, Physical Review B, **25**, 2185 (1982). Discusses edge theory.
[20] B. I. Halperin, Helvetica Physica Acta, **56**, 75 (1983).
[21] M. Hasan and C. Kane, Reviews of Modern Physics, **82**, 3045 (2010).
[22] R. Laughlin, Physical Review B, **23**, 5632 (1981).
[23] X.-L. Qi, T. L. Hughes, and S.-C. Zhang, Physical Review B, **78**, 195424 (2008).
[24] C. Nayak, Physics 243 Lectures UCLA; Boulder School for Condensed Matter and Material Physics (2013).
[25] S. A. Trugman, Physical Review B, **27**, 7539 (1983).

[26] B. I. Halperin, Physical Review Letters, **52**, 1583 (1984). Fractional statistics of quasiparticles.
[27] D. Arovas, J. Schrieffer, and F. Wilczek, Physical Review Letters, **53**, 722 (1984).
[28] D. Arovas, J. Schrieffer, F. Wilczek, and A. Zee, Nuclear Physics B, **251**, 117 (1985).
[29] X. G. Wen. International Journal of Modern Physics, **6**, 1711 (1992).
[30] J. M. Leinaas and J. Myrheim, Nuovo Cimento B, **37**, 1 (1977).
[31] F. Wilczek, Physical Review Letters, **48**, 1144 (1982).
[32] A. Lopez and E. Fradkin, Physical Review B, **44**, 5246 (1991).
[33] A. Lopez and E. Fradkin, Physical Review B, **47**, 7080 (1993).
[34] A. Lopez and E. Fradkin, Physical Review Letters, **69**, 2126 (1992).
[35] C. L. Kane, S. Kivelson, D. H. Lee, and S.-C. Zhang, Physical Review B, **43**, 3255 (1991).
[36] R. Rajaraman and S. L. Sondhi, Modern Physics Letters B, **8**, 1065 (1994).
[37] R. Shankar and G. Murthy, Physical Review Letters, **79**, 4437 (1997).
[38] G. Murthy and R. Shankar, in *Composite Fermions*, ed. O. Heinonen, World Scientific (1998).
[39] V. Kalmeyer and S. C. Zhang, Physical Review B, **46**, 9889 (1992).
[40] B. I. Halperin, P. A. Lee, and N. Read, Physical Review B, **47**, 7312 (1993).
[41] O. Heinonen (ed.), *Composite Fermions*, World Scientific (1998).
[42] N. Read, Physical Review Letters, **62**, 86 (1989).
[43] G. Murthy and R. Shankar, Reviews of Modern Physics, **75**, 1101 (2003).
[44] R. Shankar, Physical Review Letters, **83**, 2382 (1999).
[45] S. M. Girvin, A. H. MacDonald, and P. Platzman, Physical Review B, **33**, 2481 (1986).
[46] W. Kohn, Physical Review, **123**, 1242 (1961).
[47] G. Baym and L. P. Kadanoff, Physical Review, **124**, 287 (1961).
[48] L. P. Kadanoff and G. Baym, *Quantum Statistical Mechanics*, Addison-Wesley (1989).
[49] N. Read, Physical Review **58**, 16262 (1998).
[50] G. Murthy, Physical Review B, **60**, 13702 (1999).
[51] N. Read, Semiconductor Science and Technology, **9**, 1859 (1994).
[52] N. Read, Surface Science, **361/362**, 7 (1996).
[53] B. I. Halperin and A. Stern, Physical Review Letters **80**, 5457 (1998). This was a comment on our article [37].
[54] G. Murthy and R. Shankar, Physics Review Letters, **80**, 5458 (1998). This was our reply to [53].
[55] A. Stern, B. I. Halperin, F. von Oppen, and S. H. Simon, Physical Review B, **59**, 12547 (1999).
[56] C. Wang and T. Senthil, Physical Review B, **93**, 085110 (2016).
[57] G. Murthy and R. Shankar, Physical Review B, **93**, 085405 (2016).

Index